JJW McDouall.

2015.

ELECTRONIC STRUCTURE MODELING

Connections Between Theory and Software

ELECTRONIC STRUCTURE MODELING

Connections Between Theory and Software

Carl Trindle
Donald Shillady

CRC Press
Taylor & Francis Group
Boca Raton   London   New York

CRC Press is an imprint of the
Taylor & Francis Group, an **informa** business

CRC Press
Taylor & Francis Group
6000 Broken Sound Parkway NW, Suite 300
Boca Raton, FL 33487-2742

© 2008 by Taylor & Francis Group, LLC
CRC Press is an imprint of Taylor & Francis Group, an Informa business

Library of Congress Cataloging-in-Publication Data

Trindle, Carl.
   Electronic structure modeling : connections between theory and software /
Carl Trindle and Donald Shillady.
      p. cm.
   Includes bibliographical references and index.
   ISBN 978-0-8493-8406-6 (alk. paper)
   1. Electronic structure--Mathematical models. I. Shillady, Donald.
   II. Title.

QC176.8.E4T75 2008
541'.28--dc22   2008010876

Visit the Taylor & Francis Web site at
http://www.taylorandfrancis.com

and the CRC Press Web site at
http://www.crcpress.com

Contents

Chapter 9   Highly Accurate Methods: Coupled Cluster Calculations, Extrapolation to Chemical Accuracy, and Quantum Monte Carlo Methods

Chapter 10   Modeling the Coulomb Hole

Chapter 11   Density Functional Theory

Chapter 12   Calculation of Nuclear Magnetic Resonance Shielding/Shifts

Chapter 13 The Representation of Electronically Excited States

Preface

Each of the authors has studied and taught the craft of electronic structure modeling for almost four decades. Becoming conscious of this deep subject in the 1960s, we have been witness to remarkable advances in quantum chemistry. One great change is that increasingly powerful electronic structure codes are now commodities, and machines to execute the code are widely available—potentially, to every reader, on his desktop or in her carryall.

This book is a response to the growing need for those pursuing experimental investigations to add to their methods and instruments the power of electronic structure modeling. We want to ensure that those who seek it can use this power for (their own) good. This may best be accomplished by a course of study that begins with a reminder of the foundations of the field and provides a leisurely look into the operation and structure of the most common electronic structure calculations. As aids to this exercise, we include two small-scale electronic structure programs. SCF1s is very tiny, but not quite a toy—it does real SCF calculations with a real basis set. It is just powerful enough to make some important features of SCF calculations intelligible. PCLOBE includes more sophisticated methods and can be (and has been) a medium for research-level modeling. Both grew out of the pioneering work of Lee Allen and his students at Princeton.

The first half of this book rests on the capabilities of these programs, and is a stepping stone to the insightful use of powerful modern codes, many of which are proof against tinkering—a consequence of the commercial importance and wide impact of electronic structure modeling.

We were given wise (if familiar) counsel by an architect of this field, H.F. Schaefer III. "Write what you know," he said. That would have made a shorter book. The second half of our work is mainly our effort to write what we learned. We know more now of density functional theory, coupled cluster methods, and modern means of calculation of properties such as NMR shielding and coupling, and electronic excitations by response theory. Finally, our discussion turns to a subject that has proved fascinating to one of us from the beginning and to the other in its resurgence just within the past few years. Circular dichroism, optical rotatory dispersion, and magnetic circular dichroism embrace a huge range of physical chemistry. The phenomena have attracted the attention of the best theoretical and experimental chemists for over almost 80 years, and were intriguing to Goethe, Pasteur, and Faraday. PCLOBE and the most modern codes grapple with these deep subtleties of light and matter. We hope that this survey of the workings of programs and their modern use will send the reader onto independent study in electronic structure modeling, confident that powerful codes can be used in ways that are appropriate and meaningful.

For our more advanced examples we have used GAUSSIAN and ADF software, with which we have longstanding familiarity. We are grateful to those who produced results specifically for our use with GAMESS, ACES II, PSI3, and deMON. Many other powerful software suites are available, including the commercial products SCHRÖEDINGER, Q-CHEM, HYPERCHEM, and JAGUAR. Many active research groups maintain and share their own software: these include CADPAC, COLUMBUS, DALTON, MOLCAS, MOLPRO, NWCHEM, and TURBOMOL. Links to sources may be found at http://ws1.kist.re.kr among many other sites. We have found that SPARTAN software provides a particularly gentle way to gain experience with electronic structure modeling. Its excellent graphics and powerful molecule builder add to its appeal; the electrostatic potential mapped on a charge density isosurface appearing on the cover of this book is a small example.

Like everyone who writes, we are not free from debt to those who have led the way and helped us along. The study and practice of electronic structure modeling have brought us into the presence of great teachers and scientists. Some of them made a deeply personal impact, and others inspired us by example through their published work. The first group must include our mentors and colleagues. Donald Shillady thanks Walter Kauzmann, his ideal of the gentleman scientist, and John Bloor for proving that an organic chemist only needs a diagonalization routine to become a quantum chemist. Fred Richardson inspired his interest in trying to solve all the problems proposed in Kauzmann's 1940 review of optical activity, and his longtime mentor and chairman Larry Winters maintained his interest in problems of physical organic chemistry while continuing the Drexel link to Virginia Commonwealth University. The impact of Jerry Whitten's work with lobe basis sets and his encouragement of this work are also gratefully acknowledged.

The second group must include those authors of books that we have come back to again and again. Carl Trindle salutes Linus Pauling and E. Bright Wilson Jr. for a first notion of the power of quantum chemistry, Charles Coulson for a sense of adventure in applying such power, Wilson again with Decius and Cross for opening his eyes to the power of symmetry, and Frank Pilar for his example of clarity of exposition. He also thanks Abramowitz and Stegun for compiling a work of everlasting value.

We acknowledge with appreciation our colleagues who were so patient with our naïve questions, and so generous with their help. Their names appear in the chapters most relevant to their contributions. Personal thanks for making this work possible go first to Lance Wobus of Taylor & Francis, unfailing in encouragement. Barbara Body went as far as saying that writing this book could be valuable, and not only to the authors. She made time for Carl to write during vacations and weekends, lent a sympathetic ear, and supplied affectionate dogs to help console him. Nancy Shillady supported

Don throughout the work and helped in a thousand practical ways, especially in the clerical work of securing permissions. Don's son, Doug Shillady, constructed the Visual Basic® interface for PCLOBE, essential to its user's welcome. His former student Sherry Baldwin was the best possible reader. Professor Zikri Altun and the physicists at the University of Marmara heard and improved a version of this work presented in Istanbul in 2005. Not least, Toshiba made machines that survived the ordeal.

Carl Trindle
Charlottesville, Virginia

Donald D. Shillady
Ashland, Virginia

Authors

Carl Trindle studied at Grinnell College, Tufts University, Yale University, and Argonne National Laboratory. Professor K.H. Illinger of Tufts guided his dissertation on the theory of line broadening in microwave spectra. The University of Virginia has been his professional home; he is currently professor of chemistry and principal of Brown College there. He has taught the range of physical chemistry, especially applications of quantum mechanics, in the chemistry department and an assortment of seminars on scientific ethics, the history of nuclear weaponry, and more general subjects for the residential college.

Carl Trindle was named an Alfred P. Sloan fellow and was a National Academy of Sciences exchange fellow at the Institute Rudjer Bošković in Zagreb. He has lectured in India under the sponsorship of the Fulbright Foundation, spent a semester at the Technion in Haifa, and a summer at King's College (London), and has maintained a long-standing collaboration with Turkish scientists. He has served as an American Chemical Society tour speaker and, like his coauthor, was recognized by the Virginia Section of the ACS with its Distinguished Service Award.

His research interests include the computational characterization of the structures of highly reactive systems including dianions and dications, bonding in triplet carbenes and diradicals, and electronic influences on regioselectivity. Most recently he has been studying UV absorption spectra and circular dichroism of high-symmetry chiral systems.

Donald D. Shillady is an emeritus professor of chemistry at Virginia Commonwealth University in Richmond, Virginia. He has been interested in optical activity and magnetically induced optical activity since his PhD thesis in 1969 but later expanded his interests to the larger field of quantum chemistry. He enjoys teaching physical chemistry and quantum chemistry. He has carried out a few experimental spectroscopic studies of magnetic circular dichroism but has mainly devoted his career to writing programs for computation of ORD, CD, MCD, CASCI, and MCSCF results. So far he has written 77 publications and has edited two books. His past interests include the biological effects of electromagnetic waves, determination of the absolute configuration of large organic molecules, integral transform wave functions, properties of metal atom clusters, and the relationship between valence bond and configuration interaction methods in chemistry. Although retired, he occasionally teaches a chemistry course and the preparation of this text has brought him up to date with active areas of research in quantum chemistry.

His computer program PCLOBE which accompanies this text includes many features of his research over the years. This program builds orbitals from Gaussian lobes, which eases exploration of new areas of electronic structure modeling. It is a connection between theory and software.

The CD and installation instructions are located on the back inside cover and facing page respectively.

1

One-Dimensional Quantum Mechanics:
A Short Review

We want to provide a compact reminder of the framework of molecular electronic structure theory to those turning to computational chemistry. This chapter contains an account of the fundamental assumptions of quantum mechanics, including some of the rather few exact solutions to model problems, the algebra of operators, and an illustration of the uncertainty principle.

Quantum mechanics is well outside our human intuition, and we salute those adventurous scientists who were able to see so deeply into its strange world. The adventure begins with Planck's amazing account of the black-body radiation spectrum based on the idea that energy is exchanged in parcels (1901) [1] and Einstein's adoption of that principle to explain the photoelectric effect in 1905 [2]. Bohr's daring model of the hydrogen spectrum, incorporating quantization of angular momentum (1913) [3], recaptured the numerical representation of H atom absorption and emission frequencies observed by Balmer (1885) [4]. All these advances made reference to the constant h that Planck had used to fit the blackbody radiation spectrum.

This constant played a role in resolving the curious duality of wave and particle at the very small scale. Prince Louis De Broglie [5] proposed that every object with momentum had also an associated wavelength.*

$$\lambda = \frac{h}{mv} = \frac{h}{p}$$

The De Broglie wavelength λ is immeasurably small for macroscopic objects such as baseballs, ball bearings, or rifle bullets, and classical mechanics is excellent for describing these macroscopic objects. However, for subatomic particles, and in particular the electron, the De Broglie wavelength is comparable in size to chemical bonds.

---

*De Broglie arrived at this relationship by consideration of an oddity in relativity theory. [http://www.davis-inc.com/physics/]. It of course seems magical. Consider that the circular Bohr orbit with circumference $2\pi r$ accommodates one wavelength. Then the momentum is $p = h/(2\pi r)$ and the kinetic energy T is $h^2/[2m(2\pi r)^2]$. The potential V is $-Z/r$. The minimum energy $T + V = -m(2\pi Z)^2/2h^2$ is in agreement with Bohr's fit. Hey presto!

Only 3 years after the De Broglie hypothesis was published, Schrödinger [6] published a series of papers defining the wave mechanics and solving the harmonic oscillator and the hydrogen atom. Here is a way to express a link between the De Broglie waves and the Schrödinger mechanics.

The general form of a wave can be described as a complex exponential with amplitude A, frequency ω, and wavelength λ.

$$\psi(x,t) = A \exp\left[2\pi i\left(\frac{x}{\lambda} - \omega t\right)\right]$$

The first derivative of ψ with respect to x is simple

$$\frac{d\psi}{dx} = \frac{2\pi i}{\lambda} A \exp\left[2\pi i\left(\frac{x}{\lambda} - \omega t\right)\right] = \frac{2\pi i p}{h}\psi = \frac{ip}{\hbar}\psi$$

and returns the momentum as the eigenvalue. The second derivative produces the kinetic energy.

$$\frac{d^2\psi}{dx^2} = -\left(\frac{2\pi}{\lambda}\right)^2 A \exp\left[2\pi i\left(\frac{x}{\lambda} - \omega t\right)\right] = -\left(\frac{2\pi}{h}\right)^2 p^2\psi = -\frac{2mT}{\hbar^2}\psi = -\frac{2m(E-V)}{\hbar^2}\psi$$

Use of the De Broglie relation already incorporates reference to relativity. We can rearrange this further to obtain the standard form of the Schrödinger equation

$$\left[\frac{-\hbar^2}{2m}\frac{d^2}{dx^2} + V\right]\psi = H\psi = E\psi$$

The Schrödinger equation is thus a consequence of relativistic ideas.

The operator which produces the energy eigenvalue is called the Hamiltonian, after the formulation of the total energy in classical mechanics.

One may also recover the energy by evaluation of the time derivative

$$\frac{\partial\psi}{\partial t} = -2\pi i\omega A \exp\left[2\pi i\left(\frac{x}{\lambda} - \omega t\right)\right] = -i\frac{E}{\hbar}\psi$$

Now we are ready to form a strategy for using "wave mechanics."

1. Write the total energy operator H in terms of classical momenta and coordinates.
2. Insert the equivalent operator wherever a momentum occurs.
3. Consider the potential V to be a simple multiplicative operator.
4. Form the Hamiltonian and write the Schrödinger equation.
5. By whatever means solve the differential equation to find ψ and E_{tot}.

Among the few cases that have been solved exactly are the hydrogen atom, the harmonic oscillator, the rigid rotor, the particle-in-a-box, the particle-on-a-ring, and even a forced harmonic oscillator. Some seemingly simple problems such as the He atom or the diatomic hydrogen molecule have never been solved exactly. We can learn much of the structure of wave mechanics from the soluble problems, which will guide us as we construct approximations to solutions to molecular systems.

The Particle-in-a-Box

Let us consider a one-dimensional potential well extending from $x=0$ to L. The particle—perhaps an electron—can move freely inside the well (box), where $V=0$, but cannot penetrate the exterior where x is negative or greater than L, where $V=\infty$. The wave function must be zero except within the well. Inside the well, $V=0$ so the Hamiltonian is

$$H = \frac{-\hbar^2}{2m}\frac{d^2}{dx^2} + 0$$

The Schrödinger equation becomes

$$\frac{-\hbar^2}{2m}\frac{d^2}{dx^2}\psi_{Box} = E\psi_{Box}$$

If the wave function in the box is nonzero at either $x=0$ or $x=L$ then there is a discontinuity between the wave function segments. The first derivative is then infinite, the momentum is infinite, and the description of the system becomes nonsense. To avoid this we require that the wave function be exactly zero at the points $x=0$ and $x=L$. More generally, if the wave function is to be well behaved so as to give meaning to the energy:

1. ψ must be single-valued at every point.
2. ψ must be finite.
3. ψ must be a continuous function with finite first and second derivatives.

Now we can solve the Schrödinger equation; a rearrangement yields

$$\frac{d^2\psi}{dx^2} = -\frac{2m}{\hbar^2}E\psi \quad \text{and} \quad \left(\frac{d^2}{dx^2} + \frac{2mE}{\hbar^2}\right)\psi = 0$$

This equation can be factored.

$$\left(\frac{d}{dx} + i\sqrt{\frac{2mE}{\hbar^2}}\right)\left(\frac{d}{dx} - i\sqrt{\frac{2mE}{\hbar^2}}\right)\psi = \left(\frac{d}{dx} + i\alpha\right)\left(\frac{d}{dx} - i\alpha\right) = 0$$

We expect two "particular" solutions ψ_1 and ψ_2 from a second-order differential equation; either the first or second factor can be zero. The solution is straightforward

$$\left(\frac{d}{dx} \pm i\alpha\right)\psi_\pm = 0; \quad \psi_\pm = C_\pm \exp\left(\pm i\alpha x\right)$$

and

$$\psi = C_+ \exp\left(i\alpha x\right) + C_- \exp\left(-i\alpha x\right)$$

According to Euler's rule, $\exp\left(\pm ix\right) = \cos\left(x\right) \pm i \sin\left(x\right)$; we can thus reexpress the wave function in terms of the trigonometric function

$$\psi = A \cos\left(\alpha x\right) + B \sin\left(\alpha x\right)$$

Now we can apply the boundary conditions at the hard walls of the well (box) where we know $\psi = 0$. When $x = 0$, the sine is zero but the cosine is 1; thus A must vanish. When $x = L$

$$\psi = B \sin\left(\alpha L\right) = 0, \Leftrightarrow \alpha L = n\pi, n = \text{integer and so} \sqrt{\frac{2mE_n}{\hbar^2}} = \frac{n\pi}{L}$$

Therefore, E_n can only take on certain values defined by the quantum number n

$$E_n = \frac{(n\pi\hbar)^2}{2mL^2} = \frac{(n^2\pi^2\hbar^2)}{2mL^2(2\pi)^2} = \frac{n^2\hbar^2}{8mL^2}$$

Next, we need to find out the value of the constant B. Here we introduce the concept of normalizing the function

$$\int B^2 \sin^2\left(\alpha_n x\right) dx = 1$$

Normalization is a property of a probability distribution—the sum of probabilities for all possible events is unity. Here the total probability of finding the particle in the box is unity, and the square of ψ plays the role of the probability distribution.

We recommend consultation of a table of integrals as necessary. Here the integral is straightforward; use the trigonometric identity to replace $\sin^2(z)$ and complete the definite integral.

$$\int_0^L \left(B\sin\left(\frac{n\pi x}{L}\right)\right)^* \left(B\sin\left(\frac{n\pi x}{L}\right)\right)dx = B^2 \int_0^L \left[\frac{1-\cos\left(2n\pi x\right)}{2}\right]dx = \frac{B^2 L}{2} = 1$$

$$\psi_n = \sqrt{\frac{2}{L}}\sin\left(\frac{n\pi x}{L}\right)$$

Meaningful solutions for the particle in the box have positive integral quantum numbers n. If $n=0$ the wave function is zero; there is no particle in the well—a trivial solution. The negative integers merely produce a wave function of opposite sign, but with the same energy and probability distribution (Figure 1.1).

In this simple case, the energy increases (quadratically) as the quantum number increases. The lowest wave function is simply half of a sign wave with $\lambda = 2L$; higher kinetic energy is associated with shorter wavelengths, a larger number of nodes, and higher momentum.

The square of the wave function is a guide to the location of the electron; in the ground state the particle can avoid the walls, but for the higher energy states the electron is forced toward the barriers (Figure 1.2).

Notice that the higher-n wave functions have increasing numbers of nodes ($n-1$) and the higher energy wave functions and their squares tend to be spread out more over the whole length of the well (box) while the low energies are more near the middle of the well. In the classical limit, the probability distribution would be a constant from $x=0$ to L.

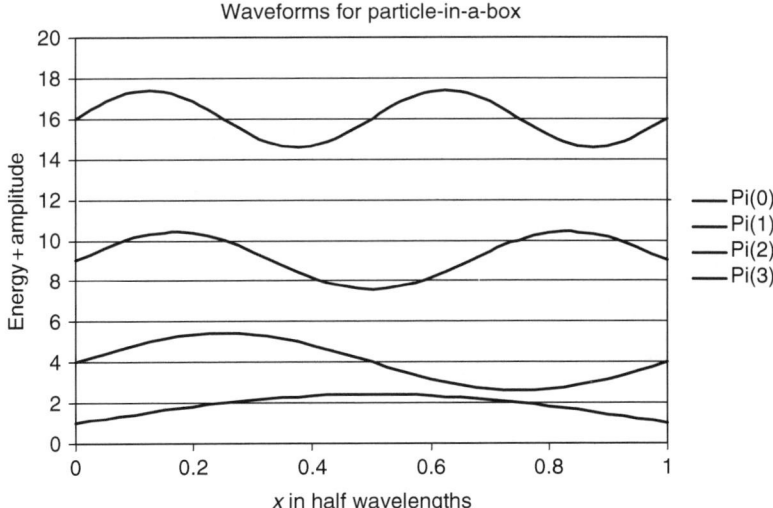

FIGURE 1.1
Normalized waveforms for the four lowest energy states of the particle in an infinite well of dimension one unit (one-half wavelength of the lowest state). The vertical axis is the dimensionless energy; state energy values $= n^2$.

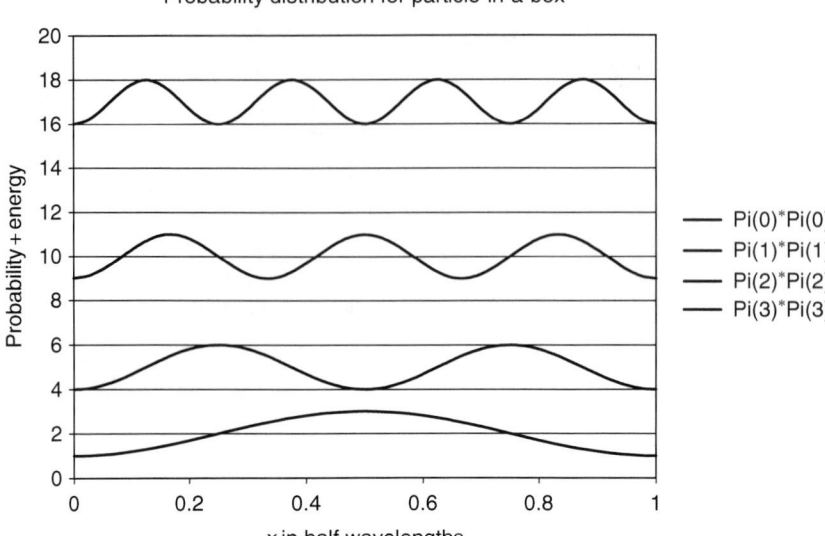

FIGURE 1.2

Normalized waveforms squared for the lowest four states of the particle in an infinite well. The axes are defined as in Figure 1.1. Note that for high-energy states, probability amplitudes migrate toward the barriers at $x = 0$ and 1.

The box model can be adapted to describe electrons in molecules. Let us consider the sigma framework of *trans*-butadiene as providing a shallow well for the four pi electrons in that compound. These electrons would occupy p orbitals on the C atoms. The box waveforms can be approximated by combinations of the four p orbitals. In Table 1.1 we show the weighting coefficients for those p orbitals, extracted from a molecular orbital (MO) calculation. The details of that calculation must await more systematic discussion, but we can anticipate some features of the description of the pi electrons. Since electrons can be placed in energy levels in pairs, the two lowest energy waveforms accept the four electrons in the model.

TABLE 1.1

Molecular Orbitals for *trans*-Butadiene

			π	π	π^*	π^*
#	At-Orb		14	15	16	17
5	C	2pz	−0.376	−0.525	0.618	−0.502
10	C	2pz	−0.488	−0.403	−0.462	0.682
15	C	2pz	−0.488	0.403	−0.462	−0.682
20	C	2pz	−0.376	0.525	0.618	0.502

* Selection of weighting coefficients for the STO-3G pi atomic orbitals produced in a SCF-MO calculation (see Chapter 4).

Now recall the nodal pattern we just derived for the particle-in-a-box (PIB) model. Column 14 has no sign changes (no nodes) in the coefficients and resembles the lowest energy waveform of the box. Column 15 has one node, column 16 has two nodes, and column 17 has three nodes. This nodal pattern agrees with the PIB model, and the relative amplitudes of the sine waves are expressed in the values of coefficients for the components in each column. Clearly this higher level all-electron calculation bears out the principle of the PIB model.

If the wave functions for the pi electrons in butadiene resemble the wave functions for the particle in the box, we may be able to relate the box properties with observable properties of the polyene. We will have to make several severe approximations. We will suppress any details of the C=C–C=C chain geometry and will assume that the electrons are independent particles. The length of the well L is adjustable; we can estimate its value, as about three times the length of a C–C bond in a delocalized pi system. In benzene, this is very close to 1.4 Å; thus, the assumed value of L would be at least 4 Å. The energy associated with removal of an electron from the second energy level and placing it in the third level would be about

$$\Delta E = h\nu = \frac{hc}{\lambda} = \frac{(3^2 - 2^2)h^2}{8mL^2}$$

Setting the box length to 5Å and rearranging the expression to isolate the wavelength.

$$\lambda \simeq \frac{8(9.11 \times 10^{-28}\text{g})(5 \times 10^{-8}\text{ cm})^2(3 \times 10^{10}\text{ cm/s})}{(3^2 - 2^2)(6.62 \times 10^{-27}\text{erg s})} = 1651 \times 10^{-8}\text{ cm} \simeq 165\text{ nm}$$

This wavelength is a little too blue, since the experimental peak is at about 217 nm [7]. To fit the transition wavelength to 217 nm, L would have to be

$$L \simeq \sqrt{\frac{(3^2 - 2^2)(6.62 \times 10^{-27}\text{erg s})(2.17 \times 10^{-5}\text{ cm})}{8(9.11 \times 10^{-28}\text{g})(3 \times 10^{10}\text{ cm/s})}} = 5.73 \times 10^{-8}\text{ cm} = 5.73\text{ Å}$$

Before we leave the PIB model we can use it to show aspects of what is meant by averaging in quantum mechanics. The "expectation value: of an operator" or the value one would expect to get after many repetitive measurements is defined in quantum mechanics as

$$\langle O \rangle = \frac{\int \psi^* O \psi \, d\tau}{\int \psi^* \psi \, d\tau}$$

If O is the Hamiltonian total energy operator H, the expectation value is precisely the eigenvalue E_n for the state ψ_n. Now suppose we want the average value of an operator for which ψ is not an eigenfunction, such as position of the particle in the PIB model. For the average position $\langle x \rangle$ of the object in the box we write

$$\langle x \rangle = \frac{\int_0^L \left(\sqrt{\frac{2}{L}} \sin\left(\frac{n\pi x}{L}\right) \right)^* (x) \left(\sqrt{\frac{2}{L}} \sin\left(\frac{n\pi x}{L}\right) \right) dx}{\int_0^L \left(\sqrt{\frac{2}{L}} \sin\left(\frac{n\pi x}{L}\right) \right)^* \left(\sqrt{\frac{2}{L}} \sin\left(\frac{n\pi x}{L}\right) \right) dx}$$

We know the integral in the denominator is 1 since we normalized the wave function. The values are available from standard tables of integrals; we find

$$\langle x \rangle = \frac{L}{2}$$

which is what we would have expected for a uniform distribution.

While the scalar operator x is nondirectional, the momentum in the x-direction is a vector operator with an associated direction. Let us evaluate the average value of the momentum operator P_x.

$$\langle P_x \rangle = \int_0^L \left(\sqrt{\frac{2}{L}} \sin\left(\frac{n\pi x}{L}\right) \right)^* \left(\frac{\hbar d}{i dx} \right) \left(\sqrt{\frac{2}{L}} \sin\left(\frac{n\pi x}{L}\right) \right) dx$$

$$\langle P_x \rangle = \left(\frac{2}{L} \right) \left(\frac{\hbar}{i} \right) \int_0^L \sin\left(\frac{2n\pi x}{L}\right) \left(\frac{2n\pi}{L} \right) \cos\left(\frac{2n\pi x}{L}\right) dx = 0$$

The zero value means that the particle has equal probability of traveling with momentum in the positive and negative directions with the net result of zero momentum.

Let us evaluate the expectation value of the square of x. One might think that all we have to do is to square the average value of $\langle x \rangle$ which would be $(L^2/4)$; however, a full evaluation of

$$\langle x^2 \rangle = \int_0^L \left(\sqrt{\frac{2}{L}} \sin\left(\frac{n\pi x}{L}\right) \right) (x^2) \left(\sqrt{\frac{2}{L}} \sin\left(\frac{n\pi x}{L}\right) \right) dx$$

shows two complications: not only is the expectation value of x^2 not the square of the expectation value of x, but the expectation value depends on the quantum number n

$$\langle n|x^2|n\rangle = \left(\frac{L^2}{3}\right) - \left(\frac{L^2}{2n^2\pi^2}\right) - \left(\frac{2}{L}\right)\left(\frac{L}{2n\pi}\right)^2(1-1) = \left(\frac{L^2}{3}\right)\left[1 - \left(\frac{3}{2n^2\pi^2}\right)\right]$$

The bracket notation at left was introduced by Dirac [8], and refers to the expectation value integral for state n. The difference between mean square and the square of the mean defines the spread in a probability distribution called the variance.

$$\text{Variance}(x) = \sigma^2(x) = \langle x^2\rangle - \langle x\rangle^2$$

Operators for which the wave function is not an eigenfunction have nonzero variance.

We can calculate the variance of the momentum with ease. The average value of the momentum of the particle is zero, as we have already seen. The square of the momentum is defined by the kinetic energy

$$\left\langle n|P_x^2|n\right\rangle = 2mE = 2m\left(\frac{n^2h^2}{8mL^2}\right) = \frac{n^2h^2}{4L^2}$$

Following Heisenberg [9] we define σ_x and σ_{P_x}. Let

$$\sigma_x^2 = (\langle x\rangle^2 - \langle x^2\rangle) \quad \text{and} \quad \sigma_{P_x}^2 = (\langle P_x\rangle^2 - \langle P_x^2\rangle)$$

Some manipulation yields

$$\sigma_x\sigma_{P_x} = \left(\frac{\hbar}{2}\right)\sqrt{1 + \left(\frac{n^2\pi^2 - 9}{3}\right)} > \left(\frac{\hbar}{2}\right)$$

Note this expression is very close to $(\hbar/2)$ when $n=1$ but increases roughly as n for large enough values of n. The variance—or "uncertainty"—in the results of simultaneous measurement of position and momentum grows larger as the energy increases.* It should not be assumed that this particular uncertainty equation is universal; in general the uncertainty depends on the shape of the potential well and the total energy. Nevertheless, we can conclude that the uncertainty is of the order of $(\hbar/2)$ and while this is small, it is not zero and at the atomic scale is significant.

Having now worked a problem we are in a position to make some postulates regarding the nature of quantum mechanics. These postulates of quantum mechanics consolidated by von Neumann [10] are generally accepted principles which capture a wide range of phenomena dealing with the physical behavior of small particles. We will briefly quote von Neumann's postulates, paraphrasing McQuarrie [11].

---

* There is considerable subtlety implicit in this statement, which we leave to the philosophically inclined reader to explore.

POSTULATE I
The state of a quantum-mechanical system is completely specified by a function $\Psi(\mathbf{r}, t)$ that depends on the coordinates of the particle and on the time. This function, called the wave function or the state function, has the important property that the product of $\Psi^*(\mathbf{r}, t)\, \Psi(\mathbf{r}, t)dx,\, dy,\, dz$ is the probability that the particle lies in the volume $dxdydz$, located at $\mathbf{r}$, at the time t.

COROLLARY
In order that the wave function be used in the Schrödinger equation, it must have several mathematical properties:

1. It must be finite.
2. It must be single-valued.
3. It must be defined for at least a first and second derivative.

These conditions assure that the kinetic energy is defined.

POSTULATE II
To every observable laboratory measurement in classical mechanics there corresponds an operator in quantum mechanics.

COROLLARY
Cartesian coordinates, (x, y, z), spherical polar coordinates, (r, θ, ϕ), or in general any set of coordinates, "q," merely become multiplicative operators, while the corresponding momentum operators, P_q, become differential operators such as $\left(\dfrac{\hbar}{i}\right)\left(\dfrac{\partial}{\partial q}\right)$.

POSTULATE III
In any measurement of the observable associated with the operator A, the only values that will ever be observed are the eigenvalues "a_j" which satisfy the eigenvalue equation

$$A\psi_j = a_j\psi_j$$

COROLLARY
If the state function is not an eigenfunction of the operator A, then only an average value can be obtained as from many measurements; see Postulate IV.

POSTULATE IV
If a system is in a state described by a normalized wave function Ψ, then the average value of the observable corresponding to the operator A is given by

$$\langle a \rangle = \int\limits_{-\infty}^{+\infty} \Psi^* A \Psi \, d\tau$$

COROLLARY

If the wave function is not normalized, then the average value of the observable corresponding to the operator A is given by

$$\langle a \rangle = \frac{\int_{-\infty}^{+\infty} \Psi^* A \Psi \, dt}{\int_{-\infty}^{+\infty} \Psi^* \Psi \, dt}$$

POSTULATE V

The wave function or state function of a system evolves in time according to the time-dependent Schrödinger equation

$$H\Psi(\mathbf{r}, t) = i\hbar \frac{\partial \Psi}{\partial t}$$

Particle-on-a-Ring

The problem of the energy levels of a particle constrained to move on a ring introduces complex wave functions and angular momentum. Let there be a particle of mass m (the electron, specifically) constrained to movement only on a ring of radius a. On the ring the potential energy of the electron is zero, $V = 0$. Otherwise the potential is infinite. Polar coordinates are the natural choice

$$x = a \sin \theta; \quad y = a \cos \theta$$

In classical mechanics we construct the Lagrangian

$$L = T - V = T = \frac{1}{2}m(\dot{x}^2 + \dot{y}^2) = \frac{1}{2}ma^2\dot{\theta}^2$$

The momentum is defined as the derivative of the Lagrangian with respect to a velocity. The kinetic energy then depends on the square of the momentum.

$$p_\theta = \frac{\partial L}{\partial \dot{\theta}} = ma^2\dot{\theta} \quad T = \frac{p_\theta^2}{2ma^2}$$

Now we can form the Hamiltonian total energy operator as

$$H = \left(\frac{-\hbar^2}{2ma^2}\right)\left(\frac{d^2}{d\theta^2}\right) + V$$

For the ring, where the potential is zero, the Schrödinger equation is

$$\left[\left(\frac{-\hbar^2}{2ma^2}\right)\left(\frac{d^2}{d\theta^2}\right) + 0\right]\psi = E\psi$$

This can be rearranged to the factored second-order differential equation.

$$\left[\left(\frac{d^2}{d\theta^2}\right) + \left(\frac{2ma^2E}{\hbar^2}\right)\right]\psi = \left[\left(\frac{d^2}{d\theta^2}\right) + \alpha^2\right]\psi = \left[\left(\frac{d}{d\theta} + i\alpha\right)\left(\frac{d}{d\theta} - i\alpha\right)\right]\psi = 0$$

The two solutions are superimposed in the general expression for the wave function:

$$\psi = A\exp(-i\alpha\theta) + B\exp(i\alpha\theta)$$

For the ring, if the function is to be continuous and single-valued

$$\psi(\theta) = \psi(\theta + 2n\pi); \quad n \text{ an integer}$$

We choose $n = 1$ for simplicity. Then

$$\exp(\pm i\alpha(\theta + 2\pi)) = \exp(\pm i\alpha\theta)\exp(\pm 2\pi\alpha i)$$

This requires that $\alpha = l$, an integer; once again confinement leads to quantization.

$$\alpha^2 = \frac{2ma^2E_l}{\hbar^2} = l^2 \quad E_l = \frac{\hbar^2 l^2}{2ma^2} = \frac{\hbar^2 l^2}{2I}$$

Here I is the moment of inertia. The wave function is complex, so the normalization requires the product of the function and its complex conjugate, obtained by replacing i by $-i$ ($i^2 = -1$).

$$\psi_n = Ce^{\pm in\theta} \text{ so set } 1 = \int_0^{2\pi} \psi_l^* \psi_l \, d\theta = \int_0^{2\pi} (C^* e^{-il\theta})(Ce^{il\theta}) \, d\theta = C^2 \int_0^{2\pi} d\theta = 2\pi C^2$$

which leads to $C = (1/\sqrt{2\pi})$. Finally we have the complete normalized solution.

$$E_n = \frac{n^2\hbar^2}{2ma^2} = \frac{n^2\hbar^2}{2I} \quad \text{and} \quad \psi_n = \frac{e^{\pm in\theta}}{\sqrt{2\pi}} \text{ for } n = 0, \pm1, \pm2, \pm3, \pm4, \ldots$$

The unique state with $l = 0$ has no angular momentum. All other pairs of states with common energy $E_{\pm l}$ have nonzero angular momentum

$$\hbar \frac{\partial [C \exp (\pm il\theta)]}{\partial \theta} = \pm l\hbar C \exp (\pm il\theta)$$

The complex exponentials are eigenfunctions of the angular momentum. The positive and negative values of the angular momentum correspond to circulation around the ring in clockwise and anticlockwise fashion, respectively.

We can model the motion of pi electrons on a ring such as benzene with the particle-on-a-ring (POR) solution. The lowest and unique energy solution with $l = 0$ has uniform amplitude around the ring. The solutions with $l = \pm 1$ have one node (change of sign), while the solutions with $l = 2$ have two nodes. We can assign electrons to these energy levels; the lowest level accommodates two electrons of opposite spin, while the higher levels accommodate as many as four electrons each. "Closed shells" are achieved when $4n + 2$ electrons are assigned in pairs to fill these energy levels. This can be related to the special stability of ring compounds such as benzene which has six pi electrons in its ring, and analogous $4n + 2$ pi electron "aromatic" systems.

Let us now compare the POR model to the description of benzene pi electrons obtained as waveforms expressed by weighted combinations of the six local p–pi atomic basis functions. We retain the labels of MOs from a description of the benzene molecule including all electrons. The energies in the list (in hartrees; 1 hartree = 27.21 electron volts) allow a ranking of the pi MOs (Table 1.2).

Orbital 17 is the lowest energy of the p–pi combinations at −0.43 hartree. The coefficients of local p–pi atomic orbitals in column 17 in Table 1.3 are all of the same sign and magnitude. Orbitals 20 and 21 are equal in energy at −0.27 hartree; the coefficients show one sign change. Orbitals 22 and 23 are also equal in energy at +0.27 hartree. This positive energy means that an electron placed in this MO is unstable relative to the free electron at zero energy.

Figure 1.3 suggests that there is no net bonding but rather net antibonding in the starred MOs, owing to the increased number of nodes.

TABLE 1.2

Energies for Waveforms for pi Molecular Orbitals in Benzene

One-Electron Energy Levels
E(17) = −0.435353305854
E(20) = −0.269941582607
E(21) = −0.269907928158
E(22) = 0.266362676286
E(23) = 0.266362976292
E(24) = 0.493982671388

TABLE 1.3

Molecular Orbitals for Benzene

			π	π	π	π^*	π^*	π^*
#	At-Orb		17	20	21	22	23	24
5	C	2pz	−0.344	−0.532	0.000	0.000	−0.645	0.511
10	C	2pz	−0.344	−0.266	0.461	−0.559	0.322	−0.511
15	C	2pz	−0.344	0.266	0.461	0.559	0.322	0.511
20	C	2pz	−0.344	0.532	0.000	0.000	−0.645	−0.511
25	C	2pz	−0.344	0.266	−0.461	−0.559	0.322	0.511
30	C	2pz	−0.344	−0.266	−0.461	0.559	0.322	−0.511

* Selection of weighting coefficients for the STO-3G pi atomic orbitals from a table produced in a SCF-MO calculation (see Chapter 4).

So far, five distinct levels have been defined; all of these follow the pattern set by the solutions of the POR. However, the set of six p–pi atomic orbitals cannot represent the solutions of the POR with shorter wavelength and higher energy. The last MO in the p–pi AO basis is unique and has a maximum number of sign changes.

We can model the spectrum of benzene, assuming the six pi electrons are assigned to the $l = 0$ level (two) and the $l = \pm 1$ level (four in all). Then, if ultraviolet light excited one of the electrons to make a transition from $l = 1$ to $l = 2$, the transition would occur at a wavelength λ.

$$E_2 - E_1 = h\nu = h\left(\frac{c}{\lambda}\right) = \left(\frac{(2^2 - 1^2)\hbar^2}{2ma^2}\right) \text{ so that } \lambda = \left(\frac{hc2ma^2}{3\hbar^2}\right) = \left(\frac{(4\pi^2)(2ma^2c)}{3h}\right)$$

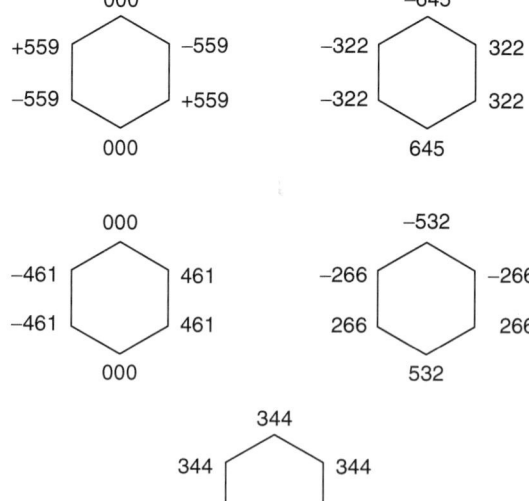

FIGURE 1.3
Amplitudes (×1000) for waveforms for a six-membered ring, as represented by basis functions at each apex. The vertical axis is energy (unscaled). The number of sign changes correlates with energy; the lowest energy wave has zero sign changes, etc.

The first pi $\rightarrow$ pi* transition of benzene is observed at 262 nm. The corresponding value of the radius a is

$$a = \sqrt{\frac{3h\lambda}{(8\pi^2)(mc)}} = \sqrt{\frac{3(6.62 \times 10^{-27}\text{erg s})(2625.3 \times 10^{-8} \text{ cm})}{8\pi^2(9.11 \times 10^{-28}\text{g})(3 \times 10^{10} \text{ cm/s})}} = 1.55 \times 10^{-8} \text{ cm}$$

The radius of a circle circumscribing benzene is the CC distance, 1.4 Å. Pretty close! Thus, the POR model gives a reasonable account of the first UV transition in benzene and also rationalizes the $4n + 2$ rule for aromatic rings. The model has guided a large number of spectroscopic studies including UV–visible absorptions and both natural and magnetically induced circular dichroism.

Operator Algebra in Quantum Mechanics

We have already noted that the momentum operator is complex and that the angular momentum wave function is complex. However, the quantities measured in a laboratory experiment are real numbers. "Hermitian" operators resolve this puzzle.

DEFINITION: Adjoint operators are defined as follows:

If $\langle\psi|\hat{A}|\psi\rangle = \langle\psi|\hat{A}^{\dagger}|\psi\rangle^*$ then $\hat{A}$ and $\hat{A}^{\dagger}$ are adjoint.

Here the asterisk refers to complex conjugation, in which every instance of i is replaced by $-i$ ($i^2 = -1$). Hermitian operators are self-adjoint.

THEOREM 1
The expectation values of Hermitian operators are real.

If$\langle\psi|\hat{A}|\psi\rangle = a + ib = \langle\psi|\hat{A}^{\dagger}|\psi\rangle^* = a - ib$; that is, $b = 0$

This implies that the eigenvalues of Hermitian operators are also real.

If $A\psi = a\psi$, then $\langle\psi|A|\psi\rangle = a$

Consider the expectation value of a Hermitian operator with a composite function

$$\int \varphi_m^* A\varphi_m \, d\tau = \langle\varphi_m|A|\varphi_m\rangle; \ \varphi_m = u + iv; \text{ real } u, v$$

$$\int (u + iv)^* A(u + iv) \, d\tau = \int u^* A u \, d\tau + \int v^* A v \, d\tau + i\left[\int u^* A v \, d\tau - \int v^* A u \, d\tau\right]$$

If the operator A is Hermitian, the expectation value must be real. Then the imaginary part vanishes, requiring

$$\int u^* A v \, d\tau = \int v^* A u \, d\tau$$

This can be rewritten as

$$\langle u | A v \rangle = \langle A u | v \rangle^*$$

This is called the turnover rule.

We often encounter collections of integrals of Hermitian operators, and this condition allows us to write an equivalence

$$A_{uv} = A_{vu}^*$$

THEOREM 2

Two different eigenfunctions of a Hermitian operator are orthogonal if their eigenvalues are not equal.

Assume $A\psi_1 = a\psi_1$, and $A\psi_2 = b\psi_2$ with $a \neq b$. Then consider $\int \psi_1 A\psi_2 d\tau$.

$$\int \psi_1^* A\psi_2 \, d\tau = b \int \psi_1^* \psi_2 \, d\tau$$

But by the turnover rule

$$\int \psi_1^* A\psi_2 \, d\tau = \int A\psi_1^* \psi_2 \, d\tau = a \int \psi_1^* \psi_2 \, d\tau$$

Subtracting the two equal expressions, we find

$$(a - b) \int \psi_1^* \psi_2 \, d\tau = 0$$

Since we initially assumed $a \neq b$ the integral must be zero.

Operator Commutation

In quantum mechanics the order in which we apply some operators makes a difference

$$AB\psi - BA\psi = [A,B]\psi$$

The quantity in square brackets is called the commutator. Consider two important operators we encountered in the PIB problem, x and P_x. Apply the operators in both orders to that function.

$$(xP_x - P_x x)\psi = x\left(\frac{\hbar}{i}\right)\frac{d\psi}{dx} - \left(\frac{\hbar}{i}\right)\frac{d}{dx}x\psi = x\left(\frac{\hbar}{i}\right)\frac{d\psi}{dx} - \left(\frac{\hbar}{i}\right)\psi - x\left(\frac{\hbar}{i}\right)\frac{d\psi}{dx} = -\left(\frac{\hbar}{i}\right)\psi$$

So, $[x, P_x]\psi = i\hbar\psi$ and we can write $[x, P_x] = i\hbar$. That is, the commutator of x and P_x is $i\hbar$. The reversed commutator is $[P_x, x] = -i\hbar$.

THEOREM 3
A single set of eigenfunctions can exist for two different Hermitian operators if the operators commute.

Assume there are two operators which have the same eigenfunction so that we have $A\psi = a\psi$ and $B\psi = b\psi$. Consider the integral where ψ is normalized and the operators commute, $[A, B] = 0$.

$$0 = \int\psi^*(0)\psi\,d\tau = \int\psi^*[A, B]\psi\,d\tau = \int\psi^*(AB - BA)\psi\,d\tau = (a^*b^* - ba)\int\psi^*\psi\,d\tau = 0$$

We know $a^*b^* = ab$ since the eigenvalues of a Hermitian operator are real (QED).

This last theorem says that there "can be" a common set of eigenfunctions for two operators that commute but does not provide a method to find them.

Commutation and Uncertainty

Consider two operators

$$A = F + i\mu G; \quad A^\dagger = F^\dagger - i\mu G^\dagger$$
$$\langle A^\dagger A\rangle = \langle F^\dagger F\rangle + \mu^2\langle G^\dagger G\rangle + \mu\langle i(F^\dagger G - G^\dagger F)\rangle$$

The first and second terms must be positive, as must be the whole expression. This requires that the roots of the quadratic form are complex. The discriminant of the quadratic equation must be negative, so

$$4\langle F^\dagger F\rangle\langle G^\dagger G\rangle \geq \langle i(F^\dagger G - G^\dagger F)\rangle^2 \equiv \langle i[F,G]\rangle^2$$

The last step required the assumption that F and G are Hermitian. Now consider that F and G are the deviations from a mean

$$F = f - \bar{f} = \Delta f; \quad G = g - \bar{g} = \Delta g$$

The equation becomes

$$(\Delta f)^2 (\Delta g)^2 \geq \tfrac{1}{4} \langle i[f,g] \rangle^2$$

That is, the uncertainty product depends on the commutator of the two operators in question. If the operators commute, the uncertainty in simultaneous measurements can be zero. For example, the energy and angular momentum of the POR can both be specified exactly. Position and momentum, however, cannot.

The Harmonic Oscillator

The harmonic oscillator was solved exactly by Schrödinger [6]; it is a model for molecular vibrations. There are several excellent accounts of the polynomial solution of the problem such as those as given by Pauling and Wilson [12] or McQuarrie [11], and of the operator method described by Dicke and Wittke [13]. Excellent discussions of molecular vibrations and especially the use of symmetry in simplifying the problem are given by Wilson et al. [14] and Cotton [15].

The problem consists of solving the Schrödinger equation for a particle of mass m constrained by a Hooke's law parabolic potential.

$$H = \left(\frac{P_x^2}{2m} \right) + \left(\frac{k}{2} \right) x^2 = \left(\frac{-\hbar^2}{2m} \right) \frac{d^2}{dx^2} + \left(\frac{k}{2} \right) x^2$$

It is useful to introduce scaled dimensionless variables. Divide the equation by half the quantum

$$H = \left(\frac{-\hbar^2}{m\hbar\omega} \right) \frac{d^2}{dx^2} + \left(\frac{k}{\hbar\omega} \right) x^2$$

Recall from classical mechanics that $k = m\omega^2$. Then defining

$$q^2 = m\omega x^2 / \hbar$$

transform the Schrödinger equation into the simple form

$$\left(-\frac{d^2}{dq^2} + q^2 \right) \psi = (p^2 + q^2)\psi = h\psi = \varepsilon\psi$$

Here $\varepsilon = 2E/\hbar\omega$ is the scaled (dimensionless) energy. Factoring the operator produces

$$(p + iq)(p - iq) = p^2 + q^2 + i[q,p] = h + i[q,p]$$

If we assume we have an eigenfunction ψ of h with eigenvalue ε, we can show that

$$h(p - iq)\psi = (\varepsilon - 2)(p - iq)\psi \quad \text{and} \quad h(p + iq)\psi = (\varepsilon + 2)(p + iq)\psi$$

That is, there is a sequence of energy levels separated by two scaled units or the vibrational quantum. There must be a lowest energy; if ψ_0 is the eigenfunction for that lowest state, the "step-down" operator must annihilate it.

$$(p - iq)\psi_0 = 0$$

The solution to this equation is

$$\psi_0 = N_0 \exp\left(-q^2/2\right)$$

The eigenvalue for this function is $\varepsilon = 1$, the half quantum. That is, in the lowest state there is still a residual energy. We noticed this in the PIB but not in the POR. From ψ_0 we can obtain higher energy states by the action of the "step-up" operator, $p + iq$. The allowed energy levels of the quantized oscillator are

$$E_n = \left(n + \frac{1}{2}\right)\hbar\omega$$

Let us revisit the idea of the uncertainty principle developed for the PIB problem. We now expect that $\langle 0|x|0\rangle^2 \neq \langle 0|x^2|0\rangle$ and $\langle 0|P_x|0\rangle^2 \neq \langle 0|P_x^2|0\rangle$. The symmetry of the problem is helpful. The harmonic oscillator potential is centered at the origin and extends over the symmetric range $(-\infty, \infty)$. Then

$$\int_{-\infty}^{\infty} \psi_0^* x \psi_0 \, dx = \int_{-\infty}^{\infty} (\text{even})(\text{odd})(\text{even}) \, dx = \int_{-\infty}^{\infty} (\text{odd})(\text{even}) \, dx = \langle 0|x|0\rangle = 0$$

Similarly, the momentum operator is antisymmetric with respect to replacement of x by $-x$, so

$$\langle 0|P_x|0\rangle = \int_{-\infty}^{\infty} (\text{even})(\text{odd})(\text{even}) \, dx = 0$$

Next, consider $\langle 0|x^2|0\rangle$. A table of integrals gives

$$\int_0^{\infty} x^2 e^{-a^2 x^2} \, dx = \frac{\sqrt{\pi}}{4a^3}$$

This leads us to

$$\frac{\hbar\omega}{2k} = \langle 0|x^2|0\rangle$$

Next, we need $\langle 0|P_x^2|0\rangle$.

$$\int_{-\infty}^{\infty} \psi_0^* \left(-\hbar^2 \frac{d^2}{dx^2}\right) \psi_0 \, dx = \langle 0|P_x^2|0\rangle = \left(\frac{\hbar^2}{2}\right)\left(\frac{k}{\hbar\omega}\right)$$

Now we can illustrate the Heisenberg uncertainty analysis of $\sigma_x\sigma_{P_x}$. Recall that

$$\sigma_x^2 = (\langle x\rangle^2 - \langle x^2\rangle) \text{ and } \sigma_{P_x}^2 = (\langle P_x\rangle^2 - \langle P_x^2\rangle);$$

we can write the product of square roots as

$$\sigma_x\sigma_{P_x} = \left[\left(\frac{\hbar\omega}{2k}\right) - (0)^2\right]^{1/2} \left[\left(\frac{\hbar^2}{2}\right)\left(\frac{k}{\hbar\omega}\right) - 0\right]^{1/2} = \sqrt{\frac{\hbar^2}{4}} = \frac{\hbar}{2}$$

For the quantized harmonic oscillator in its ground state we have the minimum uncertainty, $\sigma_x\sigma_{P_x} = (\hbar/2)$.

Summary

This preliminary chapter serves as a review of the principles of quantum mechanics by discussion of exactly soluble one-dimensional problems. Versions of this material have been used for over 30 years in lectures to undergraduates and first-year graduate students; in our experience, acquaintance with simple problems in quantum mechanics is necessary before introducing advanced topics. Exact solutions provide illustrations of the ideas of quantum mechanics and serve as models for molecular electronic structure. The principles of operator algebra carry over from these few exactly soluble problems to all methods for representing electronic structure.

In present-day electronic structure modeling, the operator algebra is replaced by an equivalent matrix algebra. We outline this alternative in our next chapter, before embarking on the study of methods of approximation.

References

1. Planck, M., *Ann. Phys.* 4:553, 1901.
2. Einstein, A., *Ann. Phys.* 17:132, 1905.

3. Bohr, N., *Phil. Mag.* 26:1, 1913.
4. Balmer, J.J., *Ann. Phys. Chem.* 25:80, 1885.
5. De Broglie, L., *Ann. Phys.* 3:22, 1925.
6. Schrödinger, E., *Ann. Phys.* 79:361, 489, 734; 80:437; 81:109, 1926.
7. Davis, J.C. Jr., *Advanced Physical Chemistry, Molecules Structure and Spectra*, Ronald Press, New York, 1965, p. 103.
8. Dirac, P.A.M., *Principles of Quantum Mechanics*, 3rd Ed., Oxford Press, Oxford, 1947.
9. Heisenberg, W., *Zeit f Phys.* 43:172, 1927.
10. von Neumann, J., *Mathematical Foundations of Quantum Mechanics*, Princeton University Press, Princeton, NJ, 1955, 1983.
11. McQuarrie, D.A., *Quantum Chemistry*, University Science Books, Mill Valley, CA, 1983.
12. Pauling, L. and Wilson, E.B., *Introduction to Quantum Mechanics with Applications to Chemistry*, McGraw Hill, New York, 1935.
13. Dicke, R.H. and Wittke, J.P., *Introduction to Quantum Mechanics*, Addison-Wesley, New York, 1960.
14. Wilson, E.B., Decius, J.C., and Cross, P.C., *Molecular Vibrations*, Dover, New York, 1955.
15. Cotton, F.A., *Chemical Applications of Group Theory*, Wiley-Interscience, New York, 1971.

2

Matrices, Representations, and Electronic Structure Modeling

Our first discussion of quantum mechanics emphasized the exact solutions of differential equation developed by Schrödinger. There are only a few such solutions and the differential equation is finally inadequate; a much more fruitful formulation, especially for molecular systems, was Heisenberg's [1] version of the theory which made use of representations of operators in matrices. von Neumann [2] reformulated quantum mechanics as an exercise in operator algebra. Matrix algebra and operator algebra are closely analogous. A formulation of quantum mechanics as an application of the linear algebra of matrices is particularly well suited to approximate calculations of electronic structure: it is the practical form of the Roothaan [3] SCF-MO equations.

After defining and illustrating the elementary operations of matrices, we will discuss the diagonalization of matrix representations. This is the most important practical procedure in quantum mechanics' matrix algebra; learn this concept and you will understand a great deal of quantum chemistry software.

Definition and Properties of Vectors and Matrices

First, we will sketch out the properties of matrices which are employed in molecular structure codes. This account of matrix algebra is based largely on a set of unpublished notes from Walter Kauzmann [4] at Princeton University.

In three-dimensional space, a point at (x, y, z) can be specified by a vector $\mathbf{p} = x\mathbf{i} + y\mathbf{j} + z\mathbf{k}$ expressed with the help of orthogonal unit vectors $\mathbf{i}, \mathbf{j}, \mathbf{k}$ along the Cartesian axes. These may be called a "basis" for representation of the vector. The unit vectors obey

$$\mathbf{u} \cdot \mathbf{u}' = 1 \text{ if } \mathbf{u} = \mathbf{u}', 0 \text{ otherwise}; \mathbf{u}, \mathbf{u}' = \mathbf{i}, \mathbf{j}, \mathbf{k}$$

Two vectors can be combined to form the scalar product

$$\mathbf{p} \cdot \mathbf{q} = (a\mathbf{i} + b\mathbf{j} + c\mathbf{k}) \cdot (r\mathbf{i} + s\mathbf{j} + t\mathbf{k}) = ar + bs + ct = |\mathbf{p}||\mathbf{q}|\cos\theta$$

which also defines the magnitude of a vector

$$|\mathbf{p}|^2 = a^2 + b^2 + c^2$$

Here θ is the angle between the two vectors.

Two vectors can also be combined as a vector product.

$$\mathbf{p} \times \mathbf{q} = \begin{vmatrix} \mathbf{i} & \mathbf{j} & \mathbf{k} \\ a & b & c \\ r & s & t \end{vmatrix} = \mathbf{i}(bt - sc) - \mathbf{j}(at - rc) + \mathbf{k}(as - rb)$$

Take note of the sign changes accompanying the expansion of the determinant. The magnitude of the "cross product" is

$$|\mathbf{p} \times \mathbf{q}| = |\mathbf{p}||\mathbf{q}|\sin\theta$$

Vectors can be reoriented in space, so that $\mathbf{v} \rightarrow \mathbf{v}'$. The reorientation can be represented by a transformation matrix

$$\begin{bmatrix} \cos\theta & -\sin\theta \\ \sin\theta & \cos\theta \end{bmatrix} \begin{bmatrix} x \\ y \end{bmatrix} = \begin{bmatrix} x\cos\theta - y\sin\theta \\ x\sin\theta + y\cos\theta \end{bmatrix} = \begin{bmatrix} x' \\ y' \end{bmatrix}$$

The matrix multiplication accomplished here is a special case of the general expression

$$\mathbf{RS} = \mathbf{T}; \quad \sum_j R_{ij}S_{jk} = T_{ik}$$

It is helpful to visualize the matrix $\mathbf{R}$ as being composed of a number of row vectors and $\mathbf{S}$ as being composed of column vectors.

$$\mathbf{R} = \begin{bmatrix} \mathbf{r}_1 \\ \mathbf{r}_2 \\ \vdots \end{bmatrix}; \quad \mathbf{S} = [\mathbf{c}_1 \quad \mathbf{c}_2 \cdots]$$

The elements of the product matrix T are scalar products of these row and column vectors.

$$T_{ik} = \mathbf{r}_i \cdot \mathbf{c}_k$$

Rotations alter the orientation of vectors but should not change the vector's magnitude. To preserve the magnitude, it is required that the rotation matrices be unitary. Unitary matrices obey the conditions

$$\det(\mathbf{U}) = 1; \quad \mathbf{U}^{\mathrm{Trans}} = \mathbf{U}^{-1}$$

where the "Trans" notation refers to the "Hermitian transpose"

$$U_{ij}^{\mathrm{Trans}} = U_{ji}^{*}$$

for which row and column labels are interchanged and the entry is replaced by its complex conjugate. Unitary matrices preserve the value of the determinant under the transform

$$\det(\mathbf{U}^{-1}\mathbf{MU}) = \det(\mathbf{U}^{-1})\det(\mathbf{M})\det(\mathbf{U}) = \det(\mathbf{M})$$

A particularly important use of the unitary transform is to reduce a matrix to diagonal form; such diagonal matrices are very simple to manipulate in computations, and the elements are generally the most meaningful results of a computation.

$$\mathbf{U}^{-1}\mathbf{LU} = \mathbf{\Lambda}; \quad \Lambda_{ij} = \lambda_{ii}\delta_{ij}$$

For a two-dimensional system

$$\begin{bmatrix} \cos\theta & \sin\theta \\ -\sin\theta & \cos\theta \end{bmatrix}\begin{bmatrix} L_{11} & L_{12} \\ L_{21} & L_{22} \end{bmatrix}\begin{bmatrix} \cos\theta & -\sin\theta \\ \sin\theta & \cos\theta \end{bmatrix} = \begin{bmatrix} \lambda_{11} & 0 \\ 0 & \lambda_{22} \end{bmatrix}$$

which permits the definition of the rotation angle.

$$\begin{aligned} P &= \begin{bmatrix} c & s \\ s & c \end{bmatrix}\begin{bmatrix} L_{11} & L_{12} \\ L_{21} & L_{22} \end{bmatrix}\begin{bmatrix} c & s \\ -s & c \end{bmatrix} = \begin{bmatrix} cL_{11} + sL_{21} & cL_{12} + sL_{22} \\ -sL_{11} + cL_{21} & -sL_{12} + cL_{22} \end{bmatrix}\begin{bmatrix} c & -s \\ s & c \end{bmatrix} \\ &= \begin{bmatrix} c^2L_{11} + csL_{21} + scL_{12} + s^2L_{22} & -scL_{11} - s^2L_{21} + c^2L_{12} + scL_{22} \\ -scL_{11} + c^2L_{21} - s^2L_{12} + csL_{22} & s^2L_{11} - scL_{21} + s^2L_{12} + c^2L_{22} \end{bmatrix} \end{aligned}$$

For symmetric matrices $L_{12} = L_{21} = B$. If $L_{11} = L_{22} = R$, the off-diagonal elements simplify to $B(c^2 - s^2)$ so the angle θ is 45°. Otherwise, the angle θ can be found from the equation which forces the off-diagonal element L_{12} to zero.

$$\tan(2\theta) = \frac{2L_{12}}{L_{11} - L_{22}}$$

One simple iterative method for diagonalizing larger matrices, the Jacobi algorithm, applies a sequence of two-dimensional rotations to the matrix to be diagonalized. In the computer program for this method, application to a given 2×2 subspace of a large matrix will, in general, change other parts of the matrix. After each pair, the largest remaining off-diagonal element defines the row and column of the new subspace. Eventually all the off-diagonal elements will become smaller than some desired threshold. The sequence of rotations acting on an initial unit matrix defines the transform which diagonalizes the matrix in question. This method is robust and has the desirable property of leaving the order of the coordinates undisturbed, but it can be slow. It is not recommended for large matrices, or when just a few eigenvalues and eigenvectors are needed.

In the algebraic formulation of quantum chemistry, we often encounter equations of the form

$$\mathbf{Fc} = \Lambda\mathbf{c}$$

Here Λ is a diagonal matrix, but $\mathbf{F}$ is not. Diagonalization accomplishes the solution. To approach the solution, we can insert the identity $\mathbf{I} = \mathbf{UU}^{-1}$ and multiply at left by $\mathbf{U}^{-1}$. $\mathbf{U}$ is such that

$$\mathbf{U}^{-1}\mathbf{FUU}^{-1}\mathbf{c} = \mathbf{fc}' = \mathbf{U}^{-1}\Lambda\mathbf{UU}^{-1}\mathbf{c} = \Lambda\mathbf{c}'$$

Here $\mathbf{f}$ is diagonal, with elements identical to those of Λ. The transformed vectors in $\mathbf{c}'$ are eigenvectors of the $\mathbf{f}$ matrix; the eigenvectors of the $\mathbf{F}$ matrix can be recovered from $\mathbf{c} = \mathbf{Uc}'$.

To make the problem simpler let us assume we are dealing with a problem in only two dimensions. Assume we have a matrix (operator) called $\mathbf{A}$ which acts on the (x, y) coordinates to produce a new vector. Once again we understand that the basis vectors are $(\mathbf{i}, \mathbf{j})$, the unit vectors in the (x, y) directions, and (c_1, c_2) as well as (d_1, d_2) are just numbers which tell us how much of each unit vector is present in the $\mathbf{c}$ or $\mathbf{d}$ vector.

$$\mathbf{Ac} = \mathbf{d} \quad \begin{bmatrix} 3 & 1 \\ 1 & 3 \end{bmatrix} \begin{bmatrix} c_1 \\ c_2 \end{bmatrix} = \begin{bmatrix} d_1 \\ d_2 \end{bmatrix}$$

Now to find the principal values (roots) of the matrix $\mathbf{A}$ we require

$$\mathbf{Ac} = \lambda\mathbf{c} \Rightarrow \begin{bmatrix} 3 & 1 \\ 1 & 3 \end{bmatrix} \begin{bmatrix} c_1 \\ c_2 \end{bmatrix} = \begin{bmatrix} \lambda & 0 \\ 0 & \lambda \end{bmatrix} \begin{bmatrix} c_1 \\ c_2 \end{bmatrix}$$

The equation is quadratic so there will be two values for λ—i.e., two roots. Notice that the matrix $\mathbf{A}$ is symmetric. In quantum physics/chemistry, matrices which represent physically measurable quantities are necessarily Hermitian-symmetric

$$A_{ij} = A_{ji}^{*}$$

The symbol * refers to complex conjugation; the imaginary number i becomes $-i$.

We can rearrange the equation to the form

$$(\mathbf{Ac} - \boldsymbol{\lambda}\mathbf{c}) = 0 \Rightarrow \left\{ \begin{bmatrix} 3 & 1 \\ 1 & 3 \end{bmatrix} - \begin{bmatrix} \lambda & 0 \\ 0 & \lambda \end{bmatrix} \right\} \begin{bmatrix} c_1 \\ c_2 \end{bmatrix} = \begin{bmatrix} 3-\lambda & 1 \\ 1 & 3-\lambda \end{bmatrix} \begin{bmatrix} c_1 \\ c_2 \end{bmatrix} = \begin{bmatrix} 0 \\ 0 \end{bmatrix}$$

Note that addition (or subtraction) of two matrices is accomplished simply by adding or subtracting the individual elements in corresponding positions. The matrices must be of the same dimension. This is a linear system of two equations in two unknowns (the c values) complicated by the fact that the vector at right is null. The way to solve this is to invoke the Cayley–Hamilton theorem [5–7] which requires that the values of λ be the roots of the polynomial

$$\begin{bmatrix} 3-\lambda & 1 \\ 1 & 3-\lambda \end{bmatrix} = 0 = (\lambda - r_1)(\lambda - r_2)$$

or $9 - 6\lambda + \lambda^2 - 1 = \lambda^2 - 6\lambda + 8 = (\lambda - 4)(\lambda - 2) = 0$, which has roots $r = 4, 2$

The roots satisfy the linear equations

$$(3 - r)c_1 + 1c_2 = 0$$
$$1c_1 + (3 - r)c_2 = 0$$

Choose a root and find the coefficients associated with that root. Consider $\lambda = 4$.

$$(3 - \lambda)c_1 + 1c_2 = 0 \quad \text{or} \quad (3 - 4)c_1 + 1c_2 = 0 \quad \text{and} \quad c_1 = c_2 = a$$

Alternatively

$$1c_1 + (3 - \lambda)c_2 = 0 \quad \text{or} \quad 1c_2 + (3 - 4)c_2 = 0 \quad \text{and} \quad c_1 = c_2 = a$$

Generally these linear equations are redundant. The last condition is that the coefficients define a normalized eigenvector. Thinking of the collection of coefficients as a vector $\mathbf{c}$, the magnitude of the vector (squared) is

$$\mathbf{c} \cdot \mathbf{c} = [c_1 \ c_2] \begin{bmatrix} c_1 \\ c_2 \end{bmatrix} = c_1^2 + c_2^2 = 1 = 2a^2$$

$$a = 1/\sqrt{2}$$

If we consider that solving the eigenvalue problem is defining a generalized rotation in a space defined by the basis, then we do not want the transform to change the determinant of the matrix. We say the transformation must be "unitary." The transformation matrix can be formed by assembling the orthogonal and normalized eigenvectors as adjacent columns in any order. Diagonalization by this transformation can be considered an uncoupling of the variables that were connected by off-diagonal elements in the original matrix.

Let us test the eigenvector; does it satisfy the eigenvalue equation? For the eigenvalue 4

$$
\begin{bmatrix} 3 & 1 \\ 1 & 3 \end{bmatrix} \begin{bmatrix} 1/\sqrt{2} \\ 1/\sqrt{2} \end{bmatrix} = \begin{bmatrix} \dfrac{3}{\sqrt{2}} + \dfrac{1}{\sqrt{2}} \\ \dfrac{1}{\sqrt{2}} + \dfrac{3}{\sqrt{2}} \end{bmatrix} = \begin{bmatrix} \dfrac{4}{\sqrt{2}} \\ \dfrac{4}{\sqrt{2}} \end{bmatrix} = \begin{bmatrix} 4 & 0 \\ 0 & 4 \end{bmatrix} \begin{bmatrix} 1/\sqrt{2} \\ 1/\sqrt{2} \end{bmatrix}
$$

$$
= 4 \begin{bmatrix} 1 & 0 \\ 0 & 1 \end{bmatrix} \begin{bmatrix} 1/\sqrt{2} \\ 1/\sqrt{2} \end{bmatrix} = 4 \begin{bmatrix} 1/\sqrt{2} \\ 1/\sqrt{2} \end{bmatrix}
$$

Let us obtain the eigenvector for the $\lambda = 2$ eigenvalue.

$$
(3 - \lambda)c_1 + 1c_2 = 0 \text{ or } (3 - 2)c_1 + 1c_2 = 0 \text{ and } c_1 = -c_2
$$
$$
1c_1 + (3 - \lambda)c_2 = 0 \text{ or } 1c_2 + (3 - 2)c_2 = 0 \text{ and } c_1 = -c_2
$$

Normalization leads to the values

$$
c_1 = \frac{1}{\sqrt{2}} \text{ and } c_2 = \frac{-1}{\sqrt{2}}
$$

Again we test this eigenvector.

$$
\begin{bmatrix} 3 & 1 \\ 1 & 3 \end{bmatrix} \begin{bmatrix} \dfrac{1}{\sqrt{2}} \\ \dfrac{-1}{\sqrt{2}} \end{bmatrix} = \begin{bmatrix} \dfrac{3}{\sqrt{2}} - \dfrac{1}{\sqrt{2}} \\ \dfrac{1}{\sqrt{2}} - \dfrac{3}{\sqrt{2}} \end{bmatrix} = \begin{bmatrix} \dfrac{2}{\sqrt{2}} \\ \dfrac{-2}{\sqrt{2}} \end{bmatrix} = \begin{bmatrix} 2 & 0 \\ 0 & 2 \end{bmatrix} \begin{bmatrix} \dfrac{1}{\sqrt{2}} \\ \dfrac{-1}{\sqrt{2}} \end{bmatrix} = 2 \begin{bmatrix} \dfrac{1}{\sqrt{2}} \\ \dfrac{-1}{\sqrt{2}} \end{bmatrix}
$$

It does produce the eigenvalue 2, as desired. The transformation matrix $\mathbf{T}$ can now be assembled.

$$
\mathbf{T} = \begin{bmatrix} \dfrac{1}{\sqrt{2}} & \dfrac{1}{\sqrt{2}} \\ \dfrac{-1}{\sqrt{2}} & \dfrac{1}{\sqrt{2}} \end{bmatrix}
$$

Let us now test whether T is unitary in two ways. First we can evaluate the determinant of the matrix. We see the value is 1.

$$\det \mathbf{T} = \begin{vmatrix} \dfrac{1}{\sqrt{2}} & \dfrac{1}{\sqrt{2}} \\ \dfrac{-1}{\sqrt{2}} & \dfrac{1}{\sqrt{2}} \end{vmatrix} = \frac{1}{2} - \left(\frac{-1}{2}\right) = 1$$

Next, we show that the product of $T^{\text{Trans}}\, T = 1$.

$$\begin{bmatrix} \dfrac{1}{\sqrt{2}} & \dfrac{-1}{\sqrt{2}} \\ \dfrac{1}{\sqrt{2}} & \dfrac{1}{\sqrt{2}} \end{bmatrix} \begin{bmatrix} \dfrac{1}{\sqrt{2}} & \dfrac{1}{\sqrt{2}} \\ \dfrac{-1}{\sqrt{2}} & \dfrac{1}{\sqrt{2}} \end{bmatrix} = \begin{bmatrix} \left(\dfrac{1}{2}+\dfrac{1}{2}\right) & \left(\dfrac{1}{2}-\dfrac{1}{2}\right) \\ \left(\dfrac{1}{2}-\dfrac{1}{2}\right) & \left(\dfrac{1}{2}+\dfrac{1}{2}\right) \end{bmatrix} = \begin{bmatrix} 1 & 0 \\ 0 & 1 \end{bmatrix}$$

and as a unitary matrix $T_{i,j}^{-1} = T_{j,i}$ so

$$\mathbf{T}^{-1} = \mathbf{T}^{\text{Trans}} = \begin{bmatrix} \dfrac{1}{\sqrt{2}} & \dfrac{-1}{\sqrt{2}} \\ \dfrac{1}{\sqrt{2}} & \dfrac{1}{\sqrt{2}} \end{bmatrix}$$

Let us construct

$$\mathbf{T}^{-1}\mathbf{MT} = \begin{bmatrix} \dfrac{1}{\sqrt{2}} & \dfrac{-1}{\sqrt{2}} \\ \dfrac{1}{\sqrt{2}} & \dfrac{1}{\sqrt{2}} \end{bmatrix} \begin{bmatrix} 3 & 1 \\ 1 & 3 \end{bmatrix} \begin{bmatrix} \dfrac{1}{\sqrt{2}} & \dfrac{1}{\sqrt{2}} \\ \dfrac{-1}{\sqrt{2}} & \dfrac{1}{\sqrt{2}} \end{bmatrix} = \begin{bmatrix} \dfrac{1}{\sqrt{2}} & \dfrac{-1}{\sqrt{2}} \\ \dfrac{1}{\sqrt{2}} & \dfrac{1}{\sqrt{2}} \end{bmatrix} \begin{bmatrix} \left(\dfrac{2}{\sqrt{2}}\right) & \left(\dfrac{4}{\sqrt{2}}\right) \\ \left(\dfrac{-2}{\sqrt{2}}\right) & \left(\dfrac{4}{\sqrt{2}}\right) \end{bmatrix}$$

$$= \begin{bmatrix} 2 & 0 \\ 0 & 4 \end{bmatrix}$$

We recover the diagonal matrix with elements equal to the eigenvalues and unchanged diagonal sum (trace) $= 6$.

Response Matrices

Rotation matrices, in particular, and unitary matrices, in general, are special cases of response matrices, which appear in many physical settings. For example, under the influence of an external field **E**, a charge distribution will deform. The deformation is described by an induced dipole **D**. The response matrix is the polarizability tensor $\boldsymbol{\alpha}$

$$\boldsymbol{\alpha}\mathbf{E} = \mathbf{D}$$

or

$$\begin{bmatrix} \alpha_{xx} & \alpha_{xy} & \alpha_{xz} \\ \alpha_{xy} & \alpha_{yy} & \alpha_{yz} \\ \alpha_{xz} & \alpha_{yz} & \alpha_{zz} \end{bmatrix} \begin{bmatrix} E_x \\ E_y \\ E_z \end{bmatrix} = \begin{bmatrix} D_x \\ D_y \\ D_z \end{bmatrix}$$

In this equation there is no explicit reference to the unit vectors, but the subscripts suggest that the Cartesian unit vectors are to be understood as the "basis" for the matrix representation of the field and dipole vectors and for the polarizability tensor.

The term "basis" is much more general than we have so far suggested. For example, a Fourier series is an expression of any periodic function as a superposition or sum of harmonic terms: for a function $F(x)$ defined for an interval $0 \leq x \leq L$

$$F(x) = \sum_{n \geq 1} A_n \sin(n\pi x/L)$$

The function $\sin(n\pi x/L)$ is the nth member of the basis for the expansion of the function. It is convenient when the members of the basis are orthogonal, which is the case here—the orthogonality is expressed as values of integrals, the analogy of the scalar products of the unit vectors in Cartesian space.

$$\left\langle \sqrt{2/L} \sin(n\pi x/L) \middle| \sqrt{2/L} \sin(m\pi x/L) \right\rangle = \delta_{nm}$$

It is no accident that we have chosen the eigenfunctions of the particle in an infinite well problem as the basis for expansion. This is a special application of the general rule that the eigenfunctions of a Hamiltonian compose a complete orthogonal and normalizable (orthonormal) basis for expression of any function in the space of that operator. Just as the Fourier coefficients are defined as

$$A_n = \left\langle F \middle| \sqrt{2/L} \sin(n\pi x/L) \right\rangle$$

the expansion coefficients in the series of eigenfunctions representing any function in the space of a Hamiltonian are given by

$$H\psi_n = E_n\psi_n; \quad G = \sum_n C_n\psi_n; \quad C_n = \langle G|\psi_n \rangle$$

A basis need not be a set of eigenfunctions of any particular operator in the space we might study. In principle, any set of functions would do. Let us consider a simple problem for which we might know the Hamiltonian and its eigenfunctions, as above. But perhaps the problem of interest, defined in this same space, has a Hamiltonian

$$H_P = H + V$$

We may construct a matrix

$$\mathbf{H_P} = (H_P)_{rs} = \langle \psi_r|H_P|\psi_s \rangle = \langle \psi_r|H|\psi_s \rangle + \langle \psi_r|V|\psi_s \rangle$$

The last two terms define matrices which are, respectively, a representation of the reference Hamiltonian in the basis of its eigenfunctions and a representation of the potential in that same basis. If the complete basis is transformed so as to produce a diagonal representation of the Hamiltonian, the diagonal elements are eigenvalues of that Hamiltonian.

$$\mathbf{U^{-1}H_P U = E}; \quad \Psi_k = \sum_\mu U_{k\mu}\psi_\mu; \quad E_k = \sum U_{k\nu}^{-1} H_{\mu l} U_{k\mu}$$

These equations are exact only if the basis is complete, but we can only fall short of the perfect completeness of an infinite basis. Finite matrix representations are approximate representations; their diagonalization yields estimates of eigenfunctions and eigenvalues.

To illustrate the use of matrix representations, we will use the exactly soluble harmonic oscillator which we have described elsewhere. In our first example we will represent the momentum and position operators in the basis of eigenfunctions. We saw from symmetry that

$$\langle n|q|n\rangle = 0; \quad \left\langle n\left|-i\frac{\partial}{\partial q}\right|n\right\rangle = 0$$

and in fact all other elements of the $\mathbf{q}$ and $\mathbf{p}$ matrices are zero except

$$\langle n|q|n \pm 1\rangle \text{ and } \left\langle n\left|-i\frac{\partial}{\partial q}\right|n \pm 1\right\rangle$$

More explicitly, the length integral is

$$\langle m|q|n\rangle = A_m A_n \int_{-\infty}^{+\infty} H_m\, q H_n \exp\left(-q^2\right) dq$$

We can evaluate this integral via the recursion relation

$$H_n = 2yH_{n-1} - 2(n-1)H_{n-2} \text{ or } yH_n = \frac{1}{2}H_{n+1} + nH_{n-1}$$

Then

$$\langle m|q|n\rangle = A_m A_n \int_{-\infty}^{+\infty} H_m\left[\frac{1}{2}H_{n+1} + nH_{n-1}\right]e^{-q^2} dq$$

Since

$$\int\limits_{-\infty}^{+\infty} H_m(y)H_n(y)e^{-y^2}\,dy = \langle m|n\rangle = \delta_{mn}2^n n!\sqrt{\pi}$$

$$\langle m|q|n\rangle = A_m A_n \left[\frac{1}{2}\delta_{m,n+1}2^{n+1}(n+1)!\sqrt{\pi} + \delta_{mn-1}2^{n-1}n!\sqrt{\pi}\right]$$

the normalization constants are

$$A_n = [2^n n!\sqrt{\pi}]^{-1/2}$$

so

$$q_{m,n} = 2^n n! (2^n 2^m n! m!)^{-1/2}\left[\delta_{m,n+1}(n+1) + \frac{1}{2}\delta_{m,n-1}\right]$$

$$q_{m,n} = \left(\frac{2^n n!}{2^m m!}\right)^{1/2}\left[\delta_{m,n+1}(n+1) + \frac{1}{2}\delta_{m,n-1}\right]$$

The only two states that mix under q are $m = n+1$ and $m = n-1$.
If $m = n+1$, then

$$q_{n+1,n} = [(n+1)/2]^{1/2}$$

If $m = n - 1$, then

$$q_{n-1,n} = [n/2]^{1/2}$$

A part of the matrix representation of q in harmonic oscillator eigenfunctions is

$$\begin{bmatrix} 0 & 1 & 0 & 0 \\ 1 & 0 & (3/2)^{1/2} & 0 \\ 0 & (3/2)^{1/2} & 0 & 2^{1/2} \\ 0 & 0 & 2^{1/2} & 0 \end{bmatrix} = \mathbf{q}$$

We will use this representation to obtain the matrices for powers of q. For example,

$$\mathbf{q}^2 = \mathbf{q} \times \mathbf{q} = \begin{bmatrix} 1/2 & 0 & \sqrt{2}/2 & 0 \\ 0 & 3/2 & 0 & \sqrt{6}/2 \\ \sqrt{2}/2 & 0 & 5/2 & 0 \\ 0 & \sqrt{6}/2 & 0 & 7/2 \end{bmatrix}$$

A similar analysis of the momentum operator produces the matrix

$$
\mathbf{p} = \begin{bmatrix} 0 & i[2]^{1/2}/2 & 0 & 0 \\ i[2]^{1/2}/2 & 0 & i[4]^{1/2}/2 & 0 \\ 0 & i[4]^{1/2}/2 & 0 & i[6]^{1/2}/2 \\ 0 & 0 & i[6]^{1/2}/2 & 0 \end{bmatrix}
$$

And $\mathbf{p}^2$ is then

$$
\mathbf{p}^2 = \begin{bmatrix} 1/2 & 0 & -\sqrt{2}/2 & 0 \\ 0 & 3/2 & 0 & -\sqrt{6}/2 \\ -\sqrt{2}/2 & 0 & 5/2 & 0 \\ 0 & -\sqrt{6}/2 & 0 & 7/2 \end{bmatrix}
$$

The uncertainty in the ground state is

$$
(U_{00})^2 = \langle 0|q^2|0\rangle \langle 0|p^2|0\rangle = (\Delta q)^2(\Delta p)^2 = (1/2)^2
$$

Reverting to unscaled (physical) units

$$
U_{00} = \hbar/2
$$

This is the familiar form of the uncertainly principle. You can verify that the uncertainty is larger for excited states, which is consistent with the more general statement

$$
U \geq (\hbar/2)
$$

Return to the scaled units and construct the commutator of operators for momentum and position

$$
pq - qp = [p, q] = i
$$

The matrix equivalent construction is

$$
\mathbf{pq} - \mathbf{qp} = [\mathbf{p}, \mathbf{q}] = \begin{bmatrix} i & 0 & 0 & 0 \\ 0 & i & 0 & 0 \\ 0 & 0 & i & 0 \\ 0 & 0 & 0 & i \end{bmatrix}
$$

This shows that the matrix representation of operators obey the same commutation laws as the operators themselves. This is, of course, required of a faithful representation.

Finally, let us construct the scaled Hamiltonian

$$\mathbf{h} = \mathbf{p}^2 + \mathbf{q}^2 = \begin{bmatrix} 1 & 0 & 0 & 0 \\ 0 & 3 & 0 & 0 \\ 0 & 0 & 5 & 0 \\ 0 & 0 & 0 & 7 \end{bmatrix}$$

Recall that the unit of energy in the scaled system is half the quantum, so the diagonal elements are

$$\mathbf{H}_{nn} = (2n + 1)\,\hbar\omega_0/2$$

Keep in mind that these 4×4 arrays are small samples of—in principle—infinite matrices. Finite matrices provide an inevitably incomplete representation except in the very special cases of angular momentum where the space of the operator is finite.

Symmetry Operations

Group theory is a powerful means of simplification of many aspects of electronic structure modeling. Accessible accounts of the theory and its applications are provided by Cotton [8] and by Harris and Bertolucci [9]. The fundamental quantities in group theory are symmetry operations. Symmetry operations can be considered to be rotations in three-dimensional Cartesian space, reflections, inversions, or combinations of these operations. To each action there corresponds a matrix. The basis may be composed of displacement vectors from reference points. For example, reflection at a particular atom affects that atom's Cartesian coordinates.

$$\text{Reflection in the } x\text{--}y \text{ plane: } \sigma_{XY}\mathbf{r} = \begin{bmatrix} 1 & 0 & 0 \\ 0 & 1 & 0 \\ 0 & 0 & -1 \end{bmatrix} \begin{bmatrix} x \\ y \\ z \end{bmatrix} = \begin{bmatrix} x \\ y \\ -z \end{bmatrix}$$

Clockwise $90°$ rotation in the x--y plane about the z-axis

$$C_4\vec{r} = \begin{bmatrix} 0 & -1 & 0 \\ 1 & 0 & 0 \\ 0 & 0 & 1 \end{bmatrix} \begin{bmatrix} x \\ y \\ z \end{bmatrix} = \begin{bmatrix} -y \\ x \\ z \end{bmatrix}$$

Clockwise 120° rotation in the $x-y$ plane about the z-axis

$$C_3\vec{r} = \begin{bmatrix} \cos(120) & \sin(120) & 0 \\ -\sin(120) & \cos(120) & 0 \\ 0 & 0 & 1 \end{bmatrix} \begin{bmatrix} x \\ y \\ z \end{bmatrix}$$

Inversion through the origin using the symbol "i" for the inversion operation

$$i\vec{r} = \begin{bmatrix} -1 & 0 & 0 \\ 0 & -1 & 0 \\ 0 & 0 & -1 \end{bmatrix} \begin{bmatrix} x \\ y \\ z \end{bmatrix} = \begin{bmatrix} -x \\ -y \\ -z \end{bmatrix}$$

There are other point group symmetry operations that can easily be represented in terms of a Cartesian matrix based on the (x, y, z) coordinates of each atom in the molecule [3].

Conclusion

While the ideas of Schrödinger's wave mechanics permeate chemistry, Heisenberg's "matrix mechanics" is actually much closer to what is used in modern computer algorithms. Almost all operations in electronic structure codes are matrix constructions and manipulations, and it is nearly impossible to comprehend the programs' structure and purpose without an acquaintance with the mathematics of matrices. We hope that the simple examples offered here are sufficient for anyone innocent of matrix algebra to proceed with relative ease through the remainder of this text which is almost entirely expressed in the language of matrices.

References

1. Heisenberg, W., *Zeit f Phys*. 43:172, 1927.
2. von Neumann, J. and Beyer, R.T. (translation), *Mathematical Foundations of Quantum Mechanics*, Princeton Press, Princeton, 1955.
3. Roothaan, C.C.J., *Rev. Mod. Phys.* 23:69, 1951.
4. Kauzman, W., Acknowledged with thanks for succinct notes sufficient for a working knowledge of linear algebra in the absence of formal training, 1962.
5. Fox, L., *An Introduction to Numerical Linear Algebra, with Exercises*, Oxford University Press, New York, 1965.
6. Kelly, L.G., *Handbook of Numerical Methods and Applications*, Addison-Wesley, Reading, MA, 1967.

7. Dahlquist, G. and Bjorck, A., *Numerical Methods*, English Translation from Swedish by N. Anderson, Dover, Mineola, Prentice-Hall, New York, 1974.
8. Cotton, F.A., *Chemical Applications of Group Theory*, 2nd Ed., Wiley-Interscience, New York, 1971.
9. Harris, D.C. and Bertolucci, M., *Symmetry and Spectroscopy: An Introduction to Vibrational and Electronic Spectroscopy*, Oxford University Press, New York, 1978, available from Dover.

3

Methods of Approximation and the SCF Method

The Variation Theorem

There are very few problems which have exact solutions in quantum mechanics. How will we proceed to describe a molecule when even the He atom does not have an exact solution? Two powerful techniques, the variation method and the perturbation method, are in wide use. The variation method is used most widely to characterize molecular energies while perturbation theory is applied to molecular properties. Almost every calculation in quantum chemistry begins with an application of the variation principle—even estimates of molecular properties, defined as a system's response to some disturbance, requires as a starting point some description of the undisturbed system. This description will always be founded on a variational method. Here we explore the variational theorem and a variety of illustrations, leaving the perturbation methods for later attention.

THEOREM
The expectation value of the Hamiltonian operator for a trial wave function is an upper bound to the exact ground state energy: $E_{\text{trial}} \geq E_0$.

PROOF

1. Assume an exact solution exists with a wave function ψ_0 and $H\psi_0 = E_0\psi_0$.
2. Assume $\{\psi_n\}$ is an orthonormal complete set and the eigenvalues are ordered so that the energy values increase with the increasing level number n, $E_{n+1} \geq E_n$ with E_0 lowest.
3. Expand the trial function ϕ as a linear combination of the exact wave functions $\{\psi_n\}$ as $\phi = \sum_n c_n\psi_n$.
4. Let the trial wave function be normalized; then

$$\langle \phi | H - E_0 | \phi \rangle = \sum_{n,m} c_n^* c_m \langle \psi_n | H | \psi_m \rangle - E_0 \sum_m c_m^* c_m = \sum_m c_m^* c_m \, (E_m - E_0)$$

Every term of the series is zero or positive, so the sum is positive. Finally, we have

$$\langle \phi | H - E_0 | \phi \rangle \geq 0 \quad \text{or} \quad \langle \phi | H | \phi \rangle \geq E_0 \text{(QED)}$$

Pauling and Wilson [1] attribute this method to Eckart [2], but he is seldom given proper credit for this very important theorem.

The Variational Treatment of the H Atom

Consider the H-like atom, a system of one electron with kinetic energy attracted electrostatically to a fixed charge of $+Z$ in electron charge units. The kinetic energy formula is

$$T = \left(\frac{-\hbar^2}{2m} \right) \left[\left(\frac{\partial^2}{\partial x^2} \right) + \left(\frac{\partial^2}{\partial y^2} \right) + \left(\frac{\partial^2}{\partial z^2} \right) \right] = \left(\frac{-\hbar^2}{2m} \right) \nabla^2(x,y,z)$$

The Laplacian ∇^2 in spherical polar coordinates is

$$\nabla^2(r,\theta,\phi) = \frac{1}{r^2} \frac{\partial}{\partial r} \left(r^2 \frac{\partial}{\partial r} \right) + \frac{1}{r^2 \sin(\theta)} \frac{\partial}{\partial \theta} \left(\sin(\theta) \frac{\partial}{\partial \theta} \right) + \frac{1}{r^2 \sin^2(\theta)} \left(\frac{\partial^2}{\partial \phi^2} \right)$$

A suitable form for the trial wave function is $\phi = Ne^{-\alpha r}$. To normalize this function in the polar coordinate system we require

$$1 = \int_0^\infty \int_0^\pi \int_0^{2\pi} N^2 e^{-2\alpha r} r^2 \sin(\theta) \, dr \, d\theta \, d\phi = (4\pi) N^2 \int_0^\infty e^{-2\alpha r} r^2 \, dr = 4\pi N^2 \left(\frac{2!}{8\alpha^3} \right)$$

The normalized trial wave function is then

$$\phi = \sqrt{\frac{\alpha^3}{\pi}} e^{-\alpha r}$$

We have a notable simplification at this point because there is no dependence on the angles (θ, ϕ); thus, the second and third terms of the ∇^2 operator have no effect. Therefore

$$H = -\frac{1}{2} \left[\frac{1}{r^2} \frac{\partial}{\partial r} \left(r^2 \frac{\partial}{\partial r} \right) \right] - \frac{Z}{r}$$

using atomic units. Then we can evaluate $\langle E \rangle$ by integrating over the radial coordinate and multiplication by the 4π angular factor.

$$\langle E \rangle = 4\pi \left(\frac{\alpha^3}{\pi} \right) \left\{ \left(\frac{-\alpha^2}{2} \right) \int_0^\infty e^{-2\alpha r} r^2 \, dr + \left(\frac{1}{2} \right) (2\alpha) \int_0^\infty e^{-2\alpha r} r \, dr - \int_0^\infty e^{-2\alpha r} r \, dr \right\}$$

which simplifies to

$$\frac{\alpha^2}{2} - \alpha = \langle E(\alpha) \rangle$$

The variation theorem assures us that the minimum of the trial energy will always be an upper bound to the exact ground state energy.

$$\frac{d}{d\alpha} \langle E(\alpha) \rangle = \alpha - 1 = 0$$

$$\alpha_{min} = 1, \quad \text{so} \quad \langle E(\alpha_{min}) \rangle = \frac{1}{2} - 1 = -\frac{1}{2} \text{ hartree}$$

This is the exact energy found by Bohr in 1913 [3] and by Schrödinger [4]. Thus, in this case the variationally optimized energy is exact. The trial wave function could express the exact ground state wave function.

Variational Treatment of the H Atom (Gaussian Trial Function)

Let us now treat the H atom variationally using a single Gaussian orbital as a trial function. We will need two general integrals:

$$\int_0^\infty x^{2n} e^{-ax^2} \, dx = \frac{\sqrt{\pi} 1 \cdot 3 \cdot 5 \cdots (2n-1)}{2^{n+1} a^{(2n+1)/2}} \quad \text{and} \quad \int_0^\pi x^{(2n+1)} e^{-ax^2} \, dx = \frac{n!}{2a^{(2n+2)/2}}$$

Assume the trial wave function $\phi = N \exp(-br^2)$. Here b has units of inverse length squared; we thus expect the total energy to contain a term proportional to b (the kinetic energy) and to $\sqrt{b}$ (the potential energy). For the spherical system

$$\frac{1}{r^2} \frac{\partial}{\partial r} \left(r^2 \frac{\partial}{\partial r} \right) e^{-br^2} = \frac{1}{r^2} \frac{\partial}{\partial r} (r^2) \left[-2bre^{-br^2} \right] = \frac{1}{r^2} \left[-6br^2 + 4b^2 r^4 \right] e^{-br^2}$$

To normalize ϕ we have

$$1 = 4\pi N^2 \int_0^\infty e^{-2br^2} r^2 \, dr = \frac{4\pi N^2 \sqrt{\pi}}{2^2 (2b)^{3/2}} \quad \text{so} \quad N = \left(\frac{2b}{\pi} \right)^{3/4}$$

and the normalized trial function becomes

$$\phi = \left(\frac{2b}{\pi}\right)^{3/4} \exp\left(-br^2\right)$$

The energy is then

$$\langle \phi|H|\phi \rangle = 4\pi \left(\frac{2b}{\pi}\right)^{3/2} \left\{ -\frac{1}{2}\int_0^\infty e^{-2br^2}[4b^2r^4 - 6br^2]\frac{r^2}{r^2}dr - Z\int_0^\infty e^{-2br^2}\frac{r^2}{r}dr \right\}$$

which simplifies to

$$E_{\text{trial}} = \frac{3b}{2} - Z\sqrt{\frac{8b}{\pi}}$$

The minimum energy is obtained from setting $\left(\dfrac{\partial E}{\partial b}\right) = 0$; we find that

$$b_{\text{OPT}} = \left(\frac{8Z^2}{9\pi}\right)$$

and the wave function is

$$\phi = \left(\frac{2b_{\text{OPT}}}{\pi}\right)^{3/4} \exp\left(-b_{\text{OPT}}r^2\right)$$

The minimum energy is evaluated using b_{OPT}

$$\langle E_{\min} \rangle = \frac{3}{2}\left(\frac{8Z^2}{9\pi}\right) - Z\left[\frac{8^2Z^2}{9\pi^2}\right]^{1/2} = Z^2\left(\frac{4}{3\pi}\right) - Z^2\left(\frac{8}{3\pi}\right) = -Z^2\left(\frac{4}{3\pi}\right)$$

Notice that the kinetic energy scales as the square of the nuclear charge Z, and the potential is linear in the nuclear charge. The viral theorem is satisfied for the optimally scaled trial function

$$\langle T \rangle = -2\langle V \rangle$$

For the H atom with $Z = 1$, the energy is -0.424413 hartree. That is to say that one Gaussian orbital can only achieve about 85% of the exact energy of H atom.

Variation Example with a Linear Combination of Basis Functions

Since it is relatively easy to use simple exponentials for atomic calculations, optimum scale factors have been collected (see, for example, Clementi and Raimondi [5]). Gaussian basis sets have been reviewed by Schaefer [6] and by Feller and Davidson [7]. For molecules, the linear combination of atomic orbitals (LCAO) approximation is central. The basis for this approximation is that an electron will be well represented by an optimized atomic orbital when it is near a nucleus. For a molecular system, this behavior is accommodated by some weighted sum of atomic functions. The weighting coefficients of the basis functions are to be determined variationally. Construct the expectation value of the energy

$$\langle \psi | H | \psi \rangle = \sum_i \sum_j c_i^* c_j \langle \phi_i | H | \phi_j \rangle$$

We assume the basis is orthonormal

$$\langle \phi_i | \phi_j \rangle = S_{i,j} = \delta_{ij}$$

and introduce an abbreviation

$$\langle \phi_i | H | \phi_j \rangle = H_{i,j}$$

The optimum energy is obtained by finding the minimum of the energy

$$E = \frac{\sum_{ij} c_i^* c_j H_{ij}}{\sum_{ij} c_i^* c_j \delta_{ij}}$$

subject to the normalization assured by the presence of the denominator.

$$\frac{\partial}{\partial c_i^*} \left[\sum_i \sum_j c_i^* c_j H_{i,j} - E \sum_i \sum_j c_i^* c_j \delta_{ij} \right] = 0$$

Then

$$\sum_j [H_{i,j} - E\delta_{ij}] c_j = 0$$

As already described in Chapter 2, we can now invoke the Cayley–Hamilton theorem [8] and set the determinant of the system of equations to zero. This produces a polynomial in E, the roots of which are the solutions to the system.

We can illustrate the method in a simple model developed by Huckel [9] in the 1930s. Consider the H_2 molecule using just a basis of two H-1s orbitals, ϕ_1 and ϕ_2. Use Huckel's approximations that the basis is orthonormal so that $S_{1,1}=1$, $S_{1,2}=S_{2,1}=0$, and $S_{2,2}=1$. Let $H_{1,1}=H_{2,2}=\alpha$ and $H_{1,2}=H_{2,1}=\beta$ where α and β are to be determined. We identify α as an approximation to the binding energy of an electron in a H atom and β as the binding energy for each electron in the diatomic. Then

$$\begin{vmatrix} (\alpha - E) & \beta \\ \beta & (\alpha - E) \end{vmatrix} = 0$$

Define $x = \dfrac{(\alpha - E)}{\beta}$; then

$$\begin{vmatrix} x & 1 \\ 1 & x \end{vmatrix} = x^2 - 1 = (x+1)(x-1) = 0$$

The solutions are $x=1$ and $x=-1$. When $x=1$, we find that $E=\alpha - \beta$ and when $x=-1$, $E=\alpha + \beta$. Substituting $x=1$ back into the system of equations leads to $c_1 + c_2 = 0$ so that $c_2 = -c_1$. The magnitude of the coefficient is defined by normalization.

$$\int N^*(c_1\phi_1 - c_2\phi_2)^* N(c_1\phi_1 - c_2\phi_2)d\tau = (N^*N)(1 - 0 - 0 + 1) = 1$$

Cross terms vanish due to orthogonality of the basis. Thus, $N = \left(\frac{1}{\sqrt{2}}\right)$; for the root $x=-1$ the same normalization constant is obtained. The energies are

$$E(x = +1) = \alpha - \beta \text{ for } \psi_{x=+1} = (\phi_1 - \phi_2)/\sqrt{2}$$

$$E(x = -1) = \alpha + \beta \text{ for } \psi_{x=-1} = (\phi_1 + \phi_2)/\sqrt{2}$$

The lower-energy state ground state (with $x=-1$) has basis functions added in phase (i.e., with like sign), which is consistent with our ideas of bonding. The first excited state transition energy of the H–H molecule would be about $\Delta E = 2\beta$; this would permit an estimate of the value of the parameter. We obtain a sensible molecular orbital interpretation of bonding, with overlapping H-1s orbitals building up charge in the region between nuclei (even though we made the approximation that the basis functions are orthonormal). The higher-energy solution has a node (a change in sign) between nuclei, which is how we visualize an antibonding orbital.

The Huckel method was widely applied to hydrocarbon pi subsystems, under the assumption that the sigma bonds and inner 1s-core orbitals were "frozen." It was quite successful in describing bonding and spectra in a large number of aromatic hydrocarbons. The value of $\beta_{C-C}=-2.39$ eV fits a number of experimental quantities. The representation of atoms other than C required specific parameters for the heteroatoms (X) in the aromatic system in terms of the carbon values for α and β [10].

$$\alpha_x = \alpha_c + h_x \beta_{c-c} \quad \text{and} \quad \beta_{c-x} = k_{c-x} \beta_{c-c}$$

For example, we can construct an approximate matrix for the pi electrons for formaldehyde. Let $h_{o=1}$ and $k_{c-o=1}$.

$$H_{\text{Huckel}}^{\text{Formaldehyde}} = \begin{vmatrix} \alpha_C + \beta - \lambda & \beta \\ \beta & \alpha_C - \lambda \end{vmatrix} \Rightarrow \begin{vmatrix} 1-x & 1 \\ 1 & -x \end{vmatrix} \Rightarrow x_{\pm} = \frac{1}{2} \pm \frac{\sqrt{5}}{2}$$

$$= +1.618 \text{ and } -0.618$$

The roots are shifted down (the unit of energy β being negative) and the stable MO is polarized. The coefficient of the O pi AO is about 0.851, and the coefficient of the C pi AO is about 0.526, accounting for a dipole moment in the CO pi manifold.

A Variational Calculation in One Dimension

Here we present a computation whereby we can estimate the eigenvalues and eigenfunctions for an arbitrary one-dimensional potential energy $V(x)$. (Extending the model to nondegenerate multidimensional problems is straightforward.) In the model called HEG defined by Harris et al. [11] we begin with a particle in a box with a zero potential between $0 \leq x \leq L$, with L a large value. The full Hamiltonian includes a potential of a flexible form, which might be a Morse potential

$$V_{\text{Morse}} = A[1 - \exp(-\beta[x - x_0])]^2$$

or might be expressed as a polynomial

$$V(x) = bx^2 + cx^3 + dx^4$$

with parameters a, b, and c. The Hamiltonian is

$$H = T + V = \frac{\hat{p}_x^2}{2m} + V(x)$$

Since all the terms in the polynomial potential are simple powers of x, $[V(x),x] = 0$. Then, since operator x commutes with whatever functional form that might be chosen for the potential $V(x)$, there exists a set of functions which are eigenfunctions for both operators. We will try to express that set of functions in a basis, the eigenfunctions for the particle in a box. The matrix representing x is then defined by

$$\langle n|x|m\rangle = \int_0^L \left(\sqrt{\frac{2}{L}}\sin\left(\frac{n\pi x}{L}\right)\right)(x)\left(\sqrt{\frac{2}{L}}\sin\left(\frac{m\pi x}{L}\right)\right)dx$$

There will be three important cases:

1. $n = m$, $\langle n|x|m\rangle = (L/2)$
2. $(n+m)$ even and $n \neq m$, $\langle n|x|m\rangle = 0$
3. $(n+m)$ odd and $m \neq n$, $\langle n|x|m\rangle = \left(\frac{-2L}{\pi^2}\right)\left[\frac{1}{(n-m)^2} - \frac{1}{(n+m)^2}\right]$

In this basis the kinetic energy will be diagonal.

$$\langle n|T|m\rangle = (n^2\pi^2\hbar^2/2mL^2)\,\delta_{nm}$$

For an accurate representation we need a large number of basis functions and a value of L large enough to span the interesting region of the potential. The experience of Dwyer [12] is that typically about 40 sine waves and L twice the range of interest in the potential should be adequate.

Now find an estimate of the eigenfunctions of x by diagonalizing its matrix representation.

$$\mathbf{U}^{-1}\mathbf{X}\mathbf{U} = \mathbf{Q} = \{x_{ii}\delta_{ij}\}$$

The eigenvectors in $\mathbf{U}$ will also diagonalize the matrix representation of the potential. That is, diagonal matrix can be computed easily.

$$\mathbf{\Lambda} = \{V(x_{ii})\delta_{ij}\}$$

There is no need to define a potential function over the whole range; one can merely compute the potential at the points x_{ii}. Then we can go back to a representation of $\mathbf{V}$ in the basis of box eigenfunctions by the transform

$$\mathbf{V} = \{\langle m|V(x)|n\rangle\} = \mathbf{U}\mathbf{\Lambda}\mathbf{U}^{-1}$$

We now have the representation of the potential energy in the box-eigenfunction basis. Adding the kinetic energy, we obtain a representation of the full Hamiltonian.

$$\mathbf{H} = \mathbf{T} + \mathbf{V}$$

Diagonalizing the matrix representing the Hamiltonian gives estimates of the system eigenvalues and vectors: the process is simplified by the fact that the basis is orthonormal.

$$\mathbf{C}^{-1}\mathbf{H}\mathbf{C} = \mathbf{E}$$

The eigenvectors correspond to approximate wave functions for the full Hamiltonian.

$$\psi_k = \sum_n C_{kn}\phi_n \text{ where } \phi_n = \sqrt{2/L}\sin\left(n\pi x/L\right)$$

This method has been tested on the vibrational potentials for symmetric vibrations of F_3PO defined by minimum basis set SCF single-point energies [12]. With 40 sine waves and well size $= 1$ Å, the authors achieved excellent agreement with experimental vibrational energies. It is now routine to evaluate harmonic frequencies for molecular vibrations, but this method is, in principle, capable of dealing with systems with highly anharmonic potentials.

Other basis sets can be used—the harmonic oscillator basis suggests itself. Shillady and Yurtsever et al. [13] extended the HEG method using Sturmian functions rather than particle-in-box eigenfunctions. These functions have the advantage of being solutions of a problem with a potential reasonably close to those found in molecules. Fewer Sturmian functions than sine waves are needed to achieve a given accuracy. The modest disadvantage is that the Sturmian functions are not orthogonal, requiring solution of the nonorthogonal eigenvalue problem $\mathbf{HC} = \mathbf{ESC}$ where the $\mathbf{S}$ matrix is the overlap between the nonorthogonal basis functions.

Determinantal Wave Functions and HF-SCF Theory

Systems with many electrons present complications and surprises. It might be our first reaction to write the eigenfunction of a system of N electrons with Hamiltonian

$$H = -\sum_j \frac{\nabla_j^2}{2m_j} - \sum_{jA} \frac{Z_A}{r_{jA}} + \sum_{i<j} \frac{1}{r_{ij}} = \sum_j h(j) + \sum_{i<j} g(ij)$$

as a product

$$\Psi = \prod_{Jj} \varphi_J(j)$$

so as to reflect the approximate separability of motion of each electron. This separability is approximate because the electron–electron repulsion couples the electrons, but the product form could still be a suitable starting point for a variational description. As reasonable as this might seem, the product form violates the fundamental principle that electrons are indistinguishable.

That is to say that swapping any two electrons cannot affect any observable. Observables depend not on the wave function but its square; so the condition can be written

$$P(i,j)\Psi^*\Psi = \Psi^*\Psi$$

Here $P(i,j)$ is the permutation operator which exchanges electron i with electron j. The effect of the permutation on Ψ can be either to leave it unchanged (the wave function is symmetric with respect to exchange) or at most change the sign (the wave function is antisymmetric with respect to exchange). Particles are either bosons (the first case) or fermions (the second case). Photons, alpha particles, and systems with integer spin in general are bosons; protons, electrons, and systems with half (odd) integer spin are fermions.

Slater [14] proposed a representation in which a determinant D is constructed with each element written as

$$D_{mn} = \varphi_m(n)$$

Here n refers to the coordinates of a particular electron including its intrinsic angular momentum or "spin" and m defines the form of a one-particle function, an "orbital." Since the sign of a determinant changes if any two rows or any two columns are interchanged, this form guarantees that the wave function for a system of electrons is antisymmetric with respect to the interchange of any two of them.

A determinantal wave function for three electrons is written as

$$\psi(1,2,3) = \frac{1}{\sqrt{3!}} \begin{vmatrix} \phi_a(1) & \phi_b(1) & \phi_c(1) \\ \phi_a(2) & \phi_b(2) & \phi_c(2) \\ \phi_a(3) & \phi_b(3) & \phi_c(3) \end{vmatrix}$$

When this is expanded one obtains an explicit but inconveniently bulky form of the wave function

$$\psi(1,2,3) = \frac{1}{\sqrt{6}} \left\{ \begin{array}{l} \phi_a(1)\phi_b(2)\phi_c(3) - \phi_a(1)\phi_c(2)\phi_b(3) + \phi_b(1)\phi_a(2)\phi_c(3) \\ -\phi_b(1)\phi_c(2)\phi_a(3) + \phi_c(1)\phi_a(2)\phi_b(3) - \phi_c(1)\phi_b(2)\phi_a(3) \end{array} \right\}$$

Here each ϕ_a, ϕ_b, or ϕ_c is actually a spin-orbital. This could be made more explicit by writing out functional forms $\phi_a\alpha$ or $\phi_a\beta$. The factor $1/\sqrt{3!}$ normalizes the wave function if the orbitals are orthonormal.

This expansion of the Slater determinant shows each of the $N!$ permutations of assignments of the N electrons to N spin-orbitals. Application of the pair permutation operation will prove that the product is antisymmetrized and thus suitable for fermions.

The unwieldy determinant can be written more compactly as the antisymmetrized product

$$\psi(1, 2, \ldots, N) = \frac{1}{\sqrt{N!}} \sum_p (-1)^p P\{\phi_a(1)\phi_b(2), \ldots, \phi_n(N)\} \equiv |\Delta\rangle$$

Here the operator P is any one of a sequence of permutations of the electrons' labels. The zeroth-order permutation changes nothing. The first-order permutation effects pair swaps and the second-order permutation shuffles three objects. Any permutation can be represented as a sequence of pair swaps; the second-order permutation requires two pair swaps. The lowercase p refers to the order of the permutation.

Clear treatments of integrals involving Slater determinants begin with the early text by Condon and Shortley [15]. Another notably clear presentation is by Pople and Beveridge [16]. We have also profited from the presentations by Pilar [17] and Szabo and Ostlund [18].

An operator in quantum chemistry can refer either to zero electrons, one electron, or two electrons. One-electron operators include the nuclear attraction, kinetic energy, dipole moment components, and angular momentum of a single electron. The most important two-electron operator is the coulomb repulsion between electrons, but we will have occasion to consider spin–spin and spin–orbit interactions. We will deal with the simpler zero-electron "overlap" case first.

$$\langle \psi_A(1, \ldots, N) | \psi_B(1, \ldots, N) \rangle \equiv \langle \Delta^A | \Delta^B \rangle$$

$$\langle \Delta^A | \Delta^B \rangle = \sum_{i=1}^n \sum_{Pp} \sum_{P'p'} \frac{(-1)^{p+p'}}{N!} \int \ldots \int P\{\phi_1(1), \ldots, \phi_N(N)\}^*$$
$$P'\{\phi_1'(1), \ldots, \phi_N'(n)\} d\tau_1, \ldots, d\tau_n$$

The turnover rule allows us to rewrite integrals of Hermitian operators

$$\langle A\Phi | B\Psi \rangle = \langle \Phi | A^\dagger B\Psi \rangle$$

Here the symbol † means "adjoint." Hermitian operators, including any constant and the permutations as well as terms in the Hamiltonian, are self-adjoint; we can thus disregard this nicety. The permutation operators form a group; thus, when any permutation in the unprimed set (call it Q) acts on any member of the unprimed set (call it Q'), the product QQ' is one of the permutations in the group. What is more, the parity of QQ' is the parity of the sum $q + q'$. If any Q is applied to the set of primed permutations, the action merely regenerates the list of members of the set (generally reordered). We have $N!$ permutations in the unprimed set so the expression simplifies to

$$\langle \Delta^A | \Delta^B \rangle = \sum_{i=1}^{n} \sum_{Rr} (-1)^r \int \cdots \int \{\phi_1(1), \ldots, \phi_N(N)\}^* R\{\phi'_1(1), \ldots, \phi'_N(n)\} d\tau_1, \ldots, d\tau_n$$

The permutation R can be the identity, the pair swap (mn), etc. Consider first the identity. Then the integral becomes a product of integrals over coordinates of a single electron

$$\langle \Delta^A | \Delta^B \rangle = \prod_j \langle \varphi_J(j) | \varphi'_{J'}(j) \rangle$$

This can be nonzero only if each function matches its primed counterpart, $J = J'$. In the orthonormal set each integral factor is unity, as is the entire product.

The pair swap (mn) acting on electronic coordinates produces a mismatch between functions at right and left.

$$\langle \Delta^A | \Delta^B \rangle = \langle \varphi_M(m) | \varphi'_{N'}(m) \rangle \langle \varphi_N(n) | \varphi'_{M'}(n) \rangle \prod_{j \neq m,n} \langle \varphi_J(j) | \varphi'_{J'}(j) \rangle$$

The first two factors are zero for orthogonal orbitals; there is thus no further contribution to the overlap. Higher permutations produce further factors of zero.

Determinantal Integrals for One-Electron Operators

Let us consider the matrix element of one-electron operators $F = \sum_i^n h(i)$.

$$\langle \Delta^A | F | \Delta^B \rangle = \sum_{i=1}^{n} \langle \phi_a(1), \ldots, \phi_n(n) | h(i) \sum_{Pp} (-1)^P P | \phi'_a(1), \ldots, \phi'_n(N) \rangle$$

Choose the zeroth-order permutation (identity) and a particular component of F referring to electron i and factor the integral. The contribution from the zeroth permutation is

$$\langle \varphi_I(i) | h(i) | \varphi'_I(i) \rangle \prod_{M \neq I} \langle \varphi_M(m) | \varphi'_M(m) \rangle$$

This can be nonzero only if all orbitals match for $M \neq I$. Then the product becomes simply unity and the remaining contribution is

$$\langle \varphi_I(i) | h(i) | \varphi'_I(i) \rangle$$

If only orbitals I and I' are mismatched (say they are K and R, respectively) then the sole nonzero part of this integral is

$$\langle \varphi_K(i)|h(i)|\varphi_R(i)\rangle = h_{KR}$$

If I and I' match, or $a = r = I$, then we obtain a nonzero result. What is more, every choice of index or i in $f(i)$ will produce a nonzero contribution. The result is

$$\sum_I \langle \varphi_I(i)|h(i)|\varphi_I(i)\rangle = \sum_I h_{II}$$

Now consider the pairwise permutation (mn). There are two possibilities:

1. Neither m nor n is equal to i.
2. Either m or n is equal to i.

In either case there is a factor of zero arising from orthogonality. For higher order permutations, more orthogonalities are forced, and there are no further contributions to the integral for the one-electron part of the Hamiltonian.

There is a subtlety hidden in the development of these results. It is assumed that the determinants at left and right have been arranged so that there is maximum coincidence; i.e., if the sets of orbitals appearing at left and right are identical, the products are written in identical order. If there is a single noncoincidence, the $N - 1$ common orbitals are written in identical order at left and right, and similarly for the common $N - 2$ orbitals if there are two noncoincidences.

In summary we have three cases for one-electron operators:

1. $\left\langle \Delta^A \left| \sum_i h(i) \right| \Delta^B \right\rangle = 0$, if Δ^A and Δ^B differ by two or more spin-orbitals

2. $\left\langle \Delta^A \left| \sum_i h(i) \right| \Delta^B \right\rangle = \int \phi_K{}^*(i)h(i)\phi_R(i)d\tau_i = h_{KR}$, if only one spin-orbital is different

3. $\left\langle \Delta^A \left| \sum_i h(i) \right| \Delta^B \right\rangle = \sum_{I=1}^{N} \int \phi_I{}^*(i)h(i)\phi_I(i)d\tau_i = \sum_I h_{II}$, when $\Delta^A = \Delta^B$

Determinantal Matrix Elements for Two-Electron Operators

Now consider matrix elements for two-electron operators. We will consider the electron–electron coulomb interaction.

$$G = \sum_{i>j} \frac{1}{r_{ij}}$$

Choose a specific term in G, the repulsion between i and j. Then the integral is

$$\langle \Delta^A | 1/r_{ij} | \Delta^B \rangle = \langle \varphi_1(1), \ldots, \varphi_N(N) | 1/r_{ij} | \Sigma(-)^P P \varphi_1'(1), \ldots, \varphi_N'(N) \rangle$$

Then begin with the zeroth-order permutation. All the factors referring to electrons other than i and j will be overlap integrals.

$$\left\langle \varphi_I(i)\varphi_J(j) \middle| r_{ij}^{-1} \middle| \varphi_I'(i)\varphi_J'(j) \right\rangle \prod_{K \neq I,J} \langle \varphi_K | \varphi_K' \rangle$$

If all the orbitals with labels K match left and right the product becomes unity and the contribution reduces to

$$\left\langle \varphi_I(i)\varphi_J(j) \middle| r_{ij}^{-1} \middle| \varphi_I'(i)\varphi_J'(j) \right\rangle$$

Note that this two-electron integral is written in "physicists'" notation, with the variables in sequence $ijij$ left to right.*

If orbitals with labels I and labels J match left and right, then we have

$$\left\langle \varphi_I(i)\varphi_J(j) \middle| r_{ij}^{-1} \middle| \varphi_I(i)\varphi_J(j) \right\rangle = J_{IJ}$$

We would find analogous results for all choices of indices; the total contribution to the two-electron energy from this case is thus

$$\sum_{I>J} J_{IJ}$$

Remember that we are still dealing with spin-orbitals.

Now return to the repulsion between electrons i and j, and consider the two-electron permutation (mn). Several choices present themselves:

1. mn and ij are identical, i.e., the pairs have both members in common.

2. mn and ij have one member in common.

3. mn and ij have no members in common.

---

*This expression assumes the Dirac convention that the "bra" at left contains all the complex conjugated functions and the "ket" at right contains all the unconjugated functions. We will refer to this as "1212" notation, referring to the order of electron coordinates. A widely used alternative used in PCLOBE program and in many other contexts emphasizes the idea that the integral expresses the Coulomb repulsion between charge distributions for two electrons. We will use the angle brackets for the 1212 and round brackets for the coulomb 1122 order. That is, $(ab|pq) = \langle ap|bq \rangle$. It is a good practice to specify the convention for two-electron integrals early in any discussion.

The first choice gives us

$$-\langle \varphi_I(i)\,\varphi_J(j)|r_{ij}^{-1}|\varphi_J'(i)\varphi_I'(j)\rangle \prod_{M\neq I,J} \langle \varphi_M|\varphi_M'\rangle$$

The negative sign is a consequence of the odd parity of the pair exchange.

Once again if all orbitals with labels M match then the product becomes unity and we are left with the expression

$$-\langle \varphi_I(i)\varphi_J(j)\big|r_{ij}^{-1}\big|\varphi_J'(i)\varphi_I'(j)\rangle$$

If the orbitals with labels I and J match, then we have the exchange integral K.

$$-\langle \varphi_I(i)\varphi_J(j)\big|r_{ij}^{-1}\big|\varphi_J(i)\varphi_I(j)\rangle = -K_{IJ}$$

There will be a term like this for every choice of i and j; the total contribution from this case is thus

$$-\sum_{I>J} K_{IJ}$$

For the second and third choices, it is inevitable that the product has at least one zero factor, arising from the electron(s) which are not held in common between mn and ij. Writing the general form

$$\langle \Delta^A|1/r_{ij}|\Delta^B\rangle = \langle \varphi_1(1),\ldots,\varphi_N(N)|1/r_{ij}|(-)P_{mn}\varphi_1'(1),\ldots,\varphi_N'(N)\rangle$$

and specifying ij and mn will make this clear.

We have not dealt with the cases where there is at least one mismatch (noncoincidence) between determinants left and right. The mismatch must not be in the product terms or else orthogonality will force the expression to be zero. Let us first assign I the form A on the left and R at right. We then have

$$\langle \varphi_I(i)\varphi_J(j)\big|r_{ij}^{-1}\big|\varphi_I'(i)\varphi_J'(j)\rangle - \langle \varphi_I(i)\varphi_J(j)\big|r_{ij}^{-1}\big|\varphi_J'(i)\varphi_I'(j)\rangle$$

$$= \langle \varphi_A(i)\varphi_J(j)\big|r_{ij}^{-1}\big|\varphi_R(i)\varphi_J'(j)\rangle - \langle \varphi_A(i)\varphi_J(j)\big|r_{ij}^{-1}\big|\varphi_J(i)\varphi_R(j)\rangle$$

Such a form would appear no matter what J ($<I$, i.e., A) we choose. Of course, we could equally have assigned J the form A at left and R at right. Then we would obtain the expression

$$\left\langle \varphi_I(i)\varphi_A(j) \middle| r_{ij}^{-1} \middle| \varphi_I(i)\varphi_R(j) \right\rangle - \left\langle \varphi_I(i)\varphi_A(j) \middle| r_{ij}^{-1} \middle| \varphi_R(i)\varphi_I(j) \right\rangle$$

Likewise we would obtain this form no matter what I ($>J$, i.e., A) we choose. Summing all those terms, we have

$$\sum_{J \neq A} \left[\left\langle \varphi_J(i)\varphi_A(j) \middle| r_{ij}^{-1} \middle| \varphi_J(i)\varphi_R(j) \right\rangle - \left\langle \varphi_A(i)\varphi_J(j) \middle| r_{ij}^{-1} \middle| \varphi_J(i)\varphi_R(j) \right\rangle \right]$$

Finally, if there were two noncoincidences (A and R in place of I; B and S in place of J), all that could survive would be

$$\left\langle \varphi_A(i)\varphi_B(j) \middle| r_{ij}^{-1} \middle| \varphi_R(i)\varphi_S(j) \right\rangle - \left\langle \varphi_A(i)\varphi_B(j) \middle| r_{ij}^{-1} \middle| \varphi_S(i)\varphi_R(j) \right\rangle$$

More than two noncoincidences force at least one orthogonality beyond the two that the two-electron operator can bridge; no further contribution to the two-electron integral can thus be made.

In summary we have four cases for two-electron operators:

1. $\langle \Delta^A | G | \Delta^B \rangle = 0$, if Δ^A and Δ^B differ by more than two spin-orbitals (there are more than two noncoincidences)

2. $\langle \Delta^A | G | \Delta^B \rangle = \langle \varphi_A(i)\varphi_B(j) | r_{ij}^{-1} | \varphi_R(i)\varphi_S(j) \rangle - \langle \varphi_A(i)\varphi_B(j) | r_{ij}^{-1} | \varphi_S(i)\varphi_R(j) \rangle$, if there are two noncoincidences (A replaced by R and B replaced by S)

3. $\langle \Delta^A | G | \Delta^B \rangle = \sum_{J \neq A} \langle \varphi_J(i)\varphi_A(j) | r_{ij}^{-1} | \varphi_J(i)\varphi_R(j) \rangle - \langle \varphi_A(i)\varphi_J(j) | | \varphi_J(i)\varphi_R(j) \rangle$,

   if there is only one noncoincidence (A replaced by R)

4. $\langle \Delta^A | G | \Delta^A \rangle = \sum_{I > J} \langle \varphi_I(i)\varphi_J(j) | r_{ij}^{-1} | \varphi_I(i)\varphi_J(j) \rangle - \langle \varphi_I(i)\varphi_J(j) | r_{ij}^{-1} | \varphi_J(i)\varphi_I(j) \rangle$, if

   there is no noncoincidence

Restricted Form of the Determinant

We want to specialize our expressions to the case of the energy expectation value of a restricted determinant in which two electrons share a single spatial form. By the Pauli principle these electron pairs are singlet-coupled, having opposite spins. It is conventional that spin-orbitals are written as $\phi_1\alpha, \phi_1\beta, \phi_2\alpha, \phi_2\beta, \ldots$ (or in some papers as $\phi_1, \bar{\phi}_1, \phi_2, \bar{\phi}_2, \ldots$); i.e., the spin functions alternate. The sum of one-electron terms over spin-orbital labels I is simplified to a sum over space-orbitals p

$$\sum_I \langle \varphi_I(i) | h(i) | \varphi_I(i) \rangle = \sum_I h_{II} = 2 \sum_p h_{pp}$$

The two-electron case is more complicated. Let us begin with the case of identical determinants

$$\langle \Delta^A | G | \Delta^A \rangle = \sum_{I>J} \left\langle \varphi_I(i)\varphi_J(j) \middle| r_{ij}^{-1} \middle| \varphi_I(i)\varphi_J(j) \right\rangle - \left\langle \varphi_I(i)\varphi_J(j) \middle| r_{ij}^{-1} \middle| \varphi_J(i)\varphi_I(j) \right\rangle$$

Following Pople and Beveridge [16], we consider the following four adjacent terms:

$I = p\alpha, J = q\alpha$

$$\left\langle \varphi_p\alpha(i)\varphi_q\alpha(j) \middle| r_{ij}^{-1} \middle| \varphi_p\alpha(i)\varphi_q\alpha(j) \right\rangle - \left\langle \varphi_p\alpha(i)\varphi_q\alpha(j) \middle| r_{ij}^{-1} \middle| \varphi_q\alpha(i)\varphi_p\alpha(j) \right\rangle$$

$I = p\alpha, J = q\beta$

$$\left\langle \varphi_p\alpha(i)\varphi_q\beta(j) \middle| r_{ij}^{-1} \middle| \varphi_p\alpha(i)\varphi_q\beta(j) \right\rangle - \left\langle \varphi_p\alpha(i)\varphi_q\beta(j) \middle| r_{ij}^{-1} \middle| \varphi_q\beta(i)\varphi_p\alpha(j) \right\rangle$$

$I = p\beta, J = q\alpha$

$$\left\langle \varphi_p\beta(i)\varphi_q\alpha(j) \middle| r_{ij}^{-1} \middle| \varphi_p\beta(i)\varphi_q\alpha(j) \right\rangle - \left\langle \varphi_p\beta(i)\varphi_q\alpha(j) \middle| r_{ij}^{-1} \middle| \varphi_q\alpha(i)\varphi_p\beta(j) \right\rangle$$

$I = p\beta, J = q\beta$

$$\left\langle \varphi_p\beta(i)\varphi_q\beta(j) \middle| r_{ij}^{-1} \middle| \varphi_p\beta(i)\varphi_q\beta(j) \right\rangle - \left\langle \varphi_p\beta(i)\varphi_q\beta(j) \middle| r_{ij}^{-1} \middle| \varphi_q\beta(i)\varphi_p\beta(j) \right\rangle$$

For the time being we assume that p and q are not identical. In this set we will find that all the positively signed coulomb integrals survive since there is no spin-disagreement. However, only two of the negatively signed exchange terms survive since orthogonality is avoided only when all spins agree. These terms can be summed for all choices of $p > q$.

$$\sum_{p>q} 4\left\langle p(1)q(2) \middle| r_{12}^{-1} \middle| p(1)q(2) \right\rangle - 2\left\langle p(1)q(2) \middle| r_{12}^{-1} \middle| q(1)p(2) \right\rangle$$

$$\sum_{p>q} 4J_{pq} - 2K_{pq} \equiv \sum_{p\neq q} 2J_{pq} - K_{pq}$$

As we scan through the indices, we see that it is possible that $I > J$ but $p = q$. Then we have $I = p\alpha, J = p\beta$

$$\left\langle \varphi_p\alpha(i)\varphi_p\beta(j) \middle| r_{ij}^{-1} \middle| \varphi_p\alpha(i)\varphi_p\beta(j) \right\rangle - \left\langle \varphi_p\alpha(i)\varphi_p\beta(j) \middle| r_{ij}^{-1} \middle| \varphi_p\beta(i)\varphi_p\alpha(j) \right\rangle$$

In this case only the coulomb term survives. We sum up all choices of p

$$\sum_p J_{pp} \equiv \sum_p 2J_{pp} - K_{pp}$$

Combining all the terms, we have

$$\langle \Delta|H|\Delta \rangle = 2\sum_p h_{pp} + \sum_{p,q} 2J_{pq} - K_{pq}$$

The 2:1 relation between J and K is a consequence of the restriction to perfect pairing.

Expansion of the RHF Energy in a Basis

One of the advances which made the SCF method applicable to molecules was the expansion of each one-electron function in a basis. The basis need not be orthogonal, but it is common to normalize each member. We represent a molecular orbital as

$$\varphi_p = \sum_\mu C_{p\mu} \chi_\mu$$

Then the one-electron energy becomes

$$2\sum_p \langle \varphi_p|h|\varphi_p \rangle = 2\sum_p \sum_{\mu\nu} C_{p\mu}^* C_{p\nu} \langle \chi_\mu|h|\chi_\nu \rangle = 2\sum_p \sum_{\mu\nu} C_{p\mu}^* C_{p\nu} h_{\mu\nu}$$

The two-electron energy is

$$\sum_{pq} \sum_{\mu\nu\rho\sigma} (2C_{p\mu}^* C_{q\nu}^* C_{p\rho} C_{q\sigma} - C_{p\mu}^* C_{q\nu}^* C_{q\rho} C_{p\sigma}) \langle \mu\nu|\rho\sigma \rangle$$

We introduce the density matrix for the restricted function

$$P_{\mu\nu} = 2\sum_{p=1}^{occ} C_{p\mu}^* C_{p\nu}$$

Following Pople, we can rewrite the energy for the restricted determinant as

$$\sum_{\mu\nu} P_{\mu\nu} h_{\mu\nu} + \frac{1}{4} \sum_{\mu\nu\rho\sigma} (2P_{\mu\rho} P_{\nu\sigma} - P_{\mu\sigma} P_{\nu\rho}) \langle \mu\nu|\rho\sigma \rangle$$

Roothaan–Hartree–Fock Self-Consistent Field Equation

The basic idea that electrons can be treated as moving in an average static field due to the nuclear attraction and the repulsion of the other electrons in an atom was exploited by Hartree [19]. His first form for the many-electron wave function was a simple product, but he soon recognized the necessity for the exchange described by Fock [20]. In 1951, Roothaan [21] derived an orbital expansion method that made it possible to treat molecules. We are indebted to Slater [22] for a clear treatment.

Assume an array of nuclei fixed in position, and write the wave function for N electrons as a single Slater determinant. We will minimize the energy for a restricted wave function

$$2\sum_p \sum_{\mu\nu} C^*_{p\mu}C_{p\nu}h_{\mu\nu} + \sum_{pq}\sum_{\mu\nu\rho\sigma}\left(2C^*_{p\mu}C^*_{q\nu}C_{p\rho}C_{q\sigma} - C^*_{p\mu}C^*_{q\nu}C_{q\rho}C_{p\sigma}\right)\langle\mu\nu|\rho\sigma\rangle$$

for which the LCAO coefficients can serve as the variational parameters. In order that the expression for the energy remains valid as we alter the coefficients, we must incorporate a condition which enforces the molecular orbitals' orthonormality.

$$S_{pq} = \sum_{\mu\nu} C^*_{p\mu}C_{q\nu}\langle\mu|\nu\rangle = \delta_{\mu\nu}$$

The method of LaGrange multipliers allows us to impose the constraint by expanding the space of variations from the NM coefficients to include dimensions for each of the N^2 constraints. The effect is to require solution of the equations

$$\frac{\partial}{\partial C^*_{r\tau}}\left[E_{\text{RHF}} - 2\sum_{pq}\lambda_{pq}(\delta_{pq} - S_{pq})\right] = 0; \quad \text{all } r, \tau$$

The λ are to be determined. The factor of 2 is inserted for later convenience.

We chose to take the derivative with respect to complex conjugated coefficient C^*. If we had evaluated the derivative with respect to C, we would obtain a very similar expression—the complex conjugate of the expression we wrote explicitly. If we solve the equation we have constructed explicitly, we will also have solved the equation's complex conjugate.

Performing the derivative, we obtain

$$2\sum_\nu h_{\tau\nu}C_{r\nu} + 2\sum_q\sum_{\nu\rho\sigma}C^*_{q\rho}C_{q\sigma}(2\langle\tau\nu|\rho\sigma\rangle - \langle\tau\sigma|\rho\nu\rangle)C_{r\rho} - 2\sum_{q\nu}\lambda_{rq}C_{q\nu}\langle r|\nu\rangle = 0$$

The "extra" factor of two multiplying the two-electron term arises because either index p or index q can take on the value r. We rearrange this to show the pseudo-eigenvalue form

$$\mathbf{FC} = \mathbf{\Lambda SC}$$

The one-electron functions seem to be coupled through the elements of $\mathbf{\Lambda}$. Recall that our wave function is a determinant of an array. From the relation

$$\| \mathbf{U}^{\dagger}\mathbf{BU} \| = \| \mathbf{U}^{\dagger} \| \| \mathbf{B} \| \| \mathbf{U} \|$$

we can deduce that any unitary transform $\mathbf{U}$ (for which the determinant is unity) of the array leaves the determinant unaffected. We can use this freedom to choose a transform which leaves the matrix of undetermined multipliers diagonal. Then

$$\mathbf{FC}_i = \lambda_i \mathbf{SC}_i$$

Now λ_i can be recognized as an energy eigenvalue, the energy of the ith MO. To complete the solution, define

$$\mathbf{C} = \mathbf{S}^{-1/2}\mathbf{C}'$$

and multiply the equation from the left by $\mathbf{S}^{-1/2}$. Then we have

$$(\mathbf{S}^{-1/2}\mathbf{FS}^{-1/2})\mathbf{C}' = \mathbf{S}^{-1/2}\lambda\mathbf{SS}^{-1/2}\mathbf{C}' = \lambda(\mathbf{S}^{-1/2}\mathbf{SS}^{-1/2})\mathbf{C}' = \lambda\mathbf{C}'$$

We need only to diagonalize the matrix $\mathbf{F}'$

$$\mathbf{F}' = (\mathbf{S}^{-1/2}\mathbf{FS}^{-1/2})$$

that gives us $\mathbf{C}'$ as the matrix that diagonalizes $\mathbf{F}'$. We recover $\mathbf{C}$ by a back transform

$$\mathbf{C} = \mathbf{S}^{-1/2}\mathbf{C}'$$

This also means we need to find $\mathbf{S}^{-1/2}$ which is easily done by diagonalizing the original $\mathbf{S}$ (overlap) matrix, replacing the diagonal elements by their inverse square roots, and then reversing the transformation.

It is possible to rewrite the energy with reference to the $\mathbf{F}$ matrix. The matrix elements of the operator $\mathbf{F}$ are

$$F_{\tau\nu} = h_{\tau\nu} + \sum_{q\rho\sigma} C^{*}_{q\rho}C_{q\sigma}(2\langle\tau\nu|\rho\sigma\rangle - \langle\tau\sigma|\rho\nu\rangle)$$

We can write its expectation value for orbital i

$$\langle i|F|i \rangle = \sum_{\tau\nu} C_{i\tau}h_{\tau\nu}C_{i\nu} + \sum_{\tau\nu}\sum_{q\rho\sigma} C_{i\tau}C_{qp}C_{q\sigma}C_{i\nu}(2\langle\tau\nu|\rho\sigma\rangle - \langle\tau\sigma|\rho\nu\rangle)$$

For later reference we write

$$\langle i|F|i \rangle = h_{ii} + \sum_{j} 2J_{ij} - K_{ij}$$

which defines the coulomb integral J and the exchange integral K. We read this equation as follows: the orbital energy is composed of an electron's kinetic energy and its interaction with the nuclear framework (h) and also contributions from repulsions with every electron in the system (J) and exchanges with those that agree in spin (K).

If we now sum the expectation value for all occupied orbitals we obtain an expression which resembles the total energy

$$\sum_{p=\text{occ}} F_{pp} = \sum_{p}\sum_{\mu\nu} C^*_{p\mu}C_{p\nu}h_{\mu\nu} + \sum_{pq}\sum_{\tau\nu}\sum_{q\rho\sigma} C_{i\tau}C_{qp}C_{q\sigma}C_{i\nu}(2\langle\tau\nu|\rho\sigma\rangle - \langle\tau\sigma|\rho\nu\rangle)$$

All that is missing is an additional one-electron term; the energy is thus

$$E_{\text{RHF}} = \frac{1}{2}\sum_{\mu\nu} P_{\mu\nu}(h_{\mu\nu} + F_{\mu\nu})$$

Take note—the energy of the $2N$ electron system is not twice the sum of orbital energies.

The Fock matrix requires a density matrix for its definition, but, of course, the coefficients that define the density matrix are precisely what we seek. A very naïve beginning would be to approximate $\mathbf{F}$ by $\mathbf{h}$. This begins an iterative solution as follows:

1. Create $H_{kl} = \langle k|h|l \rangle$ and solve $\mathbf{HC} = \lambda\mathbf{SC}$ for the initial $\mathbf{C}$ matrix.
2. Form the initial density matrix $\mathbf{P_0}$ from the initial $\mathbf{C}$ matrix.
3. Form a new $\mathbf{F}$ matrix using $\mathbf{P_0}$ and solve $FC = ESC$ for a new $\mathbf{C}$ matrix.
4. Form a new P_{mn} matrix and compare it to the previous P_{mn} matrix; if the difference is less than some small threshold value go to step 5, otherwise go back to step 3.
5. Calculate the total energy (including the nuclear repulsion) using the converged $\mathbf{F}$ and $\mathbf{P}$ matrices in the formula

$$\langle E \rangle = \sum_{m}^{M} \sum_{n}^{M} P_{mn}(H_{mn} + F_{mn})/2 + \sum_{i>}^{nuc} \sum_{j}^{nuc} \left(\frac{Z_i Z_j}{R_{ij}} \right)$$

The solution of the SCF equation is iterative, which suggests that convergence will be an issue. There are many devices to accelerate convergence ranging from such simple measures as shifting the energy of virtual (vacant) orbitals to elaborate extrapolation schemes such as Pulay's DIIS method [23]. The simple "variable-damping" scheme by Karlstrom [24] incorporated in PCLOBE works most of the time, although it is not as reliable as the DIIS method.

The SCF model describes single electrons moving in a mean field. The field is defined by all electrons in the system. Oddly, each electron sees all the other electrons including itself. The mean field does not accommodate details of the "correlation" of motions of electrons. In a full basis, the SCF energy reaches the Hartree–Fock limit, typically about 98%–99% of the total energy. Unfortunately for each atom CNO in organic molecules the 1%–2% deficit amounts to about an atomic unit of energy or about 27 eV. Chemistry, which involves energies on the order of 1–10 eV, requires a greater degree of accuracy.

Koopmans' Theorem

THEOREM
In a system of 2N electrons for occupied orbitals, λ_i are estimates of the energy necessary to remove an electron from orbital i; for vacant orbitals, λ_i are estimates of the energy attending the addition of an electron to that orbital i [25].

The second assertion is easier to demonstrate. If we add an electron to a previously empty orbital M in a system of $2N$ electrons, that electron will have an energy composed of a one-electron term h_{MM}. It will also repel all the $2N$ electrons and exchange with N of them—those who have like spin.

$$\langle M|F|M \rangle = h_{MM} + \sum_{I=1}^{M} 2J_{MI} - K_{MI}$$

This is precisely the orbital energy for the previously vacant orbital M.

If we remove an electron from one of the occupied orbitals P, it loses not only its kinetic energy and attraction for the nuclear framework (h_{PP}) but also all repulsions with the pairs of electrons in the other ($N - 1$) orbitals and exchanges with those electrons in the other orbitals that shares its spin. So far we have

$$-\left[h_{PP} + \sum_{I \neq P}^{N} 2J_{PI} - K_{PI}\right]$$

We have not taken account of the loss of a repulsion in MO P, which could be written

$$J_{PP} = 2J_{PP} - K_{PP}$$

Adding this term, we have the desired result for the ionization energy

$$IE_P = -\langle P|F|P \rangle = -\lambda_P$$

Koopmans' theorem provides a first-order approximation to ionization energies and electron affinities. There is no provision for relaxation of the charge distribution when an electron is added or removed.

Brillouin's Theorem

THEOREM
No state differing from the ground state by a single spin-orbital will mix with the ground state [26].

SCF theory optimizes one-electron functions to construct the best possible single determinant. A single excitation from an occupied MO, ϕ_a, to a virtual orbital, ϕ_m, has the effect of altering the one-electron function ϕ_a to $\phi_{a'}$. But ϕ_a is already optimized, in the sense that δE with respect to any variation in ϕ_a is already zero.

The SCF-LCAO method is so central to computational chemistry that we will look more deeply into the structure of the calculation in the succeeding chapters.

References

1. Pauling, L. and Wilson, E.B., *Introduction to Quantum Mechanics with Applications to Chemistry*, McGraw-Hill, New York, 1935.
2. Eckart, C., *Phys. Rev.* 36:878, 1930.
3. Bohr, N., *Phil. Mag.* 26:1, 1913.
4. Schrödinger, E., *Ann. d Phys.* 79:361, 489, 734; ibid 80:437; ibid 81:109, 1926.
5. Clementi, E. and Raimondi, D.L., *J. Chem. Phys.* 38:2686, 1963.
6. Schaefer, H.F., *The Electronic Structure of Atoms and Molecules: A Survey of Rigorous Quantum Mechanical Results*, Addison-Wesley, Reading, MA, 1972.
7. Feller, D. and Davidson, E.R., *Rev. Comput. Chem.* 1:1, 1990.

8. Turnbull, H.W. and Aitken, A.C., *An Introduction to the Theory of Canonical Matrices*, Dover, New York, 1961, p. 43.

9. Huckel, E., *Zeit f Phys.* 60:423, 1930.

10. Murrell, J.N., Kettle, S.F.A., and Tedder, J.M., *Valence Theory*, 2nd Ed., Wiley, London, 1970, p. 296.

11. Harris, D.O., Engerholm, G.G., and Gwinn, W.D., *J. Chem. Phys.* 43:1515, 1965.

12. Dwyer, R.W., *A Theoretical Study of Some Vibrational and Electronic Properties of Phosphoryl Fluoride*, PhD Thesis, Virginia Commonwealth University, 1975.

13. Yurtsever, E., Yilmaz, O., and Shillady, D.D., *Chem. Phys. Lett.* 85:111, 1982.

14. Slater, J.C., *Phys. Rev.* 34:1293, 1929.

15. Condon, E.U. and Shortley, G., *The Theory of Atomic Spectra*, Cambridge Press, Cambridge, England, 1951.

16. Pople, J.A. and Beveridge, D.L., *Approximate Molecular Orbital Theory*, McGraw-Hill, 1970, p. 35.

17. Pilar, F., *Elementary Quantum Chemistry*, 2nd Ed., Dover Publications, Mineola, NY, 2001.

18. Szabo, A. and Ostlund, N.S., *Modern Quantum Chemistry: Introduction to Advanced Electronic Structure Theory*, MacMillan, New York, 1982.

19. Hartree, D.R., *Proc. R. Soc. A* 141:282, ibid A143:506, 1933.

20. Fock, V., *Zeit Phys.* 61:126, 1930.

21. Roothaan, C.C.J., *Rev. Mod. Phys.* 23:69, 1951.

22. Slater, J.C., *Quantum Theory of Molecules and Solids*, Vol. 1: Electronic Structure of Molecules, McGraw-Hill, New York, Appendix 4, 1963, p. 256.

23. Pulay, P., *Chem. Phys. Lett.* 73:393, 1980; *J. Comput. Chem.* 3:556, 1982.

24. Karlstrom, G., *Chem. Phys. Lett.* 67:348, 1979.

25. Koopmans, T., *Physica* 1:104, 1933.

26. Brillouin, L. *Actualites Sci. Ind.* 71, 1933, ibid. 159, 1934.

4

Gaussian-Lobe Basis Sets

Gaussian functions do not resemble atomic orbitals very closely, but have the great advantage that integrals one encounters in a Roothaan SCF calculation are all relatively easy. The simplest case is the spherical Gaussian function, also known as the Gaussian lobe. In this chapter we detail features of the Gaussian-lobe basis, the foundation of the PCLOBE package.

The use of Gaussian basis sets was first developed by Boys [1,2]. His first examples were H and He which require only s orbitals, of the form shown below.

$$\text{GLF} = |A_x, A_y, A_z, \alpha\rangle = \left(\frac{2\alpha}{\pi}\right)^{3/4} \exp\{-\alpha[(x - A_x)^2 + (y - A_y)^2 + (z - A_z)^2]\}$$

Since two-electron integrals in an exponential basis proved to be exceedingly difficult to evaluate for the four-center cases encountered in polyatomics, Boys wrote approximations to the Slater-type orbitals (STOs) [3] (i.e., the nodeless leading terms of H-atom solutions [4])

$$\text{STO} = |n, l, m, \beta\rangle = Nr^{n-1}e^{-\beta r}Y_{lm}(\theta, \phi)$$

by linear combinations of Gaussian functions, for which the integrals were feasible. These fits are called STO-NG functions to convey that N Gaussian functions are fitted in a least-squares sense to an STO exponential function [5].

Modern quantum chemistry programs have generally adopted the so-called Cartesian–Gaussians or Gaussian-type orbitals (GTOs) shown below.

$$\text{GTO} = |x_A, y_A, z_A, a_1, a_2, a_3, a_4, \alpha\rangle = Nx_A^{a_1}y_A^{a_2}z_A^{a_3}r_A^{2a_4} \exp(-\alpha r_A^2)$$

Evaluation of integrals for these functions proved to be feasible but not simple, owing to the functions multiplying the Gaussian form. The so-called Gaussian-lobe functions (GLFs) require only very simple integral formulas.

Of course, most atoms require basis functions with higher angular momentum than s functions. The lobe basis mimic of p functions was anticipated by Parr [6] in 1957 who used two exponential 1s functions as a 2p mimic. The cusp of the exponential made this combination an imperfect match to the 2p function. In 1956, Preuss [7] suggested a combination of Gaussian lobes as an alternative without this oddity. For example, a p_z function would be represented by

$$GL(p_z) = N[GLF(\alpha: R_A + \Delta z) - GLF(\alpha: R_A - \Delta z)]$$

Here α is the scale factor for the Gaussian form, N is a normalization constant, and Δz is the displacement along the z-axis. Fletcher and Reeves [8] showed how higher angular momentum GTO functions and integrals can be obtained directly from lobes by differentiation. Most modern programs use GTO basis sets with projected "pure" spherical harmonic factors. Can lobe combinations represent GTOs' angular factors accurately? Shih et al. [9] found an energy difference of 0.1 millihartree or less in second row atoms when the same Gaussian exponents were used in pure spherical harmonic Gaussian functions and lobe mimics. Brown et al. [10] found a maximum error in the energy to be only 0.0003 hartree for light elements. The practical advantage of exact angular factors may be more apparent than real; after all, it is rare that any atomic environment is so symmetric that pure symmetry of basis functions makes a difference.

Beginning in 1963, Whitten [11,12] reported calculations which used only integrals over linear combination of Gaussian spheres. Whitten [11] developed linear combinations of Gaussian spheres that represented p- and d-type orbitals. He carried out optimization of such functions for light atoms and used these atom-optimized functions as basis functions in a Hartree–Fock–Roothaan framework [12]. Whitten's early contracted functions were designed to be of the quality of a double-zeta set of Slater functions for atoms, a major achievement. In a parallel effort, Preuss [13] laid out complete details for SCF calculations using a Gaussian-lobe basis in 1964. He used small basis sets so that he was able to treat molecules larger than benzene; such as naphthalene [14]. Whitten was initially unaware of Preuss' lobe basis representation of p functions, but upon learning of this work he acknowledged Preuss' precedence.

Shortly after Preuss' work on naphthalene [14], Buenker and Peyerimhoff [15] used Whitten's double-zeta basis to compare naphthalene to azulene. Fink [16] used the Whitten double-zeta GLF basis to study NO_2, and Harrison and Allen [17] reported a very thorough treatment of BH. These early papers established the credibility of the GLF basis.

PCLOBE uses GLF combinations of quality equivalent to the 6-31 GTO basis sets. This adaptation requires fewer primitive components than Whitten's double-zeta orbitals use.

Sambe [18] defined contractions of lobe-Gaussians which were optimized variationally to reproduce the energy levels of the H atom for 1s, 2s, 2p, 3s, 3p, and 3d orbitals rather than fit to STO AOs in a least-squares sense. The Sambe orbitals were expressed in varying numbers of Gaussian components. It is easily demonstrated that the Sambe orbitals can be adapted from the case of a nuclear charge of $Z = 1$ for the H atom to an arbitrary value of Z for some other atom by scaling the Gaussian exponents, the contraction coefficients, and the distances by which the spheres are offset from the nuclear center.

Visualizing d and f orbitals as 5 and 7 complex equivalent forms in space challenges human imagination. Consider ferrocene ($FeC_{10}H_{10}$) which is a compound of an iron atom bound to two planar cyclopentadiene moieties in a "sandwich" fashion. Efforts to describe bonding of the 3d4sp valence shell of the iron atom in this compound are less than intuitively clear if one uses the real-valued functional forms typically found in textbooks. Powell [19] addressed this problem in 1968, defining a 5×5 unitary matrix which transforms the familiar d orbitals to five with identical form. These equivalent d orbitals pointed directly to the carbon atoms in the cyclopentadienide moieties. Bonding in uranocene and lanthanide/actinide compounds of cyclooctatetraene could also be visualized with Powell's seven equivalent f orbitals pointing to eight carbons in nonplanar cyclooctatetraene.

Following Powell's idea [19], Shillady and Richardson [20] proposed a Gaussian lobe representation of the five equivalent 3d orbitals using only three spheres for the unique generator. This function is then rotated in increments of $(360/5) = 72°$ about the z-axis. The tilt angle from the z-axis was adjusted to reproduce the degenerate energy levels of the H atom 3d orbital for a completely filled (10 electrons) d shell. The orbitals are not pure d or Y_2 spherical harmonics, but contain more than 11% s-orbital character as noted by Gerloff et al. [21]. The f orbitals defined by Shillady and Talley [22] also contain substantial amounts of Y_1 (p orbital) contamination.

Later Dreissler and Ahlrichs [23] reported high quality formulations for d and f orbitals formed from spheres. The Dreissler–Ahlrichs dz^2 is constructed from only three Gaussian spheres distributed along the z-axis while the other orbitals are the usual clover-leaf shapes.

Since the PCLOBE program uses d orbitals mainly for the d-polarization of the 6-31G** basis or the simple 1G-3d orbitals for transition metals, it was deemed best to have the purest Y_2 mimics of the 3d orbitals. In the PCLOBE program the full set of five 3d orbitals are formed from lobes with the optimized 1G exponent given by Huzinaga [24] and the offset from the origin defined by Dreissler and Ahlrichs. PCLOBE incorporates a distinct $d(z^2)$ orbital formed as a linear combination of six Gaussian spheres with coefficients for the $+z$ and $-z$ spheres twice that of the (x, y) spheres. The linear combination of $[(z^2 - x^2) + (z^2 - y^2)]$ clover-leaf orbitals retains orthogonality to the other d orbitals and has no s content at all. Overall, the octahedral set standard in PCLOBE is more isotropic and a better fit to the canonical H-atom d orbitals than the Dreissler–Ahlrichs d orbitals because of the better $d(z^2)$ orbital. The Shillady–Talley f orbitals are included in PCLOBE, for possible use as polarization functions for light transition metal systems.

The question of purity of the *l*-value of a contraction of Gaussian lobes for orbitals beyond s-type has been addressed by Harrison [25]. He used a projection operator to evaluate the amount of the pure l-value in a lobe combination as well as the smaller amounts of contamination by higher l-value spherical harmonic functions. He found that lobe representation of p functions was of very high purity. A thorough mathematical analysis by Le Rouzo and Silvi [26] supported the idea that for any $Y_{1,0}$ a lobe mimic could be generated by placing spheres on a diagonal tilted away from the z-axis and then $(2l + 1)$ "equivalent orbitals" could be generated by rotations of $(360/(2l + 1))$ degrees about the z-axis.

Le Rouzo and Silvi [26] also proved analytically that $(2l + 1)$ equivalent orbitals can be formed for any desired l-value. They determine the offset distance of s-type spheres for any given l-value that meets a desired level of l-purity. The key formulas from the Le Rouzo–Silvi papers show that there is an offset parameter, λ, which controls the l-purity of the lobe-mimic orbital and this parameter determines the numerical value of the offset, R_i, according to each Gaussian exponent.

$$\lambda = a_i R_i^2, \quad R_i = \sqrt{\frac{\lambda}{a_i}}$$

Evidently Gaussian spheres can indeed represent spherical harmonic angular functions to any desired accuracy, but in practical terms implementation depends on the precision obtainable with a given digital word length in the computer used.

We will follow the notation in the review by Shavitt [27] to show how to evaluate key integrals with spherical Gaussian functions.

The key to integrals in a Gaussian basis sets is that a product of Gaussians at distinct centers is a Gaussian at an intermediate point. All integrals eventually refer to a one-center charge distribution. A normalized Gaussian sphere is

$$\varphi_A = (2a/\pi)^{3/4} e^{-ar_A^2}$$

$$(\varphi_A/\varphi_A) = 1$$

However, we plan to use linear combinations (contractions) of spherical functions to form orbital shapes close to hydrogenic functions. In that case we will have to use linear combinations of unnormalized primitive functions, ϕ_{aA}.

$$\phi_{aA} = \exp\left(-ar_A^2\right)$$

The contracted orbitals composed of n spheres will then be of the form

$$|\Phi\rangle = \sum_{i=1}^{n} c_i \phi_{aAi}$$

It will be helpful later if we also require that the linear combination is normalized.

$$\langle \Phi | \Phi \rangle = \sum_{i=1}^{n} \sum_{j=1}^{n} c_i^* c_j (\phi_{aAi} / \phi_{aAj}) = 1$$

Then the integral of any Hermitian operator, $\hat{O}$, can be evaluated over the contracted orbitals.

$$\langle \Phi | \hat{O} | \Phi \rangle = \sum_{i=1}^{n} \sum_{j=1}^{n} c_i^* c_j \langle \phi_{aAi} | \hat{O} | \phi_{aAj} \rangle$$

Here the variable r is measured from the origin of the Gaussian sphere.

$$r_A^2 = (x - A_x)^2 + (y - A_y)^2 + (z - A_z)^2$$

This means that the Gaussian can be defined by four numbers, namely (a, A_x, A_y, A_z). In the case of the product of two Gaussians another distance occurs; AB^2 is the square of the distance between the centers of the two Gaussians at the points A and B.

$$AB^2 = (A_x - B_x)^2 + (A_y - B_y)^2 + (A_z - B_x)^2$$

A similar distance can be calculated for two other Gaussian spheres at points C and D as CD^2. The Gaussian product of $\phi_{aA}\phi_{bB}$ will then be yet another Gaussian (multiplied by a factor composed of constants) centered at a point (P_x, P_y, P_z), the position of the centroid of the new Gaussian formed from the product of the original Gaussians.

$$P_x = \frac{(aA_x + bB_x)}{(a + b)}, \quad P_y = \frac{(aA_y + bB_y)}{(a + b)}, \quad P_z = \frac{(aA_z + bB_z)}{(a + b)}$$

$$\phi_{aA}\phi_{bB} = \phi_{AB} e^{-\left[\left(\frac{ab}{a+b}\right)AB^2\right]}$$

Here the product function ϕ_{AB} is a new Gaussian centered at (P_x, P_y, P_z) with a new exponent of $(a + b)$. The two normalization factors are included in the product.

Thinking ahead to programming these equations, we see that some quantities such as the sum $(a + b)$ are used repeatedly. These quantities can be evaluated once, saved, and reused to save time in evaluating the integrals within the computer program. This is probably more important with GLF basis sets because even though the formulas we show here are simple they must be used for many Gaussians in the linear combinations needed to approximate hydrogenic functions.

One slight drawback to using Gaussian orbitals, whether GLF or GTO, is that a special function occurs in the integrals related to the "Error Function, erf(x)" and higher derivatives of the lowest member of the family of functions related to F_0 [27].

$$F_0(t) = \frac{1}{\sqrt{t}} \int_0^{\sqrt{t}} e^{-v^2} dv = \frac{1}{2} \sqrt{\frac{\pi}{t}} \text{erf}(\sqrt{t})$$

$$F_{m+1}(t) = -\frac{d}{dt} F_m(t)$$

Integration by parts leads to a useful recursion relation:

$$F_m(t) = \left(\frac{1}{2m+1}\right) \left[2tF_{m+1} + e^{-t}\right]$$

Errors increase for the higher members of F_n family obtained by recursion, so the value of F_0 must be very accurate. Integrals with Gaussian spheres need F_0 and occasionally F_1. In contrast, integrals in a GTO basis require F_1 for the l-quantum number of the orbital; i.e., a d orbital would need F_2 and an f orbital would require use of F_3. This is another trade-off in efficiency attending the choice to use GLFs instead of GTOs. Although GLF basis sets require more Gaussian spheres in the linear combinations made into hydrogenic orbital shapes, the integrals only involve the simpler F_0 function while the GTO integrals require higher members of the F_n family.

The easiest way to compute the basic F_0 function is to use the special function "erf(x)" available in most computer programming languages. This intrinsic function is generally accurate to as many as 14 figures. However, due to uncertainties in the values of physical constants such as the Bohr radius and Planck's constant, the calculations are often limited to 12 significant figures or less and integrals which are accurate to 8 significant figures are often sufficiently accurate for most purposes. Thus, considerable improvement in the speed of evaluating integrals over Gaussian functions can be obtained using polynomial approximations to the F_0 function accurate to only 10 or 12 significant figures.

Here we present some significant integrals from the excellent review by Shavitt [27] and the text by Szabo and Ostlund [28].

Overlap (S_{AB})

$$(\phi_{aA}|\phi_{bB}) = \left(\frac{\pi}{a+b}\right)^{3/2} \exp\left[-\left(\frac{ab}{a+b}\right) AB^2\right] = S_{AB}$$

Kinetic Energy (T_{AB})

$$\left(\phi_{aA}\left|\left(\frac{-\nabla^2}{2}\right)\right|\phi_{bB}\right) = \left(\frac{ab}{a+b}\right)\left[3 - 2\left(\frac{ab}{a+b}\right)AB^2\right]$$

$$\times \left\{\left(\frac{\pi}{a+b}\right)^{3/2}\exp\left[-\left(\frac{ab}{a+b}\right)AB^2\right]\right\}$$

Note that the kinetic energy integral is related to the overlap integral and that the ratio of $(ab/(a+b))$ is repeated.

Nuclear Potential (for Z Nuclear Charge Value, at r_C)

$$(\phi_{aA}|(-Z/r_c)|\phi_{bB}) = -Z\left(\frac{2\pi}{a+b}\right)F_0[(a+b)CP^2]\exp\left[-\left(\frac{ab}{a+b}\right)AB^2\right]$$

This is the nuclear attraction integral of a charge density given by $\phi_{aA}\phi_{bB}$ representing a distribution of one electron attracted to a positive nuclear charge Z located at (x_c, y_c, z_c). This integral is proportional to the overlap, which allows efficient evaluation. Here CP^2 is the distance squared from the centroid position of the product of Gaussians and the position of the nucleus at position C.

$$CP^2 = (C_x - P_x)^2 + (C_y - P_y)^2 + (C_z - P_z)^2$$

So far these integrals could be evaluated easily for STOs. The multicenter two-electron integral is exceedingly difficult to evaluate with STO basis sets, but is simple for Gaussian orbitals, especially lobes.

Electron Repulsion (Coulomb's Law for Two Charge Distributions)

$$\left(\phi_{aA}(1)\phi_{bB}(1)\left|\frac{1}{r_{12}}\right|\phi_{cC}(2)\phi_{dD}(2)\right) = \left\{\frac{2\pi^{5/2}F_0(z)}{(a+b)(c+d)(a+b+c+d)^{1/2}}\right\}$$

$$\times \exp\left[-\left(\frac{ab}{a+b}\right)AB^2 - \left(\frac{cd}{c+d}\right)CD^2\right]$$

Here z is the exponent-weighted distance between centroids

$$z = \frac{(a+b)(c+d)}{(a+b+c+d)}PQ^2$$

The distance PQ^2 is the square of the distance between the two Gaussian centroids of the probability distributions of electrons 1 and 2. This integral represents coulomb repulsion between two electrons.

Note that we have used the $(11||22)$ type of charge density notation. The absence of the physical constants for the charge on an electron is due to the use of atomic units, in which system $e = |q_e| = 1$.

Dipole Moment

Although the dipole moment matrix elements are one of the simpler cases to evaluate in a lobe basis, we will treat this case in some detail, to show how one can approach a new problem in the use of a lobe basis set. Harrison and Allen [17] make the important point that a dipole moment is a vector quantity measured relative to some origin of coordinates. The dipole moment of an anion or cation is origin-dependent. In PCLOBE the origin of the coordinates is the center of mass of the molecule (X_{com}, Y_{com}, Z_{com}).

$$R_{com} = (X_{com}, Y_{com}, Z_{com}) = \frac{\sum\limits_{i=1}^{nuc}(r_i M_i)}{\sum\limits_{i=1}^{nuc} M_i} \quad (r = x, y, z)$$

$$\boldsymbol{\mu} = \mathbf{i}\mu_x + \mathbf{j}\mu_y + \mathbf{k}\mu_z$$

For each component of the dipole moment the formula is the same so we only show the derivation for the x-component.

$$\mu_x = \sum_{k=1}^{noc} 2 \sum_{i=1}^{n} \sum_{j=1}^{n} C_{k,i} C_{k,j} \sum_{l=1}^{nc(i)} \sum_{m=1}^{nc(j)} c_{i,l}\, c_{m,j} \langle a A_l | x | b B_m \rangle$$

Here capital C's refer to LCAO coefficients and lower case c's refer to contraction coefficients defining basis functions as combinations of lobes. The index k ranges over the set of noc occupied MOs and indices i and j run over the set of n basis functions. The $nc(i)$ and $nc(j)$ are the number of individual lobes in contracted basis functions i and j. These are identified by labels l and m, scale factors a or b and locations A or B. Two main types of integrals occur with Gaussian lobes, those with odd and even integrands. A simple odd case can be integrated easily as shown while cases with higher powers of x^n can be integrated by parts.

$$\int\limits_0^\infty xe^{-ax^2}\,dx = \int\limits_0^\infty \left(\frac{-2a}{-2a}\right)xe^{-ax^2}\,dx = \left(\frac{-1}{2a}\right)e^{-ax^2}\Big|_0^\infty = \left(\frac{-1}{2a}\right)[e^{-\infty} - e^0] = \frac{1}{2a}$$

The even case requires a process given by Widder [29] using the square of the integral we seek and then changing to polar coordinates in the x–y plane.

$$\int\limits_0^\infty e^{-ax^2}\,dx = I$$

$$I^2 = \int\limits_0^\infty e^{-ax^2}\,dx \int\limits_0^\infty e^{-ay^2}\,dy = \int\limits_0^\infty \int\limits_0^\infty e^{-a(x^2+y^2)}\,dx\,dy$$

$$= \int\limits_0^\infty \int\limits_0^{\pi/2} e^{-ar^2}r\,dr\,d\theta = \left(\frac{\pi}{2}\right)\left(\frac{1}{2a}\right) = \frac{\pi}{4a}$$

The original integral is the root

$$I = \frac{1}{2}\sqrt{\frac{\pi}{a}}$$

The higher powers of x can be generated by partial differentiation of the beginning forms with respect to parameter a.

Odd case

$$\int\limits_0^\infty x^{2n+1}e^{-ax^2}\,dx = \frac{n!}{2a^{n+1}}$$

Even case

$$\int\limits_0^\infty x^{2n}e^{-ax^2}\,dx = \frac{1\cdot 3\cdot 5\cdots(2n-1)\sqrt{\pi}}{2^{n+1}a^{\left(\frac{2n+1}{2}\right)}}$$

Now consider the integral for a dipole moment matrix element for, say, the x-component using unnormalized Gaussian spheres at distinct centers, with the stipulation that the normalization constants will be added as factors in the answer later.

$$(aA|x|bB) = \int\limits_{-\infty}^\infty \int\limits_{-\infty}^\infty \int\limits_{-\infty}^\infty e^{-a[(x-A_x)^2+(y-A_y)^2+(z-A_z)^2]}xe^{-b[(x-B_x)^2+(y-B_y)^2+(z-B_z)^2]}\,dx\,dy\,dz$$

As shown in Harrison's thesis [30] one can form the product Gaussian with its center at (P_x, P_y, P_z) where

$$P_\alpha = \frac{aA_\alpha + bB_\alpha}{a+b}, \quad \alpha = x, y, z$$

and shift the origin of x to

$$x' = x - P_x; \quad x = P_x + x'$$

Since P_x is a constant, the equation becomes

$$\langle aA|x|bB \rangle = P_x \langle aA|bB \rangle + \langle aA|x'|bB \rangle$$

Now x' is measured from the same origin as the centroid product Gaussian of the two former Gaussians. We recognize that the combined integrand is indeed odd relative to the x' coordinate so that $(aA|x'|bB) = 0$. The final result for the dipole matrix element over the primitive Gaussian-lobe spheres is

$$\langle aA|x|bB \rangle = P_x \langle aA|bB \rangle$$

The components of these electronic moments must still be added algebraically with the point charge positions of the nuclei. The final results in atomic units can then be multiplied by 2.541765 to obtain the dipole moment in debyes.

Quadrupole Moment

The treatment of the quadrupole tensor is given in Harrison's PhD thesis [30] and Whitten's earlier dissertation [31].

$$a_{12} = \sum_{k=1}^{noc} 2 \sum_{i=1}^{n} \sum_{j=1}^{n} C_{k,i} C_{k,j} \sum_{l=1}^{nc} \sum_{m}^{nc} c_{i,l} c_{m,j} \langle aA_l | q_1 q_2 | bB_m \rangle \quad (q = x, y, z)$$

$$\langle aA|\alpha\beta|bB \rangle = P_\alpha P_\beta \langle aA|bB \rangle; \quad (\alpha\beta = xy, xz, yz)$$

The case of the squared coordinates has also been given by Schwartz and Schaad [32].

$$(\phi_A|(3z^2 - r^2)|\phi_B) = (\phi_A|\phi_B)[2P_z^2 - P_x^2 - P_y^2]$$

An origin must be chosen since ϕ_A as well as ϕ_B are calculated relative to a fixed origin; in PCLOBE the origin is the center of mass. The full 3×3 quadrupole tensor needs to be evaluated and diagonalized to obtain the principal axes of the quadrupole moment. So far we have evaluated the electronic part of the quadrupole tensor; additional terms need to be added to each element of the tensor for the charges on the nuclei at their fixed positions. After the tensor is evaluated in the system of atomic units the answers can be reported directly as in units of 1.0e-26 esu cm^2.

Angular Momentum (Imaginary Hermitian Operator)

This integral is given in Whitten's PhD thesis [31]. In atomic units

$$\langle \phi_{aA}|(r \otimes p)|\phi_{bB}\rangle = \langle \phi_{aA}|\hat{i}L_x + \hat{j}L_y + \hat{k}L_z|\phi_{bB}\rangle$$

$$\langle \phi_{aA}|L_x|\phi_{bB}\rangle = \frac{2ab\sqrt{-1}}{(a+b)}[A_yB_z - A_zB_y]\langle \phi_{aA}|\phi_{bB}\rangle$$

$$\langle \phi_{aA}|L_y|\phi_{bB}\rangle = \frac{2ab\sqrt{-1}}{(a+b)}[A_zB_x - A_xB_z]\langle \phi_{aA}|\phi_{bB}\rangle$$

$$\langle \phi_{aA}|L_z|\phi_{bB}\rangle = \frac{2ab\sqrt{-1}}{(a+b)}[A_xB_y - A_yB_x]\langle \phi_{aA}|\phi_{bB}\rangle$$

Once again the one-electron operators are proportional to the overlap integrals. Here the components of the angular momentum are imaginary numbers and the rule for the adjoint of imaginary Hermitian operators requires that separate L_q matrices be evaluated as

$$\langle \phi_{aA}|L_q|\phi_{bB}\rangle = -\langle \phi_{bB}|L_q|\phi_{aA}\rangle \quad (q = x, y, z)$$

It is advisable to treat imaginary quantities explicitly in derivation of formulas so one does not need to use slow complex arithmetic in machine computations. Even for the imaginary elements of the angular momentum operator, one can avoid complex arithmetic entirely by suitable analysis.

Spin–Orbit and Spin–Spin Interactions

These small interactions add two additional terms to the Hamiltonian.

$$H_{so} = \left(\frac{\alpha^2}{2}\right)\sum_i\sum_\mu\left(\frac{Z_\mu}{r_{i\mu}^3}\right)(r_{i\mu} \times P_i)\bullet S_i - \left(\frac{\alpha^2}{2}\right)\sum_{i\neq j}\sum_j\left(\frac{r_{ij} \times P_i}{r_{ij}^3}\right)\bullet(S_i + 2S_j)$$

Here α is the fine structure constant, μ is the atom center, i and j are the electron labels, and $\vec{r}_{ij} = \vec{r}_i - \vec{r}_j$. The first term of the Hamiltonian is a one-electron operator while the second term is a two-electron operator. Abegg and Ha [33] evaluated the integrals for Gaussian spheres.

$$\langle \phi_a | H_{so}(1) | \phi_b \rangle = -\left(\frac{8\pi ab}{a+b}\right) \exp\left[-\left(\frac{ab}{a+b}\right)AB^2\right]$$
$$\times F_1[(a+b)VP^2]\{AB_x \bullet VP_y - AB_y \bullet VP_x\}$$

Note this expression is only for the z-component and VP is the distance from the nuclear potential at V to the centroid of the Gaussian product at P.

$$AB^2 = (A_x - B_x)^2 + (A_y - B_y)^2 + (A_z - B_z)^2$$

The two-electron term is deceptively easy but there will be as many of these integrals as there are for the $(1/r_{12})$ operator in the coulomb and exchange two-electron integral file. Again for the two-electron integrals we revert to charge density 1122 notation.

$$(\phi_a \phi_b | H_{so}(1,2) | \phi_c \phi_d)_z = \frac{8\pi^{5/2} ab}{(a+b)(a+b+c+d)^{3/2}}$$
$$\times \exp\left[-\left(\frac{ab}{a+b}\right)AB^2 - \left(\frac{cd}{c+d}\right)CD^2\right]$$
$$\times F_1\left[\frac{(a+b)(c+d)}{(a+b+c+d)}PQ^2\right]\{AB_x \bullet PQ_y - AB_y \bullet PQ_x\}$$

As above, $$CD^2 = \left[(C_x - D_x)^2 + (C_y - D_y)^2 + (C_z - D_z)^2\right]$$

with $P_k = \dfrac{aA_k + bB_k}{a+b}$ $(k = x, y, z)$ and $Q_k = \dfrac{cC_k + dD_k}{c+d}$ $(k = x, y, z)$

This is only the z-component of the two-electron operator so there are three times as many integrals for this operator as there are for the two-electron coulomb and exchange integrals. The massive number of terms required for the second term indicates the difficulty in treating metal complexes for elements heavier than the first transition row. Generally speaking, the spin–orbit terms become as large or larger than the exchange terms for elements heavier than Fe, Co, Ni, Cu, and Zn. Abegg and Ha [33] only treated BH^+ and CH which are small molecular fragments.

Frost FSGO Method

One of the venerable descriptions of bonding is the electron pair model. The idea of using a single Gaussian sphere for each pair of electrons in a closed-shell molecule's RHF determinant has been thoroughly developed by Frost and his associates. This choice of wave function, the floating spherical Gaussian orbital [34–37], or FSGO model, allows maximum flexibility in the exponent and origin of each Gaussian. All these parameters are optimized variationally to locate the minimum energy of the molecule. The FSGO model produced interesting qualitative results for molecules of light elements and is based on a rigorous mathematical model. We show this application here mainly as an illustration that one could develop a new model easily by using the Gaussian lobes' simple integrals.

The FSGO model is an almost exact paraphrase of the Lewis theory of pair structure, and defines geometries as well as the qualitative valence shell electron pair theory. Local pi bonds present complications. Frost's group resorted to "double-Gaussian" representation of p orbitals much like the 2p-orbital mimic proposed by the early work of Whitten and Preuss. Delocalized systems resist Frost representations. More flexible basis sets than Frost spheres are required for generally useful results.

Frost spheres are nonorthogonal; thus, to use the energy expression for the single determinant we require that the nonorthogonal Gaussians be Lowdin orthogonalized [38] and normalized. This can be effected by a transform $\mathbf{T} = \mathbf{S}^{-1/2}$. Then the energy can be minimized directly with a nonlinear optimization routine. The results will include the Frost orbitals, i.e., optimally located and scaled electron pair functions. We explore the Frost orbitals as a basis for SCF calculations in our next chapter.

Screened Coulomb Potential

A unique feature of the PCLOBE program is the way in which a direct inclusion of electron correlation is carried out within the lobe basis set. The original idea for the "soft coulomb hole" was pioneered by Chakravorty and Clementi [39], who suggested the screened coulomb operator

$$\frac{(1 - e^{-\eta r_{12}^2})}{r_{12}}$$

This formulation of screened electron interaction, with careful choice of the parameter, described with high accuracy the energies for atoms up to element 54. The work was extended by Clementi and Hofmann [40].

Otto et al. [41] extended this concept to molecules. In that work an important formula was derived for the screened coulomb operator in a lobe-orbital basis set.

$$\left(\phi_{aA(1)}(1)\phi_{bB}(1) \left| \frac{-e^{-\omega r_{12}^2}}{r_{12}} \right| \phi_{cC}(2)\phi_{dD(2)} \right)$$

$$= e^{-\left(\frac{ab}{a+b}\right)AB^2} e^{-\left(\frac{cd}{c+d}\right)CD^2} e^{-\left(\frac{\gamma_1\gamma_2\omega}{\gamma_{1\omega}+\gamma_2\omega+\gamma_1\gamma_2}\right)r_{PQ}^2} f(\gamma_1, \gamma_2, r_{PQ})$$

$$f(\gamma_1, \gamma_2, r_{PQ}) = \left\{ \frac{2\pi^{5/2}}{(\gamma_1\omega + \gamma_2\omega + \gamma_1\gamma_2)\sqrt{\gamma_1 + \gamma_2}} \right\}$$

$$\times F_0 \left[\frac{(\gamma_1\gamma_2)^2 r_{PQ}^2}{(\gamma_1\omega + \gamma_2\omega + \gamma_1\gamma_2)(\gamma_1 + \gamma_2)} \right]$$

$$\gamma_1 = a + b; \quad \gamma_2 = c + d$$

This is the formula for the unnormalized primitive integrals over Gaussian-lobe spheres; these integrals must be weighted by the normalized contraction coefficients and summed. These integrals are used elsewhere in this text for the "correlated-SCF" method.

Electrostatic Potential Maps

For some molecules there is interest in predicting the site of attack by a nucleophilic or electrophilic reagent. There are several levels of sophistication to this question. Consider the generation of a contour map in which a point charge such as a +1 (proton) particle is moved around the molecule under consideration in a systematic way to form a grid of values of the electrostatic potential. A contour map gives a visually arresting picture of the favored sites for attack of the electrophile. The simplest and least accurate way to do this is to place the Mulliken [42] atomic charges at the respective nuclear positions and use a simple coulombic potential based on the point charges at each nuclear site (X_i, Y_i, Z_i).

$$V(x, y, z) = \sum_{i=1}^{nuc} \frac{q_i}{\sqrt{(x - X_i)^2 + (y - Y_i)^2 + (z - Z_i)^2}}$$

The next level of treatment would be to use the three-center nuclear attraction integral formulae where r_c is the position of the test point charge at (x, y, z). This is the primitive

$$(\phi_{aA}|(-1/r_c)|\phi_{bB}) = -\left(\frac{2\pi}{a+b}\right) F_0[(a+b)CP^2] \exp\left[-\left(\frac{ab}{a+b}\right)AB^2\right]$$

integral over two spheres within the contracted basis of the calculation. Summing contributions from all molecular orbitals gives the coulomb interaction between a test point charge and the electron density. The first term is the electronic interaction and the second term accounts for effects of the nuclei.

$$V(x,y,z) = \sum_{m=1}^{occ} 2 \sum_{i=1}^{nbs} \sum_{j=1}^{nbs} C_{m,i}^{*} C_{m,j} \sum_{k=1}^{nc(k)} \sum_{l=1}^{nc(l)} c_{i,k} c_{j,l} \sum_{c=1}^{nuc} \left(\phi_{aA_i} \left| \left(\frac{-1}{r_{test}} \right) \right| \phi_{bB_j} \right)$$

$$+ \sum_{A=1}^{nuc} \left(\frac{Z_A}{|R_A - r_{test}|} \right)$$

Here occ refers to the number of occupied MOs, nbs is the number of basis functions, $nc(k)$ and $nc(l)$ refer to the number of Gaussian functions in the contracted form of the basis function labeled k and l, and nuc refers to the number of nuclei. The probe charge is at location r_{test}. A third, slower and more computationally demanding, way to generate the grid points for V (x,y,z) would be to actually do an SCF equation for the molecule in question with an extra proton at (x,y,z) but this might present difficulties in obtaining SCF convergence for geometries where the test charge is very close to one of the nuclei in the molecule.

The second way was applied by Shillady and Baldwin-Boisclair [43] to small molecules. It was found that the overall electrostatic potential is relatively insensitive to the basis set used as long as the true three-center nuclear attraction integral is used to find $V(x,y,z)$. Thus, one might expect that a small basis like STO-3G or its lobe equivalent would give qualitatively useful understanding of sites of attack by nucleophiles or electrophiles. The simple form of the GLF nuclear attraction integral makes this possible.

References

1. Boys, S.F., *Proc. R. Soc. A* 200:542, 1950.
2. Boys, S.F., *Proc. R. Soc. A* 258:402, 1960.
3. Slater, J.C., *Phys. Rev.* 36:51, 1930.
4. Pauling, L. and Wilson, E.B., *Introduction to Quantum Mechanics with Applications to Chemistry*, McGraw-Hill, New York, Chap V, 1935.
5. Hehre, W.J., Stewart, R.F., and Pople, J.A., *J. Chem. Phys.* 51:2657, 1969.
6. Parr, R.G., *J. Chem. Phys.* 26:428, 1957.
7. Preuss, H., *Z. Naturforsch.* 11:823, 1956.
8. Fletcher, R. and Reeves, C.M., *Comput. J.* Oct: 287, 1963.
9. Shih, S., Buenker, R.J., Peyerimhoff, S.D., and Wirsam, B., *Theor. Chim. Acta* 18:277, 1970.
10. Brown, R.D., Burden, F.R., and Hart, B.T., *Theor. Chim. Acta* 25:49, 1972.
11. Whitten, J.L., *J. Chem. Phys.* 39:349, 1963.
12. Whitten, J.L., *J. Chem. Phys.* 44:359, 1966.
13. Preuss, H., *Mol. Phys.* 8:157, 1964.
14. Preuss, H., *Int. J. Quantum Chem.* II:651, 1968.
15. Buenker, R.J. and Peyerimhoff, S.D., *Chem. Phys. Lett.* 3:37, 1969.
16. Fink, W.H., *J. Chem. Phys.* 49:5054, 1968.

17. Harrison, J.F. and Allen, L.C., *J. Mol. Spect.* 29:432, 1969.
18. Sambe, H., *J. Chem. Phys.* 42:1732, 1965.
19. Powell, R.E., *J. Chem. Ed.* 45:45, 1968.
20. Shillady, D.D. and Richardson, F.S., *Chem. Phys. Lett.* 6:359, 1970.
21. Gerloff, M., Ady, E., and Brickman, J., *Mol. Phys.* 26:561, 1973.
22. Shillady, D.D. and Talley, D.B., *J. Comput. Chem.* 3:130, 1982.
23. Dreissler, F. and Ahlrichs, R., *Chem. Phys. Lett.* 23:571, 1973.
24. Huzinaga, S., *J. Chem. Phys.* 42:1293, 1965.
25. Harrison, J.F., *J. Chem. Phys.* 46:1115, 1967.
26. Le Rouzo, H. and Silvi, B., *Int. J. Quantum Chem.* XIII:297, 1978.
27. Shavitt, I., *Methods Comput. Phys.* 2:1, 1963.
28. Szabo, A. and Ostlund, N.S., *Modern Quantum Chemistry: Introduction to Advanced Electronic Structure Theory*, MacMillan, New York, 1982, p. 153.
29. Widder, D.V., *Advanced Calculus*, 2nd Ed., Prentice-Hall, Englewood Cliffs, New Jersey, 1961.
30. Harrison, J.F., *The Electronic Structure and Properties of BH, NH, FH, CH_2 in Their Ground and Excited States*, PhD Thesis, Princeton University, New Jersey, 1966.
31. Whitten, J.L., *Theoretical Studies of Electronic States of Simple Polyatomic Molecules. Part I. Gaussian Expansion of Wavefunctions, Part II. Magnetic Interaction in the Triplet State of Glyoxal*, PhD Thesis, Georgia Institute of Technology, 1963.
32. Schwartz, M.E. and Schaad, L.J., *J. Chem. Phys.* 46:4112, 1967.
33. Abegg, P.W. and Ha, T.K., *Mol. Phys.* 27:763, 1974.
34. Frost, A.A., *J. Chem. Phys.* 47:3707, 1967.
35. Frost, A.A., *J. Chem. Phys.* 47:3714, 1967.
36. Frost, A.A., *J. Phys. Chem.* 72:1289, 1967.
37. Frost, A.A. and Rouse, R.A., *J. Am. Chem. Soc.* 90:1965, 1967.
38. Lowdin, P.O., *J. Chem. Phys.* 18:365, 1950.
39. Chakravorty, S.J. and Clementi, E. *Phys. Rev. A* 39:2290, 1989.
40. Clementi, E. and Hofmann, D.W.M., *Int. J. Quantum Chem.* 52:849, 1994.
41. Otto, P., Reif, H., and Hernandez-Laguna, A., *J. Mol. Struct.* 340:51, 1995.
42. Mulliken, R.S. *J. Chem. Phys.* 23:1833, 1955.
43. Shillady, D.D. and Baldwin-Boisclair, S., *Int. J. Quantum Chem., Quantum Biol. Symp.* 6:105, 1979.

5

A Very Simple MO Program

Now that we have in hand the algebraic forms necessary to compute useful wave functions for chemical systems, we are able to discuss the implementation of the methods. Here we describe a very simple program capable of illustrating features of the SCF method. It employs small basis sets composed entirely of spherical functions—"s" orbitals. Hence its name SCF1s. The choice of basis minimizes the effort to be expended in integral evaluation and permits the structure of the calculation to be seen clearly. The fact that it must apply to only a few atoms and small molecules is precisely what we need in first examples.

SCF1s is expressed in FORTRAN(77). The choice of computer language is partly historical and partly an expression of the universal familiarity of the language. The free WATCOM compiler can process the code into executable files which can run on any PC. There is no graphics interface, but we hardly need any powerful molecule-builder for the small systems which are within the reach of SCF1s. Details of its use are provided on the CD and in the listing itself.

Helium in SCF1s

The SCF1s program can describe single-center systems with exponential Slater-type orbital (STO) 1s functions. Here we describe the He atom with two STOs; the object is to solve the Roothaan equations in the small basis.

The program will read from a file called "scf1s.dat"

```
He double-zeta test                           1  2  2  1150  0
2.0          0.0        0.0       0.0
1.44608                 0.0       0.0       0.0
2.86222                 0.0       0.0       0.0
```

The codes in the first line are explained in the SCF1s program listing on the accompanying disk. The input tells us that the calculation is intended to illustrate the He atom with two basis functions. The nucleus, with charge $+2$, is situated at the origin. Two simple exponential functions (Slater 1s orbitals) are specified. Both are placed at the origin. They differ in scale factor, one with factor 1.44608 being more diffuse than the other with factor 2.86222.

Here is the output from that calculation.

```
*******************************************************************
                  Gaussian-Lobe Program for Students
                        set up by D. Shillady at
                  Virginia Commonwealth University
                        Richmond Virginia
                              2006
                        Quantum Mechanics LLC
*******************************************************************
                        He double-zeta test
*******************************************************************
*******************************************************************
     nat=1      nbs=2      nel=2      iop=1      lit=150      mul=1
*******************************************************************
```

This segment reminds us that the number of atoms is 1, the number of functions in the basis set is 2, the number of electrons is 2, the basis type is STO, and the maximum number of iterations permitted in the SCF procedure is 150. The multiplicity of the system is 1 (this is a singlet state).

```
*******************************************************************
atom(1)          zn=2.0              x=0.000000   y=0.000000   z=0.000000
*******************************************************************
1S-sphere(1)  alpha=1.44608   x=0.000000   y=0.000000   z=0.000000
1S-sphere(2)  alpha=2.86222   x=0.000000   y=0.000000   z=0.000000
*******************************************************************
```

The location and scale factor for the basis functions is quoted. The first 1s function is scaled to be slightly tighter than the best single exponential would be (alpha = 1.44 rather than 1.69) and the second is more diffuse. The relative weighting of these two functions will be chosen to optimize the SCF energy.

Next, the overlap and core matrix elements are evaluated.

```
                        S-overlap matrix

                          1              2
              1      1.000000       0.842394
              2      0.842394       1.000000

                        H-core matrix

                          1              2
              1     -1.846586      -1.885952
              2     -1.885952      -1.628288
```

In the course of the solution of the secular equation, we will need to orthogonalize the basis.

```
                         S** (-1/2) Matrix

                      1                 2
         1         1.627822         -0.891092
         2        -0.891092          1.627822
```

The starting linear combinations of the two basis functions are the eigenvectors of the core matrix.

```
              H-core Eigenvectors, by Column

                      1                 2
         1         0.642832          1.740867
         2         0.396570         -1.812894
```

Only the first column is used; there is only one (doubly) occupied MO in the RHF determinant. The P-matrix is required for construction of the first approximation to the Fock matrix.

```
                     P-matrix, noc=1

                      1                 2
         1         0.826466          0.509856
         2         0.509856          0.314535
```

So too are the two-electron integrals in the basis.

```
          Unique Two-Electron Integrals, iop=1
         1    1    1    1       0.9038000000
         1    1    1    2       0.9040457192
         1    1    2    2       1.1749298808
         1    2    1    2       0.9554023522
         1    2    2    2       1.2890820393
         2    2    2    2       1.7888875000
                ***** Iteration 1 *****

                P-matrix, noc=1

              1            2
         1  0.826466    0.509856
         2  0.509856    0.314535
```

The Fock matrix after the insertion of the repulsion terms is

```
                         F-matrix

                      1                 2
         1        -0.792870         -0.878486
         2        -0.878486         -0.113472
```

The eigenvectors of the F-matrix are the MOs, and the eigenvalues are the orbital energies.

```
Orbital Energies in hartrees, lowest first
              E(1) = -0.835243365855
              E(2) =  2.811044941327
```

The total energy is not twice the Fock eigenvalue, but

```
Electronic Energy = -2.7740974041
```

The lowest eigenvector of the Fock matrix defines the new estimate of the P-matrix.

```
            ***** Iteration 2 *****

               P-matrix, noc = 1

                  1            2
       1       0.880159     0.491905
       2       0.491905     0.291086
```

The new P-matrix defines a new F-matrix

```
               F-matrix

                  1            2
       1      -0.801185    -0.884509
       2      -0.884509    -0.120151
```

with new eigenvalues (orbital energies) and eigenvectors.

```
      Orbital Energies in hartrees, lowest first
  E(1) =                  -0.843169212850
  E(2) =                   2.802289498299
  Electronic Energy =     -2.7825055315
```

The cycling continues until the MO coefficients defined in step 65

```
               C-matrix

                  1            2
       1       0.834151     1.657722
       2       0.190601    -1.845947
```

produce a converged density matrix

```
            P-matrix, noc = 1

                  1            2
       1       1.391519     0.318016
       2       0.318016     0.072692
```

which in turn defines the final Fock matrix.

```
                              F-matrix

                         1              2
              1      -0.879575     -0.941174
              2      -0.941174     -0.183114
```

The eigenvalues of the Fock matrix define one bound state.

```
          Orbital Energies in hartrees, lowest first
                  E(1) = -0.917942509541
                  E(2) =  2.719035436217
```

The total electronic energy can now be defined.

```
          Electronic Energy = -2.8616613064

                 <<<<<ENERGY CONVERGED!>>>>>

          Nuclear Repulsion =  0.0000000000
          ***** Total Energy = -2.8616613064 hartrees *****
```

The two-member basis of STOs effects a considerable improvement over the first-order perturbation theory result (-2.75 hartree) and even the best single-STO approximation, with optimized scale factor 1.6875 and energy -2.8476 hartree. The exact energy is about -2.903 hartree; the error of 0.05 hartree or about 30 kcal/mol is significant in any chemical context. A more accurate calculation requires a more flexible wave function than the single determinant assumed in the SCF-Roothaan method. Even in a much larger basis there will be an irreducible shortfall in the Hartree–Fock energy, the "correlation" error.

Beryllium Atom in SCF1s STOs

The beryllium atom ground state might be written as $1s^2 2s^2$, and for the 1s function we can use a double-zeta expansion as we did for the He atom.

```
*********************************************************************
                  Gaussian-Lobe Program for Students
                      set up by D. Shillady at
                 Virginia Commonwealth University
                        Richmond Virginia
                              2006
                      Quantum Mechanics LLC
*********************************************************************
        Be double-zeta-1s (Huzinaga & Arnau, JCP v53, p451 (1970)
*********************************************************************
nat = 1    nbs = 3    nel = 4    iop = 1    lit = 150    mul = 1
*********************************************************************
```

```
***************************************************************
atom(1)          zn=4.0          x=0.000000   y=0.000000    z=0.000000
***************************************************************
1S-sphere(1)     alpha=3.34764   x=0.000000   y=0.000000    z=0.000000
1S-sphere(2)     alpha=5.54303   x=0.000000   y=0.000000    z=0.000000
1S-sphere(3)     alpha=0.58833   x=0.000000   y=0.000000    z=0.000000
***************************************************************
```

The scale factors are much larger for Be 1s than for He 1s, in keeping with the increase in nuclear charge. In SCF1s the valence s orbital is a more diffuse 1s function.

The sharply peaked basis functions that are intended to describe the 1s core overlap strongly, while their overlap with the diffuse basis function is relatively weak.

S-overlap matrix

	1	2	3
1	1.000000	0.909946	0.362639
2	0.909946	1.000000	0.204396
3	0.362639	0.204396	1.000000

The one-electron energies are very low for the 1s basis functions and much higher for the diffuse valence-shell basis function.

H-core matrix

	1	2	3
1	−7.787213	−7.737546	−2.497559
2	−7.737546	−6.809529	−2.173167
3	−2.497559	−2.173167	−2.180254

Solution of the first secular equation $HC = ESC$ produces core eigenvectors. The lowest energy eigenvector of the core matrix is made up of the strongly peaked and tightly bound basis functions, while the next eigenvector is dominated by the more diffuse valence basis function. Admixture of the valence basis function with the tightly bound core basis functions is required to establish orthogonality.

H-core Eigenvectors, by Column

	1	2	3
1	0.739422	0.234468	2.563235
2	0.286857	0.091682	−2.531722
3	−0.023898	−1.056796	−0.412250

The leftmost column defines the inner shell, dominated by the tightly bound basis functions, while the next column defines the valence s orbital, primarily the diffuse basis function orthogonalized to the inner eigenvector.

The vacant orbital has a node close to the nucleus, a short effective wavelength, and a very high energy. The density matrix is defined

```
                      P-matrix, noc=2

                1              2              3
    1        1.203441       0.467211      -0.530911
    2        0.467211       0.181386      -0.207490
    3       -0.530911      -0.207490       2.234778
```

and employed to refine the Fock operator so as to begin the iterative solution of the Roothaan equations.

```
    ************ delete iterations 2 to 70 **********
    ****************** Iteration 71 ****************

    Electronic Energy=-14.5552807059
            <<<<<ENERGY CONVERGED!>>>>>
    Nuclear Repulsion=   0.0000000000
    ***** Total Energy=-14.5552807059 hartrees *****
```

The converged density matrix looks very different from the initial guess above.

```
                      P-matrix, noc=2

                1              2              3
    1        1.823341       0.216916      -0.935899
    2        0.216916       0.060603       0.143126
    3       -0.935899       0.143126       2.341569
```

The final Fock matrix is strongly nondiagonal, but has eigenvalues corresponding to a strongly bound level (like 1s), a weakly bound level (like 2s), and a very high energy remnant of the variational description of the occupied levels.

```
                         F-matrix
                1              2              3
    1       -4.672569      -4.648971      -1.584083
    2       -4.648971      -2.949284      -1.642891
    3       -1.584083      -1.642891      -0.798385

    Orbital Energies in hartrees, lowest first
            E(1) =-4.736778639600
            E(2) =-0.308211159344
            E(3) = 10.601064288704
```

The final occupied SCF orbitals are (1) predominantly a combination of the tightly peaked basis functions (leftmost column of the C-matrix, below) and (2) predominantly the diffuse basis function. The SCF iteration has altered the MOs considerably from the first guess.

```
                            C-matrix
                   1             2             3
         1       0.851796      0.431432      2.502040
         2       0.160924     -0.066347     -2.543622
         3      -0.001338     -1.082029     -0.341397
```

The Koopmans ionization energy from the "2s" level is 0.3 hartree, or about 8 eV.

```
       Final Orbital Energies in hartrees, lowest first
         -4.736779          -0.308211          10.601064
```

The Koopmans electron affinity is not quantitatively meaningful, but it seems clear that Be anion is not stable.

Spherical Gaussian Contraction for Helium: 3G Expansion

Slater orbitals are very convenient for one- and two-center systems; however, owing to the great difficulty in evaluating multicenter two-electron integrals, STOs are not in wide use in descriptions of larger molecules. SCF1s provides an introduction to Gaussian basis sets using spherical Gaussian analogs to the 1s function. Now let us return to the He atom using spherical Gaussian orbitals as the basis. The input data is quoted early in the SCF1s output.

```
* * * * * * * * * * * * * * * * * * * * * * * * * * * * * * * * * * * * * * * * * * * * * * * * * * * * * *
                   Gaussian-Lobe Program for Students
                       set up by D. Shillady at
                   Virginia Commonwealth University
                          Richmond Virginia
                               2006
                        Quantum Mechanics LLC
* * * * * * * * * * * * * * * * * * * * * * * * * * * * * * * * * * * * * * * * * * * * * * * * * * * * * *
            He 3G-gaussian test
* * * * * * * * * * * * * * * * * * * * * * * * * * * * * * * * * * * * * * * * * * * * * * * * * * * * * *
nat = 1    nbs = 3    nel = 2    iop = 2    lit = 150    mul = 1
* * * * * * * * * * * * * * * * * * * * * * * * * * * * * * * * * * * * * * * * * * * * * * * * * * * * * *

* * * * * * * * * * * * * * * * * * * * * * * * * * * * * * * * * * * * * * * * * * * * * * * * * * * * * *
atom(1)          zn = 2.0             x = 0.000000   y = 0.000000   z = 0.000000
* * * * * * * * * * * * * * * * * * * * * * * * * * * * * * * * * * * * * * * * * * * * * * * * * * * * * *
1S-sphere(1)     alpha = 13.62081    x = 0.000000   y = 0.000000   z = 0.000000
1S-sphere(2)     alpha =  2.04686    x = 0.000000   y = 0.000000   z = 0.000000
1S-sphere(3)     alpha =  0.44054    x = 0.000000   y = 0.000000   z = 0.000000
* * * * * * * * * * * * * * * * * * * * * * * * * * * * * * * * * * * * * * * * * * * * * * * * * * * * * *
```

These are the basis functions which make up the Pople STO-3G contraction chosen to mimic a single STO 1s orbital of scale factor 1.69, the optimum for a single STO basis function for helium. In this calculation we allow the relative weighting of the three components to be selected variationally. Pople chose the 3G coefficients so as to match in a least-squares sense the Slater exponential 1s function. (The series can be rescaled easily to a Gaussian expansion matching any Slater 1s function.)

The now-familiar first steps are taken:

S-overlap matrix

	1	2	3
1	1.000000	0.553359	0.205660
2	0.553359	1.000000	0.667165
3	0.205660	0.667165	1.000000

H-core matrix

	1	2	3
1	8.652397	−1.989007	−1.477102
2	−1.989007	−1.495795	−1.649027
3	−1.477102	−1.649027	−1.457520

$S^{**}(-1/2)$ Matrix

	1	2	3
1	1.168225	−0.418920	0.093229
2	−0.418920	1.483353	−0.545463
3	0.093229	−0.545463	1.276891

H-core Eigenvectors, by Column

	1	2	3
1	0.096488	0.076467	1.238458
2	0.525269	1.138685	−1.049213
3	0.521792	−1.225369	0.403578

P-matrix, noc = 1

	1	2	3
1	0.018620	0.101365	0.100694
2	0.101365	0.551816	0.548163
3	0.100694	0.548163	0.544534

Nuclear Repulsion = 0.0000000000

Unique Two-Electron Integrals, iop = 2

1	1	1	1	4.1644384444
1	1	1	2	1.9692700545
1	1	1	3	0.7067131127
1	1	2	2	2.1286965388
1	1	2	3	1.1365486231
1	1	3	3	1.0424440482
1	2	1	2	0.9670674111
1	2	1	3	0.3495718918
1	2	2	2	1.1249032666
1	2	2	3	0.6103406655
1	2	3	3	0.5702845512
1	3	1	3	0.1265467924

1	3	2	2	0.4132166354
1	3	2	3	0.2250823857
1	3	3	3	0.2113082355
2	2	2	2	1.6143569354
2	2	2	3	0.9364206203
2	2	3	3	0.9608062484
2	3	2	3	0.5601189053
2	3	3	3	0.6072305841
3	3	3	3	0.7489448993

In this case the integral list consists of 21 unique values. As the basis size (N) increases, the number of two electron integrals increases drastically. The number of products of orbitals is N^4 or 81. The indistinguishability of electrons reduces this number in three ways:

$$\langle ij \| kl \rangle = \langle il \| kj \rangle = \langle kj \| il \rangle = \langle kl \| ij \rangle = \langle ji \| lk \rangle = \langle jk \| li \rangle = \langle li \| jk \rangle = \langle lk \| ji \rangle$$

This is sometimes called eightfold symmetry, but the full eightfold reduction in unique values applies only to the set of integrals for which all indices disagree.

Point group symmetry can also reduce the number of unique values of repulsion integrals. Despite these small mercies, the number of two-electron integrals increases sharply as the number of basis functions increases. This has been a continuing challenge, overcome in part by faster computers and clever algorithms.

***** Iteration 1 *****

P-matrix, noc = 1

	1	2	3
1	0.018620	0.101365	0.100694
2	0.101365	0.551816	0.548163
3	0.100694	0.548163	0.544534

F-matrix

	1	2	3
1	11.457368	−0.987722	−1.362422
2	−0.987722	0.000038	−0.933066
3	−1.362422	−0.933066	−0.412791

Orbital Energies in hartrees, lowest first

$$E(1) = -0.797315910903$$
$$E(2) = 2.168412229143$$
$$E(3) = 19.509843100378$$

A single level binds the two electrons in helium.

Electronic Energy = −2.7377589492

***** Iteration 2 *****

P-matrix, noc = 1

	1	2	3
1	0.018109	0.097632	0.101515
2	0.097632	0.527002	0.544986
3	0.101515	0.544986	0.577891

F-matrix

	1	2	3
1	11.435262	−0.990643	−1.362035
2	−0.990643	−0.005121	−0.938794
3	−1.362035	−0.938794	−0.422864

Orbital Energies in hartrees, lowest first

E(1) = −0.805776647068
E(2) = 2.162799815770
E(3) = 19.481210944928
Electronic Energy = −2.7463928881

.... delete iterations 3 to 64 ....

***** Iteration 65 *****

P-matrix, noc = 1

	1	2	3
1	0.013431	0.061781	0.110554
2	0.061781	0.284181	0.508523
3	0.110554	0.508523	0.910002

F-matrix

	1	2	3
1	11.216887	−1.019895	−1.358169
2	−1.019895	−0.055253	−0.994209
3	−1.358169	−0.994209	−0.521430

Orbital Energies in hartrees, lowest first

E(1) = −0.888518797096
E(2) = 2.106622302407
E(3) = 19.199657118057
Electronic Energy = −2.8308706943

***** Iteration 66 *****

P-matrix, noc = 1

	1	2	3
1	0.013431	0.061780	0.110555
2	0.061780	0.284176	0.508522
3	0.110555	0.508522	0.910009

F-matrix

	1	2	3
1	11.216882	−1.019896	−1.358169
2	−1.019896	−0.055254	−0.994210
3	−1.358169	−0.994210	−0.521432

Orbital Energies in hartrees, lowest first

$$E(1) = -0.888520482018$$
$$E(2) = 2.106620961213$$
$$E(3) = 19.199651077121$$
Electronic Energy = −2.8308716704

<<<<<ENERGY CONVERGED!>>>>>

Nuclear Repulsion = 0.0000000000
**** Total Energy = −2.8308716704 hartrees ****

C-matrix

	1	2	3
1	0.081947	0.064758	1.240172
2	0.376915	1.216721	−1.025126
3	0.674566	−1.155130	0.383807

F-matrix

	1	2	3
1	11.216882	−1.019896	−1.358169
2	−1.019896	−0.055254	−0.994210
3	−1.358169	−0.994210	−0.521432

Final Orbital Energies in hartrees, lowest first

−0.888520   2.106621   19.199651

In this calculation the Gaussian orbitals are the uncontracted STO-3G 1s Gaussians which have been scaled by 1.69 to match the optimum scaling of the single STO. The contracted STO-3G basis—composed of a single basis function—produces an energy for He of −2.8077 hartree. This linear combination of three Gaussians with variational weighting—a kind of 111G basis—achieves an SCF energy of −2.83087 hartree, far better than the STO-3G energy of the Pople contracted function.

Here we use six Gaussian 1s functions taken from Pople's 6G expansion of an STO-6G fit scaled for He.

	He gaussian test				1	6	2	2150	0
2.0	0.0	0.0	0.0						
102.43575		0.0	0.0	0.0					
15.410277		0.0	0.0	0.0					
3.4981788		0.0	0.0	0.0					
1.0197597		0.0	0.0	0.0					
0.2931072		0.0	0.0	0.0					
0.0500000		0.0	0.0	0.0					

The results of this input data to SCF1s are

***** Iteration 68 *****

P-matrix, noc = 1

	1	2	3	4	5	6
1	0.000101	0.000755	0.002829	0.006129	0.006485	0.000419
2	0.000755	0.005640	0.021123	0.045761	0.048420	0.003130
3	0.002829	0.021123	0.079114	0.171391	0.181347	0.011724
4	0.006129	0.045761	0.171391	0.371296	0.392864	0.025399
5	0.006485	0.048420	0.181347	0.392864	0.415701	0.026877
6	0.000419	0.003130	0.011724	0.025399	0.026877	0.001738

F-matrix

	1	2	3	4	5	6
1	124.531273	10.268532	-2.346319	-1.671275	-0.788307	-0.231027
2	10.268532	13.175187	0.535333	-1.694451	-1.113724	-0.369317
3	-2.346319	0.535333	1.007207	-0.876021	-1.111915	-0.452180
4	-1.671275	-1.694451	-0.876021	-0.565795	-0.756264	-0.413193
5	-0.788307	-1.113724	-1.111915	-0.756264	-0.424969	-0.252863
6	-0.231027	-0.369317	-0.452180	-0.413193	-0.252863	-0.056838

Orbital Energies in hartrees, lowest first

$$E(1) = -0.917260446074$$
$$E(2) = 0.119443198149$$
$$E(3) = 1.206503927743$$
$$E(4) = 6.088478065780$$
$$E(5) = 28.622496515527$$
$$E(6) = 182.732737350120$$
$$\text{Electronic Energy} = -2.8600804712$$

<<<<<ENERGY CONVERGED!>>>>>

Nuclear Repulsion = 0.0000000000

**** Total Energy = -2.8600804712 hartrees ****

C-matrix

	1	2	3	4	5	6
1	0.007112	0.001843	0.003879	0.022375	-0.027619	1.264879
2	0.053102	0.009473	0.068105	-0.022418	1.557325	-1.052578
3	0.198887	0.066756	-0.000629	1.764510	-1.805787	0.671811
4	0.430829	0.022526	1.422846	-2.222457	1.113383	-0.379727
5	0.455936	0.516540	-1.840062	1.025698	-0.427424	0.147771
6	0.029499	-1.204329	0.549051	-0.219975	0.091505	-0.032266

Final Orbital Energies in hartrees, lowest first

| -0.917260 | 0.119443 | 1.206504 | 6.088478 | 28.622497 | 182.732737 |

Extending the calculation to six Gaussians improves the variational energy to -2.86008 hartree, to be compared with the value of the He energy obtained with the two-function STO double-zeta basis, -2.86066 hartree.

The contracted STO-6G basis function yields an energy for He of -2.84629 hartree. The latter value is properly compared with the best energy obtained with a single STO basis function, -2.8476 hartree.

We see just how hard it is to describe the details of an atomic charge distribution in a Gaussian basis, and how much more efficiently the STO basis accomplishes this task.

Molecular Hydrogen in SCF1s

The simplest molecular system to which we can apply a self-consistent method is molecular hydrogen. In the 1G basis, the MO is entirely determined by symmetry so that iteration is unnecessary.

```
************************************************************************
                   Gaussian-Lobe Program for Students
                        set up by D. Shillady at
                   Virginia Commonwealth University
                          Richmond Virginia
                                2006
                         Quantum Mechanics LLC
************************************************************************
                              test on H2
************************************************************************
        nat=2    nbs=2    nel=2    iop=2    lit=99    mul=1
************************************************************************


************************************************************************
atom(1)         zn=1.0           x=0.000000   y=0.000000   z=0.000000
atom(2)         zn=1.0           x=1.400000   y=0.000000   z=0.000000
************************************************************************
1S-sphere(1)    alpha=0.40000    x=0.000000   y=0.000000   z=0.000000
1S-sphere(2)    alpha=0.40000    x=1.400000   y=0.000000   z=0.000000
************************************************************************
```

It is not necessary to place lobes at the nuclear centers but it is a widely adopted practice.

S-overlap matrix

	1	2
1	1.000000	0.675704
2	0.675704	1.000000

H-core matrix

	1	2
1	−1.068838	−0.905365
2	−0.905365	−1.068838

```
                    S**(−1/2) Matrix

                      1              2
          1        1.264262      −0.491757
          2       −0.491757       1.264262


            H-core Eigenvectors, by Column

                      1              2
          1        0.546244       1.241693
          2        0.546244      −1.241693


               P-matrix, noc = 1

                      1              2
          1        0.596764       0.596764
          2        0.596764       0.596764
        Nuclear Repulsion = 0.7142857143


     Unique Two-Electron Integrals, iop = 2
        1    1    1    1     0.7136496465
        1    1    1    2     0.4524803954
        1    1    2    2     0.5639303625
        1    2    1    2     0.3258353364
        1    2    2    2     0.4524803954
        2    2    2    2     0.7136496465
               ***** Iteration 1 *****


               P-matrix, noc = 1
                      1                    2
    1          0.596764              0.596764
    2          0.596764              0.596764


                    F-matrix

                      1                    2
    1         −0.346563             −0.511937
    2         −0.511937             −0.346563


     Orbital Energies in hartrees, lowest first
              E(1) = −0.512322251716
              E(2) =  0.509947003661
         Electronic Energy = −1.6904552584


            <<<<<ENERGY CONVERGED!>>>>>

     Nuclear Repulsion =  0.7142857143
     ***** Total Energy = −0.9761695441 hartrees *****
```

The 1G energy of H atom is 0.432 hartree, so this calculation predicts binding of about 0.112 hartree or about 70 kcal/mol.

```
                    C-matrix

                      1              2
          1        0.546244       1.241693
          2        0.546244      −1.241693
```

The coeffients are determined by symmetry but the specific numerical values depend on the geometry and scale factors.

```
                              F-matrix
                       1                         2
         1          -0.346563               -0.511937
         2          -0.511937               -0.346563

     Final Orbital Energies in hartrees, lowest first
     -0.512322  0.509947
```

LiH–Frost Spherical Gaussian Pairs

Now let us allow off-center Gaussians in an application to the LiH molecule. Here we will use the final orbitals, optimized by Frost using the FSGO method [1], as the best two Gaussians to describe the four electrons as two electron pairs, constrained to be described by a single Gaussian for each pair. Each Gaussian lobe was optimized in location and scale.

```
* * * * * * * * * * * * * * * * * * * * * * * * * * * * * * * * * * * * * * * * * * * * * * * * *
                    Gaussian-Lobe Program for Students
                         set up by D. Shillady at
                    Virginia Commonwealth University
                            Richmond Virginia
                                 2006
                         Quantum Mechanics LLC
* * * * * * * * * * * * * * * * * * * * * * * * * * * * * * * * * * * * * * * * * * * * * * * * *
                      test on LiH with Frost's orbitals
* * * * * * * * * * * * * * * * * * * * * * * * * * * * * * * * * * * * * * * * * * * * * * * * *
              nat=2  nbs=2  nel=4  iop=2  lit=99  mul=1
* * * * * * * * * * * * * * * * * * * * * * * * * * * * * * * * * * * * * * * * * * * * * * * * *

* * * * * * * * * * * * * * * * * * * * * * * * * * * * * * * * * * * * * * * * * * * * * * * * *
  atom(1)      zn=3.0     x=0.000000    y=0.000000    z=0.000000
  atom(2)      zn=1.0     x=3.226000    y=0.000000    z=0.000000
* * * * * * * * * * * * * * * * * * * * * * * * * * * * * * * * * * * * * * * * * * * * * * * * *
  1S-sphere(1)  alpha=1.99834  x=-0.007600  y=0.000000  z=0.000000
  1S-sphere(2)  alpha=0.16864  x=2.882900   y=0.000000  z=0.000000
* * * * * * * * * * * * * * * * * * * * * * * * * * * * * * * * * * * * * * * * * * * * * * * * *
```

Note the Li 1s mimic is very close to the Li nuclear center, whereas the Li–H bond pair is between the two nuclei—i.e., in the bonding region—but much closer to the H nucleus than the Li nucleus. We might be tempted to think of the system as Li cation and H anion.

S-overlap matrix

	1	2
1	1.000000	0.106952
2	0.106952	1.000000

H-core matrix

	1	2
1	−4.078638	−0.544189
2	−0.544189	−1.415783

$S^{**}(-1/2)$ Matrix

	1	2
1	1.004326	−0.053862
2	−0.053862	1.004326

H-core Eigenvectors, by Column

	1	2
1	0.994916	0.147356
2	0.040103	−1.004969

P-matrix, noc = 2

	1	2
1	2.023142	−0.216378
2	−0.216378	2.023142

Nuclear Repulsion = 0.9299442033

Unique Two-Electron Integrals, iop = 2

1	1	1	1	1.5951080710
1	1	1	2	0.1397350035
1	1	2	2	0.3381350126
1	2	1	2	0.0134351635
1	2	2	2	0.0384505338
2	2	2	2	0.4633810384

***** Iteration 1 *****

P-matrix, noc = 2

	1	2
1	2.023142	−0.216378
2	−0.216378	2.023142

F-matrix

	1	2
1	−1.824804	−0.331720
2	−0.331720	−0.284856

Orbital Energies in hartrees, lowest first

$$E(1) = -1.836599882006$$
$$E(2) = -0.225693891989$$

Electronic Energy = −7.5025403826

```
***** Iteration 2 *****

          P-matrix, noc = 2

               1                      2
  1         2.023142              -0.216378
  2        -0.216378               2.023142

               F-matrix

               1                      2
  1        -1.824804              -0.331720
  2        -0.331720              -0.284856
```

Orbital Energies in hartrees, lowest first

E(1) = −1.836599882006
E(2) = −0.225693891989

Electronic Energy = −7.5025403826

<<<<<ENERGY CONVERGED! >>>>>

Nuclear Repulsion = 0.9299442033

**** Total Energy = −6.5725961793 hartrees ****

```
               C-matrix

               1                      2
  1         0.987127               0.192745
  2         0.086065              -1.002080

               F-matrix

               1                      2
  1        -1.824804              -0.331720
  2        -0.331720              -0.284856
```

Final Orbital Energies in hartrees, lowest first

−1.836600   −0.225694

The center of mass is at xm = 0.409176   ym = 0.000000   zm = 0.000000
Dipole Moment = (−6.560500, 0.000000, 0.000000) = 6.560500 Debyes

Frost's method constructs a wave function as a determinant of nonorthogonal orbitals, each expressed as a single FSGO. His direct minimization of the energy, permitting the nuclei to find an optimum separation and the orbitals to find optimum location and scale, produced a total electronic energy of −6.572695 hartree at an internuclear separation of 3.2260 bohr. Using these orbitals as the only basis functions in SCF1s produces exactly the same energy, but orthogonalizes the FSGO functions to form delocalized MOs.

SCF1s has given us an intimate view of the operation of the Roothaan SCF calculation. Our next step is to describe small molecules in a minimum basis set, i.e., one made up only of single functions required to write configurations for each atom. H needs only a 1s function, but the series Li-F needs 1s, 2s, and 2p functions (five functions are defined for each of the atoms in this series). Calculations in this basis can be done with PCLOBE, the second software companion to this book.

PCLOBE and Sample Output from RHF Calculations

Let us now consider some specific examples using PCLOBE to treat small molecules. PCLOBE can use single Gaussian spheres, but it is more usually the case that the basis set is defined by "contracted" sets of Gaussian functions, similar to the STO-3G and STO-6G series defined by Pople. A thorough discussion of the lobe mimics of s-, p-, and d-type basis functions is provided in the chapter on lobe basis. Here we will merely define the basis used in our discussion of the elements of electronic structure in diatomics, as we encounter its description in the PCLOBE output.

The input is described elsewhere in the READpclobe.txt file available in supplementary material; PCLOBE finds the input in Li2.XYZ. The geometry of this molecule was previously optimized so that the norm of the gradient is less than 0.001 hartree/bohr. We will inspect selected portions of the output, drawn from the file called "lobe.out" in the PCLOBE.INPUT folder.

Li$_2$

```
* * * * * * * * * * * * * * * * * * * * * * * * * * * * * * * * * * * * * * * * * * * * * * * * * * * * * *
PCLOBE  PCLOBE  PCLOBE  PCLOBE  PCLOBE  PCLOBE  PCLOBE  PCLOBE  PCLOBE    PCLOBE
* * * * * * * * * * * * * * * * * * * * * * * * * * * * * * * * * * * * * * * * * * * * * * * * * * * * * *
            Gaussian-Lobe Program for Organic Molecules
          Adapted to Personal Computers by Don Shillady
                 Virginia Commonwealth University
                        Richmond Virginia
                        1978-30 May 2003
* * * * * * * * * * * * * * * * * * * * * * * * * * * * * * * * * * * * * * * * * * * * * * * * * * * * * *
          HISTORICAL FOUNDATIONS OF GAUSSIAN-LOBE BASIS SETS
 1. S.F. Boys, Proc. Roy. Soc. A200, 542 (1950)
 2. J.L. Whitten, J. Chem. Phys. v39, 349 (1963)
 3. I. Shavitt, Methods in Comp. Phys. v2, p1 (1963)
 4. H. Sambe, J. Chem. Phys. v42, 1732 (1965)
 5. H. Preuss and G. Diercksen, Int. J. Quantum Chem. I, 605 (1967)
 6. J.F. Harrison and L.C. Allen, Molec. Spectrosc. v29, 432 (1969)
 7. F. Dreisler and R. Ahlrichs, Chem. Phys. Lett. v23, 571 (1973)
 8. H. Le Rouzo and B. Silvi, Int. J. Quantum Chem. XIII, 297 (1978)
 9. S.Y. Leu and C.Y. Mou, J. Chem. Phys. v101, 5910 (1994)
10. P. Otto, H. Reif and A. Hernandez-Laguna, J. Mol. Str. (Theochem) v340,
    51 1995)
* * * * * * * * * * * * * * * * * * * * * * * * * * * * * * * * * * * * * * * * * * * * * * * * * * * * * *
```

```
                              Li2.XYZ
********************************************************************
                  Nuclear Coordinates from Input
Atomic Core          X               Y               Z          Basis
    3.           0.000000        0.000000        0.000000          1
               Z1s = 2.691     Z2s = 0.640     Z2p = 0.601
               Z3s = 0.000     Z3p = 0.000     Z3d = 0.000
               Z4sp = 0.000    Z4f = 0.000

    3.           2.838636        0.000000        0.000000          1
               Z1s = 2.691     Z2s = 0.640     Z2p = 0.601
               Z3s = 0.000     Z3p = 0.000     Z3d = 0.000
               Z4sp = 0.000    Z4f = 0.000
```

The minimal basis employed here is intended to mimic Slater orbitals. These are characterized by a location and a single scale factor. For lithium, the scale factor 2.69 is approximately the atomic number $Z = 3$ less 0.30, following Slater's rules for scaling the 1s shell. The 2s and 2p scale factors are very similar to one another and reduced from the atomic number by 2 to reflect the screening effect of the two 1s electrons, and by a further amount, roughly 0.3, to account for screening in the valence shell by other electrons.

```
        Basis Size = 10 and Number of Spheres = 62 for 6 Electrons

                     Distance Matrix in Angstroms
                            Li       Li
                 Li      0.0000    2.8386

                 Li      2.8386    0.0000

                     Spherical Gaussian Basis Set
No.   1  alpha = 0.10070791E+03  at  X =  0.000000  Y =  0.000000  Z =  0.000000  a.u.
No.   2  alpha = 0.15181710E+02  at  X =  0.000000  Y =  0.000000  Z =  0.000000  a.u.
No.   3  alpha = 0.34082354E+01  at  X =  0.000000  Y =  0.000000  Z =  0.000000  a.u.
No.   4  alpha = 0.90325404E+00  at  X =  0.000000  Y =  0.000000  Z =  0.000000  a.u.
No.   5  alpha = 0.10560932E+01  at  X =  0.000000  Y =  0.000000  Z =  0.000000  a.u.
No.   6  alpha = 0.64129564E-01  at  X =  0.000000  Y =  0.000000  Z =  0.000000  a.u.
No.   7  alpha = 0.24620285E-01  at  X =  0.000000  Y =  0.000000  Z =  0.000000  a.u.
No.   8  alpha = 0.27758703E-01  at  X =  0.266134  Y =  0.000000  Z =  0.000000  a.u.
No.   9  alpha = 0.74746090E-01  at  X =  0.266134  Y =  0.000000  Z =  0.000000  a.u.
No.  10  alpha = 0.22553946E+00  at  X =  0.266134  Y =  0.000000  Z =  0.000000  a.u.
No.  11  alpha = 0.93974774E+00  at  X =  0.266134  Y =  0.000000  Z =  0.000000  a.u.
No.  12  alpha = 0.27758703E-01  at  X = -0.266134  Y =  0.000000  Z =  0.000000  a.u.
No.  13  alpha = 0.74746090E-01  at  X = -0.266134  Y =  0.000000  Z =  0.000000  a.u.
No.  14  alpha = 0.22553946E+00  at  X = -0.266134  Y =  0.000000  Z =  0.000000  a.u.
No.  15  alpha = 0.93974774E+00  at  X = -0.266134  Y =  0.000000  Z =  0.000000  a.u.
No.  16  alpha = 0.27758703E-01  at  X =  0.000000  Y =  0.266134  Z =  0.000000  a.u.
No.  17  alpha = 0.74746090E-01  at  X =  0.000000  Y =  0.266134  Z =  0.000000  a.u.
No.  18  alpha = 0.22553946E+00  at  X =  0.000000  Y =  0.266134  Z =  0.000000  a.u.
No.  19  alpha = 0.93974774E+00  at  X =  0.000000  Y =  0.266134  Z =  0.000000  a.u.
No.  20  alpha = 0.27758703E-01  at  X =  0.000000  Y = -0.266134  Z =  0.000000  a.u.
No.  21  alpha = 0.74746090E-01  at  X =  0.000000  Y = -0.266134  Z =  0.000000  a.u.
No.  22  alpha = 0.22553946E+00  at  X =  0.000000  Y = -0.266134  Z =  0.000000  a.u.
No.  23  alpha = 0.93974774E+00  at  X =  0.000000  Y = -0.266134  Z =  0.000000  a.u.
No.  24  alpha = 0.27758703E-01  at  X =  0.000000  Y =  0.000000  Z =  0.266134  a.u.
No.  25  alpha = 0.74746090E-01  at  X =  0.000000  Y =  0.000000  Z =  0.266134  a.u.
No.  26  alpha = 0.22553946E+00  at  X =  0.000000  Y =  0.000000  Z =  0.266134  a.u.
No.  27  alpha = 0.93974774E+00  at  X =  0.000000  Y =  0.000000  Z =  0.266134  a.u.
```

```
No.  28  alpha=0.27758703E-01  at  X=  0.000000  Y=  0.000000  Z=-0.266134  a.u.
No.  29  alpha=0.74746090E-01  at  X=  0.000000  Y=  0.000000  Z=-0.266134  a.u.
No.  30  alpha=0.22553946E+00  at  X=  0.000000  Y=  0.000000  Z=-0.266134  a.u.
No.  31  alpha=0.93974774E+00  at  X=  0.000000  Y=  0.000000  Z=-0.266134  a.u.
No.  32  alpha=0.10070791E+03  at  X=  5.364245  Y=  0.000000  Z=  0.000000  a.u.
No.  33  alpha=0.15181710E+02  at  X=  5.364245  Y=  0.000000  Z=  0.000000  a.u.
No.  34  alpha=0.34082354E+01  at  X=  5.364245  Y=  0.000000  Z=  0.000000  a.u.
No.  35  alpha=0.90325404E+00  at  X=  5.364245  Y=  0.000000  Z=  0.000000  a.u.
No.  36  alpha=0.10560932E+01  at  X=  5.364245  Y=  0.000000  Z=  0.000000  a.u.
No.  37  alpha=0.64129564E-01  at  X=  5.364245  Y=  0.000000  Z=  0.000000  a.u.
No.  38  alpha=0.24620285E-01  at  X=  5.364245  Y=  0.000000  Z=  0.000000  a.u.
No.  39  alpha=0.27758703E-01  at  X=  5.630379  Y=  0.000000  Z=  0.000000  a.u.
No.  40  alpha=0.74746090E-01  at  X=  5.630379  Y=  0.000000  Z=  0.000000  a.u.
No.  41  alpha=0.22553946E+00  at  X=  5.630379  Y=  0.000000  Z=  0.000000  a.u.
No.  42  alpha=0.93974774E+00  at  X=  5.630379  Y=  0.000000  Z=  0.000000  a.u.
No.  43  alpha=0.27758703E-01  at  X=  5.098110  Y=  0.000000  Z=  0.000000  a.u.
No.  44  alpha=0.74746090E-01  at  X=  5.098110  Y=  0.000000  Z=  0.000000  a.u.
No.  45  alpha=0.22553946E+00  at  X=  5.098110  Y=  0.000000  Z=  0.000000  a.u.
No.  46  alpha=0.93974774E+00  at  X=  5.098110  Y=  0.000000  Z=  0.000000  a.u.
No.  47  alpha=0.27758703E-01  at  X=  5.364245  Y=  0.266134  Z=  0.000000  a.u.
No.  48  alpha=0.74746090E-01  at  X=  5.364245  Y=  0.266134  Z=  0.000000  a.u.
No.  49  alpha=0.22553946E+00  at  X=  5.364245  Y=  0.266134  Z=  0.000000  a.u.
No.  50  alpha=0.93974774E+00  at  X=  5.364245  Y=  0.266134  Z=  0.000000  a.u.
No.  51  alpha=0.27758703E-01  at  X=  5.364245  Y=-0.266134  Z=  0.000000  a.u.
No.  52  alpha=0.74746090E-01  at  X=  5.364245  Y=-0.266134  Z=  0.000000  a.u.
No.  53  alpha=0.22553946E+00  at  X=  5.364245  Y=-0.266134  Z=  0.000000  a.u.
No.  54  alpha=0.93974774E+00  at  X=  5.364245  Y=-0.266134  Z=  0.000000  a.u.
No.  55  alpha=0.27758703E-01  at  X=  5.364245  Y=  0.000000  Z=  0.266134  a.u.
No.  56  alpha=0.74746090E-01  at  X=  5.364245  Y=  0.000000  Z=  0.266134  a.u.
No.  57  alpha=0.22553946E+00  at  X=  5.364245  Y=  0.000000  Z=  0.266134  a.u.
No.  58  alpha=0.93974774E+00  at  X=  5.364245  Y=  0.000000  Z=  0.266134  a.u.
No.  59  alpha=0.27758703E-01  at  X=  5.364245  Y=  0.000000  Z=-0.266134  a.u.
No.  60  alpha=0.74746090E-01  at  X=  5.364245  Y=  0.000000  Z=-0.266134  a.u.
No.  61  alpha=0.22553946E+00  at  X=  5.364245  Y=  0.000000  Z=-0.266134  a.u.
No.  62  alpha=0.93974774E+00  at  X=  5.364245  Y=  0.000000  Z=-0.266134  a.u.
```

The basis is composed of spherical Gaussian functions, contracted into series representations of the 10 Slater functions required for the minimum basis. The 1s function is described by a four-Gaussian series.

```
                         Contracted Orbital No. 1
   0.41307305*(1)     0.71856839*(2)     0.81472127*(3)     0.35240992*(4)
```

The 2s function is described by a 3G series.

```
                         Contracted Orbital No. 2
    -0.00368601*(5)      0.03665051*(6)       0.02817348*(7)
```

The 2p orbitals are represented by a set of out-of-phase combinations of spherical Gaussian functions, intended to imitate the positive and negative lobes of p spherical harmonics.

```
                        Contracted Orbital No. 3

   0.20640719*(8)      0.37785315*(9)      0.19304511*(10)      0.04530796*(11)
  -0.20640719*(12)    -0.37785315*(13)    -0.19304511*(14)     -0.04530796*(15)


                        Contracted Orbital No. 4

   0.20640719*(16)     0.37785315*(17)     0.19304511*(18)      004530796*(19)
  -0.20640719*(20)    -0.37785315*(21)    -0.19304511*(22)     -0.04530796*(23)


                        Contracted Orbital No. 5

   0.20640719*(24)     0.37785315*(25)     0.19304511*(26)      0.04530796*(27)
  -0.20640719*(28)    -0.37785315*(29)    -0.19304511*(30)     -0.04530796*(31)
```

This completes the minimal set of contracted functions on the first atom.

```
                        Contracted Orbital No. 6

 0.41307305*(32)       0.71856839*(33)     0.81472127*(34)      0.35240992*(35)


                        Contracted Orbital No. 7

-0.00368601*(36)       0.03665051*(37)     0.02817348*(38)


                        Contracted Orbital No. 8

   0.20640719*(39)     0.37785315*(40)     0.19304511*(41)      0.04530796*(42)
  -0.20640719*(43)    -0.37785315*(44)    -0.19304511*(45)     -0.04530796*(46)


                        Contracted Orbital No. 9

   0.20640719*(47)     0.37785315*(48)     0.19304511*(49)      0.04530796*(50)
  -0.20640719*(51)    -0.37785315*(52)    -0.19304511*(53)     -0.04530796*(54)


                        Contracted Orbital No. 10

   0.20640719*(55)     0.37785315*(56)     0.19304511*(57)      004530796*(58)
  -0.20640719*(59)    -0.37785315*(60)    -0.19304511*(61)     -0.04530796*(62)

       ***** Nuclear Repulsion Energy in au=   1.67777583131476 *****
```

From the analogy to hydrogen molecule, we would expect to find molecular orbitals which are in-phase and out-of-phase combinations of 1s basis functions. Extending the analogy, we would expect the next level to be an in-phase combination of 2s functions. This expectation is borne out in part in the initial guesses obtained by processing the core Hamiltonian, but we also see considerable mixing of the 2p functions in the highest energy bonding level.

```
              Initial-Guess-Eigenvectors by Column
#  At-Orb    1       2       3       4       5      6      7       8
1 Li   1s  0.703  -0.703  -0.095   0.014   0.000  0.000 -0.160   0.000
2 Li   2s  0.015  -0.041  -0.023  -0.334   0.000  0.000  0.504   0.000
3 Li   2px 0.004  -0.009  -0.817   0.668   0.000  0.000  0.159   0.000
4 Li   2py 0.000   0.000   0.000   0.000   0.593  0.001  0.000  -0.709
5 Li   2pz 0.000   0.000   0.000   0.000  -0.001  0.593  0.000  -0.601
6 Li   1s  0.703   0.703   0.095   0.014   0.000  0.000 -0.160   0.000
```

```
 7 Li  2s    0.015    0.041    0.023  -0.334   0.000  0.000   0.504   0.000
 8 Li  2px  -0.004   -0.009   -0.817  -0.668   0.000  0.000  -0.159   0.000
 9 Li  2py   0.000    0.000    0.000   0.000   0.593  0.001   0.000   0.709
10 Li  2pz   0.000    0.000    0.000   0.000  -0.001  0.593   0.000   0.601
```

#	At-Orb		9	10
1	Li	1s	0.000	0.164
2	Li	2s	0.000	-2.534
3	Li	2px	0.000	-1.565
4	Li	2py	0.601	0.000
5	Li	2pz	-0.709	0.000
6	Li	1s	0.000	-0.164
7	Li	2s	0.000	2.534
8	Li	2px	0.000	-1.565
9	Li	2py	-0.601	0.000
10	Li	2pz	0.709	0.000

One-Electron Energy Levels

$$
\begin{aligned}
E(\ 1) &= -3.753657115299 \\
E(\ 2) &= -3.753142080578 \\
E(\ 3) &= -0.996921957316 \\
E(\ 4) &= -0.988659883322 \\
E(\ 5) &= -0.934032358264 \\
E(\ 6) &= -0.934032358264 \\
E(\ 7) &= -0.928720475025 \\
E(\ 8) &= -0.824496535937 \\
E(\ 9) &= -0.824496535937 \\
E(10) &= -0.400600842748
\end{aligned}
$$

This is not really a very good starting assignment for the occupied valence MO, but the SCF iteration produces an occupied valence MO dominated by the 2s atomic orbitals.

Reference State Orbitals for 3 Filled Orbitals by Column

#	At-Orb		1	2	3	4	5	6	7	8
1	Li	1s	-0.704	0.703	0.154	-0.138	0.000	0.000	0.011	0.000
2	Li	2s	-0.012	0.039	-0.514	0.787	0.000	0.000	-0.319	0.000
3	Li	2px	-0.002	0.013	-0.140	-0.274	0.000	0.000	0.670	0.000
4	Li	2py	0.000	0.000	0.000	0.000	-0.418	-0.422	0.000	0.655
5	Li	2pz	0.000	0.000	0.000	0.000	0.422	-0.418	0.000	-0.655
6	Li	1s	-0.704	-0.703	0.154	0.138	0.000	0.000	0.011	0.000
7	Li	2s	-0.012	-0.039	-0.514	-0.787	0.000	0.000	-0.319	0.000
8	Li	2px	0.002	0.013	0.140	-0.274	0.000	0.000	-0.670	0.000
9	Li	2py	0.000	0.000	0.000	0.000	-0.418	-0.422	0.000	-0.655
10	Li	2pz	0.000	0.000	0.000	0.000	0.422	-0.418	0.000	0.655

The molecular orbital coeffficients define the forms of the orbitals. We see the first two MOs are defined primarily by the Li 1s atomic orbital, only very weakly mixed with the symmetry-compatible 2s and 2pz functions. The first is an in-phase combination of the 1s orbitals on the two atoms. In German these are called "gerade" (even) orbitals, and the combination is said to be the $1\sigma_g$ orbital. Next there is an out-of-phase, i.e., odd or "ungerade" $1\sigma_u$ combination of the two Li 1s orbitals. The overlap between these basis functions is so slight that the two combinations have almost identical energies. (Their energies differ by less than 0.1 millihartree.) There is essentially no overlap or interaction between the 1s basis functions. The third MO, dominated by the in-phase combination of Li 2s functions, has higher energy (−0.17 hartree). This value is still lower than the atomic 2s function in the molecular setting as estimated by the Fock matrix element.

The energies of the Li–Li $1\sigma_g$ and $1\sigma_u$ orbitals are not far below the valence energies of the Li atom. As a result small amounts of the 2s and the symmetry-compatible 2p orbitals along the bond axis actually mix into the core MOs. In homonuclear diatomics formed from elements with higher atomic numbers we will see that 2s and 2p orbitals seldom mix into the inner 1s core orbitals; the situation here is due to the low atomic number of Li.

Since each Li atom contributes three electrons, there are a total of six electrons in the diatomic molecule; we can assign an electron pair to the valence $2\sigma_g$ orbital. The shorthand notation for the diatomic Li orbitals is $1\sigma_g^2 1\sigma_u^2 2\sigma_g^2$.

According to Koopmans' theorem, the negative binding energies of the first three lowest energy orbitals are the negative of the ionization potential of each orbital; thus, the list of one-electron energies has practical meaning. The third orbital, $2\sigma_g$, is only bound by about 0.18 hartree or about 4.9 eV. This suggests that lithium diatomic is much more easily ionized than H atom which requires 0.5 hartree (13.6 eV).

One-Electron Energy Levels

```
      E( 1) =        -2.455661029763
      E( 2) =        -2.455637544035
      E( 3) =        -0.179142682588
      E( 4) =         0.013307539116
      E( 5) =         0.048289383072
      E( 6) =         0.048289383072
      E( 7) =         0.068819901205
      E( 8) =         0.130485982682
      E( 9) =         0.130485982682
      E(10) =         0.304329131722
```

Electronic Energy =   −16.4991628465 a.u., Dif. =   0.0000000000

```
      Li      0.00000      0.00000      0.00000
      Li      2.85632      0.00000      0.00000
```

Total Energy = −14.8214281786 hartrees
Virial Ratio = −<E>/<T> = 1.010390

Dipole Moment Components in Debyes
Dx = 0.000000   Dy = 0.000000   Dz = 0.000000
Resultant Dipole Moment in Debyes = 0.000000

****

The charge distribution for the single determinant is

$$\rho = \Psi^*\Psi = 2\sum_{i=1}^{N/2}\phi_i\phi_i = 2\sum C_{i\mu}C_{i\nu}G_\mu G_\nu$$

Here $G\mu$ is a member of the set of contracted basis functions. The integral of the charge distribution is the total number of electrons N.

$$N = \sum_{\mu\nu} P_{\mu\nu}S_{\mu\nu}$$

The individual terms in the sum are called Mulliken overlap populations discussed in the account of the Roothaan method.

```
                      Mulliken Overlap Populations
#    At-Orb    1       2       3       4       5       6       7       8
1  Li   1s   2.028 -0.016   0.000  0.000  0.000   0.000  -0.012  -0.006
2  Li   2s  -0.016   0.532   0.000  0.000  0.000  -0.012   0.318   0.071
3  Li  2px   0.000   0.000   0.040  0.000  0.000  -0.006   0.071   0.008
4  Li  2py   0.000   0.000   0.000  0.000  0.000   0.000   0.000   0.000
5  Li  2pz   0.000   0.000   0.000  0.000  0.000   0.000   0.000   0.000
6  Li   1s   0.000  -0.012  -0.006  0.000  0.000   2.028  -0.016   0.000
7  Li   2s  -0.012   0.318   0.071  0.000  0.000  -0.016   0.532   0.000
8  Li  2px  -0.006   0.071   0.008  0.000  0.000   0.000   0.000   0.040
9  Li  2py   0.000   0.000   0.000  0.000  0.000   0.000   0.000   0.000
10 Li  2pz   0.000   0.000   0.000  0.000  0.000   0.000   0.000   0.000

               #    At-Orb      9      10
               1  Li   1s   0.000   0.000
               2  Li   2s   0.000   0.000
               3  Li  2px   0.000   0.000
               4  Li  2py   0.000   0.000
               5  Li  2pz   0.000   0.000
               6  Li   1s   0.000   0.000
               7  Li   2s   0.000   0.000
               8  Li  2px   0.000   0.000
               9  Li  2py   0.000   0.000
              10  Li  2pz   0.000   0.000
****************************************************************
```

It is straightforward to partition the charge into parts referring to each basis member.

$$N = \sum_\mu Q_\mu = \sum_\mu \left(\sum_\nu P_{\mu\nu}S_{\mu\nu}\right)$$

```
                        Orbital Charges
1.994564   0.892793   0.112642   0.000000   0.000000   1.994564   0.892793   0.112642
0.000000   0.000000
****************************************************************
```

This shows the limited participation of the 2p functions in the bonding.

When the basis functions are each uniquely associated with a particular nuclear center, it is tempting to partition the sum into atomic terms.

$$N = \sum_{\mu,\mu' \in M} P_{\mu\mu'} S_{\mu\mu'} + \sum_{M \neq N} \sum_{\mu \in M} \sum_{\nu \in N} P_{\mu\mu'} S_{\mu\mu'}$$

Each term in the first sum is uniquely associated with a specific atom, while each term in the second sum refers to charge shared between two nuclei. Partial sums for each atom and each pair are collected in the matrix of total overlap populations.

```
        Total Overlap Populations by Atoms
                    Li              Li
        Li        2.5674          0.4326
        Li        0.4326          2.5674
```

The simplest division of the shared charge apportions it equally between atoms. This is valid in the case of perfectly equivalent atoms and produces atomic populations of 3.0 electrons for each Li atom in this case. This cancels precisely the nuclear charge.

```
            Computed Atom Charges
        Q( 1) =   0.000  Q( 2) =   0.000
**********************************************************************
            ***** NORMAL FINISH OF PCLOBE *****
```

If the calculation is perfectly scaled and balanced between kinetic and potential energy in the system, the ratio of the total energy to the total of the kinetic energy should be exactly 1. This formulation avoids discussion of the potential energy. Since $\langle E \rangle = \langle T \rangle + \langle V \rangle$ and the virial theorem requires that $\langle V \rangle = -2\langle T \rangle$ for a perfectly scaled calculation, $\langle E \rangle = -\langle T \rangle$ and $-\langle E \rangle / \langle T \rangle = 1$. The value reported, 1.001647, should be considered "good." It is not perfect because we did not optimize the orbital scaling.

N$_2$

Next, let us consider a covalent triple bond, one of the strongest bonds in chemistry. The basis set is the same as we have used for lithium diatomic, apart from the rescaling necessary to describe the nitrogen atom in the molecular environment.

```
***********************************************************************
PCLOBE PCLOBE PCLOBE PCLOBE PCLOBE PCLOBE PCLOBE PCLOBE PCLOBE PCLOBE
***********************************************************************
            Gaussian-Lobe Program for Organic Molecules
            Adapted to Personal Computers by Don Shillady
                 Virginia Commonwealth University
                         Richmond Virginia
***********************************************************************
                 Nuclear Coordinates from Input
Atomic Core         X              Y              Z            Basis
    7.          0.000000       0.000000       0.000000          1
                Z1s  = 6.665   Z2s = 1.924    Z2p = 1.917
                Z3s  = 0.000   Z3p = 0.000    Z3d = 0.000
                Z4sp = 0.000   Z4f = 0.000

    7.          0.000000       0.000000       1.152150          1
                Z1s  = 6.665   Z2s = 1.924    Z2p = 1.917
                Z3s  = 0.000   Z3p = 0.000    Z3d = 0.000
                Z4sp = 0.000   Z4f = 0.000

    Basis Size = 10 and Number of Spheres = 62 for 14 Electrons


                   Distance Matrix in Angstroms

                         N              N
                N     0.0000         1.1522
                N     1.1522         0.0000
```

SCF calculation of the familiar form produces

```
          ##### Total Energy=   -108.3381011122 hartrees #####

              Virial Ratio = -<E>/<T> = 1.029986

                 One-Electron Energy Levels
               E ( 1) =   -15.907767987599
               E ( 2) =   -15.906176808075
               E ( 3) =    -1.436908654197
               E ( 4) =    -0.778016834589
               E ( 5) =    -0.622170328012
               E ( 6) =    -0.622170328012
               E ( 7) =    -0.611319934965
               E ( 8) =     0.158145916021
               E ( 9) =     0.158145916021
               E (10) =     0.877278715466
```

The first two levels, at -15.9 hartree, are almost undisturbed 1s orbitals as the coefficients quoted below establish. Their energy is shifted from the 1s value in atomic nitrogen (-15.5 hartree) since the Hartree–Fock orbital energy incorporates some features of the entire molecule's charge distribution. The next two levels both have negative energy so electrons in those orbitals are firmly attached to the molecule. However, the values -1.43 and -0.77 hartree are respectively lower and higher than the 2s level in atomic nitrogen, once we recognize the first-order effect of immersion of the 2s level in the molecular environment. Hence we can consider these levels to be bonding and antibonding, respectively. The next three levels, at -0.622, -0.622 (i.e., degenerate), and -0.611, are markedly stabilized compared with the atomic 2p levels. The final three values are strongly destabilized and are clearly antibonding.

```
##### Total Energy =   -108.3381011122 hartrees #####

         Virial Ratio = -<E>/<T> = 1.029986

    N          0.00000        0.00000        0.00000
    N          0.00000        0.00000        1.15215
```

The bonding MOs we just encountered differ from the sigma MOs described thus far. The atomic orbitals' p atomic orbitals carry angular momentum, which can combine into three resultants. One has zero angular momentum along the molecular axis, while the other two have a unit of angular momentum. The unit vector can be oriented in either of the two directions along the molecular axis.

The labels $\sigma, \pi, \delta, \dots$ appearing in the table refer to orbital angular momentum around the molecular axis of $0, 1, 2, \dots \hbar$ units, respectively.

Reference State Orbitals for   7 Filled Orbitals by Column

		1σ-g	1σ-u	2σ-g	2σ-u	1π-u	1π-u	3σ-g	1π-g
#	At-Orb	1	2	3	4	5	6	7	8
1	N 1s	0.704	−0.703	0.171	−0.181	0.000	0.000	−0.075	0.000
2	N 2s	0.011	−0.022	−0.478	0.812	0.000	0.000	0.403	0.000
3	N 2px	0.000	0.000	0.000	0.000	0.000	−0.630	0.000	0.000
4	N 2py	0.000	0.000	0.000	0.000	0.630	0.000	0.000	0.823
5	N 2pz	0.002	−0.007	−0.253	−0.226	0.000	0.000	−0.591	0.000
6	N 1s	0.704	0.703	0.171	0.181	0.000	0.000	−0.075	0.000
7	N 2s	0.011	0.022	−0.478	−0.812	0.000	0.000	0.403	0.000
8	N 2px	0.000	0.000	0.000	0.000	0.000	−0.630	0.000	0.000
9	N 2py	0.000	0.000	0.000	0.000	0.630	0.000	0.000	−0.823
10	N 2pz	−0.002	−0.007	0.253	−0.226	0.000	0.000	0.591	0.000

		1π-g	4σ-u
#	At-Orb	9	10
1	N 1s	0.000	0.142
2	N 2s	0.000	−1.204
3	N 2px	−0.823	0.000
4	N 2py	0.000	0.000
5	N 2pz	0.000	−1.214
6	N 1s	0.000	−0.142
7	N 2s	0.000	1.204
8	N 2px	0.823	0.000
9	N 2py	0.000	0.000
10	N 2pz	0.000	−1.214

The details of the coefficients and the MO energies show us that these three MOs (5, 6, and 7) are strongly stabilizing, which is consistent with the very strong (942 kcal/mol) bond in N_2.

```
                    Dipole Moment Components in Debyes
        Dx = 0.000000       Dy = 0.000000       Dz = 0.000000
           Resultant Dipole Moment in Debyes = 0.000000
********************************************************************
```

The dipole moment is zero by symmetry.

```
                   Mulliken Overlap Populations
#    At-Orb     1       2       3       4       5       6       7       8
1  N  1s     2.115  -0.113  0.000   0.000   0.000   0.000   0.003   0.000
2  N  2s    -0.113   2.100  0.000   0.000   0.000   0.003  -0.259   0.000
3  N  2px    0.000   0.000  0.793   0.000   0.000   0.000   0.000   0.207
4  N  2py    0.000   0.000  0.000   0.793   0.000   0.000   0.000   0.000
5  N  2pz    0.000   0.000  0.000   0.000   0.929  -0.007   0.056   0.000
6  N  1s     0.000   0.003  0.000   0.000  -0.007   2.115  -0.113   0.000
7  N  2s     0.003  -0.259  0.000   0.000   0.056  -0.113   2.100   0.000
8  N  2px    0.000   0.000  0.207   0.000   0.000   0.000   0.000   0.793
9  N  2py    0.000   0.000  0.000   0.207   0.000   0.000   0.000   0.000
10 N  2pz   -0.007   0.056  0.000   0.000   0.238   0.000   0.000   0.000

                   #    At-Orb    9      10
                   1  N  1s     0.000  -0.007
                   2  N  2s     0.000   0.056
                   3  N  2px    0.000   0.000
                   4  N  2py    0.207   0.000
                   5  N  2pz    0.000   0.238
                   6  N  1s     0.000   0.000
                   7  N  2s     0.000   0.000
                   8  N  2px    0.000   0.000
                   9  N  2py    0.793   0.000
                   10 N  2pz    0.000   0.929
********************************************************************
```

According to Mayer's analysis [1] the Mulliken overlap population is a reasonable representation of atomic charges for systems with predominantly covalent bonding represented by a minimum basis set of atom-centered functions. The Mulliken estimate of shared charge is informative in some ways—for example, the 2sA–2sB element is negative indicating that the net interaction is antibonding, while the 2pA–2pB element is positive, indicating that the pi interaction is net bonding. However, the quantitative values are harder to interpret.

The Mulliken population can be aggregated into atomic terms

$$Q_{AB} = \sum_{r \in A} \sum_{s \in B} P_{rs} S_{rs}$$

For diatomic nitrogen the result is the following matrix:

```
           Total Overlap Populations by Atoms

                      N                   N
        N          6.5041             0.4959
        N          0.4959             6.5041
```

The summed entries of each column define the total electronic charge on the associated atom; adding in the nuclear charges we obtain the atomic charges.

```
                Computed Atom Charges

            Q( 1) = 0.000   Q( 2) = 0.000
**********************************************************************
```

Another aggregation assigns charges to each member of the basis:

$$q_r = \sum_s P_{rs} S_{rs}$$

```
                  Orbital Charges

1.998258 1.786499 1.000000 1.000000 1.215242 1.998258 1.786499
1.000000 1.000000 1.215242
**********************************************************************
```

Here the seven electrons for each N atom are represented as a pair in the 1s shell, as individual electrons in the pi-symmetry 2p basis functions, and as three electrons divided between the sigma-symmetry 2s and 2pz basis functions. This may be consistent with distinct s and p combinations for the lone pair and the bond hybrid, but we do not recover detailed descriptions of those valence-bond constructs.

The off-diagonal elements of the Mulliken population matrix would seem to represent charge occupying the bonding region. The small value of charge in the bonding zone, 2×0.49 electrons, might seem inadequate to represent this very strong bond, but it is worth mentioning that chemical bonding has only a modest effect on the overall charge density.

In the diatomics we have described so far, we can assign an intuitively more appealing bond order from the count of electrons in bonding and antibonding MOs. Specifically for nitrogen, which has two-electron pairs in binding 2p–pi MOs, a bonding pair in a 2p-sigma MO and canceling pairs in 2s-sigma antibonding and 2s-sigma bonding MOs, we define a bond order of three, the (net) number of bonding pairs. Lithium diatomic would have a bond order of 1, as would fluorine diatomic.

Mayer [1] presents a bond index which estimates the order of a covalent bond between atoms R and S.

$$M_{RS} = \sum_{r \in R} \sum_{s \in S} (PS)_{rs}(PS)_{rs}$$

This is the generalization of an index introduced by Wiberg [2] which defines bond orders in the semiempirical framework wherein basis functions are orthonormal. In the case of nitrogen described in a minimum single-zeta basis, Mayer's bond index is almost exactly three.

```
                        Wiberg-Mayer W-matrix
 #   At-Orb    1       2       3       4       5       6       7       8
 1  N   1s   1.999   0.006   0.000   0.000   0.021   0.001  -0.006   0.000
 2  N   2s   0.019   1.800   0.000   0.000  -0.743  -0.019   0.200   0.000
 3  N   2px  0.000   0.000   1.000   0.000   0.000   0.000   0.000   1.000
 4  N   2py  0.000   0.000   0.000   1.000   0.000   0.000   0.000   0.000
 5  N   2pz  0.021  -0.216   0.000   0.000   1.201  -0.021   0.216   0.000
 6  N   1s   0.001  -0.006   0.000   0.000  -0.021   1.999   0.006   0.000
 7  N   2s  -0.019   0.200   0.000   0.000   0.743   0.019   1.800   0.000
 8  N   2px  0.000   0.000   1.000   0.000   0.000   0.000   0.000   1.000
 9  N   2py  0.000   0.000   0.000   1.000   0.000   0.000   0.000   0.000
10  N   2pz  0.021  -0.216   0.000   0.000  -0.799  -0.021   0.216   0.000

          #    At-Orb    9       10
          1   N   1s    0.000   0.021
          2   N   2s    0.000  -0.743
          3   N   2px   0.000   0.000
          4   N   2py   1.000   0.000
          5   N   2pz   0.000  -0.799
          6   N   1s    0.000  -0.021
          7   N   2s    0.000   0.743
          8   N   2px   0.000   0.000
          9   N   2py   1.000   0.000
         10   N   2pz   0.000   1.201

          Wiberg-Mayer Bond Orders Between Atoms
        (See Chem. Phys. Letters, v97, p270 (1983))

                       N          N
            N      11.0000     3.0000
            N       3.0000    11.0000
```

In many cases, as Mayer reports, the index corresponds to our more intuitive idea of covalent bond order (see Table 5.1).

The Mayer index for small polyatomic species also displays bond orders as the simplest ideas of bonding (Lewis models) would suggest. Oxygen in water and formaldehyde has a total bond order of very nearly 2.0, while the

TABLE 5.1

Mayer Bond Orders for Some Diatomics

Species	H2	Li$_2$	Be$_2$	N$_2$	F$_2$	
Bond order	1.0000	1.0017	0.0034	3.0000	1.0000	Nearly conventional

TABLE 5.2

Mayer Bond Orders for Polar Diatomics

Species	HF	LiF	BeO	BN	C_2	Departures
	0.9600	1.2193	2.1467	3.1173	3.2184	Slight
Species	BF	CO	BF			
	1.4207	2.4977				Serious

bonds to H in water, formaldehyde, and ethylene are close to 1.0 as expected. While the simplicity of the results is appealing, the fact that bonds identical in order differ in strength is lost in the course of the analysis.

The Mayer index can either exceed or fall short of our simplest estimates of bond order (Table 5.2).

The more exaggerated departures, at least, require some attention.

N_2, CO, BF: An Isoelectronic Series

Homonuclear diatomic molecules offer an interesting and diverse collection of bonding environments, but heteronuclear systems extend the complexity and subtlety of bonding. Here we consider an isoelectronic series based on homonuclear nitrogen diatomic. Carbon monoxide's minor departure from equivalence is a step on the way to the very different situation in BF. The table below shows the trends in orbital energies and contains brief characterizations of the orbitals. All properties show the increasing contrast between the A and B nuclei in the series, from homonuclear to highly polar. This is reflected in the increases in dipole moment, splitting and polarization of the 1s levels, and the decrease in Koopmans theorem's estimate of the ionization energy, i.e., the energy of the MO #7, a sigma-symmetry lone pair primarily on the less electronegative atom (Table 5.3).

RHF calculations produce reasonable dipole moments for polar systems. For systems with dipole moments of one Debye or more, RHF estimates are generally high by a few tenths of a Debye. Reliable estimation of the small dipole moment of the CO molecule (counterintuitively directed O+ to C− as a consequence of the diffuse lone pair on C atom) requires more accurate methods.

The Mulliken populations display the progressive charge separation in a reasonable way.

```
            Total Overlap Populations by Atoms

                     N                N
        N         6.5446          0.4554
        N         0.4554          6.5446
           Excess Charge = 0.0000  0.0000
                 Dipole = 0.000 D
```

TABLE 5.3

Orbital Properties in Diatomics N_2, CO, BF

N2	CO	BF
E(1) = −15.829336 E(2) = −15.826695 E(3) = −1.396461 E(4) = −0.726046 E(5) = −0.577425 E(6) = −0.577425 E(7) = −0.570685 E(8) = 0.214757 E(9) = 0.214757 E(10) = 0.984133	E(1) = −20.842205 E(2) = −11.396706 E(3) = −1.469712 E(4) = −0.721963 E(5) = −0.596021 E(6) = −0.596021 E(7) = −0.503065 E(8) = 0.252129 E(9) = 0.252129 E(10) = 0.945806	E(1) = −26.571702 E(2) = −7.681518 E(3) = −1.608699 E(4) = −0.733538 E(5) = −0.632916 E(6) = −0.632916 E(7) = −0.372035 E(8) = 0.237612 E(9) = 0.237612 E(10) = 0.662626
N 1s cores: in phase	O 1s core	F core
N 1s cores: out of phase	C 1s core	B core
Sigma: 2s orbitals in phase—p admixed to shift amplitude into bonding zone	Sigma: mostly O 2s with some O 2p, with a bonding admixture of C s and p	Sigma: primarily F 2s, with a minor bonding admixture of B s and p
Sigma: 2s out of phase—p admixed to shift amplitude to lone pair zone	Sigma: O 2s with some O 2p, with an antibonding admixture of C s and p	Primarily F sigma 2px
Pi: homonuclear 2p–pi MO	Pi: polarization toward O evident	Pi: polarization toward F prominent
Sigma: homonuclear, with greater weight in 2p sigma	Sigma: substantially C 2s and 2p added to shift amplitude into bond	Sigma: primarily B 2p with 2s added to shift amplitude into bond zone
Pi-antibonding (out of phase): perpendicular orientation relative to the molecular axis	Pi: slight polarization toward C	Pi: substantial polarization toward B
Dipole = 0.000 by symmetry	Dipole moment = 0.300 C to O	Dipole moment = 1.535 B to F

```
                     C              O
      C           5.4292         0.3446
      O           0.3446         7.8816
         Excess Charge = +0.2262  −0.2262
              Dipole = 0.300 D

                     B              F
      B           4.6765         0.1469
      F           0.1469         9.0297
         Excess Charge = +0.1766  −0.1766
              Dipole = 1.535
```

These isoelectronic systems have very different Wiberg–Mayer bond orders. Perfectly covalent diatomic nitrogen enjoys a bond index of 3.0000, while the isoelectronic but polar species, CO and BF, show lower values of 2.4977 and 1.4207, respectively. We might take this to be a consequence of ionic bonding; if there is a substantial component in the charge distribution of $B(+) = F(-)$ or even $B(2+) - F(2-)$, the covalent component of the bonding must decline.

The Value of Mulliken Charges and Mayer Bond Orders

While the charge distribution and certain quantities derived from it such as its moments—dipole, quadrupole, etc.—are experimentally measurable, atomic charges are not strictly related to observables. Still the notion of the atom and associated properties is such a significant part of chemical intuition that it has been given considerable attention. Correlations of NMR parameters, pK_A values, vibrational frequencies or force constants, and other properties with Mulliken charges are well established.

The Mulliken charges are mathematically as simple as can be, and despite well-known difficulties are invariably reported in widely used electronic structure modeling programs. The difficulties arise primarily in bonds involving atoms of severely different electronegativities. In the examples we have already discussed, the dipole moments are not very well represented by the temptingly simple expedient of placing the Mulliken charges at the nuclear positions. A part of the problem can be traced to the convention that the overlap population is shared equally between the two centers. For a very polar charge distribution this really cannot be true. A second aspect of the problem is that the assignment of basis functions to atoms is sometimes not clear-cut. For LiF, the very diffuse Li 2p basis members have substantial amplitude near the F atom. In a variation calculation these orbitals will be mixed with the 2s and 2p functions of the F atom and take part in the description of the F atom. The Mulliken analysis, however, assigns any charge in these orbitals to the Li atom. A system which by all reasonable measures needs to be represented as $Li(+1.0) - F(-1.0)$ will in the Mulliken analysis seem to have a dramatically reduced charge separation.

Bonding in C_2 and LiF by Natural Bond Order Analysis

Isoelectronic molecules often display strong analogies, but there are limits to their similarity. This is illustrated in the series of 12-electron diatomics from C_2 to LiF. The first must be entirely covalent in bonding, the last as close to a purely ionic interaction as we can find.

We will introduce a newer and more versatile representation of bonding developed by Frank Weinhold and collaborators [3]. This analysis is called "natural" which is a reference to key steps in the treatment of the density matrix. The fundamental assumption is still that certain members of the basis are uniquely associated with particular atoms. The density matrix is then ordered so that the n_A basis functions on each atom A form $n_A \times n_A$ blocks along the diagonal. Off-diagonal blocks of dimension $n_A \times n_B$ refer to A, B pairs of atoms.

The first step in the analysis is to define natural atomic orbitals, NAOs; these are the eigenvectors of the $n_A \times n_A$ blocks along the diagonal. The associated eigenvalues are occupation numbers for the NAOs. The NAOs are mutually orthogonal within the n_A-fold set on each atom, and have the desirable property of maximal AO population. The Gaussian implementation [4] produces the following report for diatomic carbon:

```
************************Gaussian NBO Version 3.1************************
                NATURAL ATOMIC ORBITAL AND
                NATURAL BOND ORBITAL ANALYSIS
************************Gaussian NBO Version 3.1************************
```

The first stage of the analysis assigns charge to particular atomic orbitals. We see simple results for the population in the C core 1s level (two electrons) and the p orbitals oriented perpendicular to the C–C axis (one each). Evidently there is an sp hybrid combination along the axis.

Summary of Natural Population Analysis:

Atom	No	Natural Charge	Natural Population			
			Core	Valence	Rydberg	Total
C	1	0.00000	1.99915	3.99128	0.00957	6.00000
C	2	0.00000	1.99915	3.99128	0.00957	6.00000
* Total *		0.00000	3.99830	7.98256	0.01914	12.00000

Atom	No	Natural Electron Configuration
C	1	[core]2S(1.48)2p(2.51)
C	2	[core]2S(1.48)2p(2.51)

The most detailed NBO description of the bonding is given by diagonalizing the off-diagonal block $\mathbf{D}_{AB}$ labeled by the n_A basis functions on site A and the n_B basis functions on center B. This block is generally neither symmetric nor square, but can be diagonalized indirectly. First diagonalize the square symmetric matrix formed as the product of $\mathbf{D}_{AB}$ with its adjoint (i.e., the transpose of this real matrix).

$$\mathbf{U}\mathbf{D}_{AB}\mathbf{D}^{\dagger}_{AB}\mathbf{U}^{\dagger} = \Lambda$$

Then find

$$UD_{AB}VV^\dagger D_{AB}^\dagger U^\dagger = \Lambda$$
$$(UD_{AB}V)(UD_{AB}V)^\dagger = R^2 = \Lambda$$
$$UDV = R = \Lambda^{1/2}$$

To recover **V** follow

$$D^\dagger U^\dagger UDV = D^\dagger U^\dagger \Lambda^{1/2}$$
$$D^\dagger DV = D^\dagger U^\dagger \Lambda^{1/2}$$
$$V = (D^\dagger D)^{-1}D^\dagger U^\dagger \Lambda^{1/2}$$

The matrices **U** and **V** contain hybrids taking part in AB-bonds. The eigenvalues define the relative weight of the hybrids in the two-center bonds. The report of the analysis includes the occupation number in a bond, the percent of all that population in each hybrid, and the composition of the hybrid. Here is the NBO equivalent of the C_2 $2s\sigma$ and $2s\sigma^*$ orbitals in the simplest version of MO theory for diatomics; for C_2

```
(Occupancy)   Bond orbital/ Coefficients/ Hybrids
------------------------------------------------------------
  1.   (2.000) BD ( 1) C 1 - C 2
```

This natural bond (BD) between C_1 and C_2 has two electrons in it.

```
    (79.92%)        0.8940* C       1 s(99.42%)p 0.01(0.58%)
```

The charge in this bond is about 80% situated on the first C atom; the hybrid has almost 100% s-character.

```
    0.0000      0.9971      0.0000      0.0000      -0.0764
```

The coefficient of the 2s AO is by far the largest.

```
    (20.08%)        0.4481* C       2 s(0.58%)p99.99(99.42%)
```

The second C has the remaining 20% of charge; its charge is mostly in the p manifold.

```
    0.0000      0.0764      0.0000      0.0000      0.9971
```

The z component of the p manifold has the greatest coefficient.

The second natural orbital is the reflection of the first. The two together maintain the symmetry-equivalence of the two carbons, even though symmetry of each individual is lost, perhaps owing to numerical imprecision.

```
2.      (2.000) BD ( 2) C        1 - C      2
        (20.08%)              0.4481* C       1 s(0.58%)p99.99(99.42%)
         0.0000              0.0764  0.0000   0.0000  -0.9971
        (79.92%)             0.8940* C       2 s(99.42%)p 0.01(0.58%)
         0.0000              0.9971  0.0000   0.0000  -0.0764
```

The third NBO is a doubly occupied pi MO.

```
3.      (2.000) BD          (3) C  1 - C     2
        (50.00%)             0.7071* C       1 s(0.00%)p 1.00(100.00%)
         0.0000              0.0000  1.0000   0.0000  0.0000
        (50.00%)             0.7071* C       2 s(0.00%)p 1.00(100.00%)
         0.0000              0.0000  1.0000   0.0000  0.0000
```

The fourth is its complement.

```
4.      (2.00000) BD ( 4) C      1 - C      2
        (50.00%)             0.7071* C       1 s(0.00%)p 1.00(100.00%)
         0.0000              0.0000  0.0000   1.0000  0.0000
        (50.00%)             0.7071* C       2 s(0.00%)p 1.00(100.00%)
         0.0000  0.0000      0.0000           1.0000  0.0000
```

The fifth and sixth NBOs are the carbon cores.

```
5.      (1.99978)                CR (1) C  1     s(100.00%)
         1.0000                0.0000  0.0000   0.0000  0.0000

6.      (1.99978) CR (1) C       2               s(100.00%)
         1.0000                0.0000  0.0000   0.0000  0.0000
```

All 12 electrons are accounted for; there is no sigma bonding.
  Consider the ionic extreme, LiF

Summary of Natural Population Analysis:

Atom	No	Natural Charge	Natural Population Core	Valence	Rydberg	Total
Li	1	0.52562	1.99947	0.03220	0.44271	2.47438
F	2	-0.52562	1.99995	7.52567	0.00000	9.52562
* Total *		0.00000	3.99942	7.55788	0.44271	12.00000

(Occupancy) Bond orbital/ Coefficients/ Hybrids

What were once symmetric pi bonds are now pi-symmetry lone pairs on F.

```
1.      (2.00000) BD      (1)Li  1 - F  2
                          (10.32%)  0.3213*Li  1 s(0.00%)p 1.00(100.00%)
                          0.0000  0.0000  1.0000  0.0000  0.0000
                          (89.68%)  0.9470*  F  2 s(0.00%)p 1.00(100.00%)
                          0.0000  0.0000  1.0000  0.0000  0.0000

2.      (2.00000) BD      (2)Li  1 - F  2
                          (10.32%)  0.3213*Li  1 s(0.00%)p 1.00(100.00%)
                          0.0000  0.0000  0.0000  1.0000  0.0000
                          (89.68%)  0.9470* F  2 s(0.00%)p 1.00(100.00%)
                          0.0000  0.0000  0.0000  1.0000  0.0000
```

Both Li and F have cores.

```
3. (1.99947)  CR  (1)Li        1                 s(100.00%)
                            1.0000       0.0000  0.0000  0.0000  0.0000
4. (1.99995)  CR  (1)  F        2           s(100.00%)p 0.00(0.00%)
                            1.0000       0.0000  0.0000  0.0000  0.0001
```

F also has a third lone pair—sigma in symmetry and primarily 2s.

```
5. (1.99912)  LP  (1)  F        2           s(77.57%)  p 0.29  (22.43%)
                           -0.0001     0.8807  0.0000  0.0000  0.4736
```

A fourth lone pair on F convinces us we have F anion and Li cation. The partitioning is, however, not so perfect as integer charge assignments might suggest.

```
6. (1.93936)     LP  (2)  F       2          s(22.43%)p 3.46(77.57%)
                              0.0001    0.4736  0.0000  0.0000  -0.8807
7. (0.06091)     RY*(1)Li         1          s(51.93%)p 0.93(48.07%)
                              0.0000    0.7206  0.0000  0.0000   0.6933
```

Summary and Conclusions

In this chapter we have examined some of the detail which attends any Roothaan LCAO-MO-SCF calculation, with the help of a very simple SCF program SCF1s and its bigger brother PCLOBE. In our discussion of small systems we have surveyed the foundation for most descriptions of chemical bonding. Electronic structure determines the strength of bonds which in turn define the geometry and vibrational spectrum of molecules. These are two of the primary objects of chemical experimentation. In the next chapters we will use PCLOBE to show how this kind of calculation can produce estimates of molecular structure and spectra.

References

1. Mayer, I., *Chem. Phys. Lett.* 97:270, 1983.
2. Wiberg, K., *Tetrahedron* 24:1083, 1968.
3. Reed, A.E., Curtiss, L.A., and Weinhold, F., *Chem. Rev.* 88:899, 1988.
4. Frisch, M.J., Trucks, G.W., Schlegel, H.B., et al., Gaussian, Inc., Wallingford, CT, 2004.

6

Geometry Optimization and Vibrational Frequencies by SCF

Introduction

We established the algebraic task of solving the Roothaan–Hall self-consistent field (SCF) equations [1] in a previous chapter, and illustrated the process of solving the equations for some very simple systems. Here we show how the program PCLOBE processes the SCF equations and evaluates derivatives of the SCF energy. These allow efficient search for relative minima in the potential energy and also allow estimation of the harmonic frequencies of a molecule. For the perfect pairing case the energy is

$$E = 2\sum_{i\mu\nu} C_{i\mu}C_{i\nu}\langle\mu|h|\nu\rangle + \sum_{ij}\sum_{\mu\nu\rho\sigma} C_{i\mu}C_{j\rho}C_{i\nu}C_{j\sigma} \left(2\langle\mu\rho|\nu\sigma\rangle - \langle\mu\rho|\sigma\nu\rangle\right)$$

The Fock matrix in the basis is

$$F_{\mu\nu} = \langle\mu|h|\nu\rangle + \sum_{j\rho\sigma} C_{j\rho}C_{j\sigma}(2\langle\mu\rho|\nu\sigma\rangle - \langle\mu\rho|\sigma\nu\rangle)$$

and the energy of orbitals—the diagonal elements of the Fock matrix—and the total energy are

$$F_{ii} = \sum_{\mu\nu} C_{i\mu}C_{i\nu}F_{\mu\nu}$$

$$E = \sum_{i\mu\nu} C_{i\mu}C_{i\nu}(F_{\mu\nu} + h_{\mu\nu})$$

The equations to be solved to self-consistency are

$$\sum F_{\mu\nu}C_{vi} = S_{\mu\nu}C_{vi}\lambda_{ii}$$

The SCF-Roothaan Calculation in PCLOBE

The purpose of this section is to make our way through a full SCF calculation on our central example, formaldehyde. This molecule is small and symmetric enough to treat in detail. It displays a broad array of interesting experimental properties, which we will explore further in later chapters. The structure of the calculation is, of course, similar to that we sketched for SCF1s, but we will delve a bit deeper at each step.

The HF method reviewed:

1. Choose a molecular geometry.
2. Select a basis set and locate the members of the basis.
3. Compute the matrix of the overlaps between the basis functions, S.
4. Diagonalize the S-matrix, and compute $S^{1/2}$.
5. Compute the H-core matrix.
6. Compute and save two-electron integrals.
7. Solve the nonorthogonal eigenvalue problem $\mathbf{FC} = \varepsilon\mathbf{SC}$ with an approximate $\mathbf{F}$.
8. Compute the density matrix with resulting $\mathbf{C}$.
9. Evaluate $\mathbf{F}$; then evaluate the electronic energy using $\mathbf{P.H.F}$.
10. Evaluate convergence measures; if convergence has not been achieved, return to step 7 using the new Fock matrix to obtain refined estimates of the MO coefficients $\mathbf{C}$ and orbital energies ε— otherwise go to step 11.
11. Compute aspects of the charge distributions and properties of the molecule.

We will walk through each of these steps, following PCLOBE's program structure and using illustrative quotes from detailed output. The full verbose output can be obtained using the PCLOBE program on the supplemental CD.

Geometry

Formaldehyde has useful symmetry, and it is convenient to choose its Cartesian axes to coincide with its twofold axis and planes of symmetry. Here the CO vector is placed on the x-axis and the atoms of the CH_2 fragment are confined to the x–y plane.

* *

Nuclear Coordinates from Input

Atomic Core	X	Y	Z	
6.	0.000000	0.000000	0.000000	1
	$Z1s = 5.673$	$Z2s = 1.608$	$Z2p = 1.568$	
8.	1.256472	0.000000	0.000000	1
	$Z1s = 7.658$	$Z2s = 2.246$	$Z2p = 2.227$	
1.	−0.602700	0.973800	0.000000	1
	$Z1s = 1.240$	$Z2s = 0.000$	$Z2p = 0.000$	
1.	−0.602700	−0.973800	0.000000	1
	$Z1s = 1.240$	$Z2s = 0.000$	$Z2p = 0.000$	

Symmetry is so valuable in the discussion of molecular systems that we cannot fail to use it even when a modest (or approximate) symmetry is available. Here there are slightly conflicting conventions: in point group theory the highest order rotation axis (C_2 here) is taken along the z-axis, but in the chemistry of pi-electron systems it is common to use the x–y plane as the sigma plane with the 2pz orbitals perpendicular to the x–y plane. Here we adopt the usual chemical convention for pi-electron systems. In this case the twofold axis and the two perpendicular reflection planes, with the identity, make up the group C_{2v}. We quote its character table for later reference (Table 6.1).

Basis

PCLOBE uses Gaussian lobes as the primitive members of the basis. A thorough discussion of the properties of these functions is found elsewhere in this text. For formaldehyde we will choose lobe representations of 1s, 2s, and 2p atomic functions for C and O and lobe representations of the 1s atomic orbital for H atom. These atomic functions are each characterized by a scale factor, quoted in the first section of output. Each scale factor is characteristic of a particular atomic orbital for a specific atom in a molecular environment; for example, the 1s scale factor for H atom is 1.24, reflecting the rescaling attending the incorporation of a free H atom (with scale factor of unity) into a molecule. Here we quote parameters of a portion of the set of 68 primitive Gaussian lobes; alpha is the exponential factor in the spherical Gaussian, while x, y, and z define the location of the sphere. The lobes quoted in the table are all situated on the C atom.

TABLE 6.1

Character Table for C_{2v}

C_{2v} (z)	E	C_2	σ_{XZ} (Mol)	σ_{YZ} (Vert)
A_1	1	1	1	1
A_2	1	1	−1	−1
B_1	1	−1	−1	1
B_2	1	−1	1	−1

```
                    Spherical Gaussian Basis Set
No. 1   alpha = 0.44765655E+03 at X = 0.000000 Y = 0.000000 Z = 0.000000 a.u.
No. 2   alpha = 0.67484190E+02 at X = 0.000000 Y = 0.000000 Z = 0.000000 a.u.
No. 3   alpha = 0.15149941E+02 at X = 0.000000 Y = 0.000000 Z = 0.000000 a.u.
No. 4   alpha = 0.40150529E+01 at X = 0.000000 Y = 0.000000 Z = 0.000000 a.u.
No. 5   alpha = 0.66775853E+01 at X = 0.000000 Y = 0.000000 Z = 0.000000 a.u.
No. 6   alpha = 0.40548566E+00 at X = 0.000000 Y = 0.000000 Z = 0.000000 a.u.
No. 7   alpha = 0.15567192E+00 at X = 0.000000 Y = 0.000000 Z = 0.000000 a.u.
```

Contraction and Weighting

The contraction of lobes into basis functions is made explicit in this part of the output, for the C atom's basis functions resembling 1s and 2s atomic orbitals. The coefficients quoted below define normalized basis functions.

```
                    Contracted Orbital No. 1
  1.26455545*(1),    2.19977939*(2),    2.49413566*(3),    1.07884522*(4),
                    Contracted Orbital No. 2
-0.01469754*(5),    0.14613960*(6),    0.11233844*(7),
```

Lobe Representations of p Functions

The simplest lobe representation of an atomic 2pz function requires an out-of-phase combination of two primitive Gaussians, each offset from the nucleus along the z-axis. Here is the set of eight primitives which PCLOBE uses to represent a C atom 2px basis function. The offset is very small, 0.1 bohr.

```
No.  8   alpha = 0.18879824E+00 at X =  0.102047 Y = 0.000000 Z = 0.000000 a.u.
No.  9   alpha = 0.50837859E+00 at X =  0.102047 Y = 0.000000 Z = 0.000000 a.u.
No. 10   alpha = 0.15339857E+01 at X =  0.102047 Y = 0.000000 Z = 0.000000 a.u.
No. 11   alpha = 0.63916071E+01 at X =  0.102047 Y = 0.000000 Z = 0.000000 a.u.
No. 12   alpha = 0.18879824E+00 at X = -0.102047 Y = 0.000000 Z = 0.000000 a.u.
No. 13   alpha = 0.50837859E+00 at X = -0.102047 Y = 0.000000 Z = 0.000000 a.u.
No. 14   alpha = 0.15339857E+01 at X = -0.102047 Y = 0.000000 Z = 0.000000 a.u.
No. 15   alpha = 0.63916071E+01 at X = -0.102047 Y = 0.000000 Z = 0.000000 a.u.
```

The contraction is specified below:

```
                    Contracted Orbital No. 3
  0.86930844*( 8),   1.59137354*( 9),   0.81303245*(10),   0.19081987*(11),
 -0.86930844*(12), -1.59137354*(13),  -0.81303245*(14),  -0.19081987*(15),
```

Symmetry and the Basis

The basis spans irreducible representations in C_{2v} as shown below:

C_{2v} (z)	E	C_2	σ_{XZ} (Mol)	σ_{YZ} (Vert)
Γ(basis)	12	2	8	6

This reducible representation can be resolved to $\Gamma(\text{basis}) = 7 \, A_1 + 0 \, A_2 + 2 \, B_1 + 3 \, B_2$.

Overlap Matrix

PCLOBE now computes the overlap matrix. Inspection of the sample from the matrix shows that the basis functions are normalized (the diagonal elements are all unity) but not orthogonal (the off-diagonal elements are not all zero). On any center, basis functions with different angular momenta (2s and 2p, specifically) are guaranteed to be orthogonal. Notice that the C 2s and 2px functions overlap with the O 2s and 2px functions, but the C 2py and 2pz functions do not overlap with the O 2s and 2px functions. In this way the overlap matrix encodes the symmetry of the system.

#	At-Orb	1	2	3	4	5	6	7	8
1	C 1s	1.000	0.234	0.000	0.000	0.000	0.000	0.038	−0.053
2	C 2s	0.234	1.000	0.000	0.000	0.000	0.039	0.415	−0.289
3	C 2px	0.000	0.000	1.000	0.000	0.000	0.064	0.473	−0.305
4	C 2py	0.000	0.000	0.000	1.000	0.000	0.000	0.000	0.000
5	C 2pz	0.000	0.000	0.000	0.000	1.000	0.000	0.000	0.000
6	O 1s	0.000	0.039	0.064	0.000	0.000	1.000	0.245	0.000
7	O 2s	0.038	0.415	0.473	0.000	0.000	0.245	1.000	0.000
8	O 2px	−0.053	−0.289	−0.305	0.000	0.000	0.000	0.000	1.000

Solution of the SCF Equation: Dealing with the Overlapping Basis

PCLOBE constructs roots of the overlap matrix to effect a transformation to an orthogonal basis. The first step toward evaluating this matrix is diagonalizing the overlap matrix by a unitary transform: the diagonal elements are then replaced by their inverse square root, and the transform reversed.

$$\mathbf{D} = \mathbf{U}^{-1}\mathbf{SU} = \{d_{ii}\}; \quad \mathbf{D}^{-1/2} = \{d_{ii}^{-1/2}\}; \quad \mathbf{S}^{-1/2} = \mathbf{U}\mathbf{D}^{-1/2}\mathbf{U}^{-1}$$

The inverse half power overlap reported in PCLOBE output retains the structure of the overlap matrix, as the selection shows. The roots of the overlap matrix are generally less than unity and in a large basis can approach zero. (That is why the diagonal elements of the matrix $\mathbf{S}^{-1/2}$ quoted from PCLOBE output are generally larger than unity.)

#	At-Orb	1	2	3	4	5	6	7	8
1	C 1s	1.027	−0.169	−0.002	0.000	0.000	−0.003	0.034	−0.008
2	C 2s	−0.169	1.541	0.042	0.000	0.000	0.038	−0.382	0.245
3	C 2px	−0.002	0.042	1.298	0.000	0.000	0.030	−0.403	0.262
4	C 2py	0.000	0.000	0.000	1.235	0.000	0.000	0.000	0.000
5	C 2pz	0.000	0.000	0.000	0.000	1.017	0.000	0.000	0.000
6	O 1s	−0.003	0.038	0.030	0.000	0.000	1.027	−0.156	0.019
7	O 2s	0.034	−0.382	−0.403	0.000	0.000	−0.156	1.313	−0.188
8	O 2px	−0.008	0.245	0.262	0.000	0.000	0.019	−0.188	1.122

Construct the Core Matrix Representing the Kinetic Energy and Nuclear-Electron Attraction

The elements of this matrix have energy units, called atomic units or hartrees; 1 hartree is 27.21 electron volts. (Recall that the binding energy for the electron in H atom is 13.6 eV, or 0.5 hartree.) The selection from the full PCLOBE output shows that the electron is very tightly bound in the 1s orbitals and less tightly bound in the valence 2s and 2p levels. The entries referring to valence functions on C and O can be substantial—the entry for C2s and O2s is −4.24 hartree. The structure of the core matrix is consistent with the symmetry of the molecules.

#	At-Orb	1	2	3	4	5	6	7	8
1	C 1s	−22.205	−5.080	−0.073	0.000	0.000	0.000	−0.844	1.185
2	C 2s	−5.080	−8.497	−1.003	0.000	0.000	−1.370	−4.240	3.047
3	C 2px	−0.073	−1.003	−8.210	0.000	0.000	−2.208	−5.030	3.265
4	C 2py	0.000	0.000	0.000	−7.476	0.000	0.000	0.000	0.000
5	C 2pz	0.000	0.000	0.000	0.000	−7.294	0.000	0.000	0.000
6	O 1s	0.000	−1.370	−2.208	0.000	0.000	−34.913	−8.416	0.057
7	O 2s	−0.844	−4.240	−5.030	0.000	0.000	−8.416	−11.086	0.772
8	O 2px	1.185	3.047	3.265	0.000	0.000	0.057	0.772	−9.723

Approximate the Fock Matrix and Solve the Secular Equation

The SCF equations refer to the Fock matrix, which requires the MO coefficient matrix. The simplest estimate of the Fock matrix is the core matrix itself; then with

$$\mathbf{FC} = \varepsilon\mathbf{SC}; \text{ let } \mathbf{F_0} = \mathbf{H} \text{ and define } \mathbf{C_0}, \varepsilon_0$$

The solution requires a transformation

$$\mathbf{S}^{-1/2}\mathbf{FS}^{-1/2}\mathbf{S}^{1/2}\mathbf{C} = \varepsilon\mathbf{S}^{-1/2}\mathbf{SS}^{-1/2}\mathbf{S}^{1/2}\mathbf{C}$$
$$\mathbf{F'C'} = \varepsilon\mathbf{C'}$$

The transformed secular equation can be solved by diagonalization of $\mathbf{F'}$ which produces the orbital energies ε and eigenvalues $\mathbf{C'}$. The coefficients in the original basis can be recovered by back-transformation, $\mathbf{C} = \mathbf{S}^{-1/2}\mathbf{C'}$. The roots of the core matrix are quoted below: the "aufbau" procedure would have us populate the MOs in sequence, lowest energy first. Without any inclusion of electron repulsion, all orbital energies can be negative.

One-Electron Energy Levels

E(1) = −26.277578902510
E(2) = −16.697253613040
E(3) = −7.237485506305
E(4) = −7.008502259227

```
E( 5) = -6.995680059312
E( 6) = -6.483897133334
E( 7) = -5.868487104434
E( 8) = -5.454429103406
E( 9) = -5.319250111484
E(10) = -3.628946235503
E(11) = -3.142873518255
E(12) = -2.868740145102
```

Begin the Iterative Solution of the SCF Equation—Guess MOs

To begin the iterative solution of the SCF equations, PCLOBE constructs approximate MO coefficient matrix by diagonalizing a core matrix with diagonal elements scaled by 0.75 to reflect effects of repulsion between electrons. The first MO coefficients are quoted as columns in order of ascending energy. Inspection of the coefficients reveals that the lowest energy MO is mainly the O 1s function, and the next is the C 1s function. There is little mixing between these energetically isolated basis functions and valence basis functions. MOs 3, 4, and 5 are mainly 2p basis functions on O atom (lone pairs), with modest admixtures of their symmetry-compatible counterparts on C atom. MO 4 is an approximation to the pi component for the CO bond. MO 6 is a sigma bond between C and O, while MOs 7 and 8 are dominated by basis functions of the CH_2 fragment. The initial guess MOs need considerable improvement, which is accomplished in the iterative solution of the SCF equations.

Initial-Guess-Eigenvectors by Column,

#	At-Orb	1	2	3	4	5	6	7	8
1	C 1s	0.003	-0.982	-0.067	0.000	0.000	-0.051	-0.214	0.000
2	C 2s	-0.034	-0.081	-0.040	0.000	0.000	0.007	0.697	0.000
3	C 2px	-0.027	0.018	0.071	0.000	0.000	0.234	-0.895	0.000
4	C 2py	0.000	0.000	0.000	-0.139	0.000	0.000	0.000	-1.047
5	C 2pz	0.000	0.000	0.000	0.000	-0.093	0.000	0.000	0.000
6	O 1s	0.979	0.002	0.017	0.000	0.000	-0.292	-0.058	0.000
7	O 2s	0.091	0.003	-0.068	0.000	0.000	0.893	0.315	0.000
8	O 2px	-0.017	0.015	-0.986	0.000	0.000	0.039	-0.171	0.000
9	O 2py	0.000	0.000	0.000	-0.966	0.000	0.000	0.000	0.330
10	O 2pz	0.000	0.000	0.000	0.000	-0.976	0.000	0.000	0.000
11	H 1s	0.003	0.027	0.010	0.031	0.000	0.015	-0.129	0.035
12	H 1s	0.003	0.027	0.010	-0.031	0.000	0.015	-0.129	-0.035

Symmetry Labeling the MOs

Inspection of these trial MOs allows symmetry labeling. Because the core Hamiltonian (and the Fock operator) is totally symmetric in C_{2v}, it can mix only basis function combinations of like symmetry species. We can see that the trial MOs are symmetry-adapted.

```
#      1 (A1)  2 (A1)  3 (A1)  4 (B2)  5 (B1)  6 (A1)  7 (A1)  8 (B2)
```

If we imagine that bonding in formaldehyde is specified by atomic cores, lone pairs, and local two-center bonds, we can write reducible representations for sets of such pairs:

C_{2v} (z)	E	C_2	σ_{XZ} (Mol)	σ_{YZ} (Vert)	Rep
Γ(C core)	1	1	1	1	A_1
Γ(O core)	1	1	1	1	A_1
Γ(O σ LP)	1	1	1	1	A_1
Γ(CO σ bond)	1	1	1	1	A_1
Γ(CO π bond)	1	−1	−1	1	B_1
Γ(O π' LP)	1	−1	1	−1	B_2
Γ(σ CH bonds)	2	0	2	0	$A_1 + B_2$

This adds up to exactly the symmetry representation of the occupied trial MOs.

Evaluate the Density Matrix

With the approximate MO coefficients, PCLOBE evaluates the density matrix

$$P_{\mu\nu} = \sum_{i=1}^{N} 2C_{\mu i}C_{\nu i}$$

A segment of the density matrix is shown. The diagonal elements divide themselves into core (1s) and valence sets; the former elements are near 2, consistent with the $1s^2$ full population of the core.

```
#  At-Orb    1        2        3        4       5        6       7        8
1  C   1s   2.033  -0.136   0.315  0.000  0.000   0.055  -0.221   0.172
2  C   2s  -0.136   0.992  -1.251  0.000  0.000  -0.153   0.452  -0.160
3  C  2px   0.315  -1.251   1.723  0.000  0.000  -0.084  -0.161   0.185
4  C  2py   0.000   0.000   0.000  2.230  0.000   0.000   0.000   0.000
5  C  2pz   0.000   0.000   0.000  0.000  0.017   0.000   0.000   0.000
6  O   1s   0.055  -0.153  -0.084  0.000  0.000   2.093  -0.382  -0.070
7  O   2s  -0.221   0.452  -0.161  0.000  0.000  -0.382   1.821   0.092
8  O  2px   0.172  -0.160   0.185  0.000  0.000  -0.070   0.092   2.005
```

Repulsion Integrals

The two-electron integrals will now be required; this presents a considerable problem of computation and storage. The two-electron integrals in the primitive set of lobes are all individually "easy" but the transformation to integrals in the contracted basis is cumbersome. The number of integrals in the contracted basis is formally on the order of K^4, where K is the number of contracted basis functions. In our illustrative case K is small, but still there are about 10,000 integrals in the contracted basis. Permutation symmetry reduces the total by a factor of eight, and molecular point group symmetry reduces still further the number of unique integrals. PCLOBE incorporates the eightfold permutation symmetry but makes only ad hoc use of point group symmetry. We can gain some insight into the relative magnitude of two-electron integrals by inspection of a portion of the set. In the following sample from PCLOBE output, two-electron integrals are written in the charge-density notation

$$\int d\tau_1 \, d\tau_2 \, \chi_\mu^*(1)\chi_\nu(1)\frac{1}{r_{12}}\chi_\rho^*(2)\chi_\sigma(2) = \langle\mu\rho||\nu\sigma\rangle = (\mu\nu|\rho\sigma)$$

From the segment at the beginning of the two-electron integral list quoted below we see that the repulsion in the 1s core of carbon, $(1s_C \, 1s_C \, |1s_C \, 1s_C) = (11|11)$, is very large. One-center repulsions, such as $(11|nn)$, are also substantial.

```
(1, 1/ 1,  1) =  3.521773986319   (1, 1/ 1,  2) =  0.568274648385
(1, 1/ 1,  7) =  0.089376051605   (1, 1/ 1,  8) = -0.125153023399
(1, 1/ 1, 11) =  0.131893031911   (1, 1/ 1, 12) =  0.131893031911
(1, 1/ 2,  2) =  0.769408626701   (1, 1/ 2,  6) =  0.017009594743
(1, 1/ 2,  7) =  0.245435530419   (1, 1/ 2,  8) = -0.236383828200
(1, 1/ 2, 11) =  0.309145468983   (1, 1/ 2, 12) =  0.309145468983
(1, 1/ 3,  3) =  0.778962102856   (1, 1/ 3,  6) =  0.027514372045
(1, 1/ 3,  7) =  0.255934220999   (1, 1/ 3,  8) = -0.235193996350
(1, 1/ 3, 11) = -0.142411073803   (1, 1/ 3, 12) = -0.142411073803
(1, 1/ 4,  4) =  0.778962102856   (1, 1/ 4,  9) =  0.112979733423
(1, 1/ 4, 11) =  0.230238543115   (1, 1/ 4, 12) = -0.230238543115
(1, 1/ 5,  5) =  0.778962102856   (1, 1/ 5, 10) =  0.112979733423
------------------------------------------------------------------
------------------------------------------------------------------
```

The full and bulky list of two-electron integrals in the contracted basis is not given here. Inspecting the last few values referring to H 1s functions 11 and 12, quoted below, we see that the one-center integrals are larger than all others. As symmetry dictates, values referring to the two H atoms are equivalent; $(11,11|11,11) = (12,12|12,12) = 0.775$ hartree. The next most significant integral, $(11,11|12,12) = 0.271$ hartree, corresponds to coulomb repulsion between two $1s^2$ charge distributions, one on each H atom. All other values are quite small. The full list is given in the verbose output from PCLOBE, in the supplementary disk. A substantial saving in computational effort is afforded by calculating the overlap (which appears in almost all integrals) and then deciding on the basis of its size whether or not to proceed with the full calculation of the repulsion integral.

```
------------------------------------------------------------------
------------------------------------------------------------------
                  (9, 12/ 12, 12) are done.
(10, 10/ 10, 10) = 0.871775778852   (10, 10/ 11, 11) = 0.247218846646
(10, 10/ 11, 12) = 0.031411376056   (10, 10/ 12, 12) = 0.247218846646
(10, 11/ 10, 11) = 0.000214220626   (10, 11/ 10, 12) = 0.000165030967
(10, 12/ 10, 12) = 0.000214220626
```

```
                          (10, 12/ 12, 12) are done.
 (11, 11/ 11, 11) = 0.775055675847      (11, 11/ 11, 12) = 0.055151393296
 (11, 11/ 12, 12) = 0.270883792158      (11, 12/ 11, 12) = 0.007403336866
 (11, 12/ 12, 12) = 0.055151393296
                          (11, 12/ 12, 12) are done.
 (12, 12/ 12, 12) = 0.775055675847
                          (12, 12/ 12, 12) are done.
```

Continue the Iteration

The improved Fock matrix is to be updated:

$$F_{\mu\nu} = H_{\mu\nu} + \sum_{\lambda\sigma} P_{\lambda\sigma}\left(\langle\mu\lambda|\nu\sigma\rangle - \frac{1}{2}\langle\mu\lambda|\sigma\nu\rangle\right)$$

With the Fock matrix in hand the associated electronic energy can be evaluated

$$E = \sum_{\mu,\nu} P_{\mu\nu}H_{\mu\nu} + \sum_{\mu\nu\lambda\sigma} P_{\mu\nu}P_{\lambda\sigma}\left(\langle\mu\lambda\|\nu\sigma\rangle - \frac{1}{2}\langle\mu\lambda\|\sigma\nu\rangle\right) = \frac{1}{2}\sum_{\mu\nu} P_{\mu\nu}(F_{\mu\nu} + H_{\mu\nu})$$

Test for Convergence

One can now perform convergence tests on the energy and the density matrix. If the energy value here differs from the previous estimate by more than the convergence threshold (10^{-6} hartree in PCLOBE), then one returns to the SCF equation using the improved Fock matrix to define improved MO coefficients. The sample from the final converged **F**-matrix shows that the SCF recognition of the charge distribution alters the diagonal elements most dramatically. Interactions between C and O are also weakened, but the general structure required by symmetry is, of course, maintained.

```
                          Final F-Matrix
         When the Energy converges, the F-matrix is the operator
         which solves FC = ESC in a Self-Consistent-Field sense.
         Thus the C used to form P and F is also the solution C.
```

#	At-Orb	1	2	3	4	5	6	7	8
1	C 1s	-11.470	-2.870	-0.006	0.000	0.000	0.000	-0.485	0.678
2	C 2s	-2.870	-1.639	-0.165	0.000	0.000	-0.893	-0.969	0.619
3	C 2px	-0.006	-0.165	-0.718	0.000	0.000	-1.436	-1.057	0.449
4	C 2py	0.000	0.000	0.000	-0.469	0.000	0.000	0.000	0.000
5	C 2pz	0.000	0.000	0.000	0.000	-0.216	0.000	0.000	0.000
6	O 1s	0.000	-0.893	-1.436	0.000	0.000	-20.750	-5.404	0.025
7	O 2s	-0.485	-0.969	-1.057	0.000	0.000	-5.404	-2.611	0.204
8	O 2px	0.678	0.619	0.449	0.000	0.000	0.025	0.204	-0.574

```
         Electronic Energy = -143.2779663219 a.u.

         ##### Total Energy = -113.2114345824 hartrees #####
```

A rough measure of the quality of the calculation is provided by its conformity to the virial theorem: for a system governed by Coulomb's law the kinetic energy and potential energy obey $2\langle T \rangle = -\langle V \rangle$ or, since $E = \langle T \rangle + \langle V \rangle$, the virial theorem requires that $-\langle E \rangle / \langle T \rangle = 1$.

```
Virial Ratio=-<E>/<T>=1.025599
```

Evidently the basis is not sufficiently flexible to satisfy the virial theorem, which expresses the quality of scaling of the wave function. Alternatively it may be that the geometry is not optimized, and net forces act on the nuclei.

Inspect the Output: MOs and Energies

The eigenvalues of the Fock matrix, estimates of ionization energies or electron affinities according to Koopmans' theorem [2], are listed below. The lowest energies are associated with the O and C atom cores.

```
One-Electron Energy Levels

E ( 1) = -20.756914192746
E ( 2) = -11.473757332662
E ( 3) =  -1.344589661060
E ( 4) =  -0.832659560808
E ( 5) =  -0.673507585901
E ( 6) =  -0.601856695616
E ( 7) =  -0.497583334056
E ( 8) =  -0.420825519200
E ( 9) =   0.177111638238
E (10) =   0.544086554928
E (11) =   0.627636330169
E (12) =   0.699311195798
```

The energy levels divide into bonding (negative values of the orbital energy) and antibonding levels. The final MO coefficients for the bonding set are quoted below:

```
          Alpha-spin Orbitals for   8 Filled Orbitals by Column

 #  At-Orb    1       2       3       4       5       6       7       8
 1  C   1s  -0.001  0.994  -0.107   0.187   0.000   0.022  0.000   0.000
 2  C   2s   0.008  0.027   0.227  -0.679   0.000  -0.087  0.000   0.000
 3  C  2px   0.006  0.000   0.127   0.208   0.000  -0.497  0.000   0.000
 4  C  2py   0.000  0.000   0.000   0.000  -0.563   0.000  0.000  -0.247
 5  C  2pz   0.000  0.000   0.000   0.000   0.000   0.000  0.623   0.000
 6  O   1s  -0.995  0.000  -0.228  -0.107   0.000  -0.099  0.000   0.000
 7  O   2s  -0.024 -0.005   0.801   0.474   0.000   0.525  0.000   0.000
 8  O  2px   0.005  0.000  -0.200   0.180   0.000   0.662  0.000   0.000
 9  O  2py   0.000  0.000   0.000   0.000  -0.475   0.000  0.000   0.856
10  O  2pz   0.000  0.000   0.000   0.000   0.000   0.000  0.661   0.000
11  H   1s  -0.001 -0.006   0.025  -0.208  -0.241   0.143  0.000  -0.333
12  H   1s  -0.001 -0.006   0.025  -0.208   0.241   0.143  0.000   0.333
```

Inspection of the columns makes clear that the lowest energy MOs are O and C atom 1s cores, while the next is primarily a 2s lone pair on oxygen, with some bonding stabilization with C atom. Levels #4 and #5 are primarily CH bonds, the symmetric and antisymmetric combinations, respectively. Level #6 is the CO sigma bond, level #7 is the CO pi bond, and level #8 is an oxygen lone pair dominated by the O atom's 2py orbital. There is some mixing between oxygen's s-type and p-type lone pairs with the basis orbitals on C atom, which is required to maintain orthogonality among the MOs. The overall symmetry species of the occupied MOs is unchanged, i.e.,

```
#      1  (A1)  2  (A1)  3  (A1)  4  (B2)  5  (B2)  6  (A1)  7  (B1)  8  (B2)
```

Describe the Charge Distribution

The final **C**-matrix defines the density matrix, the charge distribution, and related properties such as the dipole moment. One of the most frequently reported characterizations of the charge distribution is the Mulliken overlap population [3]; it resolves the density function or total charge N into components.

$$\rho = \sum_{\mu\nu} P_{\mu\nu} \chi_\mu \chi_\nu; \quad \int d\tau\, \rho = N = \sum_{\mu\nu} P_{\mu\nu} S_{\mu\nu}$$

Some are apparently unambiguously associated with single atoms

$$Q_A = \sum_{\mu,\mu' \text{ on } A} P_{\mu\mu'} S_{\mu\mu'}$$

but the division of elements in two-atom blocks requires a decision. Mulliken made the simplest choice. The AB population

$$Q_{AB} = \sum_{\mu \text{ on } A} \sum_{\nu \text{ on } B} P_{\mu\nu} S_{\mu\nu}$$

is divided equally between atoms A and B. The quoted section from PCLOBE shows that the charge is predominantly assigned to atomic centers, but there are important two-center populations. Note the bonding interactions between C and H, indicated by the C2s–H1s and C2py–H1s values, and the C–O pi interaction shown by the C2pz–O2pz entry.

Mulliken Overlap Populations

#	At-Orb	1	2	3	4	5	6	7	8
1	C 1s	2.071	-0.059	0.000	0.000	0.000	0.000	0.001	-0.007
2	C 2s	-0.059	1.043	0.000	0.000	0.000	0.002	-0.154	0.130
3	C 2px	0.000	0.000	0.613	0.000	0.000	-0.001	-0.057	0.194
4	C 2py	0.000	0.000	0.000	0.755	0.000	0.000	0.000	0.000

```
 5  C  2pz   0.000   0.000   0.000  0.000  0.776   0.000   0.000   0.000
 6  O  1s    0.000   0.002  -0.001  0.000  0.000   2.126  -0.128   0.000
 7  O  2s    0.001  -0.154  -0.057  0.000  0.000  -0.128   2.284   0.000
 8  O  2px  -0.007   0.130   0.194  0.000  0.000   0.000   0.000   1.021
 9  O  2py   0.000   0.000   0.000  0.024  0.000   0.000   0.000   0.000
10  O  2pz   0.000   0.000   0.000  0.000  0.175   0.000   0.000   0.000
11  H  1s   -0.005   0.134   0.057  0.182  0.000   0.000   0.000  -0.007
12  H  1s   -0.005   0.134   0.057  0.182  0.000   0.000   0.000  -0.007
```

The Mulliken population is usually aggregated into atomic and interatomic populations

```
*********************************************************************
                  Total Overlap Populations by Atoms

                 C           O           H           H
         C    5.1404      0.3043      0.3683      0.3683
         O    0.3043      7.9661     -0.0181     -0.0181
         H    0.3683     -0.0181      0.4659     -0.0239
         H    0.3683     -0.0181     -0.0239      0.4659
```

and further summed into atomic charges:

```
                     Computed Atom Charges
Q(1) = -0.181    Q(2) = -0.234    Q(3) = 0.208    Q(4) = 0.208
*********************************************************************
```

The O atom bears a substantial negative Mulliken charge, balanced, of course, by a net positive charge on the CH_2 fragment. Within the CH_2 group, H atoms are positively charged and C is negative. This is all entirely consistent with our intuition. However, placing the calculated Mulliken atomic charges at the nuclear sites will not reproduce the experimental dipole moment and will usually lead to a much larger value. The Mulliken charges are usually so exaggerated that they magnify otherwise subtle electronic effects; this may actually aid interpretation. Be forewarned, however, to use the Mulliken analysis only for qualitative judgment. The components of the charge it defines are not observables. The actual observable, the full density, can be characterized by multipole moments; i.e., expectation values of powers of the Cartesian coordinates. The simplest of these is the dipole moment

$$\mathbf{d} = \langle \psi | \mathbf{r} | \psi \rangle$$

According to PCLOBE, the dipole moment is aligned with the C–O axis, as symmetry requires.

```
            Dipole Moment Components in Debyes
       Dx = -2.011116    Dy = 0.000000    Dz = 0.000000
       Resultant Dipole Moment in Debyes = 2.011116
*********************************************************************
```

This concludes our tour of the SCF energy calculation and its outcome. We now consider how to find low-energy structures, which correspond to the equilibrium geometry of molecules.

Molecular Structure Determination by Energy Minimization

Molecules are dynamic species, but one of the most fruitful assumptions of chemistry is the molecular structure hypothesis that molecules have a persistent shape. Seeking the minimum energy geometry of a molecule is perhaps the most significant task in computational chemistry. Pulay's introduction of the analytical energy derivative [4] was a great step toward the efficient location of minima on the potential energy surface.

Derivative of the Hartree–Fock Energy

Perhaps the clearest early discussion defining the derivatives of the Hartree–Fock energy was that of Pople, Krishnan, Schlegel, and Binkley (PKSB) [5]. Following their argument and notation, we begin with the energy expression for K spin-orbitals in a single determinant

$$\langle \Phi | H | \Phi \rangle = \sum_{m}^{K} \left[\langle m|h|m \rangle \right] + \sum_{m>n}^{K} \left[\langle mn\|mn \rangle - \langle mn\|nm \rangle \right]$$

This corresponds to Equation 3 in PKSB [5]. These authors introduce the notation

$$(pq\|rs) = \langle pq\|rs \rangle - \langle pq\|sr \rangle$$

As usual spin-orbitals are written as a linear superposition of basis functions

$$\phi_i = \sum_{\mu} C_{i\mu} \chi_\mu$$

The basis functions are contractions of primitive functions

$$\chi_p = \sum_{\mu} A_{\mu p} G_p$$

The Hartree–Fock energy becomes

$$E = \sum_{i\mu\nu} C_{i\mu} C_{i\nu} \langle \mu|h|\nu \rangle + \sum_{ij} \sum_{\mu\nu\rho\sigma} C_{i\mu} C_{j\rho} C_{i\nu} C_{j\sigma} (\langle \mu\rho|\nu\sigma \rangle - \langle \mu\rho|\sigma\nu \rangle)$$

or, introducing the density matrix as a sum over spinorbitals rather than using the restricted form and writing the core matrix $H_{\mu\nu} = \langle\mu|h|\nu\rangle$,

$$E = \sum_{\mu\nu}^{\text{Basis}} P_{\mu\nu}H_{\mu\nu} + \frac{1}{2}\sum_{\mu\nu\rho\sigma}^{\text{Basis}} P_{\mu\nu}P_{\rho\sigma}(\langle\mu\rho|\nu\sigma\rangle - \langle\mu\rho|\sigma\nu\rangle)$$

Orthogonality of the spin-orbitals requires

$$\langle p|q\rangle = \sum_{\mu\nu}^{\text{Basis}} C_{p\mu}\langle\mu|\nu\rangle C_{q\nu} = \sum_{\mu\nu}^{\text{Basis}} C_{p\mu}S_{\mu\nu}C_{q\nu} = \delta_{pq}$$

The Hartree–Fock equation may be expressed as a matrix relation in the basis

$$\sum_{\mu}^{\text{Basis}} (F_{\mu\nu} - \varepsilon_p S_{\mu\nu})C_{p\mu} = 0 \text{ or } (\mathbf{F} - \varepsilon\mathbf{S})\mathbf{C} = \mathbf{0}$$

Here ε_p is the pth eigenvalue of the Fock matrix (i.e., an orbital energy), and the Fock matrix for spin-orbitals is

$$F_{\mu\nu} = \langle\mu|h|\nu\rangle + \sum_{\rho\sigma}^{\text{Basis}} P_{\rho\sigma}(\langle\mu\rho|\nu\sigma\rangle - \langle\mu\rho|\sigma\nu\rangle)$$

Differentiation of the energy produces terms referring to derivatives of integrals and also derivatives of the density matrix.

$$\frac{\partial E_{HF}}{\partial x} = \sum_{\mu\nu} P_{\mu\nu}\left(\frac{\partial H_{\mu\nu}}{\partial x}\right) + \frac{1}{2}\sum_{\mu\nu\lambda\sigma} P_{\mu\nu}P_{\lambda\sigma}\frac{\partial[\langle\mu\rho|\nu\sigma\rangle - \langle\mu\rho|\sigma\nu\rangle]}{\partial x} + \left(\frac{\partial V_{\text{nuc}}}{\partial x}\right)$$
$$+ \sum_{\mu\nu}\left(\frac{\partial P_{\mu\nu}}{\partial x}\right)H_{\mu\nu} + \sum_{\mu\nu\rho\sigma}\left(\frac{\partial P_{\mu\nu}}{\partial x}\right)P_{\rho\sigma}\left[\langle\mu\rho|\nu\sigma\rangle - \langle\mu\rho|\sigma\nu\rangle\right]$$

This, in turn, requires derivatives of the LCAO coefficients. These, however, are constrained by the required orthonormality of the MOs and by the fact that they satisfy the SCF-Roothaan equation. Differentiating the orthonormality condition subject to $p = q = i$, we find

$$\langle i|i\rangle = \sum_{\mu\nu}^{\text{Basis}} C_{i\mu}C_{i\nu} = \sum_{\mu\nu}^{\text{Basis}} C_{i\mu}S_{\mu\nu}C_{i\nu} = 1$$

$$\sum_{\mu\nu}\left[\left(\frac{\partial C_{\mu i}^*}{\partial x}\right)S_{\mu\nu}C_{\nu i} + C_{\mu i}^*\left(\frac{\partial S_{\mu\nu}}{\partial x}\right)C_{\nu i} + C_{\mu i}^*S_{\mu\nu}\left(\frac{\partial C_{\nu i}}{\partial x}\right)\right] = 0$$

Differentiation of the Roothaan equation

$$\sum_{\mu}^{\text{Basis}} F_{\mu\nu} C_{p\mu} = \varepsilon_p S_{\mu\nu} C_{p\mu}$$

produces the condition

$$\sum_{\mu} \frac{\partial F_{\mu\nu}}{\partial x} C_{p\mu} + \sum_{\mu} F_{\mu\nu} \frac{\partial C_{p\mu}}{\partial x} = \frac{\partial \varepsilon_p}{\partial x} \sum_{\mu} S_{\mu\nu} C_{p\mu} + \varepsilon_p \sum_{\mu} \frac{\partial S_{\mu\nu}}{\partial x} C_{p\mu} + \varepsilon_p \sum_{\mu} S_{\mu\nu} \frac{\partial C_{p\mu}}{\partial x}$$

Then, introducing the "energy-weighted density matrix"

$$W_{\mu\nu} = \sum_{i=1} \varepsilon_i c_{\mu i}^{*} c_{\nu i}$$

the energy derivative is cast into the form

$$\frac{\partial E}{\partial x} = \sum_{\mu\nu} P_{\mu\nu} \left(\frac{\partial H_{\mu\nu}}{\partial x} \right) + \frac{1}{2} \sum_{\mu\nu\rho\sigma} P_{\mu\nu} P_{\rho\sigma} \left(\frac{\partial [\langle \mu\rho | \nu\sigma \rangle - \langle \mu\rho | \sigma\nu \rangle]}{\partial x} \right)$$
$$+ \frac{\partial V_{NN}}{\partial x} - \sum_{\mu\nu} W_{\mu\nu} \left(\frac{\partial S_{\mu\nu}}{\partial x} \right)$$

making explicit that only derivatives of the integrals—but neither of the density matrix nor of the MO coefficients—are required to evaluate the SCF energy gradient.

Use of the gradient vector requires numerical values of derivatives of the overlap matrix, the internuclear repulsion, the core Hamiltonian matrix, and the two-electron integrals. In PCLOBE the derivatives are in every case estimated numerically, using the fundamental definition in the differential calculus.

$$\frac{\partial F(\mathbf{x})}{\partial x_i} \approx \lim_{\Delta x \to 0} \frac{[F(x_i + \Delta x) - F(x_i)]}{\Delta x}$$

This device can be criticized—its numerical stability is not guaranteed, especially in very flat regions of the potential or close to a minimum—but its simplicity recommends it.

Search Techniques Using the Gradient

The gradient vector

$$\mathbf{G} = \left\{ \frac{\partial E}{\partial x_i} \right\}$$

points in the direction of increasing energy, so one would follow $-\mathbf{G}$ to lower the energy. The length of the step is a significant question. If

$$E = E_0 + \nabla E \cdot \Delta x + \frac{1}{2} \Delta x \cdot \mathbf{K} \cdot \Delta x + \cdots$$

and we wish the derivative of E to be zero, characteristic of an extreme point, then

$$\nabla E(\mathbf{x}) \approx \nabla E(0) + \mathbf{K} \cdot \Delta x = 0$$
$$\Delta x = -\nabla E \cdot \mathbf{K}^{-1}$$

Here $\mathbf{K}$ is the matrix of second derivatives of the energy, known as the Hessian. A great variety of optimization techniques are available, reflecting the many approaches taken toward estimation of the Hessian and its inverse. Taking $\mathbf{K}$ to be the unit matrix produces the "steepest descent" method; often one searches along the line defined by the gradient vector $\mathbf{G_0}$, seeking a minimum in that direction. Recomputing the gradient at that point produces a new vector $\mathbf{G_1}$ normal to its predecessor $\mathbf{G_0}$. Conjugate gradient methods mix the new vector with its predecessor, to smooth the zigzag path characteristic of steepest descent. Actual calculation of the second derivatives, required for second-order or "Newton–Raphson" methods, is generally very demanding. To begin an optimization one may use an estimate of the Hessian obtained from a less ambitious computation, and update it with the information accumulated as gradients are evaluated in the course of a search. PCLOBE uses steepest descents, with damping, to avoid long steps along a large gradient, or a conjugate gradient method early in a minimization. As the gradient becomes small, the Newton–Raphson algorithm can be employed to hasten arrival at the extreme point. "Arrival" is recognized by an acceptably small value of Δx or reduction of the norm of the gradient below a threshold. Soft degrees of freedom (characteristic of very flexible or weakly bound molecules) tend to slow optimization, and a poor starting geometry can defeat the process entirely. In most cases, however, the gradient can be reduced to less that 0.001 hartree/bohr in a few tens of iterations.

Geometry Optimization in PCLOBE

The course of optimization of formaldehyde in PCLOBE is traced here. We begin with a structure provided perhaps from a molecular mechanics calculation.

```
*********************************************************************
     THIS PROGRAM USES FINITE-DIFFERENCE DERIVATIVES WITH
     SCHLEGEL GEOMETRY-OPTIMIZATION. USE A PREVIOUSLY OPTIMIZED
     STRUCTURE FROM A MODELING PROGRAM TO SAVE TIME. IR BANDS
     ARE CALCULATED FROM A FULL NUMERICAL ENERGY HESSIAN MATRIX.
*********************************************************************
```

```
                Nuclear Coordinates from Input
    Atomic Core        X            Y            Z
    6.              1.443750     0.000000     0.000000
    8.              0.837973    -1.047436    -0.003630
    1.              2.533745    -0.000815    -0.003265
    1.              0.899456     0.944370    -0.003267
```

The familiar minimum basis of small contractions of Gaussian lobes is employed once again.

```
     Basis Size = 12 and Number of Spheres = 68 for 16 Electrons
              Electronic Energy = −144.5487341871 a.u.,
         ##### Total Energy = −113.2016783992 hartrees #####
              Virial Ratio = -<E>/<T> = 1.023959
```

```
                    One-Electron Energy Levels

              E ( 1) =   −20.766190191407
              E ( 2) =   −11.487844780390
              E ( 3) =    −1.380722575854
              E ( 4) =    −0.846949144446
              E ( 5) =    −0.709733377773
              E ( 6) =    −0.617734776570
              E ( 7) =    −0.525020689138
              E ( 8) =    −0.426963575916
              E ( 9) =     0.187650613258
              E (10) =     0.603537615846
              E (11) =     0.698724742320
              E (12) =     0.740307228012
```

Begin with the Steepest Descents Approach to a Minimum

PCLOBE varies only the 3N-6 internal coordinates, maintaining a fixed center of mass and orientation of principal moments of inertia. Symmetry can be imposed (see the PCLOBE Readme.txt for specific capabilities), but if not imposed provides a check on the quality of the calculation. Begin with a C_{2v}-symmetric structure.

```
                Distance Matrix in Angstroms

                   C         O         H         H
         C      0.0000    1.2100    1.0900    1.0900
         O      1.2100    0.0000    1.9928    1.9928
         H      1.0900    1.9928    0.0000    1.8879
         H      1.0900    1.9928    1.8879    0.0000
```

```
     ##### Total Energy = −113.2016506817 hartrees #####
              Virial Ratio = -<E>/<T> = 1.023959
```

The root-mean-square gradient is large.

```
Minimization No. 1 RMS Gradient=0.126372 hartrees/bohr
*****************************************************************
```

The first step lengthens the CO and CH bonds.

```
             Distance Matrix in Angstroms
                 C         O         H         H
     C       0.0000    1.2435    1.1053    1.1053
     O       1.2435    0.0000    2.0368    2.0368
     H       1.1053    2.0368    0.0000    1.9117
     H       1.1053    2.0368    1.9117    0.0000
```

In this step the energy is lowered by about 0.067 hartree.

```
   ##### Total Energy=-113.2083244702 hartrees #####
           Virial Ratio=-<E>/<T>=1.024909
```

After the fifth step, the gradient is substantially reduced.

```
Minimization No. 5 RMS Gradient=0.012161 hartrees/bohr
*****************************************************************
```

The geometry is near its optimum, but the search slows as the minimum is approached.

```
             Distance Matrix in Angstroms
                 C         O         H         H
     C       0.0000    1.2583    1.1389    1.1389
     O       1.2583    0.0000    2.0832    2.0832
     H       1.1389    2.0832    0.0000    1.9604
     H       1.1389    2.0832    1.9604    0.0000
   ##### Total Energy=-113.2112171688 hartrees #####
           Virial Ratio=-<E>/<T>=1.025527


Minimization No. 10 RMS Gradient=0.004216 hartrees/bohr
*****************************************************************
```

```
             Distance Matrix in Angstroms
                 C         O         H         H
     C       0.0000    1.2575    1.1442    1.1442
     O       1.2575    0.0000    2.0907    2.0907
     H       1.1442    2.0907    0.0000    1.9624
     H       1.1442    2.0907    1.9624    0.0000
```

```
              The Two-Electron Integrals Have Been Computed
          ##### Total Energy = -113.2113542941 hartrees #####
                 Virial Ratio = -<E>/<T> = 1.025572


      Minimization No. 25 RMS Gradient = 0.001271 hartrees/bohr
**********************************************************************
```

```
                   Distance Matrix in Angstroms
                 C         O         H         H
          C   0.0000    1.2568    1.1452    1.1452
          O   1.2568    0.0000    2.0965    2.0964
          H   1.1452    2.0965    0.0000    1.9527
          H   1.1452    2.0964    1.9527    0.0000

          ##### Total Energy = -113.2114255580 hartrees #####

      Minimization No. 29 RMS Gradient = 0.000932 hartrees/bohr
**********************************************************************
```

The optimized geometry is achieved at last. The norm of the gradient is less than 10^{-3}.

```
                        GEOMETRY CONVERGED

**********************************************************************

                   Distance Matrix in Angstroms
                 C         O         H         H
          C   0.0000    1.2567    1.1452    1.1452
          O   1.2567    0.0000    2.0971    2.0971
          H   1.1452    2.0971    0.0000    1.9513
          H   1.1452    2.0971    1.9513    0.0000

          ##### Total Energy = -113.2114293246 hartrees #####
                 Virial Ratio = -<E>/<T> = 1.025595
```

Second Derivatives of the Hartree–Fock Energy and Vibrational Spectra

In their pioneering paper, Pople et al. [5] defined second derivatives of the Hartree–Fock energy as well as the gradient. Second derivatives of the potential energy define the force matrix

$$K_{rs} = \frac{\partial^2 V}{\partial q_r \partial q_s}$$

For a system at an extreme point, where the gradient is zero, the roots of the force matrix tell us if the extreme point is a relative minimum (all roots positive), a saddle point (a single negative root, one displacement which is stabilizing), or a maximum (two or more roots negative and several paths stabilizing).

PCLOBE defines the force matrix in a way that is very simple to program, by a second numerical differentiation of the energy. Here is the form of the diagonal elements of the force constant matrix:

$$K_{rr} = \frac{E(q_1, \ldots, q_r + \Delta, \ldots) - 2E(q_1, \ldots) + E(q_1, \ldots, q_r - \Delta, \ldots)}{\Delta^2}$$

This is an adaptation of a method used by Ransil [6] to deal with strongly nonlinear problems such as the optimization of basis function scale factors. The accuracy of numerical differentiation is notoriously poor, but for the purpose of diagnosis of the nature of the extreme point the method has some practical value.

If the coordinates are mass-weighted Cartesian coordinates, the classical equations of motion for a harmonic oscillator can be written in compact form. Defining

$$q_i = \sqrt{m_A} \Delta C_{Aj}$$

where

A is an atom label

ΔC_{Aj} is a Cartesian displacement coordinate in x, y, or z directions as identified by index j

The Newtonian kinetic energy is

$$T = \frac{1}{2} \sum_i \dot{q}_i^2$$

and the potential is

$$V = \frac{1}{2} \sum_{ij} K_{ij} q_i q_j$$

which leads to classical equations of motions for a system of harmonic oscillators

$$\frac{d(\partial L / \partial \dot{q}_i)}{dt} = \partial L / \partial q_i; \quad L = T - V$$

This reduces to

$$\ddot{q}_i = -\sum_j K_{ij} q_j$$

which can be decoupled by defining normal modes $\mathbf{Q} = \mathbf{Uq}$ such that the force matrix is diagonal:

$$\ddot{\mathbf{Q}} = \mathbf{UKU}^{-1}\mathbf{Uq} = \Lambda\mathbf{Q}$$

The diagonal elements are proportional with the square of the frequency of each mode, and the associated eigenvector describes the oscillation in terms of the mass-weighted displacement coordinates. For the formaldehyde calculation with the small basis, PCLOBE finds the matrix $\mathbf{K}$ quoted below.

Mass-Weighted Force Constant Matrix

	1	2	3	4	5	6	7
1	0.965E-01	0.258E-04	0.457E-05	−0.641E-01	0.101E-04	0.109E-04	−0.392E-01
2	0.258E-04	0.594E-01	0.255E-05	0.782E-05	−0.271E+00	0.581E-05	0.369E-01
3	0.457E-05	0.255E-05	0.136E-01	0.803E-05	0.460E-05	−0.395E-02	−0.382E-04
4	−0.641E-01	0.782E-05	0.803E-05	0.594E-01	−0.732E-04	−0.916E-05	−0.852E-02
5	0.101E-04	−0.271E+00	0.460E-05	−0.732E-04	0.412E-02	0.572E-05	0.829E-02
6	0.109E-04	0.581E-05	−0.395E-02	−0.916E-05	0.572E-05	0.857E-03	−0.258E-04
7	−0.392E-01	0.369E-01	−0.382E-04	−0.852E-02	0.829E-02	−0.258E-04	0.155E+00
8	0.370E-01	−0.888E-01	−0.282E-04	0.732E-03	0.356E-02	−0.279E-04	−0.147E+00
9	−0.329E-04	−0.174E-04	−0.156E-01	−0.625E-04	−0.137E-04	0.411E-02	−0.637E-04
10	−0.392E-01	−0.368E-01	−0.371E-04	−0.852E-02	−0.861E-02	−0.263E-04	0.115E-01
11	−0.370E-01	−0.887E-01	−0.380E-04	−0.113E-02	0.381E-02	−0.210E-04	−0.149E-01
12	−0.352E-04	−0.335E-04	−0.156E-01	−0.626E-04	−0.187E-04	0.411E-02	−0.713E-04

	8	9	10	11	12
1	0.370E-01	−0.329E-04	−0.392E-01	−0.370E-01	−0.352E-04
2	−0.888E-01	−0.174E-04	−0.368E-01	−0.887E-01	−0.335E-04
3	−0.282E-04	−0.156E-01	−0.371E-04	−0.380E-04	−0.156E-01
4	0.732E-03	−0.625E-04	−0.852E-02	−0.113E-02	−0.626E-04
5	0.356E-02	−0.137E-04	−0.861E-02	0.381E-02	−0.187E-04
6	−0.279E-04	0.411E-02	−0.263E-04	−0.210E-04	0.411E-02
7	−0.147E+00	−0.637E-04	0.115E-01	−0.149E-01	−0.713E-04
8	0.297E+00	−0.110E-03	0.147E-01	−0.595E-02	−0.737E-04
9	−0.110E-03	0.178E-01	−0.944E-04	−0.693E-04	0.172E-01
10	0.147E-01	−0.944E-04	0.155E+00	0.146E+00	−0.362E-04
11	−0.595E-02	−0.693E-04	0.146E+00	0.297E+00	−0.370E-04
12	−0.737E-04	0.172E-01	−0.362E-04	−0.370E-04	0.178E-01

Examination of the off-diagonal elements will show that symmetry has been lost, probably a consequence of the roundoff error attending numerical differentiation. Despite modest departures from C_{2v} symmetry, the normal modes (below) are easily identifiable (#1–#6 in order of decreasing frequency) as CH and CO stretches, HCH and HCO bends, and the O–CH$_2$ pyramidalization, i.e., familiar internal motions.

Normal Modes Relative to Optimized Geometry

Ordered Atom-1(x,y,z),Atom-2(x,y,z), ...
Including Rotational and Translational False Modes

	1	2	3	4	5	6	7
1	−0.001	−0.240	0.000	−0.737	0.006	−0.002	−0.003
2	0.458	−0.002	−0.567	0.000	0.001	0.000	0.058
3	0.000	0.000	0.000	0.000	0.000	−0.526	−0.004
4	0.001	0.024	0.000	0.619	0.292	−0.002	−0.003
5	−0.266	0.001	0.638	0.000	0.000	0.000	−0.094
6	0.000	0.000	0.000	0.000	0.000	0.142	0.000
7	0.305	0.365	0.257	0.036	−0.573	−0.002	0.584
8	−0.519	−0.579	−0.264	0.188	−0.356	−0.001	0.393
9	0.000	0.000	0.000	0.000	0.002	0.593	0.002
10	−0.302	0.366	−0.258	0.038	−0.576	0.003	−0.582
11	−0.516	0.582	−0.264	−0.188	0.358	−0.002	0.391
12	0.000	0.000	0.000	0.000	0.001	0.593	0.002

	8	9	10	11	12
1	−0.003	−0.054	0.254	0.576	0.000
2	0.000	0.000	0.000	0.000	0.682
3	0.000	0.506	−0.608	0.314	0.000
4	−0.004	−0.063	0.293	0.665	0.000
5	0.000	0.000	0.000	−0.001	0.716
6	0.002	−0.736	−0.628	0.208	0.000
7	−0.001	−0.015	0.075	0.173	−0.040
8	0.000	0.000	0.001	0.004	0.095
9	0.707	0.313	−0.191	0.119	0.000
10	−0.001	−0.016	0.075	0.169	0.041
11	0.000	0.000	−0.001	0.000	0.094
12	−0.707	0.311	−0.196	0.113	0.000

The frequencies for the internal motions are reasonable, but the translations and rotations should have frequencies very near zero. These results depart seriously from this prescribed value.

Normal Mode Eigenvalues (hartrees/(amu*bohr∧2))
0.466501 0.401375   0.248914   0.135406   0.089365   0.049817   0.037166
0.000561 0.000083 −0.000424 −0.000564 −0.254378

Harmonic Vibrational Frequencies and Corrections

J.A. Pople et al. Int. J. Quantum Chem. S15, 269 (1981)
We apply 3-21G factor of 0.89 to any basis set initially.

(K dyne/(amu*angstrom);	frequency 1/cm;	0.89×1/cm)
(0.007263;	3511.00;	3124.79)
(0.006249;	3256.72;	289848)
(0.003875;	2564.66;	2282.55)
(0.002108;	1891.57;	1683.50)
(0.001391;	1536.70;	1367.66)
(0.000776;	1147.34;	1021.13)

```
                  The following are false frequencies.
( 0.000579;                       991.02;              882.00)
( 0.000009;                       121.80;              108.40)
( 0.000001;                        46.84;               41.69)
(-0.000007;            Imag       105.85;   Imag        94.20)
(-0.000009;            Imag       122.05;   Imag       108.63)
(-0.003960;            Imag      2592.65;   Imag      2307.46)

Imaginary frequencies (Imag) indicate either an
insufficient optimization or a less complete basis set.
Try further optimization and/or a larger basis set.
Fortunately, imaginary frequencies are usually in the
range where they are not easily measured below 400 cm⁻¹.
```

These results give us high motivation to find alternatives that preserve symmetry and separate the overall translations and rotations from internal motions. Analytical second derivatives were defined by Pople et al., but the pragmatic compromise for many years was to evaluate numerical differences of analytic first derivatives of the energy, since those first derivatives required only expressions for the derivatives of one- and two-electron integrals. Higher derivatives of the energy required derivatives of the wave function as well. The work of Handy and Schaefer [7] addressed this point and is the foundation of all derivative techniques in HF and more sophisticated methods as well. The compilation by Yamaguchi et al. [8] contains detailed accounts of advances in derivative methods.

SCF Calculation Revisited: Alternatives and Points of Contention

Having made our way through a full SCF calculation and geometry optimization for formaldehyde, we can now consider some of the subtleties of each step. Many of the purely computational steps need not be revisited, but there are alternatives worth more detailed discussion for steps 1 (specification of geometry) and 2 (choice of basis). The question of management of the large number of two-electron integrals needs to be addressed, and convergence of the iterative solution of the SCF equations needs to be encouraged sometimes.

Geometry specification. Formaldehyde has useful symmetry and as we saw above it is convenient to choose the symmetry axis to coincide with one of the Cartesian axes (x was chosen) and the molecular plane to coincide with one of the Cartesian planes (x–y was chosen in the example calculation). Many programs accept internal coordinates as well as Cartesian coordinates, and effect the transformation automatically. Internal coordinates are used in the "Z-matrix" specification of geometry. Here the first atom is by convention situated at the Cartesian origin of coordinates, the second atom is placed at bond length along one axis (say x), and the third atom is located in a Cartesian plane (say *x*–*y*). These conventions set values of three coordinates for the first atom, two for the second atom, and one for the third atom. Thus, 3N-6 coordinates will specify all the internal degrees of freedom for a nonlinear molecule. An example of a Z-matrix for formaldehyde can be borrowed from GAUSSIAN's treatment of formaldehyde.

```
        Symbolic Z-matrix:
C
H    1    B1
H    1    B2    2    A1
O    1    B3    2    A2    3    D1    0

  Variables:
B1      1.10114
B2      1.10114
B3      1.21655
A1    114.50841
A2    122.7458
D1    180.
```

Here the first atom is carbon. An H atom is connected to atom 1 (C) and separated by the bond length B1 = 1.101 Å. Atom 3, a hydrogen, is also connected to the carbon, with bond length 1.101 Å. Atom 2 defines an angle A1 = 3-1-2 of 114.5°. The oxygen, atom 4, is separated from Carbon by 1.216 Å. With atom 2, the angle A2 (4-1-2) is defined as 122.7°. Finally, oxygen is located by the dihedral angle D1 (4-1-2-3) defined as 180.0°. The program generates Cartesian coordinates with these Z-matrix parameters and shifts the origin to the center of mass. The rotational constants are inverses of the principal moments of inertia.

Center Number	Atomic Number	Atomic Type	Coordinates (Angstroms)		
			X	Y	Z
1	6	0	0.000000	0.000000	−0.533822
2	1	0	0.000000	0.926149	−1.129445
3	1	0	0.000000	−0.926149	−1.129445
4	8	0	0.000000	0.000000	0.682727

Rotational constants (GHZ): 292.3077530 38.0571398 33.6730605

It is not apparent that C_{2v} symmetry is incorporated in the geometry specification of the Z-matrix, but the Cartesian coordinates derived from the Z-matrix do show full symmetry. Imposing high symmetry on molecules is sometimes a challenge for graphic builders, and can require clever use of dummy atoms in a manually generated Z-matrix. It is not an enormous challenge to impose symmetry in an approximate Z-matrix in this case; here is one that accomplishes C_{2v} symmetry for formaldehyde.

```
                Numeric Z-matrix:
X
O    1    1.0
C    2    1.22    1    090.0
H    3    1.11    2    120.0    1    090.0
H    3    1.11    2    120.0    1    -090.0
```

The dummy atom permits easy definition of symmetrically disposed H atoms and maintenance of a molecular plane. The CH bond distances are explicitly set equal to maintain symmetry, as are the OCH angles.

The Choice of basis. PCLOBE uses Gaussian lobes as the primitive members of the basis [7]. This is now a rare choice in spite of its clear advantages; most molecular codes employ atom-centered Gaussians with not only s-type lobes but forms with higher angular momentum, i.e., p and d. PCLOBE uses contractions of Gaussian lobes to mimic basis functions of higher angular momentum, as we have already seen for the p-functions used in formaldehyde. Further discussion can be found in the technical chapter on the lobe basis.

PCLOBE combines (contracts) primitive lobes into a series of representations intended to resemble atomic functions of 1s, 2s, and 2p (etc.) types. For our first treatment of formaldehyde we chose lobe representations only of 1s, 2s, and 2p atomic functions for C and O and lobe representations of the 1s atomic orbital for H atom. This constitutes a minimum basis set. The main advantage of minimal basis sets is simplicity and low computational demands. If even modest accuracy is required, a more extensive and flexible basis is required. It is possible to enhance the description of the valence shell by "splitting the basis"—i.e., using two contractions for the 2s shell and a distinct set of two contractions for the 2p shell. This doubles the number of valence-shell contractions and improves the energy considerably. This kind of basis set is often termed a double-zeta valence (DZV) [9] set, with zeta standing for a scale factor. One could imagine defining triple-, quadruple-, and quintuple-zeta basis sets, and a standard series of such basis functions has in fact been developed by Dunning, using GTO primitives [10,11].

Besides enhancing the flexibility of a basis set by incorporating more and more functions with distinct radial scaling, one could add to the basis functions with higher angular momentum than would be contained in a minimal basis. When d-type functions are added to a basis for C or O, or p-type functions to the basis for H atoms they serve to increase the angular flexibility of the basis; they are thus called polarization functions. Finally for systems with very weakly held electrons, it may be advantageous (or even essential for valid estimates of energetics) to include in the basis very diffuse functions capable of a realistic representation of Rydberg states. This is especially important for anions and excited states.

Impure Symmetry of Properties Computed in the Contracted Lobe Basis

The lobe representation of basis functions with high angular momentum—d and above—is a considerable challenge. This is discussed in the technical section on the lobe basis. While the angular momentum composition of individual basis functions is a subtlety over which we need not linger, a molecular basis should be able to represent the molecular symmetry perfectly. The contractions used in PCLOBE fall short of this ideal. For example, the dipole moment of formaldehyde is not perfectly aligned with the symmetry axis. This is of little significance; the symmetry error in the magnitude of the dipole moment is less than 0.05%, while the overall value is in error by about 10%, owing to the inflexibility of the basis.

Completeness and Linear Dependence in the Lobe Basis

Increasing the size of basis sets will eventually present practical problems of overcompleteness or "linear dependence" in the basis even for GTO or STO functions. This means that one (or more) of the roots of the overlap matrix will approach zero. Then the matrix $S^{-1/2}$ becomes singular and the solution of the secular equation becomes indeterminate. Any basis set composed of a collection of subsets on atomic centers can become overcomplete, since a complete basis can be defined at each center. In contrast, a set of spherical Gaussians will always be "undercomplete." Paradoxically, linear dependence can occur when spherical Gaussians are contracted into representations of sets of functions with angular momentum higher than s.

Management of Two-Electron Integrals

Disregarding symmetry, the number of unique orbital products for N basis orbitals is $N(N+1)/2 = M$. Then, the number of unique integrals will be $M(M+1)/2$. As N increases this number becomes very large, increasing as N's 4th power. As we see from the discussion of integrals in the lobe basis, these integrals have two distinct factors, one exponential and another depending on the error function. In PCLOBE a screening subroutine called PREINT terminates evaluation of the integral if the exponent guarantees that the integral is less than 10^{-12}. In molecules of even moderate size this device can lead to neglect of many two-electron integrals. For large molecules the number of retained integrals scales as N^3 rather than N^4.

Assembly of the Fock Matrix

There many two-electron integrals; they need to be read (or regenerated) as the Fock matrix is constructed in each step of the iteration. PCLOBE stores unique integrals, which limits its practical scope to perhaps 100 basis functions. Many codes use direct methods, in which the two-electron integrals are generated as needed and never stored, but the problem of using the integrals efficiently however they are produced remains a major challenge for efficient data processing in the program.

PCLOBE uses the method of Preuss and Diercksen [12] for construction of the **F**-matrix. The two-electron integrals are specified by their packed labels (I,J | K,L). FILLF recognizes that any specific integral is used in up to eight places in the Fock matrix (eightfold symmetry applies) as a cofactor of the P-matrix, either in the coulomb or exchange terms of the Fock matrix. FILLF can be used for either restricted or unrestricted SCF calculations, and for direct methods in which two-electron integrals are computed as needed to assemble the Fock matrix.

Convergence of the Iterative Solution

Convergence of the SCF procedure is not guaranteed. A variety of convergence-forcing measures have been devised, including:

Level shifting. The energy of the virtual orbitals plays no role in the definition of the ground state energy, which is defined entirely by the properties of the occupied MOs. A small gap between the highest energy occupied MO and the lowest energy virtual MO can produce large HO-LU mixing which would retard or prevent convergence. Simply adding a few tenths of a hartree to the virtual orbital energies can accelerate convergence.

Damping. If successive SCF iterations produce oscillating energies and density matrix values, damping may force convergence. One damping scheme produces the new guess of the density matrix by admixture of a portion of the newly calculated matrix with a portion of the density matrix from the previous step:

$$\mathbf{P}_i^{\text{damped}} = x\mathbf{P}_i + (1 - x)\mathbf{P}_{i-1}$$

The extent of damping, x, can be varied depending on how rapidly the energy changes, so to calm the most drastic changes. This is coded into PCLOBE.

DIIS. The most powerful method to force convergence is an extrapolation technique called "direct inversion in iteration space (DIIS)". As described by Pulay [13], it employs the sequence of Fock matrices and density matrices generated in several steps of the iteration

$$\mathbf{F}_0, \mathbf{F}_1, \mathbf{F}_2, \dots \quad \text{and} \quad \mathbf{P}_0, \mathbf{P}_1, \mathbf{P}_2, \dots$$

Assume that the exact value can be expressed as a sum of members of the sequence.

$$\mathbf{F} = \sum_j c_j \mathbf{F_j} \quad \text{or} \quad \sum_j c_j (\mathbf{F} + \mathbf{e_j}) = \sum_j c_j \mathbf{F} + \sum_j c_j \mathbf{e_j}$$

where $\mathbf{e_j}$ are (as yet undetermined) error matrices. Then we require that $\sum_j c_j = 1$ and attempt to force the last term to zero. We approximate the error matrices by $\mathbf{e_j} \approx (\mathbf{F_{j+1}} - \mathbf{F_j})$ and minimize the norm $\sum c_i c_j B_{ij}$ where $B_{ij} = \sum_{nm} (\mathbf{e_i})_{mn} (\mathbf{e_j})_{nm}$ subject to the constraint on the weighting coefficients. This is accomplished by the method of Lagrange multipliers; one is to minimize

$$L = \mathbf{cBc} - \lambda \left(1 - \sum_i c_i \right)$$

with respect to each of the coefficients, which leads to the matrix equation

$$\begin{pmatrix} B_{11} & \cdots & B_{1m} & -1 \\ \vdots & & \vdots & \vdots \\ B_{m1} & \cdots & B_{mm} & -1 \\ 1 & \cdots & 1 & 0 \end{pmatrix} \begin{pmatrix} c_1 \\ \vdots \\ c_m \\ \lambda \end{pmatrix} = \begin{pmatrix} 0 \\ \vdots \\ 0 \\ -1 \end{pmatrix}$$

which is Pulay's Equation 6. This method requires storage of several small matrices and a matrix inversion to recover the refined Fock matrix at each step, but has proved to be a good compromise, the investment in computational effort at each step repaid by acceleration of the approach toward convergence.

Reoptimization of Formaldehyde in an Extended Basis

The single-zeta minimum basis set with lobe representations produces an approximate geometry for formaldehyde, with overlong CO and CH distances relative to experimental values: (1.25 vs. 1.21 Å; 1.14 vs. 1.12 Å, respectively).

Distance Matrix in Angstroms

	C	O	H	H
C	0.0000	1.2565	1.1452	1.1452
O	1.2565	0.0000	2.0988	2.0988
H	1.1452	2.0988	0.0000	1.9476
H	1.1452	2.0988	1.9476	0.0000

The OCH angle is better estimated (122° vs. 121°). We might wonder whether improvement of the basis set would allow the SCF method to produce a better description of geometry. PCLOBE incorporates a basis set with split-valence and polarization flexibility. The basis, defined in the selections from the output quoted below, represents a C 1s cores by a single-zeta 6-Gaussian contraction.

Spherical Gaussian Basis Set

```
No.1  alpha=0.30475249E+04 at X=0.000000 Y=0.000000 Z=0.000000 a.u.
No.2  alpha=0.45736952E+03 at X=0.000000 Y=0.000000 Z=0.000000 a.u.
No.3  alpha=0.10394868E+03 at X=0.000000 Y=0.000000 Z=0.000000 a.u.
No.4  alpha=0.29210155E+02 at X=0.000000 Y=0.000000 Z=0.000000 a.u.
No.5  alpha=0.92866630E+01 at X=0.000000 Y=0.000000 Z=0.000000 a.u.
No.6  alpha=0.31639270E+01 at X=0.000000 Y=0.000000 Z=0.000000 a.u.
```

The contraction is shown here:

Contracted Orbital No. 1

```
0.53634519*(1),  0.98945214*(2),  1.59728255*(3),  2.07918728*(4),
1.77417427*(5),  0.61257974*(6),
```

In the valence shells of these first-row atoms, the 2s subshell is described by two contractions, one made up from three lobes of the following set:

```
No.  7 alpha=0.78682723E+01 at X=0.000000 Y=0.000000 Z=0.000000 a.u.
No.  8 alpha=0.18812885E+01 at X=0.000000 Y=0.000000 Z=0.000000 a.u.
No.  9 alpha=0.54424926E+00 at X=0.000000 Y=0.000000 Z=0.000000 a.u.
No.10 alpha=0.16871448E+00 at X=0.000000 Y=0.000000 Z=0.000000 a.u.
```

specifically

Contracted Orbital No. 2

```
-0.39955639*(7),  -0.18415517*(8),  0.51639033*(9),
```

and one composed of a single Gaussian lobe

Contracted Orbital No. 3

```
0.18761794*(10),
```

The valence p subshell is likewise split—i.e., represented by two contractions for each p component; the first is constructed from three sets of six lobes

```
No. 11  alpha=0.78685871E+01 at X= 0.003565 Y= 0.000000 Z= 0.000000 a.u.
No. 12  alpha=0.78685871E+01 at X=-0.003565 Y= 0.000000 Z= 0.000000 a.u.
No. 13  alpha=0.78685871E+01 at X= 0.000000 Y= 0.003565 Z= 0.000000 a.u.
No. 14  alpha=0.78685871E+01 at X= 0.000000 Y=-0.003565 Z= 0.000000 a.u.
No. 15  alpha=0.78685871E+01 at X= 0.000000 Y= 0.000000 Z= 0.003565 a.u.
No. 16  alpha=0.78685871E+01 at X= 0.000000 Y= 0.000000 Z=-0.003565 a.u.
```

```
No. 17  alpha=0.18813638E+01 at  X=  0.007291  Y=  0.000000  Z=  0.000000  a.u.
No. 18  alpha=0.18813638E+01 at  X=-0.007291  Y=  0.000000  Z=  0.000000  a.u.
No. 19  alpha=0.18813638E+01 at  X=  0.000000  Y=  0.007291  Z=  0.000000  a.u.
No. 20  alpha=0.18813638E+01 at  X=  0.000000  Y=-0.007291  Z=  0.000000  a.u.
No. 21  alpha=0.18813638E+01 at  X=  0.000000  Y=  0.000000  Z=  0.007291  a.u.
No. 22  alpha=0.18813638E+01 at  X=  0.000000  Y=  0.000000  Z=-0.007291  a.u.

No. 23  alpha=0.54427103E+00 at  X=  0.013555  Y=  0.000000  Z=  0.000000  a.u.
No. 24  alpha=0.54427103E+00 at  X=-0.013555  Y=  0.000000  Z=  0.000000  a.u.
No. 25  alpha=0.54427103E+00 at  X=  0.000000  Y=  0.013555  Z=  0.000000  a.u.
No. 26  alpha=0.54427103E+00 at  X=  0.000000  Y=-0.013555  Z=  0.000000  a.u.
No. 27  alpha=0.54427103E+00 at  X=  0.000000  Y=  0.000000  Z=  0.013555  a.u.
No. 28  alpha=0.54427103E+00 at  X=  0.000000  Y=  0.000000  Z=-0.013555  a.u.
```

These are contracted as we find later in the output.

```
                    Contracted Orbital No. 4

11.55227699*(11),-11.55227699*(12),18.11447535*(17),-18.11447535*(18),
16.80800160*(23),-16.80800160*(24),
```

```
                    Contracted Orbital No. 5

11.55227699*(13),-11.55227699*(14),18.11447535*(19),-18.11447535*(20),
16.80800160*(25),-16.80800160*(26),
```

```
                    Contracted Orbital No. 6

11.55227699*(15),-11.55227699*(16),18.11447535*(21),-18.11447535*(22),
16.80800160*(27),-16.80800160*(28),
```

The second component of the split valence p-shell basis is more simply described, by the spherical Gaussians listed here

```
No. 29  alpha=0.16872123E+00 at  X=  0.024345  Y=  0.000000  Z=  0.000000  a.u.
No. 30  alpha=0.16872123E+00 at  X=-0.024345  Y=  0.000000  Z=  0.000000  a.u.
No. 31  alpha=0.16872123E+00 at  X=  0.000000  Y=  0.024345  Z=  0.000000  a.u.
No. 32  alpha=0.16872123E+00 at  X=  0.000000  Y=-0.024345  Z=  0.000000  a.u.
No. 33  alpha=0.16872123E+00 at  X=  0.000000  Y=  0.000000  Z=  0.024345  a.u.
No. 34  alpha=0.16872123E+00 at  X=  0.000000  Y=  0.000000  Z=-0.024345  a.u.
```

and contracted as shown:

```
                    Contracted Orbital No. 7

9.38164736*(29),  -9.38164736*(30),
```

```
                    Contracted Orbital No. 8

9.38164736*(31),  -9.38164736*(32),
```

```
                    Contracted Orbital No. 9

9.38164736*(33),  -9.38164736*(34),
```

Similar expressions define the 3:1 split for the valence shell of O and H atoms.

Polarization—addition of functions with higher angular momentum—in a basis can have an impact on bond lengths, especially for bonds to atoms with lone pairs as is the case for formaldehyde. PCLOBE the d-set for C atom is composed of lobes listed here:

```
No. 35   alpha=0.75359483E+00 at  X= 0.325819  Y= 0.000000  Z= 0.000000  a.u.
No. 36   alpha=0.75359483E+00 at  X=-0.325819  Y= 0.000000  Z= 0.000000  a.u.
No. 37   alpha=0.75359483E+00 at  X= 0.000000  Y= 0.325819  Z= 0.000000  a.u.
No. 38   alpha=0.75359483E+00 at  X= 0.000000  Y=-0.325819  Z= 0.000000  a.u.
No. 39   alpha=0.75359483E+00 at  X= 0.000000  Y= 0.000000  Z= 0.325819  a.u.
No. 40   alpha=0.75359483E+00 at  X= 0.000000  Y= 0.000000  Z=-0.325819  a.u.
No. 41   alpha=0.75359483E+00 at  X= 0.230389  Y= 0.000000  Z= 0.230389  a.u.
No. 42   alpha=0.75359483E+00 at  X=-0.230389  Y= 0.000000  Z= 0.230389  a.u.
No. 43   alpha=0.75359483E+00 at  X=-0.230389  Y= 0.000000  Z=-0.230389  a.u.
No. 44   alpha=0.75359483E+00 at  X= 0.230389  Y= 0.000000  Z=-0.230389  a.u.
No. 45   alpha=0.75359483E+00 at  X= 0.000000  Y= 0.230389  Z= 0.230389  a.u.
No. 46   alpha=0.75359483E+00 at  X= 0.000000  Y=-0.230389  Z= 0.230389  a.u.
No. 47   alpha=0.75359483E+00 at  X= 0.000000  Y=-0.230389  Z=-0.230389  a.u.
No. 48   alpha=0.75359483E+00 at  X= 0.000000  Y= 0.230389  Z=-0.230389  a.u.
No. 49   alpha=0.75359483E+00 at  X= 0.325819  Y= 0.000000  Z= 0.000000  a.u.
No. 50   alpha=0.75359483E+00 at  X=-0.325819  Y= 0.000000  Z= 0.000000  a.u.
No. 51   alpha=0.75359483E+00 at  X= 0.000000  Y= 0.325819  Z= 0.000000  a.u.
No. 52   alpha=0.75359483E+00 at  X= 0.000000  Y=-0.325819  Z= 0.000000  a.u.
No. 53   alpha=0.75359483E+00 at  X= 0.230389  Y= 0.230389  Z= 0.000000  a.u.
No. 54   alpha=0.75359483E+00 at  X=-0.230389  Y= 0.230389  Z= 0.000000  a.u.
No. 55   alpha=0.75359483E+00 at  X=-0.230389  Y=-0.230389  Z= 0.000000  a.u.
No. 56   alpha=0.75359483E+00 at  X= 0.230389  Y=-0.230389  Z= 0.000000  a.u.
```

These are contracted into five components.

```
                    Contracted Orbital No. 10
```

$$-2.16440680*(35), -2.16440680*(36), -2.16440680*(37), -2.16440680*(38),$$
$$4.32881360*(39), 4.32881360*(40),$$

```
                    Contracted Orbital No. 11
```

$$3.74886255*(41), -3.74886255*(42), 3.74886255*(43), -3.74886255*(44),$$

```
                    Contracted Orbital No. 12
```

$$3.74886255*(45), -3.74886255*(46), 3.74886255*(47), -3.74886255*(48),$$

```
                    Contracted Orbital No. 13
```

$$3.74886255*(49), 3.74886255*(50), -3.74886255*(51), -3.74886255*(52),$$

```
                    Contracted Orbital No. 14
```

$$3.74886255*(53), -3.74886255*(54), 3.74886255*(55), -3.74886255*(56),$$

Orbital 10 is a d_{z^2} function with zero s-character. The H atom 1s shell is split 3:1, using the lobes listed here

```
No. 123  alpha=0.18731137E+02 at  X=-1.138938  Y=-1.840215  Z=0.000000  a.u.
No. 124  alpha=0.28253944E+01 at  X=-1.138938  Y=-1.840215  Z=0.000000  a.u.
No. 125  alpha=0.64012169E+00 at  X=-1.138938  Y=-1.840215  Z=0.000000  a.u.
No. 126  alpha=0.16127776E+00 at  X=-1.138938  Y=-1.840215  Z=0.000000  a.u.
```

to construct contractions #34 and #35.

```
                        Contracted Orbital No. 34
        0.21493545*(123),   0.36457120*(124),   0.41505143*(125),

                        Contracted Orbital No. 35
         0.18138065*(126),
```

The polarization of the basis on H atom is accomplished by adding a p-set made up of lobes as listed.

```
No. 127  alpha=0.11000440E+01 at  X=-1.129404  Y=-1.840215  Z= 0.000000  a.u.
No. 128  alpha=0.11000440E+01 at  X=-1.148472  Y=-1.840215  Z= 0.000000  a.u.
No. 129  alpha=0.11000440E+01 at  X=-1.138938  Y=-1.830681  Z= 0.000000  a.u.
No. 130  alpha=0.11000440E+01 at  X=-1.138938  Y=-1.849750  Z= 0.000000  a.u.
No. 131  alpha=0.11000440E+01 at  X=-1.138938  Y=-1.840215  Z= 0.009534  a.u.
No. 132  alpha=0.11000440E+01 at  X=-1.138938  Y=-1.840215  Z=-0.009534  a.u.
```

and contracted as shown:

```
                        Contracted Orbital No. 36
  38.27889158*(127),   -38.27889158*(128),

                        Contracted Orbital No. 37
  38.27889158*(129),   -38.27889158*(130),

                        Contracted Orbital No. 38
  38.27889158*(131),   -38.27889158*(132),
```

These are diffuse 2p functions based on the Sambe 2p orbitals for H.

The overlap, core, and related matrices are now much larger (38×38); it is thus not convenient to examine them in any detail. There are 38 eigenvalues of the core matrix, and a like number of eigenvectors; however, only the eight occupied MOs are used to construct the Fock matrix. The SCF proceeds as we have seen before. The better basis improves the energy and the virial ratio considerably, and shifts the orbital energies noticeably. The starting geometry, which we found in the optimization in the single-zeta basis, is not however the equilibrium structure in this new double-zeta polarization basis.

```
        ##### Total Energy=-113.8552102209 hartrees #####
            Virial Ratio=-<E>/<T>=1.005636

              One-Electron Energy Levels
            E( 1)=   -20.594826887395
            E( 2)=   -11.360292685747
            E( 3)=    -1.370771834195
            E( 4)=    -0.861400163191
```

```
E ( 5) =    -0.669440984728
E ( 6) =    -0.630323335406
E ( 7) =    -0.510780415485
E ( 8) =    -0.438886769617
E ( 9) =     0.123945477233
E (10) =     0.228672589748
E (11) =     0.319109153746
E (12) =     0.372580585162
E (13) =     0.736691537375
E (14) =     0.830315338128
E (15) =     0.834825653637
            (etc.)
```

The SCF MOs are more elaborate descriptions of the familiar entities we found in the single-zeta calculation: 1s cores for C and H, the CO sigma bond, the CH_2 symmetric and antisymmetric combinations of CH sigma bonds, the oxygen sigma lone pair, and finally the pi bond and the pseudo-pi oxygen lone pair. Exactly the same symmetry species is spanned by the occupied MOs in this larger basis as was found in the small basis.

Alpha-spin Orbitals for   8 Filled Orbitals by Column

#	At-Orb	1	2	3	4	5	6	7	8
1	C	0.000	-0.996	-0.105	-0.164	0.000	-0.017	0.000	0.000
2	C	0.000	-0.023	0.201	0.347	0.000	0.055	0.000	0.000
3	C	0.005	0.007	0.072	0.311	0.000	-0.050	0.000	0.000
4	C	0.000	-0.001	0.165	-0.165	0.000	0.377	0.000	0.000
5	C	0.000	0.000	0.000	0.000	0.401	0.000	0.000	-0.195
6	C	0.000	0.000	0.000	0.000	0.000	0.000	0.318	0.000
7	C	0.003	-0.002	-0.011	-0.081	0.000	0.087	0.000	0.000
8	C	0.000	0.000	0.000	0.000	0.165	0.000	0.000	-0.042
9	C	0.000	0.000	0.000	0.000	0.000	0.000	0.218	0.000
10	C	0.000	0.001	-0.014	-0.013	0.000	-0.002	0.000	0.000
11	C	0.000	0.000	0.000	0.000	0.000	0.000	0.037	0.000
12	C	0.000	0.000	0.000	0.000	0.000	0.000	0.000	0.000
13	C	-0.001	0.000	0.029	-0.012	0.000	0.026	0.000	0.000
14	C	0.000	0.000	0.000	0.000	-0.008	0.000	0.000	0.059
15	O	0.996	0.000	-0.198	0.086	0.000	0.065	0.000	0.000
16	O	0.022	0.000	0.451	-0.206	0.000	-0.157	0.000	0.000
17	O	-0.011	0.002	0.404	-0.252	0.000	-0.275	0.000	0.000
18	O	-0.002	0.000	-0.125	-0.153	0.000	-0.503	0.000	0.000
19	O	0.000	0.000	0.000	0.000	0.337	0.000	0.000	0.561
20	O	0.000	0.000	0.000	0.000	0.000	0.000	0.498	0.000
21	O	0.003	-0.002	-0.050	-0.076	0.000	-0.286	0.000	0.000
22	O	0.000	0.000	0.000	0.000	0.187	0.000	0.000	0.422
23	O	0.000	0.000	0.000	0.000	0.000	0.000	0.351	0.000
24	O	0.000	0.000	-0.007	-0.003	0.000	-0.013	0.000	0.000
25	O	0.000	0.000	0.000	0.000	0.000	0.000	-0.039	0.000
26	O	0.000	0.000	0.000	0.000	0.000	0.000	0.000	0.000
27	O	0.000	0.000	0.012	0.009	0.000	0.024	0.000	0.000
28	O	0.000	0.000	0.000	0.000	-0.022	0.000	0.000	-0.018
29	H	0.000	0.000	0.028	0.173	0.185	-0.095	0.000	-0.180
30	H	0.000	-0.002	-0.001	0.075	0.134	-0.082	0.000	-0.237

31	H	0.000	0.000	0.005	0.006	0.007	0.002	0.000	−0.003
32	H	0.000	0.000	−0.004	−0.013	−0.009	0.006	0.000	0.005
33	H	0.000	0.000	0.000	0.000	0.000	0.000	0.005	0.000
34	H	0.000	0.000	0.028	0.173	−0.185	−0.095	0.000	0.180
35	H	0.000	−0.002	−0.001	0.075	−0.134	−0.082	0.000	0.237
36	H	0.000	0.000	0.005	0.006	−0.007	0.002	0.000	0.003
37	H	0.000	0.000	0.004	0.013	−0.009	−0.006	0.000	0.005
38	H	0.000	0.000	0.000	0.000	0.000	0.000	0.005	0.000

We embark on the reoptimization of geometry.

```
*************************************************************************
        Ransil-McIver-Komornicki Numeric Geometry Optimization
              B. J. Ransil, Rev. Mod. Phys. v32, p239 (1960)
        J. W. McIver and A. Komornicki, Chem Phys. Lett. v10, p303 (1971)
                       Quasi-Newton Search No. = 1
*************************************************************************
              But first, a faster Schlegel-Steepest-Descent!
*************************************************************************

*************************************************************************
                  Using Schlegel-Gradient Equations, see:
                   Pople, Krishnan, Schlegel and Binkley,
              I.J.Q.C. Symp. 13, p225 (1979); equation (21).
*************************************************************************

                  Using Cs symmetry w.r.t. the X-Y plane.
*************************************************************************
Varying 5 coordinates by 0.000100 until RMS-norm = 0.0010000 hartrees/bohr
*************************************************************************
*************************************************************************
```

Distance Matrix in Angstroms

	C	O	H	H
C	0.0000	1.2565	1.1452	1.1452
O	1.2565	0.0000	2.0988	2.0988
H	1.1452	2.0988	0.0000	1.9476
H	1.1452	2.0988	1.9476	0.0000

(several optimization steps ensue)

```
        Minimization No. 14 RMS Gradient = 0.0009972 hartrees/bohr
*************************************************************************
```

Distance Matrix in Angstroms

	C	O	H	H
C	0.0000	1.1858	1.0941	1.0941
O	1.1858	0.0000	1.9950	1.9950
H	1.0941	1.9950	0.0000	1.8545
H	1.0941	1.9950	1.8545	0.0000

The CO and CH bond distances are considerably shortened to values actually less than experimental estimates. This is characteristic of the

RHF–SCF model chemistry. The MOs are changed very little in general form by geometry optimization.

Total Energy = −113.8677137321 hartrees

Virial Ratio = −<E>/<T> = 1.001957

Reference State Orbitals for   8 Filled Orbitals by Column

#	At–Orb	1	2	3	4	5	6	7	8
1	C	0.000	0.996	0.114	0.161	0.000	0.020	0.000	0.000
2	C	0.000	0.024	−0.213	−0.336	0.000	−0.056	0.000	0.000
3	C	−0.005	−0.008	−0.069	−0.290	0.000	0.056	0.000	0.000
4	C	0.000	0.001	−0.185	0.187	0.000	−0.370	0.000	0.000
5	C	0.000	0.000	0.000	0.000	−0.404	0.000	0.000	0.203
6	C	0.000	0.000	0.000	0.000	0.000	0.000	0.329	0.000
7	C	−0.003	0.002	0.013	0.083	0.000	−0.068	0.000	0.000
8	C	0.000	0.000	0.000	0.000	−0.146	0.000	0.000	0.046
9	C	0.000	0.000	0.000	0.000	0.000	0.000	0.203	0.000
10	C	0.000	−0.001	0.015	0.012	0.000	0.002	0.000	0.000
11	C	0.000	0.000	0.000	0.000	0.000	0.000	0.040	0.000
12	C	0.000	0.000	0.000	0.000	0.000	0.000	0.000	0.000
13	C	0.001	0.001	−0.032	0.014	0.000	−0.025	0.000	0.000
14	C	0.000	0.000	0.000	0.000	0.003	0.000	0.000	−0.064
15	O	−0.996	0.000	0.196	−0.086	0.000	−0.068	0.000	0.000
16	O	−0.023	0.000	−0.442	0.206	0.000	0.167	0.000	0.000
17	O	0.011	−0.002	−0.377	0.250	0.000	0.296	0.000	0.000
18	O	0.002	0.000	0.144	0.148	0.000	0.516	0.000	0.000
19	O	0.000	0.000	0.000	0.000	−0.353	0.000	0.000	−0.551
20	O	0.000	0.000	0.000	0.000	0.000	0.000	0.497	0.000
21	O	−0.003	0.002	0.046	0.075	0.000	0.281	0.000	0.000
22	O	0.000	0.000	0.000	0.000	−0.192	0.000	0.000	−0.419
23	O	0.000	0.000	0.000	0.000	0.000	0.000	0.338	0.000
24	O	0.000	0.000	0.007	0.003	0.000	0.013	0.000	0.000
25	O	0.000	0.000	0.000	0.000	0.000	0.000	−0.041	0.000
26	O	0.000	0.000	0.000	0.000	0.000	0.000	0.000	0.000
27	O	0.000	0.000	−0.013	−0.008	0.000	−0.026	0.000	0.000
28	O	0.000	0.000	0.000	0.000	0.025	0.000	0.000	0.017
29	H	0.000	−0.001	−0.030	−0.182	−0.182	0.092	0.000	0.192
30	H	0.000	0.002	0.003	−0.075	−0.124	0.076	0.000	0.253
31	H	0.000	0.000	−0.005	−0.006	−0.007	−0.002	0.000	0.004
32	H	0.000	0.000	0.004	0.013	0.009	−0.006	0.000	−0.004
33	H	0.000	0.000	0.000	0.000	0.000	0.000	0.005	0.000
34	H	0.000	−0.001	−0.030	−0.182	0.182	0.092	0.000	−0.192
35	H	0.000	0.002	0.003	−0.075	0.124	0.076	0.000	−0.253
36	H	0.000	0.000	−0.005	−0.006	0.007	−0.002	0.000	−0.004
37	H	0.000	0.000	−0.004	−0.013	0.009	0.006	0.000	−0.004
38	H	0.000	0.000	0.000	0.000	0.000	0.000	0.005	0.000

The dipole moment computed in the bigger basis has increased considerably; the more flexible basis permits both polarization of the charge distribution and a greater shift of charge density to the oxygen.

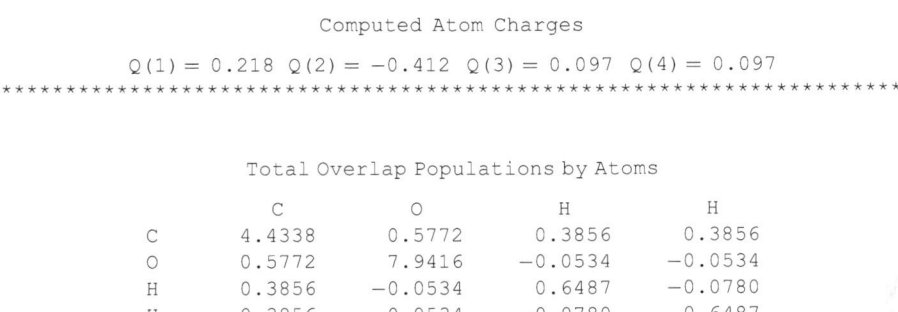

Dipole Moment Components in Debyes

Dx = −2.621357   Dy = 0.000020   Dz = 0.000000

Resultant Dipole Moment in Debyes =   2.621357
****

The Mulliken population is most compatible with a minimum basis, but it can still define local charges on atoms and in bonding regions.

Computed Atom Charges

Q(1) = 0.218 Q(2) = −0.412 Q(3) = 0.097 Q(4) = 0.097
****

Total Overlap Populations by Atoms

	C	O	H	H
C	4.4338	0.5772	0.3856	0.3856
O	0.5772	7.9416	−0.0534	−0.0534
H	0.3856	−0.0534	0.6487	−0.0780
H	0.3856	−0.0534	−0.0780	0.6487

********** NORMAL FINISH OF PCLOBE **********

Historical Landmark: The Accomplishment of Boys

Formaldehyde was the first illustration of the power of ab initio MO theory. Foster and Boys [14] treated the polyatomic formaldehyde by the Hartree–Fock–Roothaan model in 1960. Most ab initio work in the 1950s had been restricted to diatomics, owing to the two-electron integral bottleneck—i.e., the difficulty of three- and four-center coulomb integrals in a basis of Slater orbitals. Formaldehyde displays a variety of subtle and interesting physical and spectroscopic properties. It illustrates a range of bonding, with a pi-bond as well as sigma bonds and also contains lone pairs.

Geometry and basis. Foster and Boys chose the values of CH distance at 2.0 bohr or 1.058 Å, the CO distance at 2.3 bohr or 1.217 Å, and the HCO angle at 120° for their simplicity. The model structure bore a resemblance to the experimental geometry, but since the authors were skeptical that their method could produce a reliable geometry they did not seek a local minimum in the energy. The minimum basis of Slater-type orbitals was employed, with scale factors deduced from Slater's rules, as shown below.

Nuclear Coordinates (Foster and Boys)

Atomic Core	X	Y	Z	
6.	0.0000	0.0000	0.0000	1
	Z1s = 5.70	Z2s = 1.625	Z2p = 1.625	
8.	1.2167	0.0000	0.0000	1
	Z1s = 7.700	Z2s = 2.275	Z2p = 2.227	
1.	−0.5292	0.9165	0.0000	1
	Z1s = 1.200	Z2s = 0.000	Z2p = 0.000	
1.	−0.5292	−0.9195	0.0000	1
	Z1s = 1.200	Z2s = 0.000	Z2p = 0.000	

The functions px, py, pz were defined as the Cartesian coordinates x, y, or z, respectively multiplying a simple exponential, as is the definition for a Slater-type 2p set. The authors constructed 9-Gaussian contractions representing these STOs solely to evaluate integrals; theirs was not an ab initio calculation in a Gaussian basis in the sense we have employed, but a calculation in a Slater basis with convenient approximations to difficult integrals, although parts of this calculation could be likened to what would be called an STO-9G basis. It is of historical interest to realize how much credibility was attached to the exponential STO functions that even when they were expanded in terms of Gaussian functions, the calculation was described as an STO calculation.

The SCF energy for formaldehyde in the single-zeta Gaussian-lobe basis was −113.211 hartree, while Foster and Boys report −113.450 hartree for their single-zeta basis set. The lower (variationally superior) value is to be attributed to their basis set's better representation of the wave function very near nuclear centers, whether that be due to the use of STOs or the 9-Gaussian approximation to the simple exponential. Standard Gaussian basis sets generally use a smaller STO-6G contraction to represent core atomic orbitals. This may impose some small energy penalty, but since the error is a phenomenon of the representation of charge near the nuclei it is generally not chemically significant.

We present the C-matrix of Foster and Boys below; a comparison with the PCLOBE output assures us that the two calculations produce very nearly equivalent descriptions of MOs.

Foster-Boys [14] SCF Orbitals for Formaldehyde (STO basis functions)

#	At-Orb	1	2	3	4	5	6	7	8
1	C 1s	0.0003	−0.9957	−0.1113	0.1665	0.0000	0.0256	0.0000	0.0000
2	C 2s	−0.0058	−0.0233	0.2789	−0.5752	0.0000	−0.0981	0.0000	0.0000
3	C 2px	0.0000	0.0000	0.0000	0.0000	0.0000	0.0000	−0.6543	0.0000
4	C 2py	0.0000	0.0000	0.0000	0.0000	0.5540	0.0000	0.0000	0.1896
5	C 2pz	0.0052	0.0004	−0.1589	−0.2391	0.0000	0.4593	0.0000	0.0000
6	O 1s	0.9961	0.0000	−0.2112	−0.0979	0.0000	−0.0869	0.0000	0.0000
7	O 2s	0.0186	0.0044	0.7580	0.4396	0.0000	0.5016	0.0000	0.0000
8	O 2px	0.0000	0.0000	−0.0000	0.0000	0.0000	0.0000	−0.6280	0.0000
9	O 2py	0.0000	0.0000	0.0000	0.0000	0.4246	0.000	0.0000	−0.8829
10	O 2pz	0.0048	0.0012	0.1708	−0.1716	0.0000	−0.6847	0.0000	0.0000

| 11 | H | 1s | 0.0002 | 0.0054 | 0.0338 | −0.2460 | 0.2769 | 0.1474 | 0.0000 | 0.3425 |
| 12 | H | 1s | 0.0002 | 0.0054 | 0.0338 | −0.2460 | −0.2769 | 0.1474 | 0.0000 | −0.3425 |

#	At-Orb		9 = a1	10 = b2	11 = a1	12 = b1
1	C	1s	−0.0814	0.0000	−0.1679	0.0000
2	C	2s	0.6599	0.0000	1.4385	0.0000
3	C	2px	0.0000	0.0000	0.0000	−0.7878
4	C	2py	0.0000	−1.2548	0.0000	0.0000
5	C	2pz	−1.1776	0.0000	0.4864	0.0000
6	O	1s	0.1103	0.0000	0.0240	0.0000
7	O	2s	−0.9111	0.0000	−0.1523	0.0000
8	O	2px	0.0000	0.0000	0.0000	0.8090
9	O	2py	0.0000	0.3195	0.0000	0.0000
10	O	2pz	−0.9165	0.0000	−0.2127	0.0000
11	H	1s	0.1320	0.9555	−0.9863	0.0000
12	H	1s	0.1320	−0.9555	−0.9863	0.0000

Results from PCLOBE

(lobe contractions mimicking STO-4G 1s, STO-3G 2s, STO-4G 2p, STO-3G H1s)

Alpha-spin Orbitals for 8 Filled Orbitals by Column

#	At-Orb		1	2	3	4	5	6	7	8
1	C	1s	−0.001	0.994	−0.107	0.187	0.000	0.022	0.000	0.000
2	C	2s	0.008	0.027	0.227	−0.679	0.000	−0.087	0.000	0.000
3	C	2px	0.006	0.000	0.127	0.208	0.000	−0.497	0.000	0.000
4	C	2py	0.000	0.000	0.000	0.000	−0.563	0.000	0.000	−0.247
5	C	2pz	0.000	0.000	0.000	0.000	0.000	0.000	0.623	0.000
6	O	1s	−0.995	0.000	−0.228	−0.107	0.000	−0.099	0.000	0.000
7	O	2s	−0.024	−0.005	0.801	0.474	0.000	0.525	0.000	0.000
8	O	2px	0.005	0.000	−0.200	0.180	0.000	0.662	0.000	0.000
9	O	2py	0.000	0.000	0.000	0.000	−0.475	0.000	0.000	0.856
10	O	2pz	0.000	0.000	0.000	0.000	0.000	0.000	0.661	0.000
11	H	1s	−0.001	−0.006	0.025	−0.208	−0.241	0.143	0.000	−0.333
12	H	1s	−0.001	−0.006	0.025	−0.208	0.241	0.143	0.000	0.333

#	At-Orb		9 = b1	10 = a1	11 = b2	12 = a1
1	C	1s	0.000	0.212	0.000	−0.075
2	C	2s	0.000	−1.451	0.000	0.549
3	C	2px	0.000	0.304	0.000	1.263
4	C	2py	0.000	0.000	1.170	0.000
5	C	2pz	0.812	0.000	0.000	0.000
6	O	1s	0.000	−0.051	0.000	0.127
7	O	2s	0.000	0.331	0.000	−0.918
8	O	2px	0.000	−0.321	0.000	0.901
9	O	2py	0.000	0.000	−0.319	0.000
10	O	2pz	−0.781	0.000	0.000	0.000
11	H	1s	0.000	0.964	−0.881	0.244
12	H	1s	0.000	0.964	0.881	0.244

Inspection of the MO coefficients may be mildly complicated by a difference in Cartesian coordinate systems. Foster and Boys took the x direction perpendicular to the molecular y–z plane. In the PCLOBE program the out-of-plane direction is chosen as z, and the atoms are located in the x–y plane. This is illustrated in orbital #7 which in PCLOBE output has nonzero coefficients only for the 2pz orbitals on C and O. In the orbitals reported by Foster and Boys, orbital #7 has nonzero amplitudes for the px basis functions on C and O. The choice is, of course, entirely a matter of convention, which has changed over the years.

The order of the virtual orbitals #9–#12 is the same in the two methods, though the orbital energies differ seriously. This has no impact on the ground state energy, which is defined entirely and only by the occupied MOs. And even though the virtual orbital energies are, according to Koopmans' theorem, to be interpreted as electron affinities, their energies cannot claim such accuracy that one should make much of disagreements. This is, in part, a consequence of the inflexibility of the basis. The SCF calculation that chooses occupied orbitals' form to minimize the system energy does not impart any meaning or value to virtual orbitals.

One-Electron Energy Levels

Orbital label	(PCLOBE)	STO-6G	Foster-Boys
A1 (Oxygen core)	−20.756914192746	−20.57476	−20.59
A1 (Carbon core)	−11.473757332662	−11.27949	−11.36
A1 (C-O sigma bond)	−1.344589661060	−1.34544	−1.370
A1 CH2 symmetric	−0.832659560808	−0.81414	−0.8363
B2 CH2 antisymmetric	−0.673507585901	−0.63832	−0.6768
A1 O sigma Lone Pair	−0.601856695616	−0.55091	−0.5701
B1 CO pi bond	−0.497583334056	−0.44779	−0.4698
B2 O pseudo-pi LP	−0.420825519200	−0.36143	−0.3829
B1 CO pi antibond	0.177111638238	0.27360	0.2488
A1 CH2 antibond	0.544086554928	0.62414	0.6410
B2 CH2 antibond	0.627636330169	0.72771	0.7615
A1 CO antibond	0.699311195798	0.89375	0.8303

The energies of occupied MOs are in fair agreement, predicting that the highest lying MO is the pseudo-pi lone pair on oxygen atom, with a Koopmans' ionization energy (KIE) of 0.42 atomic units (about 11.4 eV) and the next is the pi bond with a KIE of 0.50 atomic units (about 13.6 eV). Improving the basis to 6-31G(d) yields Koopmans' estimates of 0.54 and 0.45 atomic units, or 11.8 and 14.7 eV. The experimental values are 10.9 and 14.5 eV. The agreement is pleasing, but one must suspect that good fortune plays a role here. Much more sophisticated and difficult calculations are required to do substantially better.

References

1. Roothaan, C.C.J., *Rev. Mod. Phys.* 23:69, 1951; *Rev. Mod. Phys.* 32:179, 1960.
2. Koopmans, T., *Physica* 1:104, 1933.
3. Mulliken, R.S., *J. Chem. Phys.* 23:1833, 1955.
4. Pulay, P., *Mol. Phys.* 17:197, 1969.

5. Pople, J.A., Krishnan, R., Schlegel, H.B., and Binkley, J.S., *Int. J. Quantum Chem.* 14:545, 1978.
6. Ransil, B.J., *Rev. Mod. Phys.* 32:239, 1960.
7. Handy, N.C. and Schaefer, III H.F., *J. Chem. Phys.* 81:5031, 1984.
8. Yamaguchi, Y., Osamura, Y., Goddard, J.D., et al., *A New Dimension in Quantum Chemistry: Analytic Derivative Methods in Ab Initio Molecular Electronic Structure Theory*, Oxford University Press, Oxford, UK, 1994.
9. Whitten, J.L., *J. Chem. Phys.* 44:359, 1966.
10. Dunning, T., *J. Chem. Phys.* 53:2823, 1970; *J. Chem. Phys.* 55:716, 1971.
11. Wilson, A., van Mourik, T., and Dunning, Jr. T.H., *J. Mol. Struct. (Theochem)* 388:339, 1997.
12. Preuss, H. and Diercksen, G., *Int. J. Quantum Chem.* I:605, 1967.
13. Pulay, P., *J. Comput. Chem.* 3:556, 1982.
14. Foster, J.M. and Boys, S.F., *Rev. Mod. Phys.* 32:303, 1960.

7

Configuration Interaction and Potential Curves

To this point we have focused our attention on single-determinant wave functions. Such wave functions are of course approximate, but for the conventionally bound molecules we have studied, the single determinant is a reasonable starting point. There are, however, important chemical contexts when the single determinant is inadequate for our purposes. Whenever the correlation error of the HF method is quantitatively significant, or when the single determinant is not even qualitatively suited to a particular system, we cannot be content with the single-determinant approximation. It is possible to write a more flexible and competent wave function by mixing several—perhaps many—determinants. This is called configuration interaction.

Configuration Interaction in General

Configuration interaction (CI) employs a wave function of the general form

$$\Phi_{CI} = \sum_{K \geq 0} A_K \Delta_K$$

Δ_K is a single determinant (or in some formulations, a symmetry-adapted combination of determinants) and A_K are weighting coefficients. Orbitals which make up the determinants are either taken from SCF calculations or refined in the course of the variational optimization of Φ. The latter more elaborate calculation is called multiconfigurational SCF. As our H_2 example will demonstrate, CI is particularly valuable in defining a wave function flexible enough to describe a chemical system in the course of bond breaking, or in representing systems in general with low-lying excited states, that is, near-degeneracies. A simple system with substantial configuration mixing is in the Be atom, for which the $1s^2 2s^2$ and $1s^2 2p^2$ determinants are almost equal in energy. The energy increment associated with mixing near-degenerate states is called "nondynamical correlation energy," which is to be distinguished from the "dynamical correlation energy" increment associated with mixing of many high-energy states into the ground determinant. The terms allude to the characteristic scale of the correction. Dynamical correlation is associated with the details of short-range (i.e., short wavelength or high momentum) motion of electrons in close proximity.

The CI calculation is an example of the linear variational optimization of weighting coefficients; thus, a secular equation must be solved

$$|\langle \Delta_K|H|\Delta_L\rangle - E\langle \Delta_K|\Delta_L\rangle| = 0$$

The roots of the secular determinant are state energies. Generally only a few roots are required since chemistry deals with the ground state and a few excited states. Eigenvectors are obtained by solving the associated linear equations, as we have seen before.

In principle, the exact wave function can be expressed in a CI expansion. Its approach to the exact energy has proved to be frustratingly slow. Limited CI suffers another flaw which can be serious for description of potential surfaces. Incomplete CI fails the simple size-extensivity condition, which requires that the energy of N independent identical molecules be equal to $N \times$ the energy of a single such system. This is demonstrated elsewhere in this book in the discussion of the coupled cluster model, which does not suffer from this shortcoming.

Slater Determinant MO-CI

CI can use the RHF single determinant as the reference Δ_0, and generate excited states by substitution of occupied MOs by virtual MOs. Schaefer [1] writes simple but powerful equations for the matrix elements between determinants. If a single substitution is made $(p \rightarrow q)$ then

$$\langle \Delta|H|\Delta'\rangle = \langle p|h|q\rangle + \sum_{r \neq p}^{n} \left[\langle pr|qr\rangle - \langle pr|rq\rangle \right]$$

$$h = \frac{-\nabla^2}{2} - \sum_N \frac{Z_N}{R_N}$$

In Schaefer's notation the one-electron integrals are

$$I\langle p|q\rangle = \left\langle p \left| \frac{-\nabla^2}{2} - \sum_N \frac{Z_N}{R_N} \right| q \right\rangle$$

For a double substitution $(p \rightarrow q; r \rightarrow s)$ a particularly simple formula occurs

$$\langle \Delta|H|\Delta'\rangle = \langle pr|qs\rangle - \langle pr|sq\rangle$$

The simplicity is only apparent, because the "AO-to-MO" transformation is necessary.

$$\langle pq|rs \rangle = \sum_i \sum_j \sum_k \sum_l c_{ip}\, c_{jq}\, c_{kr}\, c_{ls}\, \langle ij|kl \rangle$$

This transformation of the two-electron interactions from the basis set representation to the molecular orbital basis is often called for and is inescapable in any form of CI calculation. A great deal of research has gone into making this transformation efficient; the simplest summation scales as N^8 (for N basis functions) but clever algorithmic analysis has reduced this to N^5—still formidable. Modern techniques to simplify the transform still further include resolution of the identity, use of auxiliary basis sets, and exploitation of the sparseness of the two-electron supermatrix [2]. The PCLOBE program does not incorporate such subtleties but does take advantage of permutation equivalences in the indices. This has the modest effect of reducing the scaling to N^7.

In CI studies it is almost always observed that a few configurations have substantial coefficients in the linear expansion. Often the coefficient of the RHF determinant is 0.90 or greater. However, the CI series is slowly convergent, and the number of configurations increases sharply with the degree of excitation. Important contributions to energy are made by single and double excitations. For this reason, many CI investigations truncate the series at this stage, as CID (doubles only) or CISD (singles are mixed with the doubles). The size consistency error of this incomplete CI can be estimated by a device of Langhoff and Davidson [3]. Pople et al. [4] add estimates of the energy terms arising from triple excitations in their quadratic (Q) CISD(T) method, to assure size-extensivity.

Full CI is out of reach for all but the smallest systems; the number of configurations in a system of N electrons generated at a given order of excitation K grows as the binomial coefficient. We can learn some things from the few systems which are amenable to full CI. Molecular hydrogen in the STO-3G basis has, according to Szabo and Ostlund [5], the exact electronic energy

$$E_{\text{exact}} = 2h_{\sigma\sigma} + J_{\sigma\sigma} + \Delta - \sqrt{\Delta^2 + K^2_{\sigma\sigma^*}}$$
$$2\Delta = 2(\varepsilon_{\sigma^*} - \varepsilon_\sigma) + J_{\sigma\sigma} + J_{\sigma^*\sigma^*} - 4J_{\sigma\sigma^*} + 2K_{\sigma\sigma^*}$$

Consider the limit as the interatomic distance grows. Then both h integrals become $E(H)$, i.e., the energy of H atom in the basis; all two-electron integrals approach a common value, and Δ approaches zero. The exact energy then approaches $2E(H)$, the proper limit.

The CI wave function has a well-defined S_z spin component quantum number since each determinant is an eigenfunction of that operator. However, the CI function is not automatically adapted into eigenfunctions of the total spin operator S^2. Schaefer [1] refers to an early paper by Johnson [6] which shows how to diagonalize the matrices of operators L^2 and S^2 in a basis of single determinants so that the determinants will be grouped into symmetry-adapted linear combinations of determinants called "configurations"—hence the name CI. Such symmetry-adaptation reduces the size of the CI matrix considerably, a major advantage. Schaefer cites a case for the CI treatment of the Ne atom by Viers et al. [7] in which the problem involving 8393 Slater determinants was reduced to 434 configurations of ^1S spin symmetry. Molecular symmetry, when present, may be included as well. The symmetry-adapted cluster CI method [8] makes more extensive use of point group symmetry than many codes.

Modules for CI calculations for ground states are available in most widely applicable codes, including GAUSSIAN, GAMESS, DALTON, and MOLPRO among others. These are often limited to CI with singles and doubles, but multiconfigurational SCF and complete active space codes allow for full CI for a subset of electrons within defined orbital subspaces. The MOLE-CULE-SWEDEN [9] CASSCF/MRI code and another program for full CI calculations by Knowles and Handy [10] have been used by Bauschlicher and his colleagues [11,12]. We will have occasion to mention this work below.

CI Without Canonical Orbitals

Determinants need not be formed from canonical RHF orbitals. Any set of orbitals could be used. In fact, one could carry out a CI calculation using the lowest occupied eigenfunctions of the core-Hamiltonian and include only (all) single excitations in the CI list to find the RHF result since, in principle, the Hartree–Fock scheme consists of annihilation of single excitations so as to satisfy Brillouin's theorem. This looks like an attractive way to guarantee SCF convergence until one starts to count the many single excitations required for that approach.

Pauling Valence Bond and CI

Pauling [13] defined the valence bond (VB) method, which starts from nonorthogonal valence shell hybrids directed toward bonded neighbors. The VB method has been held in high regard for a long time, has been the medium for fruitful qualitative description of bonding, and has the merit of properly describing bond dissociation [14]. The method has been refined by Shull [15] among others. The name derives from the bond orbitals formed from nonorthogonal valence shell hybrids. These hybrids are localized linear combinations of basis functions on a given atom. They are not orthogonal to hybrids on other atoms; indeed, the overlap was considered as the source of bonding. Bond functions are formed as products of local hybrids. In the nomenclature of Pauling, these bond functions are termed A, B, $C,\ldots$. We can write the Slater determinant as

$$\Psi_I = \frac{1}{\sqrt{N!}} \sum_P (-1)^P P[A(1)B(2)C(3)\cdots E(N)]$$

P is the permutation operator that generates all the orbital products which result from expansion of the Slater determinant. The two-electron integrals would reduce to the form we discussed previously if the bond functions were orthonormal. However, the simplifications we relied upon are no longer working in our favor. In evaluating the integral

$$\left\langle \prod_I F_I^\alpha(2i-1)\, F_I^\beta(2i) \left| r_{mn}^{-1} \right| \sum_P (-)^P P \prod_J G_J^\alpha(2j-1) G_I^\beta(2j) \right\rangle$$

we could previously count on the simplification that factors of the form

$$\langle H(h)|M(h)\rangle \cdots \langle K(k)M(m)|K(k)H(m)\rangle \cdots \langle R(r)|R(r)\rangle$$

arising from the swap of electrons m and h would vanish owing to the zero overlap between functions M and H. This is no longer the case; all $N!$ terms in the expansion of the determinant must be processed.*

Returning to the hydrogen problem, the bond functions can be written

$$\psi_{co} \propto \left| 1s_A^\alpha(1)1s_B^\beta(2) \right|$$

In this case the "hybrid" is particularly simple. The complication arises from the nonorthogonality of the one-electron functions. To increase the flexibility of the wave function, one may add ionic terms such as

$$\left| 1s_A^\alpha(1)1s_A^\beta(2) \right|$$

and the equivalent for nucleus B. This is a version of configuration inter-action. Unfortunately, configurations as well as the bond functions are nonorthogonal, and further complications ensue.

The computational burden of the VB method prevented its rigorous application until the 1960s. In a landmark study, Harrison and Allen [16] applied the VB method to the study of the low-lying states of the BH molecule. This was one of the first all-electron ab initio calculations to

---

*When the PCLOBE program is running the Boys–Reeves CI, the real time foreground output display pauses while the diagonal matrix elements are computed since all the electrons must be treated. The off-diagonal elements only require a few integrals; their computation is much faster.

show calculated potentials for excited states. Although calculations based on the VB method are rare, accurate results are reported even now [17].

Counting bond functions in VB theory is a challenge. As in other connections such as the many-body perturbation theory [18] and coupled cluster theory [19] where complex algebra can be replaced by diagrammatic techniques, there is also a diagrammatic formulation for the VB method. This is the Rumer method [20] of generating bond functions. In VB nomenclature a set of linearly independent VB configurations can be set up by arranging number labels for the basis functions in a ring (regardless of the molecular structure). This is most convenient for the minimum basis sets first used to define valence shell hybrid orbitals. Rumer showed that if arrows were used to connect pairs of labels, a canonical set of diagrams could be formed from those diagrams in which no bond lines crossed. These are linearly independent "bonded" configurations, termed "canonical." The canonical bonded diagrams look exactly like the Kekule and Dewar bond lines in the usual VB representation of the pi bonding in benzene. The Rumer diagrams generated in this way include only the neutral covalent diagrams. Ionic and nonbonding functions exist which could be included in a CI-VB calculation (Figure 7.1).

The diagrams have some numerical meaning and value beyond their use as an aid to enumeration, in that simply superposing two canonical diagrams can be related to the energy matrix element for the interaction of the two structures. Looking at the second row of diagrams from Pauling's paper we can see enclosed areas—"islands." The coulomb part J_{AB} of an energy integral involving two Rumer structures A and B, i.e.,

$$H_{AB} = \langle A|H|B \rangle = I_{AB} + J_{AB} + K_{AB}$$

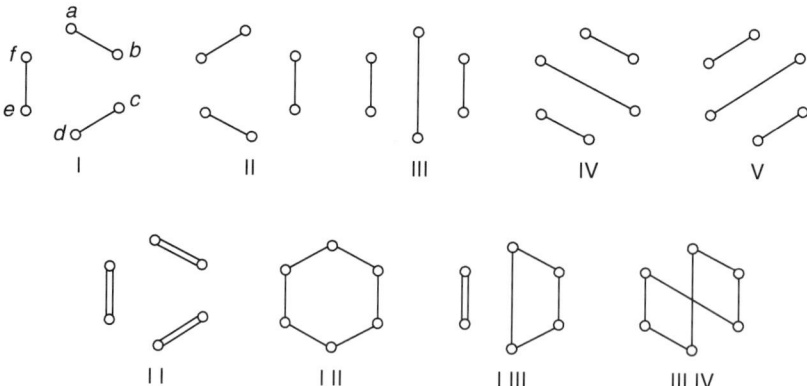

FIGURE 7.1
The five canonical valence-bond structures for a six-membered ring, as defined by Rumer. (From Pauling, L., *J. Chem. Phys.* 1, 280, 1933. With permission.)

has a coefficient of $(-1)^r [2^{-(n-i)}]$. Here "i" is the number of islands (rings) in the pattern and "r" is the number of reversals of "n" $\alpha \to \beta$ bond arrows required to make the islands (rings) alternate as $\alpha, \beta, \alpha, \beta$, etc. Pauling notes that when canonical structures are used, the Coulomb coefficient is even simpler, $[1/2^{(n-i)}]$, and in fact bond lines can be used without directional arrows provided only canonical structures are used. Allowing for permutation of orbitals within the Slater determinants, Pauling [13] then gives the coefficient for the exchange integral as $[2^{-(n-i)}]f(p)$ where $f(p) = -\frac{1}{2}$, $p = 0, f(p) = +1, p = 1, 3, 5, \ldots$, odd and $f(p) = -2$, $p = 2, 4, 6, \ldots$, even. The number p is the number of bonds in the superposition pattern along the path between the two orbitals which are interchanged with "i" as the number of islands (rings) in the superposition pattern and "n" as the number of bonds.

MOLE, the pioneering ab initio lobe basis program by Whitten and Allen, included a VB-CI program developed by Erdahl [21].

Boys–Reeves CI (MOVB)

Considering the determinantal-CI as the first method and the VB method as a second form of CI, we find a third form of CI developed by Boys and his students from 1956 to 1963 [22,23]. The Boys–Reeves algorithm [24] lent itself to efficient coding; a detailed explanation was provided by Cooper and McWeeny [25].

The Boys–Reeves algorithm has advantages of both MO-CI and VB-CI methods. First, configurations are automatically formed as eigenfunctions both of S^2 and S_z—a form called a "codetor" by Boys—which serves to reduce the scale of the problem. One slight limitation is that for a given spin the method only forms the codetor with the highest S_z component of the given multiplet. The second and major advantage of the Boys–Reeves algorithm is its use of orthogonal orbitals. All those nonzero overlap integrals that appeared as factors to the VB integrals are now Kroneker delta functions, δ_{ij}, and most are zero. Since the Boys–Reeves "projective reduction" [22,23] method uses orthogonal molecular orbitals within a VB algorithm we have called this the "MOVB" method [26].

Resolution of an MOVB-CI Wave Function into Leading Excitations

We will step through a Boys–Reeves analysis for our favorite molecule, formaldehyde. The molecular geometry is the result of optimization with RHF in the 6-31GL** basis, but the CI calculation used the STO-4G basis.

The Boys–Reeves CI report allows us to characterize the most important admixtures into the ground RHF configurations. In this example we will see that a pairwise $\pi \to \pi^*$ doubly excited codetor represents about 7.1% of the entire density, as deduced from the square of its coefficient. The third most important codetor in this case is negligible at only about 0.09%. The excitation reduces the net dipole moment and alters the strength of the CO bond; both these effects are corrections to the predictions derived from the ground state RHF determinant.

The output for an MOVB calculation on formaldehyde has many familiar elements, including a quote of the molecular geometry, the minimal basis of 12 AOs represented by lobe contractions, the ground state MOs, and their energies. All these quantities can be found elsewhere in this text, so we begin with the 20 codetors developed by removing electrons from the two highest energy occupied MOs (#7 and #8) and placing them in the lowest energy virtual MOs.

Automatic CAS List of SINGLET Codetors

Configuration Interaction HC = ESC

Top Spin-Orbital Codetor Labels

```
 1)  1  1  2  2  3  3  4  4  5  5  6  6  7  7   8   8  0  0  0  0  0  0  0  0  0  0
 2)  1  1  2  2  3  3  4  4  5  5  6  6  7  7   8   9  0  0  0  0  0  0  0  0  0  0
 3)  1  1  2  2  3  3  4  4  5  5  6  6  7  7   8  10  0  0  0  0  0  0  0  0  0  0
 4)  1  1  2  2  3  3  4  4  5  5  6  6  7  7   9  10  0  0  0  0  0  0  0  0  0  0
 5)  1  1  2  2  3  3  4  4  5  5  6  6  7  8   8   9  0  0  0  0  0  0  0  0  0  0
 6)  1  1  2  2  3  3  4  4  5  5  6  6  7  9   8  10  0  0  0  0  0  0  0  0  0  0
 7)  1  1  2  2  3  3  4  4  5  5  6  6  7  8   8  10  0  0  0  0  0  0  0  0  0  0
 8)  1  1  2  2  3  3  4  4  5  5  6  6  7  7   9   9  0  0  0  0  0  0  0  0  0  0
 9)  1  1  2  2  3  3  4  4  5  5  6  6  7  7  10  10  0  0  0  0  0  0  0  0  0  0
10)  1  1  2  2  3  3  4  4  5  5  6  6  8  8   9   9  0  0  0  0  0  0  0  0  0  0
11)  1  1  2  2  3  3  4  4  5  5  6  6  8  8   9  10  0  0  0  0  0  0  0  0  0  0
12)  1  1  2  2  3  3  4  4  5  5  6  6  8  8  10  10  0  0  0  0  0  0  0  0  0  0
13)  1  1  2  2  3  3  4  4  5  5  6  6  7  8   9   9  0  0  0  0  0  0  0  0  0  0
14)  1  1  2  2  3  3  4  4  5  5  6  6  7  9   9  10  0  0  0  0  0  0  0  0  0  0
15)  1  1  2  2  3  3  4  4  5  5  6  6  8  9   9  10  0  0  0  0  0  0  0  0  0  0
16)  1  1  2  2  3  3  4  4  5  5  6  6  9  9  10  10  0  0  0  0  0  0  0  0  0  0
17)  1  1  2  2  3  3  4  4  5  5  6  6  7  8  10  10  0  0  0  0  0  0  0  0  0  0
18)  1  1  2  2  3  3  4  4  5  5  6  6  7  9  10  10  0  0  0  0  0  0  0  0  0  0
19)  1  1  2  2  3  3  4  4  5  5  6  6  8  9  10  10  0  0  0  0  0  0  0  0  0  0
20)  1  1  2  2  3  3  4  4  5  5  6  6  7  8   9  10  0  0  0  0  0  0  0  0  0  0
```

Codetor 1 is the ground state; excitations are limited to removals from orbitals 7 and 8 and population of MOs 9 and 10. The excitations include single, pairwise double, split double, triples, and quadruples.

Remember Reeves said: To restrict ourselves to a linearly independent set of bonded functions . . . allocate the remaining left (alpha) and right (beta) brackets subject to the requirement that there must be more left brackets than right brackets to the left of every right bracket. (‼ ? ‼)

Configuration Overlap Matrix
(Off-diagonal 1.000 indicates a repeated Codetor)

The configuration overlap is the unit matrix.

H = Configuration Interaction Matrix
(Only upper left corner for large matrices.)

	1	2	3	4	5	6	7
1	-113.2114	0.0000	0.0000	0.0000	0.0000	0.0000	0.0000
2	0.0000	-113.0684	0.0000	0.0000	0.0000	0.0000	0.0000
3	0.0000	0.0000	-112.5243	0.0000	0.0000	-0.2466	0.0000
4	0.0000	0.0000	0.0000	-112.3118	0.0000	0.0000	0.0194
5	0.0000	0.0000	0.0000	0.0000	-112.9371	0.0000	0.0000
6	0.0000	0.0000	-0.2466	0.0000	0.0000	-111.9928	0.0000
7	0.0000	0.0000	0.0000	0.0194	0.0000	0.0000	-112.4315
8	0.0190	0.0000	0.0000	0.0000	0.0274	0.0000	0.0000
9	0.0461	0.0000	0.0000	0.0000	0.0000	0.0000	0.0000
10	0.1952	0.0000	0.0000	0.0000	0.0150	0.0000	0.0000
11	0.0000	0.0000	0.0000	0.0178	0.0000	0.0000	0.0885
12	0.0248	0.0000	0.0000	0.0000	0.0179	0.0000	0.0000
13	0.0000	0.0769	0.0000	0.0000	0.0000	0.0000	0.0000
14	0.0000	0.0000	0.0000	0.1900	0.0000	0.0000	0.0190
15	0.0000	0.0000	-0.0976	0.0000	0.0000	0.1270	0.0000
16	0.0000	0.0000	0.0000	0.0000	0.0000	0.0000	0.0000
17	0.0000	0.0253	0.0000	0.0000	0.0000	0.0000	0.0000
18	0.0000	0.0000	0.0000	0.0000	-0.0230	0.0000	0.0000
19	0.0000	0.0248	0.0000	0.0000	0.0000	0.0000	0.0000
20	0.0000	0.0000	0.1646	0.0000	0.0000	55.9895	0.0000

The ground state has symmetry A_1, as do the codetors 8, 9, 10, and 12. Its mixing with codetor 10, which is the MO 7 → MO 9 double excitation, is strongest. The effective CI for the ground state is 5×5 owing to the simplifications of symmetry. Codetor 2, the first singlet excitation (MO 8 → MO 9), mixes with 13, 17, and 19. The rest of the 20×20 matrix can be found in the full output for this calculation, in the supplementary CD.

State Energy Levels, Iteration No. 1

E(1) =	-113.26754957 au	Transition =	0.000 ev	lambda =	0.00 nm
E(2) =	-113.08607719 au	Transition =	4.938 ev	lambda =	251.08 nm
E(3) =	-112.94409747 au	Transition =	8.802 ev	lambda =	140.87 nm
E(4) =	-112.81439184 au	Transition =	12.331 ev	lambda =	100.55 nm
E(5) =	-112.72639065 au	Transition =	14.726 ev	lambda =	84.20 nm
E(6) =	-112.63727252 au	Transition =	17.151 ev	lambda =	72.29 nm
E(7) =	-112.50989718 au	Transition =	20.617 ev	lambda =	60.14 nm

Eigenvectors for Spin Multiplicity = 1 by Columns
(Only first 7 for large matrices.)

	1	2	3	4	5	6	7
1	0.9629	0.0000	-0.0251	0.0023	0.0000	0.0000	0.2662
2	0.0000	0.9764	0.0000	0.0000	0.2142	0.0000	0.0000

3	0.0000	0.0000	0.0000	0.0000	0.0000	0.9066	0.0000
4	0.0000	0.0000	0.0000	0.0000	0.0000	0.0000	0.0000
5	0.0146	0.0000	0.9759	0.2135	0.0000	0.0000	0.0374
6	0.0000	0.0000	0.0000	0.0000	0.0000	0.1776	0.0000
7	0.0000	0.0000	0.0000	0.0000	0.0000	0.0000	0.0000
8	−0.0287	0.0000	−0.2141	0.9722	0.0000	0.0000	0.0691
9	−0.0278	0.0000	0.0134	−0.0386	0.0000	0.0000	−0.0183
10	−0.2664	0.0000	−0.0152	−0.0807	0.0000	0.0000	0.9591
11	0.0000	0.0000	0.0000	0.0000	0.0000	0.0000	0.0000
12	−0.0081	0.0000	−0.0118	0.0004	0.0000	0.0000	−0.0391
13	0.0000	−0.2154	0.0000	0.0000	0.9754	0.0000	0.0000
14	0.0000	0.0000	0.0000	0.0000	0.0000	0.0000	0.0000
15	0.0000	0.0000	0.0000	0.0000	0.0000	0.1186	0.0000
16	0.0069	0.0000	0.0071	−0.0152	0.0000	0.0000	−0.0310
17	0.0000	−0.0139	0.0000	0.0000	−0.0425	0.0000	0.0000
18	−0.0039	0.0000	0.0214	−0.0321	0.0000	0.0000	−0.0178
19	0.0000	−0.0124	0.0000	0.0000	−0.0285	0.0000	0.0000
20	0.0000	0.0000	0.0000	0.0000	0.0000	−0.2858	0.0000

Density Matrix in M.O. Basis
Iteration No. 1

	1	2	3	4	5	6	7
1	2.0000	0.0000	0.0000	0.0000	0.0000	0.0000	0.0000
2	0.0000	2.0000	0.0000	0.0000	0.0000	0.0000	0.0000
3	0.0000	0.0000	2.0000	0.0000	0.0000	0.0000	0.0000
4	0.0000	0.0000	0.0000	2.0000	0.0000	0.0000	0.0000
5	0.0000	0.0000	0.0000	0.0000	2.0000	0.0000	0.0000
6	0.0000	0.0000	0.0000	0.0000	0.0000	2.0000	0.0000
7	0.0000	0.0000	0.0000	0.0000	0.0000	0.0000	1.8576
8	0.0000	0.0000	0.0000	0.0000	0.0000	0.0000	0.0000
9	0.0000	0.0000	0.0000	0.0000	0.0000	0.0000	−0.0071
10	0.0000	0.0000	0.0000	0.0000	0.0000	0.0000	0.0000
11	0.0000	0.0000	0.0000	0.0000	0.0000	0.0000	0.0000
12	0.0000	0.0000	0.0000	0.0000	0.0000	0.0000	0.0000

	8	9	10	11	12
1	0.0000	0.0000	0.0000	0.0000	0.0000
2	0.0000	0.0000	0.0000	0.0000	0.0000
3	0.0000	0.0000	0.0000	0.0000	0.0000
4	0.0000	0.0000	0.0000	0.0000	0.0000
5	0.0000	0.0000	0.0000	0.0000	0.0000
6	0.0000	0.0000	0.0000	0.0000	0.0000
7	0.0000	−0.0071	0.0000	0.0000	0.0000
8	1.9967	0.0000	0.0000	0.0000	0.0000
9	0.0000	0.1439	0.0000	0.0000	0.0000
10	0.0000	0.0000	0.0018	0.0000	0.0000
11	0.0000	0.0000	0.0000	0.0000	0.0000
12	0.0000	0.0000	0.0000	0.0000	0.0000

The density matrix would be diagonal if the MOs were "natural orbitals" (by definition). The important $7 \rightarrow 9$ $\pi \rightarrow \pi^*$ double excitation is the source of the population in MO 9 and the depletion of the population in MO 7.

The eigenvectors of the density matrix—called pseudo-natural orbitals since we do not have the exact density matrix—can be interpreted as mixtures of cores, lone pairs, and local bonds.

Pseudo-Natural Orbitals by Column

#	At-Orb		1	2	3	4	5	6	7	8
1	C	1s	0.001	−0.994	−0.117	0.181	0.000	0.026	0.000	0.000
2	C	2s	−0.010	−0.027	0.215	−0.650	0.000	−0.089	0.000	0.000
3	C	2px	0.000	0.000	0.000	0.000	0.568	0.000	0.000	0.253
4	C	2py	0.000	0.000	0.000	0.000	0.000	0.000	−0.622	0.000
5	C	2pz	−0.007	0.000	0.136	0.245	0.000	−0.475	0.000	0.000
6	H	1s	0.001	0.006	0.030	−0.214	0.220	0.126	0.000	0.351
7	H	1s	0.001	0.006	0.030	−0.214	−0.220	0.126	0.000	−0.351
8	O	1s	0.995	0.000	−0.226	−0.106	0.000	−0.107	0.000	0.000
9	O	2s	0.025	0.005	0.778	0.462	0.000	0.571	0.000	0.000
10	O	2px	0.000	0.000	0.000	0.000	0.479	0.000	0.000	−0.850
11	O	2py	0.000	0.000	0.000	0.000	0.000	0.000	−0.647	0.000
12	O	2pz	−0.006	0.001	−0.237	0.166	0.000	0.681	0.000	0.000

#	At-Orb		9	10	11	12
1	C	1s	0.000	−0.212	0.000	−0.088
2	C	2s	0.000	1.555	0.000	0.753
3	C	2px	0.000	0.000	1.224	0.000
4	C	2py	0.822	0.000	0.000	0.000
5	C	2pz	0.000	−0.396	0.000	1.336
6	H	1s	0.000	−1.032	−0.927	0.192
7	H	1s	0.000	−1.032	0.927	0.192
8	O	1s	0.000	0.048	0.000	0.142
9	O	2s	0.000	−0.323	0.000	−1.102
10	O	2px	0.000	0.000	−0.359	0.000
11	O	2py	−0.803	0.000	0.000	0.000
12	O	2pz	0.000	0.262	0.000	0.950

These pseudo-natural orbitals can serve as a basis for a second CI. From the density matrix defined by that CI, new estimates of the natural orbitals could be obtained. The capability for this calculation is built into PCLOBE. Exact natural orbitals allow most rapid convergence of the CI series, and it would seem that this iteration could allow approach to those valuable constructs. Bender and Davidson [27] explored this possibility; convergence difficulties in the iterative process have limited the value of this approach, but approximations to natural orbitals are still favored for large CI studies.

Dipole Moment Components in Debyes
Dx = 0.000000   Dy = 0.000000   Dz = −1.757097
Resultant Dipole Moment in Debyes =   1.757097

****

Computed Atom Charges
Q(1) = −0.148   Q(2) = 0.205   Q(3) = 0.205   Q(4) = −0.262   Q(

****

Total Overlap Populations by Atoms

	C	H	H	O
C	5.1908	0.3669	0.3669	0.2723
H	0.3669	0.4672	−0.0237	−0.0180

| H | 0.3669 | −0.0237 | 0.4672 | −0.0180 |
| O | 0.2723 | −0.0180 | −0.0180 | 7.9821 |

Wiberg-Mayer Bond Orders Between Atoms
(See Chem. Phys. Letters, v97, p270 (1983))

	C	H	H	O
C	8.6218	0.8883	0.8883	1.7275
H	0.8883	0.6279	0.0212	0.0454
H	0.8883	0.0212	0.6279	0.0454
O	1.7275	0.0454	0.0454	14.3480

* *

Orthonormal Resonance Analysis: Configuration No. 1 Contributes 92.7088 % to the Ground State, if Configuration Overlap is diagonal.

* *

Boys-Codetor:
1 1 2 2 3 3 4 4 5 5 6 6 7 7 8 8 0 0 0 0 0 0 0 0 0 0

* *

Dipole Moment Components in Debyes
Dx = 0.000000   Dy = 0.000000   Dz = 1.960109
Resultant Dipole Moment in Debyes = 1.960109

Note the large dipole of the leading term, the RHF ground state

* *

Computed Atom Charges
Q(1) = −0.190   Q(2) = 0.208   Q(3) = 0.208   Q(4) = −0.226   Q(

* *

	C	H	H	O
C	8.6081	0.8895	0.8895	1.9925
H	0.8895	0.6277	0.0215	0.0459
H	0.8895	0.0215	0.6277	0.0459
O	1.9925	0.0459	0.0459	14.3670

These are properties of the charge distribution of the dominant codetor. It has a conventional double bond $C = O$.

* *

Orthonormal Resonance Analysis: Configuration No. 10 Contributes 7.0974 % to the Ground State, if Configuration Overlap is diagonal.

* *

Boys-Codetor:
1 1 2 2 3 3 4 4 5 5 6 6 8 8 9 9 0 0 0 0 0 0 0 0 0 0

* *

This is the pairwise double excitation $7 \rightarrow 9$; this populates the π^* MO and alters the dipole moment. Its contribution of 7.1% is modest but detectable.

Dipole Moment Components in Debyes
Dx = 0.000000   Dy = 0.000000   Dz = 0.820829
Resultant Dipole Moment in Debyes = 0.820829

* *

Computed Atom Charges
Q(1) = −0.272   Q(2) = 0.208   Q(3) = 0.208   Q(4) = −0.143

* *

This configuration has a lower dipole moment owing to the shift of charge from the O atom to the CH_2 fragment.

```
****************************************************************
               Wiberg-Mayer Bond Orders Between Atoms
               (See Chem. Phys. Letters, v97, p270 (1983))

          C          H          H          O
  C     8.7728     0.8895     0.8895     1.9925
  H     0.8895     0.6277     0.0215     0.0459
  H     0.8895     0.0215     0.6277     0.0459
  O     1.9925     0.0459     0.0459    14.2023
****************************************************************
Orthonormal Resonance Analysis: Configuration No. 8 Contributes 0.0824 % to
the Ground State, if Configuration Overlap is diagonal.

****************************************************************
Boys-Codetor:
1 1 2 2 3 3 4 4 5 5 6 6 7 7 9 9 0 0 0 0 0 0 0 0 0 0
****************************************************************
```

This is the $8 \to 9\, n \to \pi^*$ double excitation. It makes such a small contribution to the ground state charge distribution that we will not discuss it further.

Sydnone CI

Sydnone is an unusual molecule. It appears in the Merck index, represented by the shorthand shown below.

which captures the large dipole moment, and conveys the notion that the ring may be aromatic. The remarkably large dipole moment of its methyl derivative suggests that the VB form at right (below) with the formal charges well separated must be a significant aspect of the bonding picture.

Our RHF/STO-4GL geometry and the subsequent 20×20 MOVB-CI calculation reproduces the large dipole moment. The alternating distances of the bond lengths in the ring argue against 6π ring aromaticity; the Wiberg–Mayer bond indices and bond lengths are consistent in suggesting a conventional carbonyl bond coupled to a CN double bond as shown above (left). Both leading configurations suggest a strong component of the cross-ring structure shown below. The easy photochemically induced extrusion of CO_2 helps make this seemingly unlikely aspect of the bonding more plausible.

CI and Potential Curves

Now we turn our attention to the task of tracing potential curves for chemical reactions. In certain cases the single determinant will still serve us well. If all species have closed-shell structures throughout the course of the reaction, then the single determinant is appropriate at all stages. Protonation of a base, association of ions, and most concerted cyclo-additions are examples of such amenable processes. It is also possible that single-determinant representations of open-shell systems can be effective throughout the course of dissociation. We will illustrate these by tracing the dissociation of ketene to CO and methylene, sketched in Figure 7.2.

The figure displays curves for the simplest possible treatment of ketene dissociation in its ground state singlet state and also the lowest triplet state. Anticipating a later discussion, we also include curves resulting from a much more advanced treatment, a coupled cluster expansion through single and double excitations. The simpler calculations employ the restricted (RHF) single-determinant wave function for the singlet and the unrestricted (UHF) single-determinant wave function for the triplet. The basis is a very small split-valence set called 3-21G. The crossing at about 1.7 Å for the simpler calculation confirms that the triplet methylene and singlet CO are the low-energy products while the singlet ketene is the stable reactant. The estimate of the energy difference between singlet and triplet products— about 1.5 eV or over 30 kcal/mol—is a serious overestimate of the splitting between singlet and triplet states of methylene [28], which suggests that there is an imbalance in the small basis RHF and UHF single-determinant descriptions of the two spin states. The coupled cluster calculation, using a superior but still modest 6-31G(d) basis, produces curves broadly similar in

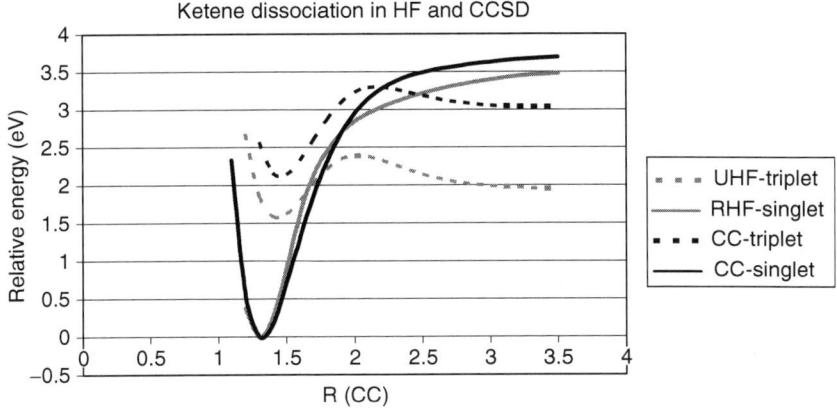

FIGURE 7.2
Representations of dissociation of ketene to fragments CH_2 and CO. The horizontal axis is the CC distance in angstroms. The vertical axis is energy in electron volts relative to the minimum energy for ketene in CCSD/6-31G(d) and RHF/6-31G(d). Dashed lines refer to the triplet state. The RHF and UHF descriptions of the dissociations are qualitatively correct, but correlation correction is quantitatively significant.

form, but in significant quantitative disagreement. Both calculations agree generally with the detailed CI studies of Allen and Schaefer [29] who showed that the C_{2v} symmetry-preserving dissociation of ketene is forbidden on the ground state surface, in the sense that one observes different orbital occupations in reactant and product. (Whereas the singlet molecule has 8 electrons in a_1 MOs, 4 in b_1 MOs, and 4 in b_2 MOs, the singlet products have 10, 2, and 4 electrons in such MOs, respectively.) In-plane distortion (CCO < 180°) does not help, but pyramidalization leads to allowed dissociation. The pyramidalization also allows the triplet A_1 state of the ketene to lead smoothly to the ground state triplet B_1 products. These authors locate the transition state for the ketene triplet dissociation at R(CC) about 2.07 Å, very close to our correlation-corrected value.

When bonds are broken and the number of electron pairs changes, the single-determinant representation is no longer of even qualitative value. Consider the dissociation of diatomic hydrogen:

$$\Psi_{gg} = \frac{1}{2} \left| \left(\frac{1s_A + 1s_B}{(1+S)^{1/2}} \right)^\alpha \left(\frac{1s_A + 1s_B}{(1+S)^{1/2}} \right)^\beta \right| = N_g^2 \left| 1s\sigma_g^\alpha(1) 1s\sigma_g^\beta(2) \right|$$

In the limit of large separation of the hydrogen atoms, the wave function approaches

$$\Psi_{\infty,gg} = \left| 1s_A^\alpha(1) 1s_A^\beta(2) + 1s_A^\alpha(1) 1s_B^\beta(2) + 1s_B^\alpha(1) 1s_A^\beta(2) + 1s_B^\alpha(1) 1s_B^\beta(2) \right|$$

Notice that the first and last terms place both electrons on a single nucleus. Symmetry requires that these ionic terms have equal weight with the covalent terms in which one electron is found on each nucleus. This mixture cannot be the proper dissociation limit, two neutral H atoms, and is an artifact of the single-determinant constraint. One can fix this problem by adopting a wave function which is flexible enough to vary the relative weighting of ionic and covalent terms. One form which accomplishes this is

$$\Phi_{CI} = \lambda \Psi_{gg} + (1 - \lambda^2)^{1/2} \Psi_{uu}$$

Here λ is a weighting coefficient which can be chosen variationally, and

$$\Psi_{uu} = \frac{1}{2} \left| \left(\frac{1s_A - 1s_B}{(1-S)^{1/2}} \right)^\alpha \left(\frac{1s_A - 1s_B}{(1-S)^{1/2}} \right)^\beta \right| = N_u^2 \left| 1s\sigma_u^\alpha(1) 1s\sigma_u^\beta(2) \right|$$

In the limit of large separations between the hydrogen nuclei

$$\Psi_{\infty,uu} = \frac{1}{2} \left| 1s_A^\alpha(1)1s_A^\beta(2) - 1s_A^\alpha(1)1s_B^\beta(2) - 1s_B^\alpha(1)1s_A^\beta(2) + 1s_B^\alpha(1)1s_B^\beta(2) \right|$$

The energies of the two limiting forms of Ψ are identical; i.e., the *gg* and *uu* states are nearly degenerate. Note that in this limit the covalent and ionic terms have opposite signs. Adding together the limiting forms, we see the undesirable ionic terms being canceled

$$\frac{1}{\sqrt{2}} \{ \Psi_{gg,\infty} + \Psi_{uu,\infty} \} = \frac{1}{\sqrt{2}} \left\{ \left| 1s_A^\alpha(1)1s_B^\beta(2) \right| + \left| 1s_B^\alpha(1)1s_A^\beta(2) \right| \right\}; \quad \lambda = \frac{1}{\sqrt{2}}$$

This is the proper dissociation limit, a special case of the two-determinant wave function. The second determinant is obtained by a double excitation from the reference RHF determinant, so this is an example of a CI(Doubles) or CID calculation.

Three Descriptions of Dissociation of the Hydrogen Molecule

Figure 7.3 shows the dissociation of hydrogen as described in RHF, MP2, and CID. RHF is unsuited to the task since it is not able to represent the dissociation limit of two neutral H atoms. MP2, a perturbative method correcting the single-determinant wave function a perturbative method which we will discuss later, adds double excitations in a way that is valid only for small mixing, and cannot represent the dissociation limit where the reference RHF determinant and the doubly excited determinant have equal weight. CID is able to describe the system properly over the whole range.

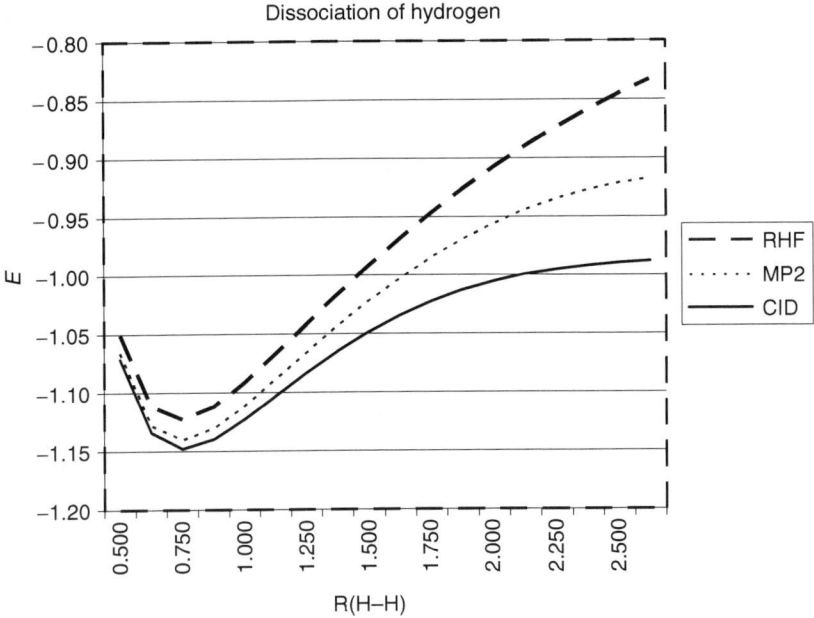

FIGURE 7.3
Dissociation curves for diatomic hydrogen. The horizontal axis is the H–H distance in angstroms. The vertical axis is energy in hartrees. Configuration interaction with double excitations (CID—solid line) produces the correct dissociation products, neutral H atoms. Restricted Hartree–Fock (RHF—dashed line) is constrained to produce a mixture of ionic and neutral products. The perturbative mixture of double excitations (MP2—dotted line) cannot correct the RHF wave function fully.

We mentioned that the dissociation curve for systems that form closed-shell fragments is well represented by RHF. For LiF in the 6-31G(d) basis, the RHF curve for dissociation to ions resembles the CID curve. The double excitations stabilize the system; the CID curve is lower than the RHF curve by about 5 eV regardless of interatomic distance (Figure 7.4).

MOVB-CI for BH

Harrison and Allen [16] described dissociation of the BH molecule, using the VB-CI program developed by Ehrdahl [21] and the state-of-the-art double-zeta quality lobe basis developed by Whitten [40]. They described the lowest four singlet and the lowest four triplet states of the system. Their full CI included all possible configurations in the nonorthogonal basis of bond functions. With their achievement as a standard for comparison, we will describe BH states using a modest 6-31GL** lobe basis with Boys–Reeves CASCI with 175 functions for singlets and only 62 for triplets. The size of the CI is defined by the number of spin symmetry–adapted

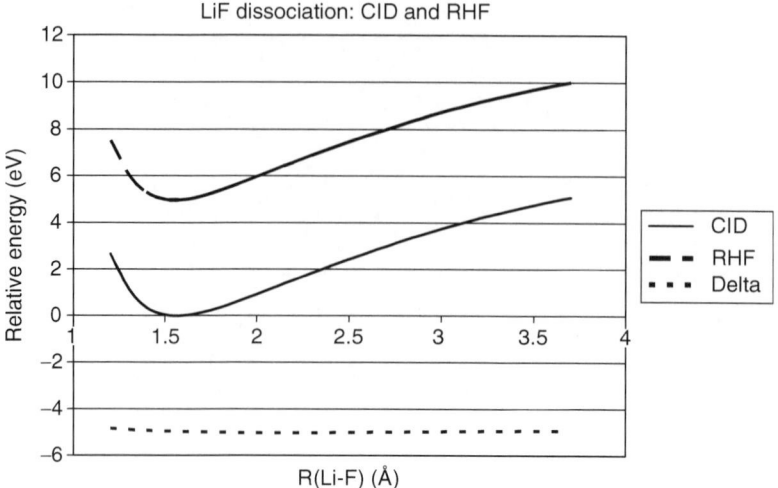

FIGURE 7.4

Dissociation of LiF. The horizontal axis is the Li-F distance in angstroms; the vertical axis is energy in electron volts relative to the minimum energy in CID/6-31G(d). The difference in RHF and CID energies (dashed line) is almost constant. Since the dissociation fragments Li cation and F anion are closed-shell species, RHF produces a qualitatively correct dissociation curve. CID lowers the energy but does not change the form.

excitations from the ground state singlet's determinant obtained by removing electrons from the four highest occupied MOs and populating the lowest energy virtual orbitals.

From Table 7.1 it appears that the VB-CI captures more correlation energy than does the MOVB-CI. The lower energy from the PCLOBE calculation may be due mainly to its better basis which gave a lower RHF energy. Note also that the full MP2 method (which includes double excitations from the B 1s core) captures more than twice the correlation energy than does either CI method. MP2 approximates "dynamical" correlation, while small CI calculations can capture only "nondynamical" correlation. The small CI

TABLE 7.1

Comparison of Energy Calculated for BH Singlet Ground State

Method	R eq (Å)	Energy (hartrees)	Correlation (hartrees)
RHF(lobe DZ) [16]	1.236	−25.0751	
RHF (6-31GL**)	1.256	−25.11865277	
VB-CI [16]	1.236	−25.1426	0.0675
MOVB-CI	1.256	−25.14635149	0.02769872
MP2 (frozen core)	1.256	−25.17803140	0.05937863
MP2 (incl. core)	1.256	−25.17979843	0.06114566
Corr.-SCF	1.256	−25.28181615	0.16316338
QMC [30]		−25.2879	

calculation, however, overcomes the limitations of the single-determinant starting point. Some slight differences can be noted between the 1969 "full" VB-CI and the CAS-MOVB-CI description of equilibrium BH. The bond length found with the Boys–Reeves CI is slightly longer (1.256 Å) than the value found by Harrison and Allen (1.236 Å). It is not clear if this difference is due to the use of (1) a different basis set, (2) a smaller number of configurations/codetors, or (3) the orthogonal molecular orbitals rather than the VB nonorthogonal valence shell hybrids. The energy comparison in Table 7.1 shows that the PCLOBE 6-31GL** basis leads to a lower RHF energy, but the full VB-CI captures more correlation energy than the MOVB-CI. While the 175 codetor CASCI treatment is complete within its space, the VB-CI calculations are said to be complete within the basis set.

The dissociation fragment B atom and H atom have ground configuration $1s^2 2s^2 2p^1$ and $1s^1$; or in state symbols, ^2P and ^2S. An excited state of B atom $1s^2 2s^1 2p^2$ can span $S(^{1,3}\Sigma)$ and $^{1,3}D(^{1,3}\Delta)$; these can couple with H to form doublet and quartet spins. Results of our small Boys–Reeves calculations are plotted in Figure 7.5. We have defined the energies in eV rather than hartrees and plotted the energy relative to the minimum value for the ground state singlet. The dashed lines are triplets and the solid lines are singlets.

Open-shell systems require special attention. In a small CI it is advantageous to choose virtual orbitals that mix effectively with the valence set.

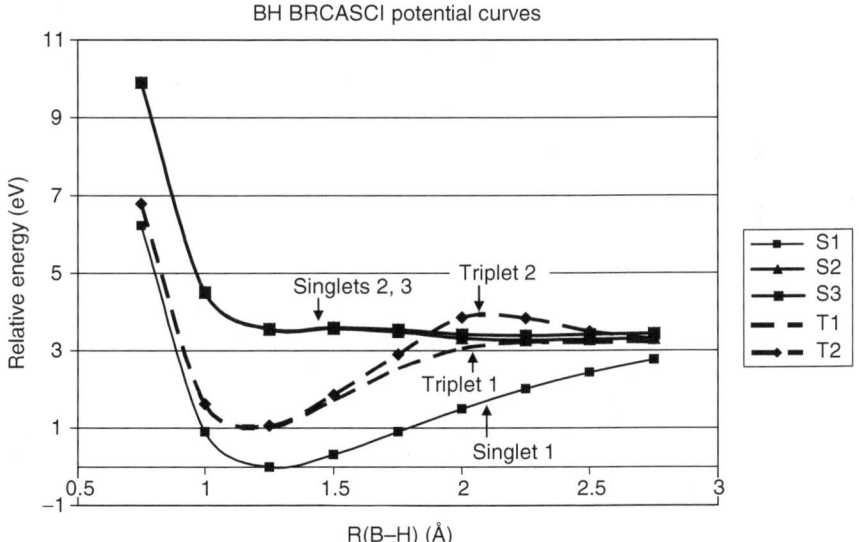

FIGURE 7.5
BH dissociation curves for low-lying singlet states (solid lines) and triplet states (dashed lines) computed by Boys–Reeves configuration interaction in a small active space and 6-31GL** basis. The horizontal axis is the B–H distance in angstroms; the vertical axis is energy in electron volts relative to the minimum energy of the ground state S1. All these states approach the dissociation products ^2S ($1s^1$) H atom and ^2P K$2s^2 2p^1$ B atom.

Virtual orbitals from SCF calculations are generally too diffuse to satisfy this condition. According to Koopmans' theorem, they describe electrons moving in the field of all N electrons in the system, whereas suitable virtual orbitals would describe the Nth electron in a field defined by the remaining $N - 1$ electrons. PCLOBE generates orbitals for open-shell systems by SCF computations on the cation resulting from removal of all unpaired electrons, and conducts the MOVB-CI in that MO basis.*

The results for the first few excited singlet potentials calculated using the MOVB-CI method are quite similar to those obtained by Harrison and Allen [16]. Notice that the $^3\Pi$ equilibrium bond length is shorter than the more stable $^1\Sigma^+$ state's equilibrium bond length, and these states approach a common asymptote and large B–H distances. The splitting of the $^3\Pi$ state at intermediate distances is attributed to small departures from perfect symmetry in the basis. We do not report triplet states above the degenerate $^1\Pi$ state in the MOVB-CI plot. MacDonald's theorem [31] states that the nth eigenvalue of the secular determinant is an upper bound for the nth exact level. In practice we find that the very lowest eigenvalue is the most accurate, while in a small space there can be serious error in the upper levels. The small CASCI space containing 62 states in the representation of the triplet here is adequate to describe only the low-lying states.

Unfortunately, the original Erdahl [21] VB-CI program is no longer available, but Harrison was able to use MOLPRO 2006, a multireference-CI program, to produce a modern set of potentials in less than a day [32]. His calculation uses a considerably larger basis and active space, plus all single and double excitations from each of the reference states. The PCLOBE MOVB-CI uses only the ground state as the reference state.

Harrison and Allen's calculations show a curious inflection in the $^3\Sigma^+$ potential (hinting at interaction with still higher levels) but produce smooth curves with proper dissociation limits for the lowest states (Figure 7.6).

One important feature present in both calculations is the small energy barrier to dissociation in the $^1\Pi$ state. This feature has been noticed by Bauschlicher and Langhoff [33] who estimate the height of the barrier to be 1.9 ± 0.2 kcal/mol, lessened by tunneling to about 0.8–0.9 kcal/mol. Other similar estimates have been given by Curtiss and Pople [34] and by Martin et al. [35,36]. The MOVB-CI yields an estimate of 2.04 kcal/mol, and the CASSCF (plus singles and doubles from the active space) calculation by Harrison gives a slightly higher value, about 3 kcal/mol. Thus, the results from the MOVB-CI seem to be in reasonable agreement with much more extensive calculations for the lowest state of the sigma and pi manifolds. But once again the hard lesson of using CI is that you need a lot of

---

* Kelly, H.P. (1968) private communication at the University of Virginia; although he used Hartree orbitals for the MBPT treatment of Be, he recommended using virtual orbitals of a cation to lower the energy of the virtual orbitals in an MBPT application based on Hartree–Fock orbitals.

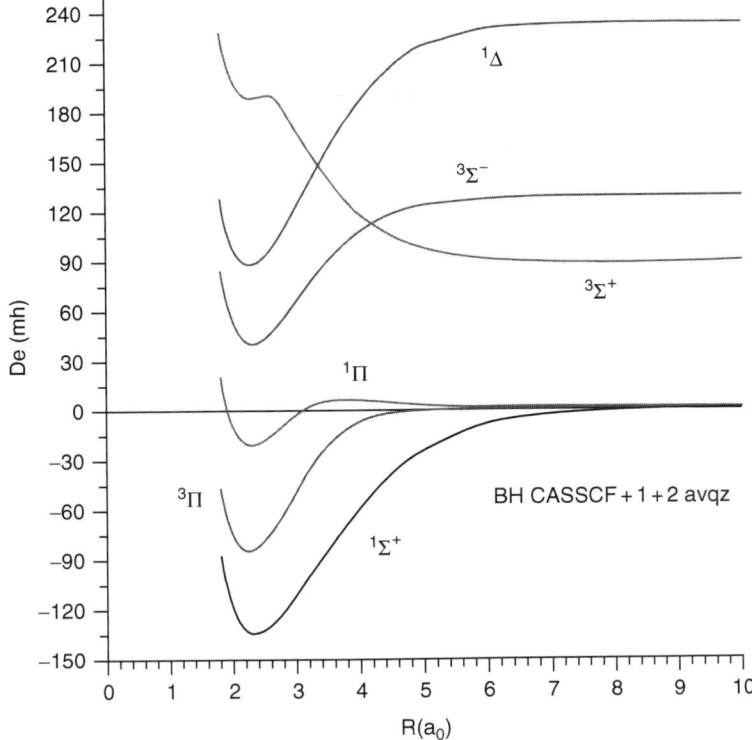

FIGURE 7.6

Harrison's representation of the low-lying states of BH. The horizontal axis is the B–H distance in bohrs; the vertical axis is energy in millihartrees relative to ground states of the B- and H-atom dissociation products. The computational model was a complete active space (all electrons) MCSCF plus a perturbative estimate of the effect of single and double excitations from the active space. Dunning's augmented correlation-consistent quadruple-zeta basis set was used. There is a general resemblance between the Boys–Reeves description and this very complete characterization. (From Harrison, J.F., Private communication; www.molpro.net, 2007. With permission.)

configurations and a full CI calculation is a very fine thing that is not easy to achieve.

MOVB-CI of Formaldehyde Dissociation

The Boys–Reeves MOVB-CI method allows us to describe bond breaking for compounds with double or triple bonds, since we can include excitations beyond pairwise doubles. We examine the singlet, triplet, and quintet states of formaldehyde as the molecule dissociates into ground state triplet CH_2 and triplet O atom. These open-shell systems are treated by MOVB-CI in a basis of MOs of the associated closed-shell cations.

TABLE 7.2

Comparison of Ground State Energy
for Formaldehyde

Method	Energy (hartrees)
RHF–SCF	−113.86771541
MOVB-CI	−113.89223071
MP2 (frozen core)	−114.17457191

Generating the codetor list by excitations from two occupied MOs into two vacant MOs would produce only one quintet

$$\varphi_a^2 \varphi_b^2 \rightarrow \varphi_a^\alpha \varphi_b^\alpha \varphi_m^\alpha \varphi_n^\alpha$$

To somewhat improve the calculation, we expand the active space to the highest four occupied MOs and the lowest four virtual MOs. Still, this gives us only 36 codetors for the quintets, 62 for the triplets, and 175 for the singlets. We must expect that our result will be biased in favor of the singlet manifold.

Table 7.2 displays several estimates of the energy of formaldehyde at the RHF/6-31GL(d) optimized geometry. Although the amount of correlation energy recovered using the MOVB-CI method is not impressive relative to other methods for the ground state, we can obtain information about the excited states. When we stretch the carbonyl bond to the breaking point we expect to get a triplet methylene fragment and a triplet oxygen atom for a combined spin state of four unpaired electrons.

In the ground state optimization, the CH bond length was found to be 1.0937 Å. This bond length was frozen while HCH bond angle was optimized for each point on the graph. At each R(CO) distance an MOVB-CI calculation was run for three different HCH angles and a parabola fitted to the three energies. This was followed by running the MOVB-CI calculation at the predicted angle for the minimum energy of that value of the R(CO) bond length. This is a simple form of optimization in the absence of an automated routine for optimization of excited state geometry (discussed elsewhere in this text). Throughout the process symmetry is maintained as C_{2v}.

Although the HCH angles are only optimized probably to the nearest degree, the energy reported is for the exact angle given in Table 7.3 and the R(CH) bond length of 1.0937 Å. With R(CH) = 1.0937 Å, the HCH angle of the dissociated methylene fragment is 132.1° after only a single pass through the parabolic optimization, and it is well known that the HCH angle in triplet methylene is due to a broad shallow potential [37,38] (Figure 7.7A).

At about 2.2 Å, the energy of the quintet drops below that of the singlet state while the energy of the singlet state is still increasing. This appears to be an artifact of the symmetry constraint.

If we relax the constraint of C_{2v} symmetry and extend the active space to 6 electrons in 8 MOs we obtain a more refined potential, displayed in Figure 7.7B.

TABLE 7.3

CASCI Energies for Dissociation of Formaldehyde
(6-31GL**), R(CH) = 1.0937 Å

R(CO) (Å)	No. of Codetors	Energy (a.u.)	HCH Angle (°)
Quintet			
1.00	36	−113.094801	42
			(H–H = 0.7839 Å)
1.1855	36	−113.332556	51
1.50	36	−113.397586	103.2
2.00	36	−113.603529	125.7
2.50	36	−113.658430	130.1
3.00	36	−113.651134	130.4
Triplet			
1.00	76	−113.566858	106.1
1.1855	76	−113.723411	122.3
1.50	76	−113.723173	131.1
2.00	76	−113.615135	118.0
2.50	76	−113.617447	105.0
3.00	76	−113.616734	105.0
Singlet			
1.00	196	−113.782616	110.0
1.185517	196	−113.867154	115.7 (RHF)
1.50	196	−113.807623	116.4
2.00	196	−113.661682	120.0
2.50	196	−113.626327	148.2
3.00	196	−113.621961	130.0
Energy of HCH as a triplet			
∞	66	−38.918745	130.5
Energy of O atom as a triplet			
∞	56	−74.789658	—
Energy of separate triplet fragments			
		−113.708403	—

MCSCF and CASSCF: Achieving Proper Dissociation for Larger Systems

CI with double excitations should be capable of describing any dissociation which consists in one bond pair being disrupted. For multiple-bonded systems, such as diatomic oxygen or nitrogen, further complications emerge owing to the larger number of pairs to be decoupled during dissociation and the open-shell character of the species produced.

$$O_2(^3\Sigma) \rightarrow 2O(^3P); \quad N_2 \rightarrow 2N(^5S)$$

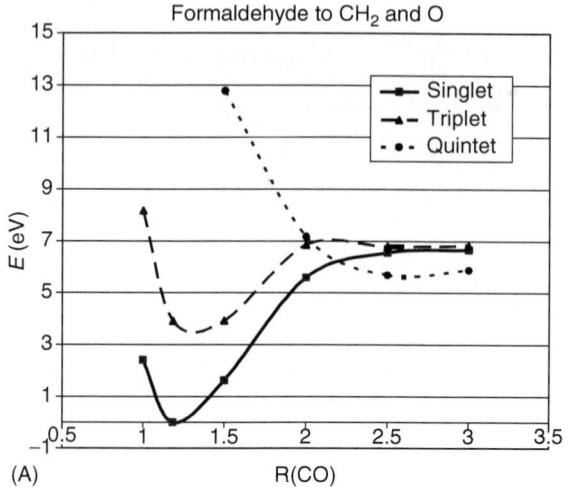

(A)

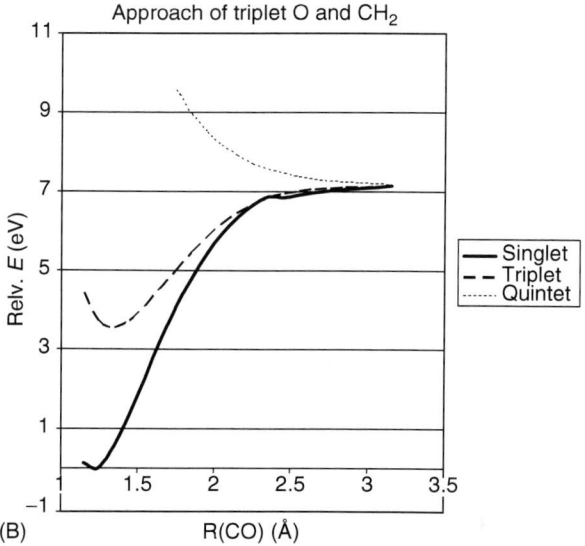

(B)

FIGURE 7.7

(A) Dissociation of formaldehyde to CH_2 and O atom, described by a small Boys–Reeves MO-CI calculation (see text). The horizontal axis is the C–O distance in angstroms; the vertical axis is energy in electron volts relative to the minimum energy of the singlet molecule (solid line). The dissociation products are triplets ^3P O and $^3B_1$ CH_2. These can couple to singlet, triplet (dashed line), or quintet (dotted line) states. The calculation shows that the quintet is unbound, while the singlet is more strongly bound than the triplet. C_{2v} symmetry was maintained throughout. (B) Dissociation of formaldehyde to oxygen atom and methylene, in CAS (10,8)/6-31G(d). The horizontal axis is the C–O distance in angstroms and the vertical axis is the energy in electron volts relative to the minimum energy of the singlet ground state. For each curve the geometry was optimized at each point. (The triplet is pyramidalized.) This active space is large enough to represent a smooth dissociation to triplet fragments. Once again the singlet molecule is most stable, the triplet is more weakly bound, and the quintet is unbound.

A complete CI within the subspace required to represent the reactants and products would afford a qualitatively correct dissociation of dissociation. A complete active space SCF calculation [39] begins with a full CI of the N electrons assigned to the active space of K orbitals.

$$\Psi_{CAS}(N,K) = \sum_{k=1}^{N} C_k \Delta(k \in K)$$

Here the index k refers to the number of electrons (up to N) involved in the excitation within the manifold of $K = 0$ orbitals of the active space. Both the weighting coefficients for the configurations and the LCAO coefficients are optimized variationally. For diatomic oxygen, a natural active set would be the 2p sigma and pi bonding and antibonding MOs which as the system dissociates can span the set of 2p orbitals on each atom. We choose to correlate 8 electrons in the 6-orbital space.

Near the molecular oxygen minimum energy distance, the CAS(6,8)/ 3-21G triplet wave function is obtained; the CI series is dominated by the lead term, and the electron assignment agrees with the aufbau for diatomics.

$$\Delta_0 = \left\| 1s\sigma^2 1s\sigma^{*2} 2s\sigma^2 2s\sigma^{*2} \left[2p\sigma^2 2p\pi^4 2p\pi_x^{\alpha 1} 2p\pi_y^{\alpha 1} 2p\sigma^0 \right] \right\|$$

In this expression the active space is bound by square brackets [].

EIGENVALUES AND EIGENVECTORS OF CI MATRIX
EIGENVALUE   −148.8774393171

```
(1)    0.9442967    (44)−0.2089558    (37)−0.1235637    (46) 0.1044523
(45)−0.1044523    (35) 0.0971592    (36)−0.0971592
(102) 0.0831349    (63)−0.0242890    (85)−0.0242890    (39)−0.0166395
(41) 0.0166395    (70)−0.0037225    (91)−0.0037225
```

Final one electron symbolic density matrix:

```
      1          2          3          4          5          6
1   1.92759
2  −0.00000    1.92759
3  −0.00000    0.00000    1.91439
4   0.00000    0.00000    0.00000    1.07120
5   0.00000    0.00000    0.00000    0.00000    1.07120
6   0.00000    0.00000    0.00000    0.00000    0.00000    0.08803
```

The wave function passes smoothly to two ^3P atoms as the limit at 5.0 Angstroms shown below:

Final one electron symbolic density matrix:

```
      1          2          3          4          5          6
1   1.50000
2   0.00000    1.50000
```

```
3 −0.00000   0.00000   1.00001
4   0.00001   0.00005   0.00000   1.50000
5   0.00005   0.00001   0.00000   0.00000   1.50000
6   0.00000   0.00000   0.00000   0.00000   0.00000   0.999991
```

	EIGENVALUE	−148.7840482520	

```
(39)  0.4330127      (41)−0.4330127       (1)  0.3535569      (44)−0.3535566
(37)−0.3535502      (102) 0.3535499      (36)−0.2500000
(35)  0.2500000       (20) 0.0000087       (6)−0.0000087      (84)  0.0000087
(62)−0.0000087       (16) 0.0000087       (5)−0.0000087
(90)  0.0000087       (66)−0.0000087      (74)  0.0000033      (10)  0.0000029
(12)  0.0000023       (11) 0.0000023      (78)  0.0000023
(79)  0.0000023        (8) 0.0000017
```

In order to represent the pair of separate O atoms at the dissociation limit, there must be equal population (or amplitude) in the bonding and anti-bonding MOs. The populations matrix shows this clearly, but the wave function form is less transparent.

Choosing a 10-orbital space in which to correlate 8 of the 12 valence electrons of formaldehyde, we obtain refined dissociation curves for the singlet, triplet, and quintet coupled states of (^{3}P) O atom and $^3B_1$ (see Figure 7.7B).

The quintet cannot bind; the singlet combination passes to the molecule's ground state. The triplet couple forms the weakly bound triplet state of the molecule. The optimization allowed symmetry-breaking and full geometry optimization.

It can generally be assured that a complete active space consisting of all the valence electrons and the equivalent bonding, nonbonding, and antibonding MOs would constitute an active space adequate to the description of any chemical process. The craft in selecting the electrons to be correlated and the space in which the description is expressed consists in balancing the requirement for flexibility (large active spaces) with feasibility (minimum active spaces).

Conclusions

While configuration interaction suffers from the lack of size extensiveness and slow convergence, it is still very useful for chemical research in several ways. A particular niche that remains for CI is in studying excited states and chemical transitions as long as a dissociation process is limited to the reaction coordinate near a transition state. While full-CI results are obtainable for very small molecules, it would seem that CI is not a method suited to the study of very large molecules. However, we have shown that it may be qualitatively useful to employ small CI lists which are complete to model bonding in molecules that have unusual bonding as from what would be ionic resonance structures in Pauling VB theory.

We wish to acknowledge useful discussions with Professor Jim Harrison and the calculations he carried out on BH. Comments from Dr Charlie Bauschlicher have also been quite helpful.

References

1. Schaefer, H.F. III, *The Electronic Structure of Atoms and Molecules: A Survey of Rigorous Quantum Mechanical Results*, Addison-Wesley, Reading, MA, 1972, p. 30.
2. Jung, Y., Shao, Y., and Head-Gordon, M., *J. Comp. Chem.* 28:1953, 2007.
3. Langhoff, S.R. and Davidson, E.R. *Int. J. Quantum Chem.* 8:61, 1974.
4. Pople, J.A., Head-Gordon, M., and Raghavachari, K., *J. Chem. Phys.* 87:5968, 1987.
5. Szabo, A. and Ostlund, N.S., *Modern Quantum Chemistry: Introduction to Advanced Electronic Structure Theory*, Macmillan, New York, NY, 1982.
6. Johnson, M.H., *Phys. Rev.* 39:197, 1932.
7. Viers, J.W., Harris, F.E., and Schaefer, H.F., *Phys. Rev. A* 1:24, 1970.
8. Toyota, K., Ehara, M., and Nakatsuji, H., *Chem. Phys. Lett.* 356:1, 2002 and *www.sbchem.kyoto-u.ac.jp/nakatsuji-lab*.
9. Almlöf, J., Bauschlicher, C.W., Blomberg, M.R.A., Chong, D.P., Heiberg, A., Langhoff, S.R., Malmqvist, P.A., Rendell, A.P., Roos, B.O., Siegbahn, P.E.M., and Taylor, P.R., MOLECULE-SWEDEN described by Almölf, J. and Taylor, P.R., Modern Techniques in Computational Chemistry: MOTECC-91 ed Clementi E, 1991.
10. Knowles, P.J. and Handy, N.C., *Chem. Phys. Lett.* 111:315, 1984; Siegbahn, P.B.M., ibid. 109:417.
11. Bauschlicher, C.W., Langhoff, S.R., Taylor, P.R., and Partridge, H., *Chem. Phys. Lett.* 126:436, 1986.
12. Bauschlicher, C.W., Langhoff, S.R., Taylor, P.R., Handy, N.C., and Knowles, P.J., *J. Chem. Phys.* 85:1469, 1986.
13. Pauling, L., *J. Chem. Phys.* 1:280, 1933.
14. Newton, M.D., Boer, F.P., and Lipscomb, W.N., *J. Am. Chem. Soc.* 88:2353, 1966; ibid 88:2357.
15. Shull, H., *Int. J. Quantum Chem.* III:523, 1969.
16. Harrison, J.F. and Allen, L.C., *J. Mol. Spect.* 29:432, 1969.
17. Thorsteinsson, T. and Cooper, D.L., *Int. J. Quantum Chem.* 70:637, 1998.
18. Kelly, H.P., *Adv. Chem. Phys.* 14:129, 1969.
19. Bartlett, R.J. and Musial, M., *Rev. Mod. Phys.* 79:291, 2007.
20. Rumer, G., Teller, E., and Weyl, H. 1932 *Nachr. Ges. Wiss.* Göttingen, p. 499.
21. Erdahl, R.M., *Valence Bond Configuration Interaction and the Electronic Structure of Molecules*, PhD Thesis, Princeton University, 1965.
22. Boys, S.F., Cook, G.B., Reeves, C.M., and Shavitt, I., *Nature (London)* 178:1207, 1956.
23. Boys, S.F. and Cook, G.B., *Rev. Mod. Phys.* 32:285, 1960.
24. Reeves, C.M., *Comm. ACM* 9:276, 1966.
25. Cooper, I.L. and McWeeny, R., *J. Chem. Phys.* 45:226, 1966; ibid 45:3484.
26. Alston, P.V., Shillady, D.D., and Trindle, C., *J. Am. Chem. Soc.* 97:469, 1975.
27. Bender, C.F. and Davidson, E.R., *J. Phys. Chem.* 70:2675, 1966.
28. Leopold, D.G., Murray, K.K., and Lineberger, W.C., 81:1084, 1984.

29. Allen, W.D. and Schaefer, III, H.F., *J. Chem. Phys.* 89:329, 1988.
30. Lüchow, A. and Anderson, J.B., *J. Chem. Phys.* 105:7573, 1996.
31. MacDonald, J.K.L., *Phys. Rev.* 43:830, 1933.
32. Harrison, J.F., Private communication; www.molpro.net, 2007.
33. Bauschlicher, C.W. and Langhoff, S.R., *J. Chem. Phys.* 93:502, 1990.
34. Curtiss, L.A. and Pople, J.A., *J. Chem. Phys.* 90:2522, 1989.
35. Martin, J.M.L., Francois, J.P., and Gijbels, R., *Chem. Phys. Lett.* 163:387, 1989.
36. Martin, J.M.L., Francois, J.P., and Gijbels, R., *J. Chem. Phys.* 91:4425, 1989.
37. Bender, C.F., Schaefer, H.F., Franceschetti, D.R., and Allen, L.C., *Am. Chem. Soc.* 94:6888, 1972; Harrison, J.F. and Allen, L.C., *J. Am. Chem. Soc.* 91:807, 1969.
38. Saxe, P., Schaefer, H.F., and Handy, N.C., *J. Phys. Chem.* 85:745, 1981; Cole, S.J., Purvis, G.D., and Bartlett, R.J., *Chem. Phys. Lett.* 113:271, 1985.
39. Hegarty, D. and Robb, M.A., *Mol. Phys.* 38:1795, 1979.
40. Whitten, J., *J. Chem. Phys.* 39:349, 1963.

8

Perturbation Theory

It is often the case that we can consider properties of molecules to be defined by their response to a disturbance. For example, the dipole moment of a molecule is defined by a change in energy of the system as an electric field is applied. Much of spectroscopy is simply the study of responses of a molecular system to a periodic perturbation provided by external fields. Scattering experiments similarly can be considered the study of the effects of molecular collisions, momentary disturbances of the reference systems. Perturbation theory permits a systematic representation of the effects of small disturbances on a reference system, and is the natural language for many molecular phenomena.

First-Order Correction to a Nondegenerate Reference System

We may seek solution for $H\psi = E\psi$. Assume we can solve a similar reference system $H^0\psi_n^0 = E_n^0\psi_n^0$ for which the Hamiltonian H^0 closely resembles H.

$$H = H^{(0)} + \lambda V \text{ where } \lim_{\lambda \to 0} H = H^0$$

Here the term V is the perturbation. The key idea of perturbation theory is to represent the unknown wave function as the reference system's wave function with a series of corrections. The wave function is then

$$\psi_n = \psi_n^{(0)} + \lambda\psi_n^{(1)} + \lambda^2\psi_n^{(2)} + \lambda^3\psi_n^{(3)} + \cdots \text{ where again } \lim_{\lambda \to 0} \psi_n = \psi_n^{(0)}$$

Furthermore, the associated eigenvalues are each represented as the reference system's energies with a sequence of corrections.

$$E_n = E_n^{(0)} + \lambda E_n^{(1)} + \lambda^2 E_n^{(2)} + \lambda^3 E_n^{(3)} + \cdots$$

Note that, for instance, in $\lambda^3 E_n^{(3)}$ 3 is the power of λ and in $E_n^{(3)}$ 3 refers to the third-order correction to the nth energy value. Now if we insert these expressions into $H\psi_n - E_n\psi_n = 0$, we obtain a power series in the ordering parameter λ as follows:

$$(H^{(0)}\psi_n^{(0)} - E_n^{(0)}\psi_n^{(0)})\lambda^0 + (H^{(0)}\psi_n^{(1)} + V\psi_n^{(0)} - E_n^{(0)}\psi_n^{(1)} - E_n^{(1)}\psi_n^{(0)})\lambda$$
$$+ (H^{(0)}\psi_n^{(2)} + V\psi_n^{(1)} - E_n^{(0)}\psi_n^{(2)} - E_n^{(1)}\psi_n^{(1)} - E_n^{(2)}\psi_n^{(0)})\lambda^2 + \cdots = 0$$

We have included λ^0 as the factor of the first term to emphasize that this is the zeroth-order term. For the polynomial to vanish, it is required that the coefficients of each power of λ be zero, as follows:

$$(H^{(0)}\psi_n^{(0)} - E_n^{(0)}\psi_n^{(0)}) = 0$$
$$(H^{(0)}\psi_n^{(1)} + V\psi_n^{(0)} - E_n^{(0)}\psi_n^{(1)} - E_n^{(1)}\psi_n^{(0)}) = 0$$
$$(H^{(0)}\psi_n^{(2)} + V\psi_n^{(1)} - E_n^{(0)}\psi_n^{(2)} - E_n^{(1)}\psi_n^{(1)} - E_n^{(2)}\psi_n^{(0)}) = 0$$

The first equation merely reminds us that we have a soluble reference problem. Consider the equation for the first-order effect. It can be rearranged to

$$(H^{(0)} - E_n^{(0)})\psi_n^{(1)} = (E_n^{(1)} - V)\psi_n^{(0)}$$

The first-order correction to the wave function is necessarily orthogonal to the zeroth-order function when we represent the correction in the basis of unperturbed orthogonal eigenfunctions of the reference Hamiltonian excluding $\psi_n^{(0)}$.

$$\psi_n^{(1)} = \sum_{n \neq k} a_{nk}\psi_k^{(0)}$$

Inserting this form, we have

$$(H^{(0)} - E_n^{(0)}) \sum_{k \neq n} a_{nk}\psi_k^{(0)} = (E_n^{(1)} - V)\psi_n^{(0)}$$

or, since $H^{(0)}\psi_k^{(0)} = E_k^{(0)}\psi_k^{(0)}$

$$\sum_{k \neq n} (E_k^{(0)} - E_n^{(0)})a_{nk}\psi_k^{(0)} = (E_n^{(1)} - V)\psi_n^{(0)}$$

Now we can multiply the equation by $\langle \psi_n^{(0)}|$ from the left and integrate. Then, since $\langle \psi_n^{(0)*}|\psi_k^{(0)}\rangle = \delta_{nk}$ the left-hand side vanishes, and

$$\sum_{k \neq n} (E_k^{(0)} - E_n^{(0)})a_{nk}\langle \psi_n^{(0)}|\psi_k^{(0)}\rangle = 0 = E_n^{(1)} - \langle \psi_n^{(0)}|V|\psi_n^{(0)}\rangle$$

That is, the first-order correction to the energy is simply given by

$$E_n^{(1)} = \langle \psi_n^{(0)} | V | \psi_n^{(0)} \rangle$$

In order to find the expression for the first-order correction to the wave function $\psi_n^{(1)}$ we multiply the first-order equation by $\langle \psi_k^0 |$ from the left and integrate. The orthonormality allows simplification to

$$a_{nk}(E_k^{(0)} - E_n^{(0)}) = -\langle \psi_k^{(0)} | V | \psi_n^{(0)} \rangle$$

or

$$a_{nk} = -\frac{\langle \psi_k^{(0)} | V | \psi_n^{(0)} \rangle}{E_k^{(0)} - E_n^{(0)}}, \quad k \neq n$$

To summarize the case for perturbation to first order with nondegenerate energy levels we have

$$H = H^{(0)} + V, \quad \psi_n = \psi_n^{(0)} + \psi_n^{(1)} \quad \text{and} \quad E_n = E_n^{(0)} + E_n^{(1)}$$

$$E_n^{(1)} = \langle \psi_n^{(0)} | V | \psi_n^{(0)} \rangle \text{ and } \psi_n^{(1)} = -\sum_{k \neq n} \frac{\langle \psi_k^{(0)*} | V | \psi_n^{(0)} \rangle}{(E_k^{(0)} - E_n^{(0)})}$$

We remind ourselves that state n is being corrected.

Second-Order Correction—Nondegenerate Case

We can recover the second-order energy correction to the energy from the third equation in the series, equating terms with quadratic λ dependence.

$$H^{(0)}\psi_n^{(2)} + V\psi_n^{(1)} - E_n^{(0)}\psi_k^{(2)} - E_n^{(1)}\psi_n^{(1)} - E_n^{(2)}\psi_n^{(0)} = 0$$

Multiply at left by $\langle \psi_n^{(0)} |$ and recognize again that all corrections to the nth-order wave function are orthogonal to the zeroth-order approximation. Then, from

$$\langle \psi_n^{(0)} | H^{(0)} | \psi_n^{(2)} \rangle + \langle \psi_n^{(0)} | V | \psi_n^{(1)} \rangle = E_n^{(0)} \langle \psi_n^{(0)} | \psi_n^{(2)} \rangle + E_n^{(1)} \langle \psi_n^{(0)} | \psi_n^{(1)} \rangle + E_n^{(2)}$$

all that remains is

$$E_n^{(2)} = \langle \psi_n^{(0)} | V | \psi_n^{(1)} \rangle = -\sum_{k \neq n} \frac{\langle \psi_n^{(0)} | V | \psi_k^{(0)} \rangle \langle \psi_k^{(0)} | V | \psi_n^{(0)} \rangle}{E_k^{(0)} - E_n^{(0)}}$$

This shows that the perturbation represented in the basis set of zeroth-order wave functions is all that is needed to construct not only the first-order correction to the wave function, but also the first- and second-order corrections to the energy.

The size of the second-order energy correction is defined by the extent of admixture of zeroth-order eigenfunctions under a perturbation. This depends on the interaction integral of the perturbation and the energy gap between states. Substantial mixing occurs when states of similar energy are strongly coupled by the perturbation.

Example: Cubic and quartic perturbations of the harmonic oscillator

Recall our treatment of the harmonic oscillator: in one dimension

$$\left(-\frac{\hbar^2}{2m} \frac{\partial^2}{\partial x^2} + \frac{kx^2}{2} \right) \psi = E\psi$$

Recall that $k = m\omega_0^2$ and divide both sides of the equation by the unit of energy, $\hbar\omega_0/2$. Then,

$$\frac{-\hbar\omega_0}{m} \frac{\partial^2}{\partial x^2} \psi + \frac{mx^2}{\hbar\omega_0} \psi = \varepsilon\psi \text{ or } -\frac{\partial^2}{\partial q^2}\psi + q^2\psi = \varepsilon\psi$$

introduces the dimensionless length q, defined by

$$q^2 = \frac{mx^2}{\hbar\omega_0} = \frac{mx^2}{h\nu}$$

Now assume the perturbation has the form

$$V(q) = Bq^3 + Cq^4$$

The unperturbed energy states are $\varepsilon_n = 2n + 1$ and the key integral is

$$\langle n|q|n+1 \rangle = [(n+1)/2]^{1/2}$$

The matrix of these integrals serves as a representation of variable q in the unperturbed basis. We encountered this idea in the discussion of representations and matrices; consulting that section would be helpful. In principle, this is a matrix of infinite dimensions because there are an infinite number of energy levels. We can generate matrix representations of the powers of q by raising this matrix to the desired power. Practically we are required to use a truncated form of the matrix. This approximation is accurate as long as the matrix is large enough so that it includes states well above the energy levels of interest. Here we are mainly interested in the $0 \rightarrow 1$ transition; thus, it is possible that a matrix representation up to $n = 8$ is adequate for this application.

The matrix representing $\langle n|x^2|m \rangle$ has diagonal elements which are $(2n + 1)/2$; thus, the ground state has half of the unit of energy, or one quarter of the quantum. This is a consequence of equipartition which divides the total energy equally between kinetic and a harmonic potential. The total ground state energy of half a quantum has one quarter of the quantum as potential, one quarter of a quantum as kinetic.

The values in the matrices can be used in estimating first- and second-order perturbations. For the first-order corrections we need only the diagonal elements. The diagonal elements of the cubic term vanish; thus, no first-order correction to the energy can arise from that odd power of q. The quartic term in the perturbation contributes in first order. For the ground vibrational state the first-order correction is 0.75 and for the first excited vibrational state the first-order correction is 3.75, in scaled units of half the fundamental quantum. The first excited state explores more of the steep quartic potential, so it is more severely elevated. In first order, the vibrational transition is blue-shifted by $3C$.

In second order the cubic term makes a contribution to the energy of the state $n = 0$, through terms

$$\frac{\langle 0|Bq^3|1\rangle\langle 1|Bq^3|0\rangle}{E_0^{(0)} - E_1^{(0)}} + \frac{\langle 0|Bq^3|3\rangle\langle 3|Bq^3|0\rangle}{E_0^{(0)} - E_3^{(0)}} = -\frac{(3\sqrt{2}/4)^2}{2} - \frac{(\sqrt{3}/2)^2}{6} = -\frac{11}{16}B^2$$

and the quartic term has the second-order effect

$$\frac{\langle 0|Cq^4|2\rangle\langle 2|Cq^4|0\rangle}{E_0^{(0)} - E_2^{(0)}} + \frac{\langle 0|Cq^4|4\rangle\langle 4|Cq^4|0\rangle}{E_0^{(0)} - E_4^{(0)}} = -\frac{(3\sqrt{2}/2)^2}{4} - \frac{(\sqrt{6}/2)^2}{8} = -\frac{21}{16}C^2$$

The analogous expressions for the $n = 1$ state are

$$\frac{\langle 1|Bq^3|0\rangle\langle 0|Bq^3|1\rangle}{E_1^{(0)} - E_0^{(0)}} + \frac{\langle 1|Bq^3|2\rangle\langle 2|Bq^3|1\rangle}{E_1^{(0)} - E_2^{(0)}} + \frac{\langle 1|Bq^3|4\rangle\langle 4|Bq^3|1\rangle}{E_1^{(0)} - E_4^{(0)}}$$

$$= +\frac{(3\sqrt{2}/4)^2}{2} - \frac{(3)^2}{2} - \frac{(\sqrt{3})^2}{6} = -\frac{71}{16}B^2$$

and

$$\frac{\langle 1|Cq^4|3\rangle\langle 3|Cq^4|1\rangle}{E_1^{(0)} - E_3^{(0)}} + \frac{\langle 1|Cq^4|5\rangle\langle 5|Cq^4|1\rangle}{E_1^{(0)} - E_5^{(0)}} = -\frac{(5\sqrt{6}/2)^2}{4} - \frac{(\sqrt{30}/2)^2}{8} = -\frac{165}{16}C^2$$

The energy difference between the two states becomes

$$E_1 - E_0 = 2 + \left(\frac{15}{4} - \frac{3}{4}\right)C + \left(-\frac{71}{16} + \frac{11}{16}\right)B^2 + \left(-\frac{165}{16} + \frac{21}{16}\right)C^2$$

$$= 2 + 3C - \frac{15}{4}B^2 - 9C^2$$

Both second-order terms red-shift the transition.

The Morse Potential

$$V_{\text{Morse}} = D_e[1 - \exp(-a[r - r_0])]^2 = D_e[1 - \exp(-bx)]^2;$$
$$b = ar_0, \ x = \frac{r - r_0}{r_0}$$

is commonly used to model anharmonicity. The Schroedinger equation is exactly soluble for the Morse potential [1], with energy levels (in wave numbers)

$$G_n = \bar{\nu}(n + 1/2) - \chi_e\bar{\nu}(n + 1/2)^2$$

The fundamental frequency and the anharmonicity parameters are related to the Morse constants by

$$2D_e a^2 = \mu\omega_0^2 \ \text{ and } \ \chi_e = \frac{\hbar\omega_0}{4D_e}$$

Here $\omega_0 = hc\bar{\nu}$ is the fundamental frequency and μ is the reduced mass for the effective particle. In our example, HCl, the reduced mass is very nearly the unit of atomic mass and r_0 is about 1.13 Å [2]. The fundamental frequency is 2990 cm^{-1}, D_0 is 4.43 eV, D_e is about 4.60 eV, and the anharmonicity parameter χ_e is about 0.0838. The anharmonicity shifts the first absorption frequency to the red by about 104 cm^{-1}.

The Morse potential can be approximated by a polynomial form if we construct the Taylor series in the scaled distance x.

$$V_{\text{Morse}} = D_e\left[b^2x^2 - b^3x^3 + \frac{7}{12}b^4x^4 + \cdots\right]$$

The scaling is similar to what we have already seen; dividing by half the quantum

$$V_{\text{scaled}} = \left(\frac{2D_e b^2}{\hbar\omega_0}\right)\left[x^2 - bx^3 + \frac{7}{12}b^2x^4 + \cdots\right] = q^2 + Cq^3 + Dq^4$$

The scaled variable is defined by

$$q^2 = \frac{2D_e b^2 x^2}{\hbar \omega_0} = \alpha^2 b^2 x^2$$

and the parameters of the potential are defined by

$$B = -b \left(\frac{2D_e b^2}{\hbar \omega_0} \right)^{-1/2} = -\alpha; \quad C = \frac{7b^2}{12} \left(\frac{2D_e b^2}{\hbar \omega_0} \right)^{-1} = \frac{7\alpha^2}{12}$$

The perturbation expansion is likely to be useful when the perturbation parameter, α, is small, i.e., when the dissociation energy is much larger than the vibrational quantum.

For HCl, the parameter $\alpha \sim 0.2$; thus,

$$B = -0.200; \quad C = +0.023$$

The first transition becomes

$$E_1 - E_0 = 2 + 3C - \frac{15}{4}B^2 - 9C^2 \sim 2 + 0.069 - 0.15 - 0.05$$
$$\sim 1.87 \text{ scaled units}$$

The estimated Morse transition is

$$\Delta E_{0 \to 1} \approx 1.87 \left(\frac{2990}{2} \right) \text{cm}^{-1} = 2796 \text{ cm}^{-1}$$

This is quite accurate.

Most appreciations of anharmonicity in polyatomic molecules begin with the harmonic oscillator reference point, since the quadratic form of the potential allows separation into terms for each of the normal modes

$$V = \frac{1}{2} \sum_{ij} \left(\frac{\partial^2 V}{\partial x_i \partial x_j} \right)_0 x_i x_j = \sum K_{mm} \left(\frac{\partial^2 V}{\partial q_m^2} \right)_0 q_m^2$$

and separation of the vibrational Schroedinger equation into independent equations, one for each normal mode. Then, if one can define perturbation terms for each mode and neglect mode–mode coupling, one needs only to solve a sequence of single-mode corrections as defined above.

The Degenerate Case

It will often happen that the reference system is degenerate: i.e., some sets of solutions will have identical energies:

$$H^{(0)}\psi_{n\alpha}^{(0)} = E_n^{(0)}\psi_{n\alpha}^{(0)}$$

Then, the perturbation expansion must take a slightly more elaborate form; the perturbed system's Hamiltonian has the familiar form

$$H = H^{(0)} + \lambda V$$

but the perturbed wave function must reflect the substantial mixing which takes place within the originally degenerate set upon application of the perturbation. To first order

$$\psi_{n\beta} = \sum_{\alpha} C_{\alpha\beta}^n \psi_{n\alpha}^{(0)} + \lambda \sum_{m'} a_{nm'} \psi_{m'}^{(0)} + \cdots$$

Here the index m' refers to all unperturbed functions outside the degenerate set with common energy $E_n^{(0)}$. Constructing the Schroedinger equation, we find the zeroth- and first-order equations as

$$H^{(0)} \sum_{\alpha} C_{\alpha\beta}^n \psi_{n\alpha}^{(0)} = E_n^{(0)} \sum_{\alpha} C_{\alpha\beta}^n \psi_{n\alpha}^{(0)}$$

$$V \sum_{\alpha} C_{\alpha\beta}^n \psi_{n\alpha}^{(0)} + H^{(0)} \sum_{m'} a_{nm'} \psi_{m'}^{(0)} = E_n^{(1)} \sum_{\alpha} C_{\alpha\beta}^n \psi_{n\alpha}^{(0)} + E_n^{(0)} \sum_{m'} a_{nm'} \psi_{m'}^{(0)}$$

$$V \sum_{m'} a_{nm'} \psi_{m'}^{(0)} + H^{(0)} \psi_n^{(2)} = E_n^{(2)} \sum_{\alpha} C_{\alpha\beta}^n \psi_{n\alpha}^{(0)} + E_n^{(1)} \sum_{m'} a_{nm'} \psi_{m'}^{(0)} + E_n^{(0)} \psi_n^{(2)}$$

The first of these is solved in the reference problem. The second equation can be simplified by multiplication at left by $\langle \psi_{n\gamma}^{(0)} |$ which upon integration yields

$$\sum_{\alpha} C_{\alpha\beta}^n \langle \psi_{n\gamma}^{(0)} | V | \psi_{n\alpha}^{(0)} \rangle = E_n^{(1)} \sum_{\alpha} C_{\alpha\beta}^n \langle \psi_{n\gamma}^{(0)} | \psi_{n\alpha}^{(0)} \rangle$$

This is a linear equation which we have met before; definition of the roots requires solution of the secular determinant

$$\| V_{\gamma\alpha} - E_n^{(1)} S_{\gamma\alpha} \| = 0$$

Since the basis of unperturbed functions within the degenerate set can always be defined as orthonormal, the task is simply to diagonalize the representation of the perturbation.

Recovery of the coefficients $a_{nm'}$ can be accomplished by multiplication of the first-order equation by $\langle \psi_{k'}^{(0)} |$. Integration yields

$$\langle \psi_{k'}^{(0)} | V \sum_{\alpha} C_{\alpha\beta}^n | \psi_{n\alpha}^{(0)} \rangle + E_{k'}^{(0)} a_{nk'} = E_n^{(0)} a_{nk'}$$

Rearranging this equation yields the expression for the mixing coefficients for states outside the degenerate set.

$$a_{nk'} = \sum \frac{C_{\alpha\beta}^n \langle \psi_{k'}^{(0)} | V | \psi_{n\alpha}^{(0)} \rangle}{E_n^{(0)} - E_{k'}^{(0)}}$$

This is very similar to the result for the nondegenerate case. The difference is that the perturbation has effected a combination (sometimes called "polarization") within the degenerate set before mixing of states outside that set can occur.

The second-order energy correction can be recovered by multiplying the second-order equation by one of the polarized solutions to the secular equation above and integrating

$$\sum_{\alpha'} C^n_{\alpha'\beta} \sum_{m'} a_{nm'} \langle \psi^{(0)}_{n\alpha'} | V | \psi^{(0)}_{m'} \rangle = E^{(2)}_n \sum_{\alpha,\alpha'} C^n_{\alpha\beta} C^n_{\alpha'\beta} \langle \psi^{(0)}_{n\alpha'} | \psi^{(0)}_{n\alpha} \rangle$$

Orthogonality between the polarized functions and those outside the degenerate set with label n has already removed several terms. Now recalling that the polarized functions are normalized and inserting the expressions for the coefficients a_{nm}

$$E^{(2)}_n = \sum_{\alpha} C^n_{\alpha\beta} C^n_{\alpha\beta} \frac{\langle \psi^{(0)}_{n\alpha} | V | \psi^{(0)}_{k'} \rangle \langle \psi^{(0)}_{k'} | V | \psi^{(0)}_{n\alpha} \rangle}{E^{(0)}_{k'} - E^{(0)}_n}$$

A simple example of a degenerate reference system is provided by the bending modes of a linear molecule. The displacements of the central atom in the x and y directions, which are symmetry equivalent, define the potential

$$V = \frac{1}{2} k(x^2 + y^2) \equiv \frac{1}{2} k r^2$$

The energy levels are

$$E(n, m) = \hbar \omega_0 (n + m + 1)$$

The degeneracy is considerable: for energy of n quanta the index is n.

Total energy in quanta	1	2	3	4
States	0,0	1,0; 0,1	2,0; 1,1; 0,2	3,0; 2,1; 1,2; 0,3

A perturbation proportional to a coordinate product Axy has selection rules for each of the two vibrational modes; $\Delta n = \pm 1$ and $\Delta m = \pm 1$. Individual integrals for scaled functions are

$$\langle j | q | j + 1 \rangle = [(j + 1)/2]^{1/2}$$

The set of degenerate states with energy 2, subjected to this kind of potential, is described by the matrix

$$\begin{bmatrix} \langle 1,0|Axy|1,0 \rangle - \lambda & \langle 1,0|Axy|0,1 \rangle \\ \langle 0,1|Axy|1,0 \rangle & \langle 0,1|Axy\langle 0,1| - \lambda \end{bmatrix} = \begin{bmatrix} -\lambda & C \\ C & -\lambda \end{bmatrix} = 0$$

The roots are $\pm A$ and the combinations are

$$(|1,0\rangle \pm |0,1\rangle)/\sqrt{2}$$

The selection rules limit mixing of $\{1,0; 0,1\}$ to the members 2,1 and 1,2 of the degenerate state with energy 4. The second-order energy effect is

$$E_{2j}^{(2)} = -\frac{\langle 1,0|Ayx|2,1\rangle\langle 2,1|Axy|1.0\rangle}{2} - \frac{\langle 0,1|Ayx|1,2\rangle\langle 2,1|Axy|1.0\rangle}{2}$$

The value of the second-order stabilization is $-A^2$.

Perturbation Theory in Approximate MO Theory

Perturbation theory is invaluable in chemical spectroscopy since the field effects are typically small, but is also valuable as a means of discussions of interactions and structures of molecules even when the central presumptions of the theory, that only small disturbances are present, cannot be defended. The argument is based on some simple outcomes of perturbation theory already developed. The familiar results can be reexpressed as rules:

1. Unperturbed levels are mixed pairwise. The effect is largest for unperturbed levels that are close in energy.
2. The result of a pair mixing is that the lower energy level is further stabilized, while the higher energy level is further elevated (destabilized).
3. The less stable orbital is mixed with the more stable orbital in a bonding way, and vice versa.

Of course, it is often the case that energy levels are degenerate. Then, we consider linear combinations within a degenerate set and solve a secular determinant, as above. Once this polarization within the degenerate set is complete, the rules can still be applied.

Example 1: Electron donor substituent on ethylene

Ethylene has a pi bond between the carbons, and we might consider the effects of a substitution $-OCH_3$ on the pi bond. The oxygen pi lone pair is rather strongly bound, perhaps comparably stable to the pi electrons in the CC bond. The pi CC antibonding level is considerably higher in energy, so the major mixing will involve the O lone pair with the pi bonding level of ethylene. The mixing produces a stabilized combination with new bonding between O and its attachment point on ethylene, and also a destabilized combination antibonding between O and its attachment point on ethylene. The latter is the new, elevated HOMO.

PREDICTION 1
The HOMO is destabilized. Approximate calculation (PM3) produces values of the HOMO energy of -10.64 eV before substitution, -9.51 eV after substitution.

PREDICTION 2
There is little admixture of the CC pi antibonding level, and little polarization of the HOMO. Calculation bears this out.

Example 2: Electron withdrawing substituent on ethylene
Substitution with CN, which has a low-lying vacant pi antibonding MO, will involve mixing of ethylene pi and CN pi (bonding); the levels will split, and the waveforms will mix in and out of phases but there will be little net stabilization. Stabilization derives from mixing a filled level with an empty level. Here the CN pi antibonding level will play a role.

PREDICTION 1
The CN pi bonding level is below the CC bonding level (-13.67 vs. -10.55 eV); these two levels mix, stablizing the in-phase combination. The CN and CC pi antibonding levels, destablizing the out of phase combination (from 1.44 to 2.19 ev).

PREDICTION 2
The HO-LU gap is diminished (CC pi to CC pi* is 12 in ethylene, 10.69 in cyanoethylene; the prediction fails—contrary to expectation, the HOMO is depressed in value).

Example 3: Approach of ethylene to butadiene
Using the simplest Huckel orbitals, we find butadiene has levels at 1.618 ill-defined "β" units (an estimate of the energy representing the pi bonding between adjacent carbons is used throughout Huckel theory), 0.618, -0.618, and -1.618; ethylene has pi levels at 1.0 and -1.0 β. In ethylene the lower level is filled; in butadiene the two lower levels are filled, as shown in the figures of Chapter 1.

Stabilization of the whole system occurs when there is mixing between vacant levels on one fragment and filled levels on the other. There will be significant mixing between those levels close in energy. The LU (butadiene) and HO (ethylene), and the HO (butadiene) and LU (ethylene) are both separated by 1.618 β. The symmetries are compatible in each case and the occupied levels are stabilized by bonding interactions between the terminal carbons of butadiene and counterparts on ethylene. This analysis suggests that the formation of the six-membered ring should proceed smoothly.

MP2 as Perturbation Theory

MP2 theory begins with the Hartree–Fock result, and constructs a zeroth-order Hamiltonian as the sum of Fock operators.

$$H_0 = \sum_i F_i = \sum \left(h_i + \sum_j (J_{ij} - K_{ij}) \right) = \sum_i h_i + 2\overline{V}_{ee}$$

$$H_1 = H - H_0 = V_{ee} - 2\overline{V}_{ee}$$

Then, the first-order correction $<\Delta|V_{ee} - 2\overline{V}_{ee}|\Delta> = -\overline{V}_{ee}$ cancels the double-counting of the electron–electron repulsion. The first-order corrected energy is precisely the Hartree–Fock energy. The second-order correction is

$$E^{(2)} = \sum_{i<j} \sum_{a<b} \frac{<\Delta_0|H - H_0|\Delta_{ij}^{ab}><\Delta_{ij}^{ab}|H - H_0|\Delta_0>}{E_0 - E_{ij}^{ab}}$$

Here Δ_{ij}^{ab} is a doubly excited configuration; electrons are removed from orbitals i and j, and placed in virtual orbitals a and b. The denominator simplifies to $\varepsilon_i + \varepsilon_j - \varepsilon_a - \varepsilon_b$. The time-consuming step is the transformation of the two-electron integrals from atomic to molecular bases, but since not all integrals need to be transformed, this is not as demanding as it might be.

MP2 improves the energy considerably and typically captures about 80% or more of the correlation energy. It is not variational (it may produce ground state energies below the exact value) but is size-extensive. It is particularly important in strained systems.

A severe test of any method is the representation of the energies of atomization of a standard set of thermodynamically well-characterized molecules, the G2 set [8–10]. The error of a method is expressed in the mean absolute deviation. To get an idea of the value of MP2 calculations, we reproduce a segment of the table on page 147 of *Exploring Chemistry with Electronic Structure Methods* by Foresman and Frisch (Table 8.1).

MP2 removes a portion of the systematic error RHF suffers in its description of dissociation (see Chapter 7), and this is expressed in its superior performance in representation of fundamental vibrational frequencies. Typically one is obliged to scale frequencies computed by HF or MP2 methods, by factors of about 0.89 and 0.94, respectively. However, there is still scatter. As an illustration we reproduce a portion of Tables 7.3 and 7.4 in Hehre's "*A Guide to Molecular Mechanics and Quantum Chemical Calculations.*" These values are errors in estimates of shifts in frequencies relative to the value for a reference molecule, methylamine (Tables 8.2 and 8.3).

TABLE 8.1

Performance of Various Model Chemistries

Model Chemistry	MAD (kcal/mol) in Heats of Atomization
HF/STO-3G//HF/STO-3G	93
HF/3-21G//HF/3-21G	58
HF/6-31+G(d,p)//HF/6-31+G(d,p)	47
MP2/6-31+G(d,p)//MP2/6-31+G(d,p)	11

MAD is the mean absolute deviation of calculated heats of atomization from experimental values.

TABLE 8.2

CX Stretching in CH_3X Molecules

Method	MAD (wave numbers cm^{-1})
HF/STO-3G//HF/STO-3G	216
HF/3-21G//HF/3-21G	43
HF/6-31G(d)//HF/6-31G(d)	83
HF/6-311+G(d,p)//HF/6-311+G(d,p)	70
MP2/6-31G(d)//MP2/6-31G(d)	49
MP2/6-311+G(d,p)//MP2/6-311+G(d,p)	38

MAD is the mean absolute deviation of computed frequencies from experimental values.

TABLE 8.3

Methyl Symmetric Stretch in CH_3X Molecules

Method	MAD (wave numbers cm^{-1})
HF/STO-3G//HF/STO-3G	85
HF/3-21G//HF/3-21G	21
HF/6-31G(d)//HF/6-31G(d)	20
HF/6-311+G(d,p)//HF/6-311+G(d,p)	24
MP2/6-31G(d)//MP2/6-31G(d)	30
MP2/6-311+G(d,p)//MP2/6-311+G(d,p)	41

MAD is as defined for Table 8.2.

Time-Dependent Perturbation Theory

The time-dependent version of perturbation theory is an indispensable guide to spectroscopic events of all kinds. It also lends itself to any setting in which a part of the Hamiltonian is not constant, as for molecular motions including collisions and even reactions. The time-dependent Schroedinger equation is the starting point:

$$i\hbar \frac{\partial \Psi^0(r,t)}{\partial t} = H^0 \Psi^0$$

Here H^0 is the time-independent reference Hamiltonian. This equation can be separated

$$\Psi^0(r,t) = f(t)\psi^0(r)$$

and terms referring to the time can be isolated

$$\frac{1}{f}\frac{\partial f}{\partial t} = i\frac{H\psi^0}{\psi^0} = E^0$$

Assuming we are studying bound states, we can assign quantum numbers to E values.

$$\Psi_n^0(r,t) = \exp\left(-i\omega_n t\right)\psi_n^0(r); \quad \omega_n = E_n^0/\hbar$$

Consider that the reference Hamiltonian is augmented by a time-dependent perturbing term

$$H(r,t) = H^0 + \lambda V(r,t)$$

The symbol λ plays a formal role, defining "orders" as in the time-independent case already discussed. The Schroedinger equation becomes

$$-i\hbar \frac{\partial \Psi(r,t)}{\partial t} = (H^0 + V)\Psi$$

We write the time-dependent wave function as a sum of eigenfunctions of H^0

$$-i\hbar \frac{\partial}{\partial t} \sum_k C_k(t) \exp\left(i\omega_k t\right)\psi_k^0(r) = (H^0 + V) \sum_k C_k(t) \exp\left(i\omega_k t\right)\psi_k^0(r)$$

Conducting the time derivative we obtain

$$-i\hbar \sum_k \left[\frac{\partial}{\partial t} C_k(t)\right] \exp(i\omega_k t)\psi_k^0 + \sum_k \hbar\omega_k C_k(t) \exp(i\omega_k t)\psi_k^0$$
$$= \sum_k C_k(t) \exp(i\omega_k t) H^0 \psi_k^0 + V \sum_k C_k(t) \exp(i\omega_k t)\psi_k^0$$

The second term at left cancels the first term at right; thus,

$$-i\hbar \sum_k \left[\frac{\partial}{\partial t} C_k(t)\right] \exp(i\omega_k t)\psi_k^0 = V \sum_k C_k(t) \exp(i\omega_k t)\psi_k^0$$

Isolate a particular term in the sum by multiplying at left by

$$\psi_m^{0*} \exp(-i\omega_m t)$$

After integration over the space coordinates

$$-i\hbar \sum_k \left[\frac{\partial}{\partial t} C_k(t)\right] \langle\psi_m^0|\psi_k^0\rangle = \sum_k C_k(t) \exp(i[\omega_k - \omega_m]t)\langle\psi_m^0|V|\psi_k^0\rangle$$

Only the $k = m$ term survives at left, but V may mix other states with m. The equation can be written more compactly

$$-i\hbar\frac{\partial}{\partial t} C_m(t) = \sum_k C_k(t) \exp(i\omega_{km} t) V_{km}$$

An incompletely defined integrated form

$$-i\hbar \int dC_m = \sum_k \int C_k(t) \exp(i\omega_{km} t') V_{km} \, dt'$$

suggests that the perturbation is being subjected to a Fourier transform. That is, the Fourier component of V which matches the energy gap $E_k - E_n$ is being selected. The following initial conditions are convenient:

$$C_k(0) = 1 \text{ if } n = k; \quad 0 \text{ otherwise}$$

Then, for short times so that C_k is only slightly changed we have

$$-i\hbar C_{m\neq n}(t) = \int_0^t V_{nm}(t') \exp(i\omega_{nm} t') \, dt'$$

We might consider a very gradual application of a perturbation. If V is almost constant, the dominant Fourier component is a very low frequency and V will not be effective in populating any excited state m with ω_{mn} very different from zero. The energy of the original state may change, but the system will stay in the original state. This is the "adiabatic" limit.

The opposite extreme is the "sudden" limit. In this case, the perturbation is a pulse or sharp blow of duration $\Delta\tau$, which has Fourier components up to frequency $\omega_{\text{Pulse}} \sim 2\pi/\Delta\tau$. In the limit, the Fourier components are uniform and every state within the range of the pulse can be populated.

The most important perturbation is oscillatory with a well-defined frequency, ω_F; this has a single Fourier frequency, and the potential can populate only a state that lies at an energy $\hbar\omega_F$ away from the origin n. We will return to this case in the discussion of the interaction of light and matter.

Hamiltonian for Matter in an Electromagnetic Field

The force $\mathbf{F}$ on a particle with charge e moving with velocity $\mathbf{v}$ according to Maxwell's equations is

$$\mathbf{F} = e\left[\mathbf{E} + \frac{1}{c}(\mathbf{v} \times \mathbf{H})\right]$$

The electric field $\mathbf{E}$ and the magnetic field $\mathbf{H}$ are derived from a scalar field Φ and a vector field $\mathbf{A}$:

$$\mathbf{E} = -\frac{1}{c}\frac{\partial \mathbf{A}}{\partial t} - \nabla\Phi \quad \text{and} \quad \mathbf{H} = \nabla \times \mathbf{A}$$

In classical mechanics $\mathbf{F} = m\mathbf{a} = m\dfrac{d^2\mathbf{r}}{dt^2} = m\ddot{\mathbf{r}}$ so that we have

$$\mathbf{F} = e\left[\mathbf{E} + \frac{1}{c}(\mathbf{v} \times \mathbf{H})\right] = m\ddot{\mathbf{r}} = e\left[-\frac{1}{c}\frac{\partial \mathbf{A}}{\partial t} - \nabla\Phi + \frac{1}{c}(\mathbf{v} \times \nabla \times \mathbf{A})\right]$$

$$m\ddot{x} = -e\frac{\partial\Phi}{\partial x} - \left(\frac{e}{c}\right)\frac{\partial A_x}{\partial t} + \left(\frac{e}{c}\right)\left[\dot{y}\left(\frac{\partial A_y}{\partial z} - \frac{\partial A_z}{\partial y}\right) + \dot{z}\left(\frac{\partial A_z}{\partial x} - \frac{\partial A_x}{\partial z}\right)\right]$$

The equations for y and z can be written by cyclic permutation of (x, y, z).

The Lagrangian equations of motion recapture Newton's laws in a compact way

$$\frac{d}{dt}\left(\frac{\partial L}{\partial \dot{q}_j}\right) - \frac{\partial L}{\partial q_j} = 0; \quad q_j = x, y, z$$

The expression $L = T - V$ solves this equation:

$$L = \sum_j \frac{1}{2} m\dot{q}_j^2 - e\phi + \left(\frac{e}{c}\right)(\dot{\mathbf{r}} \cdot \mathbf{A})$$

The classical momentum is $p_j = \dfrac{\partial L}{\partial \dot{q}_j}$; the classical Hamiltonian is $H = \sum_j p_j \dot{q}_j - L$. Then,

$$H = \left(\frac{1}{2m}\right) \sum_{j=1}^{3} \left[p_j + \left(\frac{e}{c}\right) A_j \right]^2 + e\Phi$$

To obtain the quantum mechanical Hamiltonian, replace the momentum with the operator $p_j = -i\hbar \frac{\partial}{\partial q_j}$; for a charged particle in an electromagnetic field the result is

$$H = \left(\frac{1}{2m}\right) \left[-\hbar^2 \nabla^2 + i\hbar \left(\frac{e}{c}\right)(\nabla \cdot \mathbf{A}) + i\hbar \left(\frac{e}{c}\right)(\mathbf{A} \cdot \nabla) + \left(\frac{e^2}{c^2}\right)|A|^2 \right] + e\Phi$$

Using the Coulomb gauge ($\nabla \cdot \mathbf{A} = 0$ and $\Phi = 0$) we obtain

$$H = \left(\frac{1}{2m}\right) \left[-\hbar^2 \nabla^2 + i\hbar \left(\frac{e}{c}\right)(\mathbf{A} \cdot \nabla) + \left(\frac{e^2}{c^2}\right)|A|^2 \right]$$

Assuming $|A^2|$ is negligibly small we neglect the last term. The Schroedinger equation becomes

$$H|\psi\rangle \cong \sum_j \left[\left(\frac{-\hbar^2 \nabla_j^2}{2m_j}\right) + V + \left(\frac{i\hbar e_j}{m_j c}\right)(\vec{A}_j \cdot \nabla_j) \right] |\psi\rangle = i\hbar \left(\frac{\partial}{\partial t}\right)|\psi\rangle$$

Since the reference Hamiltonian is

$$H^0 = \sum_j \left(\frac{-\hbar^2 \nabla_j^2}{2m_j}\right) + V$$

The perturbation arising from the electromagnetic field is

$$V_R = \frac{i\hbar e}{mc} \sum_i (\mathbf{A}_j \cdot \nabla_j)$$

Time-Dependent Perturbation Theory for Charged Particles in the Electromagnetic Field

Following Penner [3] we can carry out a quantum mechanical treatment of the light absorption process. The Hamiltonian for a particle in a radiation field is

$$H_R = \left(\frac{1}{2m}\right)\left[-\hbar^2\nabla^2 + i\hbar\left(\frac{e}{c}\right)(\nabla\cdot\mathbf{A}) + i\hbar\left(\frac{e}{c}\right)(\mathbf{A}\cdot\nabla) + \left(\frac{e^2}{c^2}\right)|A|^2\right] + e\Phi$$

Again using the Coulomb gauge ($\nabla\cdot\mathbf{A}=0$ and $\Phi=0$) we obtain

$$H_R = \left(\frac{1}{2m}\right)\left[-\hbar^2\nabla^2 + i\hbar\left(\frac{e}{c}\right)(\mathbf{A}\cdot\nabla) + \left(\frac{e^2}{c^2}\right)|A|^2\right]$$

Assuming $|A^2|$ is negligibly small we neglect the last term as well. For a system of particles the reference Hamiltonian and the perturbation become

$$V_R = \sum_j \left(\frac{i\hbar e_j}{m_j c}\right)(\mathbf{A}\cdot\nabla) \text{ for all } j \text{ particles} \quad \text{and} \quad H^0 = \sum_j \left(\frac{-\hbar^2\nabla_j^2}{2m_j}\right) + V$$

Then,

$$\frac{dC_m(t)}{dt} = \left(\frac{1}{i\hbar}\right)\sum_{n\neq m} C_n(t)\exp(i\omega_{nm}t)\left\langle \psi_m^0 \left| \sum_j \frac{i\hbar e_j}{m_j c}(\mathbf{A}_j\cdot\nabla_j) \right| \psi_n^0\right\rangle$$

Canceling $(i\hbar)$ and assuming $\mathbf{A}$ is roughly constant over the dimensions of the molecule (compare the 3000 Å wavelength vs. 10 Å molecule size), we can factor the average $\mathbf{A}$ out of the integral to obtain

$$\frac{dC_m(t)}{dt}\bar{\mathbf{A}}\cdot\left[\sum_{n\neq m} C_n(t)\exp(i\omega_{nm}t)\left\langle \psi_m^0 \left| \sum_j \frac{e_j}{m_j c}\nabla_j \right| \psi_n^0\right\rangle\right]$$

Now let the vector potential of the light wave be written as an oscillation

$$\mathbf{A} = \mathbf{A}^0\cos(\omega t)$$

The differential equation for the amplitude becomes

$$\frac{dC_m(t)}{dt} = \{\exp[i(\omega_{nm}+\omega)t] + \exp[i(\omega_{nm}-\omega)t]\}\left(\frac{\mathbf{A}^0}{2}\right)\cdot\left\langle \psi_m^0 \left| \sum_j \frac{e_j}{m_j c}\nabla_j \right| \psi_n^0\right\rangle$$

The integrated form is

$$C_m(t) \cong \frac{\mathbf{A}^0}{2}\cdot\left\langle \psi_m^0 \left| \sum_j \left(\frac{e_j}{m_j c}\right)\nabla_j \right| \psi_n^0\right\rangle$$
$$\times \left\{\frac{(\exp[i(\omega_{nm}+\omega)t]-1)}{i(\omega_{nm}+\omega)} + \frac{(\exp[i(\omega_{nm}-\omega)]-1)}{i(\omega_{nm}-\omega)}\right\}$$

The second term will dominate if $\omega \sim \omega_{nm}$; then

$$C_m(t) \approx \frac{1}{2} \mathbf{A}^0 \cdot \left\langle \psi_m^0 \left| \sum_j \left(\frac{e_j}{m_j c} \right) \nabla_j \right| \psi_n^0 \right\rangle \left\{ \frac{(\exp\left[i(\omega_{nm} - \omega)t\right] - 1)}{i(\omega_{nm} - \omega)} \right\}$$

Next we can factor out

$$\exp\left[i(\omega_{nm} - \omega)t/2\right]$$

(noting the 1/2 factor in the exponent) and use the definition

$$\sin(x) = \left[\exp(ix) - \exp(-ix)\right]/2i$$

to obtain

$$C_m(t) \approx \frac{1}{2} \mathbf{A}^0 \cdot \left\langle \psi_m^0 \left| \sum_j \left(\frac{e_j}{m_j c} \right) \nabla_j \right| \psi_n^0 \right\rangle e^{i(\omega_{nm} - \omega)t/2}$$

$$\times \left\{ \frac{\exp\left[i(\omega_{nm} - \omega)t/2\right] - \exp\left[-i(\omega_{nm} - \omega)t/2\right]}{\left[(\omega_{nm} - \omega)t\right](i/t)} \right\}$$

Note the multiplication by $(t/t) = 1$ in the denominator. This allows the change of variable to

$$x = \left[(\omega_{nm} - \omega)t/2\right]$$

Then,

$$C_m(t) \approx \mathbf{A}^0 \cdot \left\langle \psi_m^0 \left| \sum_j \left(\frac{e_j}{m_j c} \right) \nabla_j \right| \psi_n^0 \right\rangle \exp\left[i(\omega_{nm} - \omega)t/2\right] \left\{ \frac{\sin(x)}{(x)} \right\} t$$

The probability that the system will be in the state m with energy E_m at time $t > 0$ is

$$|C_m(t)|^2 = C_m^*(t) C_m(t)$$

In this product the complex exponential becomes 1 so that we find

$$|C_m(t)|^2 \approx \left| \left\langle \psi_m^0 \left| \sum_j \left(\frac{e_j}{m_j c} \right) \nabla_j \right| \psi_n^0 \right\rangle \cdot \mathbf{A}^0 \right|^2 \left\{ \frac{\sin^2(x)}{x^2} \right\} t^2$$

This is the probability of population of a specific state m by the effect of electromagnetic radiation in the neighborhood of resonance (i.e., where $x = 0$). In practice, the radiation is not monchromatic and there are many states near resonance. The number of states is expressed by a density ρ (states per unit energy) multiplied by a measure of the breadth of the radiation. If the density of states varies slowly over this interval then the square of the matrix element can be replaced by the constant

$$|C_m(t)|^2 \approx \left| \left\langle \psi_m^0 \left| \sum_j \left(\frac{e_j}{m_j c} \right) \nabla_j \right| \psi_n^0 \right\rangle \cdot \mathbf{A}^0 \right|^2 \left\{ \frac{\sin^2(x)}{x^2} \right\} t^2$$

Now we must integrate

$$[\sin^2(x)/x^2]$$

which is sharply peaked near the resonant frequency. The function takes on the value 1 at $x = 0$, and reaches 0 when $x = \pm\pi$ (Figure 8.1).

After a certain elapsed time t the width of the form is

$$\Delta\omega = 2\pi/t$$

The area under the curve is approximately $4\pi/t$. After the integration over x we have

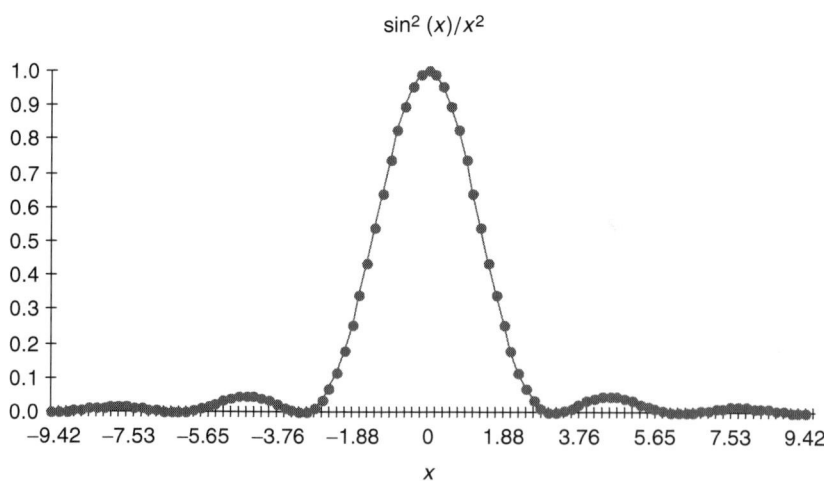

$$\sin^2(x)/x^2$$

FIGURE 8.1

This function appears in the definition of the intensity of absorption and the rate of population of the excited state in first-order time-dependent perturbation theory. The area under the curve depends on peak height (always unity) and the points where the form goes to zero (see text).

$$|C_m(t)|^2 \approx \left| \left\langle \psi_m^0 \left| \sum_j \left(\frac{e_j}{m_j c} \right) \nabla_j \right| \psi_n^0 \right\rangle \cdot \mathbf{A}^0 \right|^2 \frac{4\pi}{t} t^2$$

This is the probability of occupancy of initially vacant state m—or, alternatively, the fraction of population in state m—after elapsed time t. The rate of passage into state m from the initial state W can be expressed as the probability of occupancy divided by the elapsed time. Defining $dE = 2\hbar dx$, we obtain the probability per unit time as "W" where

$$W \approx \frac{|C_m(t)|^2}{t} \approx 4\pi \left| \left\langle \psi_m^0 \left| \sum_j \left(\frac{e_j}{m_j c} \right) \nabla_j \right| \psi_n^0 \right\rangle \cdot \mathbf{A}^0 \right|^2 \rho(\hbar \omega_{nm})$$

This is sometimes called "Fermi's golden rule" and is a well-known and widely used approximation.

One may wonder about the effect of the several approximations. Fortunately, Lick [4] has tested this approximation for one of the few exact solutions in quantum mechanics (a forced oscillator). The graph reproduced from the paper by Lick [4] shows that for short times (small epsilon) the time-dependent perturbation approximation is very good. Even after longer times the evolution of the state coefficients is close to the exact result (Figure 8.2).

Lick's analysis acknowledges the interdependence of the coefficients, while we have assumed in the derivation that each C_m is independent of the other coefficients. The system must after all maintain normalization; over all time we require

$$\sum_m |C_m(t)|^2 = 1$$

Length–Velocity Relationship

We see above that the probability per unit time of a transition occurring due to an electric dipole perturbation is proportional to the square of the dipole velocity matrix element. However, there are other equivalent forms of matrix elements. Here we trace the proof that a dipole length integral can replace the dipole velocity integral.

Consider the relationship between the dipole length and the dipole velocity operators. Given the Hamiltonian

$$H = \frac{-\hbar^2}{2m} \left[\frac{\partial^2}{\partial x^2} + \frac{\partial^2}{\partial y^2} + \frac{\partial^2}{\partial z^2} \right] + V(x, y, z)$$

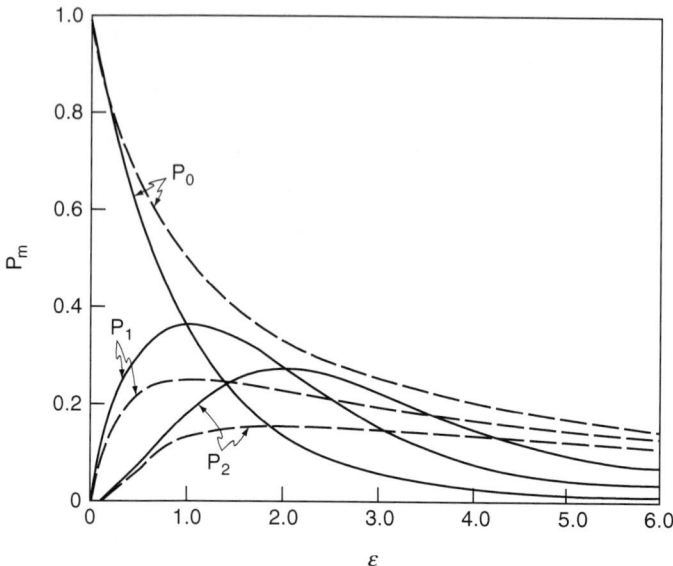

FIGURE 8.2
Lick's exact solution to the time-dependent equation for population of excited states of the forced oscillator [4]. The horizontal axis is the total energy scaled by the vibrational quantum; the vertical axis is the population of each state according to the exact solution of the coupled equations (solid line) and the solution for uncoupled equations (dashed line). The perturbative solutions are accurate only very early in the process. (From Lick, W., *J. Chem. Phys.*, 47, 2438, 1967. With permission.)

we compute its commutator with the x (position) operator.

$$[H, x]\Psi = (Hx - xH)\Psi = \frac{-\hbar^2}{2m} \left\{ \frac{\partial^2}{\partial x^2} x - x \frac{\partial^2}{\partial x^2} \right\} \Psi = \frac{-\hbar^2}{m} \frac{\partial \Psi}{\partial x}$$

Thus,

$$[H, x] = \frac{-\hbar^2}{m} \frac{\partial}{\partial x}$$

Therefore, the expectation value of the commutator is

$$\int \Psi^* (Hx - xH)\Psi \, d\tau = \frac{-\hbar^2}{2m} \int \Psi^* (Hx - xH)\Psi \, d\tau = \left(\frac{-\hbar^2}{m}\right) \int \Psi^* \left(\frac{\partial}{\partial x}\right) \Psi \, d\tau$$

Then, for exact functions

$$\int \Psi_i^* (xH) \Psi_j \, d\tau = E_j \int \Psi_i^* x \Psi_j \, d\tau$$

and using the turnover rule proved earlier we can write

$$\int \Psi_i^* (Hx) \Psi_j \, d\tau = E_i \int \Psi_i^* x \Psi_j \, d\tau$$

so that

$$(E_i - E_j) \int \Psi_i^* x \Psi_j \, d\tau = \left(\frac{-\hbar^2}{m} \right) \int \Psi_i^* \left(\frac{\partial}{\partial x} \right) \Psi_j \, d\tau$$

The left-hand side refers to a dipole length, while the right-hand side refers to the dipole velocity.

More explicitly

$$\int \Psi_i^* r \Psi_j \, d\tau = \left\{ \frac{(-\hbar^2/m)}{(E_i - E_j)} \right\} \int \Psi_i^* \nabla \Psi_j \, d\tau = \left(\frac{\hbar^2}{m} \right) \frac{\int \Psi_i^* \nabla \Psi_j \, d\tau}{(E_j - E_i)}$$

We note in passing that there is a third equivalent form called the dipole acceleration. In principle, if Ψ is an exact wave function and the energies are exact energies, all three of these forms of the transition moment would give the same result. However, for any less-than-exact wave function and energies the expressions will disagree. The results would tend to converge as the basis set increases in size.

Response Theory

Response theory deals directly with changes in a system induced by a perturbation, and expresses that change as a partial derivative. Jensen [5] collects some 23 electronic properties, each of which can be expressed as a partial derivative of the energy. In his notation, F is an electric field, B is a magnetic field, I is a nuclear magnetic moment, R is a coordinate, and P is a property.

$$P \propto \frac{\partial^{n_F + n_B + n_I + n_R} E}{\partial F^{n_F} \partial B^{n_B} \partial I^{n_I} \partial R^{n_R}}$$

The order of derivatives with respect to a particular agent is specified by the integers n_F, n_B, n_I, and n_R. Even an abbreviated form of Jensen's table [5] shows the range of this formulation (Table 8.4).

TABLE 8.4

Electronic Properties as Energy Derivatives

n_F	n_B	n_I	n_R	Property
0	0	0	0	Energy
1	0	0	0	Electric dipole moment
0	1	0	0	Magnetic dipole moment
1	1	0	0	Circular dichroism
0	0	0	2	Harmonic vibrational frequencies
0	0	0	3	Cubic anharmonic vibrational correction
2	1	0	0	Magnetic circular dichroism: Faraday effect
0	0	0	4	Quartic anharmonic vibrational correction
2	2	0	0	Cotton–Mouton effect

Source: From Jensen, F., *Introduction to Computational Chemistry*, Wiley, New York, 1999.

Many investigators have contributed to the growing influence and power of methods based on derivatives. We should salute the pioneering work of Pulay [6] who saw that computation of derivatives of the Hartree–Fock energy was feasible and effective, and the breakthrough by Handy and Schaefer [7], who showed that efficient evaluation of derivatives was possible for correlation-corrected energies as well. We will have more to say on this subject in later chapters.

References

1. Morse, P.M., *Phys. Rev.* 34:57, 1929.
2. McHale, J.L., *Molecular Spectroscopy*, Pearson Education–Prentice Hall, Upper Saddle River, NJ, 1999.
3. Penner, S.S., *Quantitative Spectroscopy and Gas Emissivities*, Addison-Wesley, London, pp. 115–176, 1959.
4. Lick, W., *J. Chem. Phys.* 47:2438, 1967.
5. Jensen, F., *Introduction to Computational Chemistry*, Wiley, New York, p. 239, 1999.
6. Pulay, P., *Mol. Phys.* 17:197, 1969; *Mol. Phys.* 18:473, 1970.
7. Handy, N.C. and Schaefer, H.F., *J. Chem. Phys.* 81:5031, 1984.
8. Curtiss, L.A., Raghavachari, K., Redfern, P.C., and Pople, J.A., *J. Chem. Phys.* 93:1063, 1997.
9. Curtiss, L.A., Redfern, P.C., Raghavachari, K., and Pople, J.A., *J. Chem. Phys.* 109:42, 1998.
10. Curtiss, L.A., Raghavachari, L., Redfern, P.C., Rassolov, V., and Pople, J.A., *J. Chem. Phys.* 109:7764, 1998.

9

Highly Accurate Methods: Coupled Cluster Calculations, Extrapolation to Chemical Accuracy, and Quantum Monte Carlo Methods

For most of the 80 years that quantum mechanics has been employed to characterize chemical systems, its value has been conceptual and qualitative. This is, of course, familiar territory to chemistry, and invaluable to our task of developing insight. In 1960, a division was evident; there were those who were content with qualitative theory and hastened to apply its concepts to large chemical systems. But there were those who were not satisfied with this level of description and sought ever more accurate accounts of small chemical systems. Recently, the persistence of this latter group has generated methods of great value. We will explore two branches of their endeavor. The more thoroughly developed and widely adopted approach focuses on the systematic treatment of correlation by introducing refinements on the Hartree–Fock approximation either entirely rigorously (coupled cluster theory) or by empirical estimates of small effects not captured directly in other ways (extrapolation methods). The second branch, which may now be poised to make its impact on chemistry, deals more directly with the motion of electrons in a field of nuclear charges. The first approach aspires to "chemical accuracy" while the second, called quantum Monte Carlo, aspires to exact solution of the Schroedinger equation.

Aspiration to Chemical Accuracy

A sign of the maturity of electronic structure modeling is its engagement with the most thoroughly reviewed and reliable thermochemical data. The pioneering work in this ambitious project was the definition of data sets by which the performance of model chemistries could be judged. For that critique, only the most trustworthy data would be valuable; the systems chosen would have to be within reach of the best computational methods. Pople et al. [1–3] devised the first of the series of Gn data sets in 1989, and development of these standard resources [4–6] guided the evaluation of several kinds of methods aspiring to chemical relevance—not just for their explanatory power and qualitative value, but numerical reliability so well established that the computation would be able to help judge the validity of thermochemical data. Two major strands of investigation can be discerned in this project: strictly ab initio treatment of small systems by the coupled cluster expansion, and hybrid extrapolation schemes which employed a mixture of ab initio modeling and empirical correction. Pople et al. [1,2] led the way with an approach called G1 theory, beginning with modestly demanding MP2 calculations to establish structures, and employing empirical scaling to estimate vibrational frequencies. Higher-level correlation corrections were added in a complex but consistent way, well described in Foresman and Frisch [7] and in Jensen [8]. Later extrapolation schemes incorporated DFT steps and most recently coupled cluster calculations to define structures and correlation corrections. Before discussion of these extrapolation techniques, the coupled cluster method needs to be described.

An Aerial View of CC

Widely available electronic structure modeling software suites invariably have an implementation of coupled cluster calculations. The coupled cluster expansion presents formidable obstacles to understanding and is quite demanding of computer resources and time. Still, since it is widely regarded as the best available way to treat difficult systems, it must be part of the armamentarium of any modeler.

We begin with a survey of the landscape before turning to discussion of details of the theory and its performance as implemented in GAUSSIAN, GAMESS, and other prominent codes. This broad summary is borrowed from Martin Head-Gordon's discussion of computations of chemical accuracy delivered at the ACS-PRF summer school on computational chemistry organized by Prof. Jack Simons at the University of Utah [9].

The starting point for the coupled cluster expansion is the expression of the exact wave function as developed from a reference function.

$$|\Psi_{\text{exact}}\rangle = \exp{(T_1 + T_2 + \cdots + T_n)}|\Phi\rangle$$

The operators T_k refer to excitations from the reference function of a single electron, pairs, etc. up to excitations of all n electrons in the system. Truncating the sequence of T operators produces approximate wave functions

$$|\Psi_{\text{CCSD}}\rangle = \exp{(T_1 + T_2)}|\Phi\rangle$$

There is a variational version of CCSD in which we consider Ψ_{CCSD} the trial function, but the awkwardness of the calculation tilts us toward an alternative obtained by "subspace projections." This gives "tractable" equations but sacrifices the lower bound guaranteed by the variational principle. This is common to all perturbation methods. We quote the equation which defines the energy

$$\langle \Phi | H - E | \exp{(T_{CCSD})} \Phi \rangle = 0$$

$$E = \left\langle \Phi | H | \left(1 + T_1 + T_2 + \frac{1}{2} T_1^2 \right) \Phi \right\rangle$$

The equations defining the amplitudes are

$$\left\langle \Phi | H - E | \left(1 + T_1 + T_2 + \frac{1}{2} T_1^2 + T_1 T_2 + \frac{1}{3!} T_1^3 \right) \Phi \right\rangle = 0$$

$$\left\langle \Phi_{ij}^{ab} | H | \left(1 + T_1 + T_2 + \frac{1}{2} T_1^2 + T_1 T_2 + \frac{1}{3!} T_1^3 + \frac{1}{2} T_2^2 + \frac{1}{3} T_1^2 T_2 + \frac{1}{4!} T_1^4 \right) \Phi \right\rangle = 0$$

The CCSD model chemistry has a number of desirable features: it is exact for a two-electron system and it is "size consistent"—i.e.,

$$E_{CCSD}(N \text{ isolated systems}) = N \, E_{CCSD}(\text{single system})$$

Broadly, CCSD is more powerful and complete than MP2, containing some higher excitations, specifically products of pair excitations. Its advantage is shown more clearly in radicals than closed-shell systems. Extending the expansion to higher order than doubles is difficult, owing to the scaling of the method, about N^6 and higher as the degree of excitation is increased (MP2 scales as N^5). CCSD(T) includes triples, estimated in a perturbative sense and added on after the CCSD is done.

CCSD(T) gives excellent structures ($|\Delta R|$ about 0.002 Å), vibrational frequencies ($|\Delta \omega|$ about 2%), and energies for chemical processes ($|\Delta E| < 2$ kcal/mol in bond energies)—for closed-shell systems. Results for open-shell systems can be erratic, and bond-breaking is not well described. Though it is computationally expensive—it is reasonable to limit the molecules in question to fewer than 10 heavy (Li, . . . F) atoms—CCSD(T) is taken to be the current standard of excellence, and results of cheaper methods such as the variants of DFT are often judged by their agreement with CCSD(T) estimates.

Theoretical Foundations

This aerial view must motivate us to look deeper into the structure of the theory, to judge the quality of its modeling in more detail, and to see how it is deployed in popular software packages. Quantum chemistry texts beginning with Szabo and Ostlund [10] are never without a description of CCD, CCSD, CCSD(T), and some of their predecessors. We recommend Jensen's presentation [8] as a model for a first introduction. Crawford and Schaefer [11] have given what may be the best compromise between clarity and depth in their tutorial review, and Bartlett's periodic technical reviews [12a,b,c] give a thorough picture of the advances of coupled cluster theory since it was introduced to the chemical community by Čižek and Paldus [13]. The following sketch of the theory is drawn from Crawford and Schaefer.

We are accustomed to writing a single determinant of one-electron functions as our Hartree–Fock starting point:

$$\Phi = \det\left(\varphi_1\varphi_2\varphi_3\varphi_4\right)$$

It may be that electrons in some pairs of orbitals i and j call for more detailed description by a pair function f_{ij}. Then, we could imagine augmenting an orbital product thus:

$$\varphi_i(1)\varphi_j(2) + f_{ij}(1,2)$$

The purpose of the correction function f is to increase the flexibility in our description of the relative motions of the two electrons it helps to describe. This extended form for the pair function leads to an expansion of the wave function

$$\begin{aligned}\Psi_{\text{PAIR}} = \left|\varphi_i\varphi_j\varphi_k\varphi_l\right\rangle &+ \left|f_{ij}\varphi_k\varphi_l\right\rangle - \left|f_{ik}\varphi_j\varphi_l\right\rangle + \left|f_{il}\varphi_j\varphi_k\right\rangle + \left|\varphi_i f_{jk}\varphi_l\right\rangle - \left|\varphi_i f_{jl}\varphi_k\right\rangle \\ &+ \left|\varphi_i\varphi_j f_{kl}\right\rangle + \left|f_{ij}f_{kl}\right\rangle - \left|f_{ik}f_{jl}\right\rangle + \left|f_{il}f_{jk}\right\rangle\end{aligned}$$

We could imagine incorporating pair functions to describe details of interactions of electrons. The refinement could include excitations from an occupied pair i, j to the virtual orbitals a, b.

$$f_{ij}(r_m, r_n) = \sum_{ab} t_{ij}^{ab}\, \varphi_a(r_m)\phi_b(r_n)$$

Here we have introduced the t amplitudes which convey the relative importance of specific excitations. Then if all pairs are improved, the wave function would become

$$\Psi = \Phi + \sum_{\substack{a>b \\ i<j}} t_{ij}^{ab} \|\, \varphi_a(r_1)\varphi_b(r_2)\varphi_k(r_3)\varphi_l(r_4)\|$$

Antisymmetrizing the new function forces equivalences among the t amplitudes.

$$t_{ij}^{ab} = -t_{ji}^{ab} = -t_{ij}^{ba} = t_{ji}^{ba}$$

A Notational Convenience: Creation and Annihilation Operators

It is a part of the vocabulary of coupled cluster theory to refer to excited states as if they were constructed from the HF determinant by removal of electrons from the HF manifold, and insertion of electrons in the virtual manifold. Removal is accomplished by an "annihilation operator" a:

$$a_3 |\varphi_1 \varphi_2 \varphi_3 \varphi_4\rangle = |\varphi_1 \varphi_2 \varphi_4\rangle$$

Insertion is done by a "creation" operator $a^\dagger$:

$$a_7^\dagger |\varphi_1 \varphi_2 \varphi_4\rangle = |\varphi_1 \varphi_2 \varphi_7 \varphi_4\rangle$$

Creation operators can make something out of nothing. Beginning with the "vacuum state" with no electrons at all

$$a_i^\dagger a_j^\dagger a_k^\dagger a_l^\dagger |0\rangle = |\varphi_i \varphi_j \varphi_k \varphi_l\rangle$$

you cannot, however, annihilate nothing:

$$a_p |0\rangle = 0$$

Permuting the order of a sequence of creation operators changes the sign of the result:

$$a_i^\dagger a_j^\dagger |0\rangle = |\varphi_i \varphi_j\rangle = -|\varphi_j \varphi_i\rangle = -a_j^\dagger a_i^\dagger |0\rangle$$
$$\text{or } a_i^\dagger a_j^\dagger = -a_j^\dagger a_i^\dagger$$

A similar commutation relation holds for the annihilation operators; for the mixed expression

$$a_p^\dagger a_q + a_p^\dagger a_q = \delta_{pq}$$

So far, we have considered that our pair functions are formed by double excitations. Before we return to that topic, let us write an expression for a single excitation

$$|\varphi_a \varphi_j \varphi_k \varphi_l\rangle = a_a^\dagger a_i |\varphi_i \varphi_j \varphi_k \varphi_l\rangle$$

Then, the coupled cluster expression for the collection of single-excitation corrections to the wave function

$$\sum_{ia} t_i^a \left| \varphi_a \varphi_j \varphi_k \varphi_l \right\rangle = \sum_{ia} t_i^a a_a^\dagger a_i \left| \varphi_i \varphi_j \varphi_k \varphi_l \right\rangle$$

helps us define a single-excitation operator, thus:

$$\hat{t}_i = \sum_a t_i^a a_a^\dagger a_i$$

$$\hat{T}_1 = \sum_i \hat{t}_i$$

Consider the fourth term in Ψ_{PAIR} above:

$$\left| f_{il} \varphi_j \varphi_k \right\rangle = \sum_{ab} t_{il}^{ab} \left| \varphi_a \phi_b \varphi_k \varphi_l \right\rangle$$

We can envision this term as if it were constructed from the reference state by annihilation operators a which remove electrons from orbitals i and l, and creation operators $a^\dagger$ which place electrons in orbitals a and b.

$$\left| f_{il} \varphi_j \varphi_k \right\rangle = \sum_{ab} t_{il}^{ab} a_a^\dagger a_b^\dagger a_i a_l \left| \varphi_i \varphi_j \varphi_k \varphi_l \right\rangle$$

Coupled cluster theory defines a second-level cluster operator:

$$\hat{t}_{il} = \sum_{ab} t_{il}^{ab} a_a^\dagger a_b^\dagger a_i a_l$$

Second-level t operators are collected into

$$\hat{T}_2 \equiv \frac{1}{2} \sum_{il} \hat{t}_{il} = \frac{1}{4} \sum_{ab,il} t_{il}^{ab} a_a^\dagger a_b^\dagger a_i a_l$$

We will dispense with the caret to minimize clutter, but will keep in mind that the T's are operators. Analogous higher excitations are also representable by creation and annihilation operators, but this will serve our purpose.

Specification of the *T* Operators

Let us make explicit what the operators T_j accomplish. We assume that our reference function Φ is a single determinant of Hartree–Fock SCF molecular orbitals; the convention is that labels $i, j, \ldots$ refer to MOs occupied in Φ, while $a, b, \ldots$ refer to MOs vacant in Φ. Then,

$$T_1 \Phi = \sum_{ia} t_i^a \Phi_i^a$$

is how all single excitations are generated. Analogously

$$T_2\Phi = \sum_{i<j,a>b} t_{ij}^{ab} \Phi_{ij}^{ab}$$

generates all double excitations. Triples, etc. are obtained similarly with higher order t amplitudes.

In contrast to configuration interactions with singles and doubles, the coupled cluster expansion includes terms such as

$$\frac{1}{2}T_2^2\Phi = \sum_{ab,ij} \sum_{cd,kl} t_{ij}^{ab} t_{kl}^{cd} \Phi_{ijkl}^{abcd}$$

$$\frac{1}{2}T_1^2\Phi = \sum_{ia} \sum_{jb} t_i^a t_j^b \Phi_{ij}^{ab}$$

$$T_1 T_2 \Phi = \sum \sum t_i^a t_{kl}^{cd} \Phi_{ikl}^{acd}$$

A welcome simplification is that these higher excitations are defined by two- and four-index amplitudes rather than the distinct six- and eight-index amplitudes that CI for triple and quadruple excitations would require. CISD cannot include triple and quadruple excitations (the last and first of these) although it does accommodate simultaneous single excitations. This has significant consequences for extended systems, which we explore in the next section.

The Size Consistency Issue in CC and CI

A key distinction between truncated CI and CC is the coupled cluster expansions' size consistency. Consider two separated species X and Y. The reference function factors naturally into a portion for X and a portion for Y:

$$\Phi = \Phi_X \Phi_Y$$

The exponential operator also divides naturally into factors

$$\exp(T) = \exp(T_X + T_Y) = \exp(T_X)\exp(T_Y)$$

The energy then becomes

$$H \exp(T)\Phi = (H_X + H_Y)\exp(T_X)\Phi_X \exp(T_Y)\Phi_Y = (E_X + E_Y)\Phi$$

If the coupled cluster sequence is terminated at doubles, we will still obtain simultaneous double excitations in each subsystem. This cannot happen with CI (doubles) in which only two electrons in the entire system can be excited. To illustrate this point, we borrow from Szabo and Ostlund [10] a treatment in CI (doubles) of a model system, H_2, in a minimal basis. Consider two molecules separated by a large distance. The HF ground state is

$$|\psi\rangle = |1s\sigma_A^\alpha 1s\sigma_A^\beta 1s\sigma_B^\alpha 1s\sigma_B^\beta\rangle$$

Here the occupied MO is

$$1s\sigma_A = N(1s_{A1} + 1s_{A2})$$

The sole virtual orbital is

$$1s\sigma_A^* = N^*(1s_{A1} - 1s_{A2})$$

and similarly for molecule B. Single excitations vanish owing to g–u symmetry disagreement in this case, and more generally owing to Brillouin's theorem [14]. The only double excitations which can mix with ground state are local to one of the molecules.

$$D_A = |1s\sigma_A^{\alpha *} 1s\sigma_A^{\beta *} 1s\sigma_B^\alpha 1s\sigma_B^\beta\rangle$$
$$D_B = |1s\sigma_A^\alpha 1s\sigma_A^\beta 1s\sigma_B^{\alpha *} 1s\sigma_B^{\beta *}\rangle$$

The secular equation for the mixing is

$$\begin{bmatrix} 0 & K & K \\ K & 2\Delta & 0 \\ K & 0 & 2\Delta \end{bmatrix} \begin{bmatrix} 1 \\ c \\ c \end{bmatrix} = E_{CID}^{n=2} \begin{bmatrix} 1 \\ c \\ c \end{bmatrix}$$

The solution is

$$E_{CID}^{n=2} = \Delta - \left(\Delta^2 + 2K^2\right)^{1/2}$$

The energy of a single H_2 in CID is

$$E_{CID}^{n=1} = \Delta - \left(\Delta^2 + K^2\right)^{1/2}$$

We see that the system of two hydrogen molecules does not have twice the energy of a single molecule. For a system of N molecules

$$E_{CID}^{n=N} = \Delta - \left(\Delta^2 + NK^2\right)^{1/2}$$

The energy of N diatomics each treated by CID is

$$NE_{CID}^{n=1} = N\left[\Delta - \left(\Delta^2 + K^2\right)^{1/2}\right]$$

Consider the perturbation equivalent of the CID expression, valid for large Δ and small K.

$$E_{CID}^{n=N} = \Delta - \Delta\left(1 + NK^2/\Delta^2\right)^{1/2} \approx -NK^2/2\Delta$$

This is just N times the perturbation approximation for the single molecule

$$NE_{CID}^{n=1} = N\left[\Delta - \Delta(1 + 1K^2/\Delta^2)^{1/2}\right] \approx -NK^2/2\Delta$$

Perturbation theory is size consistent.

Solving the CC Equations

To find the amplitudes t we will use the similarity-transformed Hamiltonian

$$\exp(-T)H\exp(T) = \bar{H}$$

A similarity-transformed operator retains all its eigenvalues; thus,

$$\langle\Phi|e^{-T}He^T|\Phi\rangle = \langle\Phi|\bar{H}|\Phi\rangle = E$$

The expansion produces the first few terms

$$\bar{H} = H + (HT - TH) + \frac{1}{2}\{(HT - TH)T - T(HT - TH)\} + \cdots$$

This is the beginning of the Hausdorff expansion in commutators

$$[A,B] = AB - BA$$

The Hausdorff expansion is well described by Merzbacher [61]. The expansion of the Hamiltonian is

$$\bar{H} = H + [H,T] + \frac{1}{2!}[[H,T],T] + \frac{1}{3!}[[[H,T],T],T]\cdots$$

which truncates at the fourth-level commutator owing to the fact that H refers to no more than two electrons.

CC-Doubles Approximation

To illustrate in the simplest way the key features of the coupled cluster expansion we consider the CC-doubles approximation

$$T = T_2$$

Then, the wave function is

$$\Psi_{CCD} = \exp(T_2)\Phi$$

The similarity-transformed Hamiltonian is

$$\bar{H}_{CCD} = \exp(-T_2)H\exp(T_2)$$

Expanding the exponentials as above

$$\bar{H} = H + (HT_2 - T_2H) + \tfrac{1}{2}\{(HT_2 - T_2H)T_2 - T_2(HT_2 - T_2H)\} + \cdots$$

and inserting the expression for T_2 above

$$T_2\Phi = \sum_{i<j,a>b} t_{ij}^{ab}\Phi_{ij}^{ab}$$

we can see that the leading term in the Hausdorff series yields the reference HF energy, and the remainder of the energy must be expressed by two-electron integrals. The energy expression is

$$E_{\mathrm{CCD}} = E_0 + \frac{1}{4}\sum_{ijab}[\![ij \parallel ab]\!]t_{ij}^{ab}$$

We use the indices i, j to refer to occupied orbitals, a, b to refer to virtual orbitals (occupied only after excitation) and adopt the notation

$$[\![pr|qs]\!] = \langle pq \parallel rs \rangle - \langle ps \parallel qr \rangle$$

which expresses the repulsion of and possible exchange with the second electron. Then, the amplitude equation, expressing the fact that all excitations are orthogonal to the reference function Φ, is

$$0 = \left\langle \Phi_{ij}^{ab}|\bar{H}|\Phi \right\rangle$$

Let us write this a little more explicitly.

$$0 = \left\langle \Phi_{ij}^{ab}|\bar{H}|\Phi \right\rangle = \left\langle \Phi_{ij}^{ab}|\exp(-T_2)H\exp(T_2)|\Phi \right\rangle$$

Expanding the exponentials we encounter

$$\left\langle \Phi_{ij}^{ab}\left|\left(1 - T_2 + \frac{1}{2}T_2^2 + \cdots\right)H\left(1 + T_2 + \frac{1}{2}T_2^2 + \cdots\right)\right|\Phi \right\rangle$$

$$= \left\langle \Phi_{ij}^{ab}|H|\Phi \right\rangle + \left\langle \Phi_{ij}^{ab}|HT_2|\Phi \right\rangle - \left\langle \Phi_{ij}^{ab}|T_2H|\Phi \right\rangle + \cdots$$

$$= (\varepsilon_a + \varepsilon_b - \varepsilon_i - \varepsilon_j) + [\![ab \parallel ij]\!] + \left\langle \Phi_{ij}^{ab}\Big|H\sum_{kl,cd} t_{kl}^{cd}\Big|\Phi \right\rangle - \left\langle \Phi_{ij}^{ab}\Big|\sum_{kl,cd} t_{kl}^{cd}H\Big|\Phi \right\rangle + \cdots$$

We have written out the first two terms which make no reference to the amplitudes t, and have not expressed the terms quadratic in amplitudes. It is possible to enumerate all the nonzero two-electron integrals, although this effort becomes unwieldy and has been replaced by diagrammatic techniques. After manipulation, the equation for the four-index t amplitudes becomes

$$0 = [\![ab \| ij]\!] - D_{ijab}t_{ij}^{ab} + \sum_{c>d} [\![ab \| cd]\!]t_{ij}^{cd} + \sum_{k>l} [\![kl \| ij]\!]t_{kl}^{ab}$$

$$+ \sum_{kc} \left\{ -[\![kb \| jc]\!]t_{ik}^{ac} + [\![ka \| jc]\!]t_{ij}^{bc} + [\![kb \| ic]\!]t_{jk}^{ac} - [\![ka \| ic]\!]t_{jk}^{bc} \right\}$$

$$+ \sum_{k>l,c>d} [\![kl \| cd]\!] \left\{ t_{ij}^{cd}t_{kl}^{ab} - 2\left(t_{ij}^{ac}t_{kl}^{bd} + t_{ij}^{bd}t_{kl}^{ac} \right) - 2\left(t_{ik}^{ab}t_{jl}^{cd} + t_{ik}^{cd}t_{jl}^{ab} \right) \right.$$

$$\left. + 4\left(t_{ik}^{ac}t_{jl}^{bd} + t_{ik}^{bd}t_{jl}^{ac} \right) \right\}$$

Here

$$D_{ijab} = \varepsilon_i + \varepsilon_j - \varepsilon_a - \varepsilon_b$$

The iterative solution of this nonlinear equation begins with all $t = 0$ except for the multiplier of D; then

$$t_{ij}^{ab}(\text{1st}) = \frac{[\![ab \| ij]\!]}{D_{ijab}}$$

We have managed something like an MP2 calculation. The energy is now

$$E(\text{2nd}) \approx \sum_{\substack{a>b \\ i>j}} [\![ab \| ij]\!]t_{ij}^{ab}(\text{1st}) = \sum_{\substack{a>b \\ i>j}} \frac{[\![ab \| ij]\!][\![ij \| ab]\!]}{D_{ijab}}$$

Using these first estimates to the t amplitudes, a refined value becomes

$$t_{ij}^{ab}(\text{2nd}) = \sum_{c>d} [\![ab \| cd]\!]t_{ij}^{cd}(\text{1st}) + \sum_{k>l} [\![kl \| ij]\!]t_{kl}^{ab}(\text{1st})$$

$$+ \sum_{kc} \left\{ -[\![kb \| jc]\!]t_{ik}^{ac}(\text{1st}) + [\![ka \| jc]\!]t_{ij}^{bc}(\text{1st}) + [\![kb \| ic]\!]t_{jk}^{ac}(\text{1st}) \right.$$

$$\left. - [\![ka \| ic]\!]t_{jk}^{bc}(\text{1st}) \right\}$$

This has involved only the terms linear in the approximate t amplitudes. Later iterations, which include the quadratic terms, produce further estimates of the t amplitudes.

The CCSD equations are substantially more complex, and are nonlinear in both the four-index t amplitudes and the two-index t amplitudes. Simplifications are obtained by symmetry-adaptations, spin adaptation, and factoring nonlinear expressions into products of linear forms. But even so, ACES II [15] reports upon conclusion of the iterative treatment of the equations for singles and doubles, "A miracle happened: The CCSD Equations Converged!"

Imminent Developments

Crawford and Schaefer [11] identify several lines of inquiry which pre-occupy students of coupled cluster formalisms and algorithms. While the closed-shell problem has been solved, the open-shell problem is much more demanding. Spin restriction can remove some of the difficulty, which arises in part because some quantities which are zero in the closed-shell case (such as the Fock matrix elements, f_{ai}) are nonzero for open shells. The key seems to be the nature of the reference function Φ. For closed-shell systems, the HF Slater determinant of canonical orbitals is the convenient choice; a comparable choice for open-shell systems is still sought. Current implementations use an unrestricted (UHF) Φ, but much attention is now focused on the possibility of use of a spin-restricted open-shell reference.

It appears than in CC expansions the first-order t operators have the task of improving the orbitals in the reference function Φ. Brueckner orbitals [16] have the desirable property that all the first-order corrections are already incorporated, and all the first-order t amplitudes vanish. This may counter certain pathologies in HF reference functions, in which spin (or other) instability produces a singularity in the Hessian.

In general the (in)adequacy of the reference function is expressed as a value of a parameter $\mathcal{T}$

$$\mathcal{T} = \left[\frac{\sum \left(t_i^a\right)^2}{N}\right]^{1/2}$$

$\mathcal{T}$ expresses the extent of the first-order correction in orbitals comprising the reference function. Crawford has kindly provided us a report on the amplitudes of formaldehyde generated in the software suite PSI3. In this table we see that many important excitations originate in orbital 5—i.e., the b_1 CO pi bonding MO. We can also see that the most significant excitations, both single and double, involve rather few of the virtual orbitals; numbers 127–130 and 83–85 are important. This suggests that restricted active spaces may be employed to good effect; choices could be guided by the MP2 amplitudes.

```
              Largest TIA Amplitudes:

          5      127      -0.0306478960
          5      128      -0.0242201569
          3       85       0.0126866182
          4       83      -0.0096674932
          5      136       0.0092790802
          4       85       0.0091720044
          5      132      -0.0091064572
          5      131      -0.0086120409
          5      129       0.0082642537
          4       84      -0.0082104180
```

Largest TIjAb Amplitudes:

5	5	127	127	−0.0995342765
5	5	127	129	−0.0349754191
5	5	129	127	−0.0349754191
5	5	127	128	−0.0347666625
5	5	128	127	−0.0347666625
1	1	127	127	−0.0264939004
3	3	85	85	−0.0228837421
4	4	85	85	−0.0228507601
5	5	127	130	−0.0222233701
5	5	130	127	−0.0222233701

The magnitudes of the single-excitation amplitudes are modest; thus, we are assured that the single determinant is a reasonable starting point. Results of calculations for which the measure is more than 0.1 are to be mistrusted. Such a large value of $\mathcal{T}$ suggests that a multireference starting point is required. This need spurs formal studies intended to accommodate more flexible reference functions.

This effort will also help coupled cluster theory to extend its reach to regions of potential energy surfaces where the single-determinant reference state is a very poor starting point.

Linear response and equations of motion variants of coupled cluster methods describe excited states well, so long as those states are primarily achieved by single excitations from the reference state. We can be sure that more flexible descriptions of excited states within the coupled cluster theory will be forthcoming.

Finally, the N^6 (or worse) scaling properties of coupled cluster methods will apparently always put some important systems out of reach. The large basis sets now considered necessary to achieve chemical accuracy-1 kcal/mol or better-cannot be managed for molecules with more than a few heavy atoms. Localization ideas are appealing. If for an N-fold problem we can identify $K(N/K)$-fold subproblems, the N^6-scaled problem becomes K $(N/K)^6$-scaled problems. Even if $K = 2$, the computational demand falls to $N^6/2^5$, 32-fold smaller. Composite methods such as the ONIOM hierarchy can also focus the use of highly demanding computational methods on subsections of a molecule, with other portions treated by more approximate methods.

Beyond CCSD

The scaling problem just mentioned is a considerable challenge, increasing for every level of excitation included in the coupled cluster expansion. For practical purposes this puts out of reach full description of triple excitations of even medium-sized molecules. CCSDT scales as N^8, with N being a measure of size of the system, often the number of basis functions. There is considerable advantage, however, to inclusion of at least approximate consideration of triple excitations. An effective measure has been proposed by Raghavachari et al. [17], in which the effects of triples are recognized by a perturbative correction using quantities obtained in a fully converged CCSD calculation. The proposal is expressed in somewhat schematic form as

$$E_{\text{Triples}} = \sum_{s=S,D} \sum_{t=T} \sum_{u=D} (E_0 - E_s)^{-1} a_s V_{st} V_{tu} a_u$$

Here the indices s, t, u refer to levels of excitation, i.e., S = singles, D = doubles, T = triples; the as are evidently composites of t amplitudes found in a CCSD calculation, E_s is the energy of a particular state of excitation level $s = S$ or D, and the Vs are matrix elements involving the various excited states. Evaluation of this correction is demanding—it scales as the seventh power of the size of the basis—but need be done only once.

Performance of CCSD(T)

The coupled cluster expansion is now incorporated in readily available software suites: Lee and Scuseria [18] took note already in 1995 that GAUSSIAN, CADPAC, ACES-II, PSI, and other packages had implementations of various forms of coupled cluster procedures. This is a tribute to the performance of (primarily) the CCSD(T) formulation, wherein the effect of triple excitations is acknowledged and estimated by a perturbative addition to a CCSD calculation. It is equally a tribute to those who transformed coupled cluster methods from theoretical arcana to efficient and practical computational implementations. This advance has been formal—the use of symmetry had its usual salutary effect, and development of means to evaluate gradients for correlated methods was critical to the use of CC methods for geometry optimizations, for example—as well as more practical, as software and hardware improvements flourished.

There are two ways to judge the power of a quantum chemical computational scheme: to refer to even more advanced calculations, or to appeal to experiment. Multireference configuration interaction has served as the more complete mode of calculation, and the databases of reliable experimental data—G2, G3, etc.—assembled by Pople's group [1–6] have provided the grounding in observation. Hard cases may make bad law but they are a critical means of testing the power of computational methods: alkaline earth clusters, the strong bond of Cr_2, the vibrations of the ozone molecule, and the geometry and especially the torsional potential of FOOF are among these intransigents.

Quantities of central interest to chemists include ground state properties such as equilibrium geometries and vibrational frequencies. Lee and

Scuseria [18] survey geometries of small polyatomics (at most three heavy atoms). Errors in CCSD(T) bond lengths are typically less than 0.01 Å. Basis sets including f orbitals seem to be "saturated" for this purpose, and cc-pVTZ performs well. Errors in bond angles less than 1° are obtained by CCSD(T)/cc-pVTZ. To mention specifically our familiar species, we notice that the HCH angle in C_{2v} formaldehyde is within 0.01° of the experimental value of 116.18° for CCSD(T) in the valence triple-zeta basis, but 0.26° for the basis cc-pVQZ. It is not easy to know what point such small differences make. It does appear that small differences in geometry have an impact on properties such as NMR shielding constants. Vibrational averaging is probably essential to any final judgment.

CCSD(T) in triple-zeta basis sets will typically produce harmonic frequencies for small molecules within about 10 cm^{-1}. The harmonic frequencies for formaldehyde in particular are recaptured within 10 cm^{-1}. Even the hard case of the ozone OO asymmetric stretch is located within 35 cm^{-1} if a sufficiently large basis is employed. For the family of molecules studied by Lee and Scuseria, the MAD is halved if g-functions are included in the basis. For the few systems reported in detail—N_2 and other diatomics and alkaline earth M_3 and M_4 clusters—CCSD(T) dissociation energies agreed well (<2 kcal/mol, but often better) with MR-CI values. Proton affinities and heats of formation—especially of marginally stable species such as *cis*-FONO—were reliably described by CCSD(T), so that in some cases corrections to experimental data were possible.

Electrical properties—dipole moments and polarizabilities—are a test of the quality of the charge distribution in a region which is not critical to the energy value. For small polyatomics predicted dipoles were typically within 0.1 debye, if TZ2P basis sets are used. More diffuse functions are required for reliable polarizabilities.

A linear response variant within coupled cluster theory, or an "equations of motion" formulation, is the basis for estimation of vertical excitation energies for small systems. Most examples contained two heavy atoms, but larger symmetric systems were also accessible. The goal of accuracy of 0.1–0.2 eV seemed attainable for spectra dominated by one-electron excitations.

More than a decade after Lee and Scuseria's 1995 summary [18], Bartlett and Musial [12c] presented charts showing that geometry optimization with CCSD(T)/cc-pVTZ achieved values of bond distances with standard errors relative to experiment of ca. 0.02 Å (2 pm). Heats of atomization are more difficult to estimate; CCSD(T)/cc-pVQZ estimates for a modest set of small molecules showed a range of errors of −13 to +6 kcal/mol but a MAD of only about 2 kcal/mol. For the easier task of estimating single bond energies, CCSD(T) displays a range of ±6 kcal/mol and a mean error less than 1 kcal/mol.

Feller and Peterson [19] used the GAUSSIAN-2 data set to judge the performance of CCSD and CCSD(T) relative to MP2, MP3, and MP4 methods. They employed the series of correlation-consistent basis sets developed by Dunning's group and described by Martin et al. [20], and also estimated the "complete basis set" limit errors for atomization energies, ionization energies, electron affinities, and bond lengths. This estimate rested on the extrapolation

$$E(x) = E_{CBS} + b \exp(-cx)$$

Here $E(x)$ refers to the energy computed in a Dunning basis of split-valence order x, in the correlation-consistent series aug-cc-pVxZ. The mean absolute deviation of heats of atomization achieved with G-2 theory, 1.4 kcal/mol (max errors ca. ± 5 kcal/mol), can serve as a reference point. As the basis set size increases, the MAD for CCSD(T) (frozen-core) declines. The MAD for aug-cc-pVQZ, 2.7 kcal/mol, becomes 1.3 kcal/mol at the basis set limit. Electron affinities and ionization potentials show the same general pattern, with CCSD(T) at the basis set limit achieving a MAD of 0.7 kcal/mol for electron affinities and comparable accuracy for ionization potentials. Proton affinities seem to be much easier to describe quantitatively, with errors as low as a few tenths kcal/mol. Dunning [21] provides us more refined vocabulary for description of errors, distinguishing three types:

Basis Set Convergence Error

$$DQ_{bs}^M(n) = Q(M,n) - Q(M,\infty)$$

Intrinsic Error

$$DQ^M = Q(M,\infty) - Q(expt'l)$$

Calculational Error

$$DQ_{calc'd}^M(n) = Q(M,n) - Q(expt'l)$$
$$= DQ_{bs}^M(n) + DQ^M$$

"Infinity" refers to the complete basis set, and n refers to the order of the Dunning basis aug-cc-pVnZ. Q is any computed quantity, and DQ is a departure of a computed value from an experimental value. The superscript M refers to a particular model chemistry. The intrinsic error is a property of the computational model, in a presumed-to-be perfectly flexible and complete basis set. It may be positive or negative, and could be approached from above or below. In a fortunate—but misleading—case, the intrinsic error may be positive but approached from below. In such circumstances, the errors of a small basis set can cancel with the errors of a computational model and the calculational error can be fortuitously small. An example is the bond lengths in HF theory; the intrinsic error is negative, but use of a small basis may produce an accurate value by error cancellation. Unfortunately, one cannot know if a pleasing result is coincidental without the context provided by a sequence of calculations in a series of basis sets.

CCSD(T) studies on larger molecules must be done in smaller basis sets. For aug-cc-pVDZ, the smallest basis set studied, the MAD in atomization energies for CCSD(T) was about 18 kcal/mol. The step to aug-cc-pVTZ was well rewarded; the MAD for atomization energies became 5 kcal/mol. Analogously, the MAD for electron affinities dropped from 4.5 kcal/mol to less than 2.0 kcal/mol, and the MAD for vertical ionization potentials dropped from about 6.0 kcal/mol to less than 3.0 kcal/mol. CCSD(T) is unambiguously better than CCSD only for basis sets aug-cc-pVTZ or better. MP2 proves to provide the best value for computational effort, but since the quality of the virtual orbitals is important to the success of a correlation model, use of a basis set at level aug-cc-pVTZ or better is advisable, according to Dunning [21].

Figure 9.1 shows the distribution of errors in the AH dissociation energy with the best basis possible—aug-cc-pV6Z—for use in molecules with a few first-row atoms.

Here the mean errors and standard deviations illustrate the importance of triple excitations in the estimation of dissociation energies. CCSD is not obviously better than MP2, but CCSD(T) is substantially better (Table 9.1).

Basis set convergence is systematic and regular for H–X bond energies, as is shown in Figure 9.2. The best correlation model will still not produce excellent results without a suitably flexible basis. The systematic trend of errors

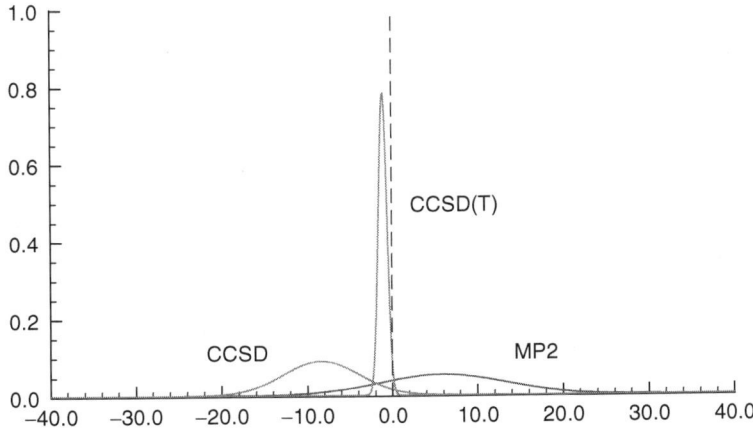

FIGURE 9.1

Normalized error distribution for dissociation energies for small molecules according to Bartlett and Musial. Horizontal axis is the signed error in kilocalories per mole, and the vertical axis is the normalized frequency. CCSD(T)/aug-cc-pV6Z displays a much smaller mean error and variance than either MP2 or CCSD in the same basis. (From Bartlett, R.J. and Musial, M., *Rev. Mod. Phys.*, 79, 291, 2007. With permission.)

TABLE 9.1

Mean Error (Signed) and Variance of Dissociation Energies
According to Three Post-HF Methods (kcal/mol)

MP2	MP2	CCSD	CCSD(T)
Mean error	6.0	−8.3	−1.0
Breadth	7.5	4.5	0.5

allows extrapolation to the complete basis set limit. This is the foundation of
the CBS series of extrapolation schemes which we discuss below.

Other molecular properties are also well described by post-HF methods
[22]. Bond lengths and angles, well defined by the HF model chemistry, are
less sensitive to flexibility of basis sets and completeness of correlation
models than are energetic quantities; MP2/aug-cc-pVDZ gives accurate
A-H distances (MAD <2 pm) and HAH angles (MAD <1°) though A-B
distances are less well modeled (MAD ca. 4 pm for MP2/aug-cc-pVDZ) and
there is an advantage to better basis sets. CCSD(T) has a major advantage
over MP2 in vibrational motions, displaying a MAD of <50 cm^{-1} for aug-
cc-pVTZ, about half of the value for MP2; CCSD is comparable in quality.
The main advantage of improved basis sets is that the range of errors in
computed frequencies is narrowed.

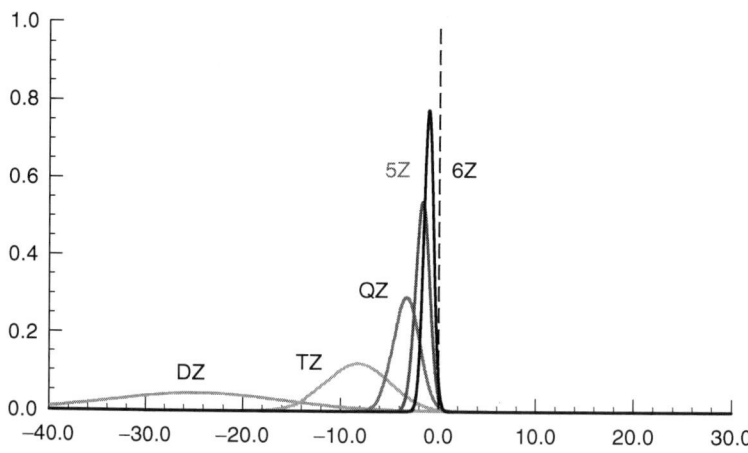

FIGURE 9.2

Normalized error distribution for dissociation energies for small hydrides according to Bartlett
and Musial. The horizontal axis is the signed error in kilocalories per mole, and the vertical axis
is the normalized frequency. The terms DZ, TZ, QZ, 5Z, and 6Z refer to the degree of flexibility
in the valence shell afforded by Dunning aug-cc-pVnZ. CCSD(T) calculations seem to achieve
satisfactory accuracy for quadruple zeta or better basis sets. (From Bartlett, R.J. and Musial, M.,
Rev. Mod. Phys., 79, 291, 2007. With permission.)

As we look toward investigation of chemical reactions, we can take as a rule of thumb that MP2 calculations in DZ basis sets give good structures, and if CCSD(T) calculations are to serve as a check on the unpredictable errors of density functional methods, a basis set of TZ quality should be employed. In the nomenclature used for Pople basis sets this probably means that 6-311+G(d,p) is a reasonable choice.

As a familiar example, we consider the energy of formaldehyde and its alternative products closed-shell CO and H_2 and open-shell CH_2 and O. Similarly, we consider the ketene products CO and singlet or triplet methylene. Gaussian's CCSD(T) is invoked [23], and we quote portions of the output.

```
Cite this work as:
Gaussian 03, Revision D.01,
M. J. Frisch, G. W. Trucks, H. B. Schlegel, G. E. Scuseria,
M. A. Robb, J. R. Cheeseman, J. A. Montgomery, Jr., T. Vreven,
K. N. Kudin, J. C. Burant, J. M. Millam, S. S. Iyengar, J. Tomasi,
V. Barone, B. Mennucci, M. Cossi, G. Scalmani, N. Rega,
G. A. Petersson, H. Nakatsuji, M. Hada, M. Ehara, K. Toyota,
R. Fukuda, J. Hasegawa, M. Ishida, T. Nakajima, Y. Honda, O. Kitao,
H. Nakai, M. Klene, X. Li, J. E. Knox, H. P. Hratchian, J. B. Cross,
V. Bakken, C. Adamo, J. Jaramillo, R. Gomperts, R. E. Stratmann,
O. Yazyev, A. J. Austin, R. Cammi, C. Pomelli, J. W. Ochterski,
P. Y. Ayala, K. Morokuma, G. A. Voth, P. Salvador, J. J. Dannenberg,
V. G. Zakrzewski, S. Dapprich, A. D. Daniels, M. C. Strain,
O. Farkas, D. K. Malick, A. D. Rabuck, K. Raghavachari,
J. B. Foresman, J. V. Ortiz, Q. Cui, A. G. Baboul, S. Clifford,
J. Cioslowski, B. B. Stefanov, G. Liu, A. Liashenko, P. Piskorz,
I. Komaromi, R. L. Martin, D. J. Fox, T. Keith, M. A. Al-Laham,
C. Y. Peng, A. Nanayakkara, M. Challacombe, P. M. W. Gill,
B. Johnson, W. Chen, M. W. Wong, C. Gonzalez, and J. A. Pople,
Gaussian, Inc., Wallingford CT, 2004.

*****************************************************************
Gaussian 03:    IA32W-G03RevD.01 13-Oct-2005
                13-Jul-2007
*****************************************************************

%chk=formalccsd.chk
%mem=500MB
```

```
# rccsd(t)/6-311+g(d,p) guess=read geom=connectivity
```

The familiar quote of structure and basis is omitted: Then,

```
SCF Done:  E(RHF)  =   -113.901270061 A.U. after  15 cycles
             Convg  =     0.9316D-08          -V/T =   2.0015
             S**2   =     0.0000
. . . . . . . . . .
```

```
Range of M.O.s used for correlation:  3  56
NBasis =      56 NAE =       8 NBE =      8 NFC =        2 NFV =       0
NROrb =      54 NOA =       6 NOB =      6 NVA =       48 NVB =      48
Estimate disk for full transformation  7490906 words.
Spin components of T(2) and E(2):
alpha-alpha T2 =         0.1251984026D-01 E2 =      -0.4233497015D-01
alpha-beta  T2 =         0.7722480329D-01 E2 =      -0.2556031431D+00
beta-beta   T2 =         0.1251984026D-01 E2 =      -0.4233497015D-01
ANorm =   0.1049887843D+01
E2 =      -0.3402730834D+00 EUMP2 =   -0.11424154314411D+03
Iterations =  50 Convergence =  0.100D-06

Iteration Nr.  1
* * * * * * * * * * * * * * * *
MP4 (R+Q) =   0.56033334D-02
E3 =          -0.37660243D-02   EUMP3 =        -0.11424530917D+03
E4 (DQ) =     -0.39960819D-02   UMP4 (DQ) =    -0.11424930525D+03
E4 (SDQ) =    -0.96230231D-02   UMP4 (SDQ) =   -0.11425493219D+03
DE (Corr) = -0.33809322        E (CORR) =   -114.23936328
NORM (A) =    0.10519024D+01

Iteration Nr.  2
* * * * * * * * * * * * * * * *
DE (Corr) =  -0.35058512  E (CORR) =   -114.25185518  Delta = -1.25D-02
NORM (A) =    0.10579943D+01
```

Iteration
Result of first iteration
Notice that the MP2 energy is obtained in the zeroth iteration, as we saw in the introduction to iterative solution of the CCD equations. Then, the first full iteration produces MP3 and portions of the MP4 increments. The solution of the CCSD equations proceeds to convergence.

```
Iteration Nr.  11
* * * * * * * * * * * * * * * *
DE (Corr) =   -0.35249696  E (CORR) =   -114.25376702  Delta = -2.31D-07
NORM (A) =     0.10607005D+01

Iteration Nr.  12
* * * * * * * * * * * * * * * *
DE (Corr) =   -0.35249701  E (CORR) =   -114.25376707  Delta = -4.59D-08
NORM (A) =     0.10607006D+01
Largest amplitude =  6.26D-02
Time for triples =  4.00 seconds.
T4 (CCSD) =   -0.14011052D-01
T5 (CCSD) =    0.11422316D-02
CCSD (T) =    -0.11426663589D+03
```

Perturbative correction
The triple-excitation perturbative correction is added to conclude the calculation. That value appears in Table 9.2.

The basis set 6-311+G(d,p) used in the MP2 and CC calculations may not be adequate to the purpose. Crawford [24] has provided us CCSD(T)

TABLE 9.2

Post-HF Energies for Ketene, Formaldehyde, and Fragments

Species	CCSD/ 6-311+G(d,p)	CCSD(T)/ 6-311+G(d,p)	B3LYP/ 6-31G(d)	MP2/ 6-311+G(d,p)
Formaldehyde	−114.2537671	−114.2666359	−114.5004726	−114.2416893
CO	−113.0827293	−113.0783872	−113.3094543	−113.0781692
H$_2$	−1.1683403	−1.1683403	−1.1754824	−1.1602723
CH$_2$ singlet	−39.0330702	−39.0365717	−39.1282494	−39.0064771
CH$_2$ triplet	−39.0536753	−39.0558982	−39.1500196	−39.0360944
O singlet	−74.8457696	−74.8540800	−74.9573979	−74.8135191
O triplet	−74.9366457	−74.9381521	−75.0606231	−74.9217813
Ketene	−152.2466221	−152.2676349	−152.5894707	−152.2353194

computed values for formaldehyde with 170 basis functions. Here is a fragment of his report:

```
Nuclear Rep. energy (chkpt)      =      31.165323797697845
   SCF energy (chkpt)            =    -113.919768024639041
   Reference energy (file100)    =    -113.919768024638842
   CCSD energy (file100)         =      -0.430531303434135
   Total CCSD energy (file100)   =    -114.350299328072978

   (T) energy                    =      -0.018489881954704
   Total CCSD(T) energy          =    -114.368789210027685
```

The 58-member basis we used is adequate to the RHF, falling short by "only" 0.0185 hartree, but is sorely tested by the CCSD step. It achieves only 0.3525 hartree of correlation energy, to be compared with the 0.4305 hartree accessible in the more extensive basis. The triples increment in the small basis is 0.0129 hartree, cf. 0.0185 for the larger basis.

The chemically relevant quantity is not the absolute energy relative to separated electrons and nuclei, but the energies of reaction. In forming the energy difference we may find that errors cancel to a helpful degree. The sequence of energies shows that for the quoted reaction energies MP2, CCSD, and CCSD(T) form an oscillating perturbation sequence, while the DFT results are outside the sequence. The triples correction usually improves agreement with the results of the CBS-QB3 extrapolation scheme designed for faithful thermodynamic estimates (Table 9.3). (We will describe this and related methods later.)

The singlet–triplet gap for methylene estimated by perturbation theory is rather higher than its best variational of 9.2 kcal/mol [25]; CBS-QB3 (described below) places the gap at 7.85 kcal/mol. The PT sequence seems to be declining toward this limit. However, B3LYP already provides a decent estimate in a small basis. The same could be said—though with less conviction—for the singlet–triplet gap for O atom. The CBS-QB3 estimate is 49.46 kcal/mol. All other estimates are higher in value, but not ludicrously in error. The dissociation energies show considerable scatter, up to 30 kcal/mol for formaldehyde

TABLE 9.3

Correlation-Corrected Estimates of Dissociation Energies of Formaldehyde and Ketene

Species	MP2/ 6-311+G(d,p)	CCSD/ 6-311+G(d,p)	CCSD(T)/ 6-311+G(d,p)	CBS-QB3[a]	B3LYP/ 6-31G(d)
Dissociation of Formaldehyde to Methylene and Oxygen Atom (kcal/mol)					
Singlets	264.61	235.27	235.93	234.64	260.30
Triplets	178.09	165.31	171.05	178.11	181.87
Dissociation of Ketene into Methylene and CO (kcal/mol)					
Singlets	94.55	82.09	95.80	85.96	100.88
Triplets	75.96	69.16	83.68	78.11	87.22
Singlet–Triplet Gap (kcal/mol)					
Methylene	18.58	12.93	12.13	7.85	13.66
O atom	69.02	57.93	53.60	49.46	65.81

[a] Energy at 0 K, includes ZPE correction.

and 20 kcal/mol for ketene. The CBS-QB3 and CCSD(T) estimates for formaldehyde dissociation are within about 7 kcal/mol, and the values for ketene dissociation are within 10 kcal/mol. This is apparently a hard case for calculations with aspirations toward chemical accuracy. This may arise from the fact that in the course of these reactions the bonding has changed drastically from a more or less conventional collection of CH, CC, and CO sigma bond pairs, oxygen sigma and pseudo-pi lone pairs, and CC and CO pi double bonds, to produce the CO triple bond and methylene and O-atom open shells.

Application to Isomer Energies

The dissociation and atomization energies used as indices of the performance of choices of model chemistry methods present perhaps the most severe possible tests of the flexibility and power of electronic structure models. Estimates of reaction energetics are most reliable for isodesmic reactions, in which the number of bonds of each kind remains unchanged.

$$\underset{H_3C}{\overset{H}{\diagdown}}C{=}C{=}O + 2CH_4 \longrightarrow H_3C{-}CH_3 + H_2C{=}CH_2 + H_2C{=}O$$

The error cancellation is particularly extensive for this kind of process. A slightly more severe test is set by isogyric reactions, in which the same number of electron pairs is found in reactants and products. Many computational studies of the reactions of medium-sized systems fall into this category. Jensen [8] treats the series of C_4H_6 isomers, in which we find a variety of bond types including strained rings, within the class of isogyric species. The energetic order of the C_4H_6 isomers is a suitable measure of

TABLE 9.4

Electronic Energies of C_4H_6 Systems (Hartrees) by GAUSSIAN-03

Species: Model	HF/3-21G	HF/6-31G(d)	MP2/6-31G(d)
2-Methylcyclopropene	−153.9842317	−154.8606616	−155.3750712
1-Methylcyclopropene	−153.9901810	−154.8680520	−155.3812960
Bicyclo[1.1.0]butane	−153.9866377	−154.8717672	−155.3899740
Methylenecyclopropane	−154.0187262	−154.8873446	−155.3966988
Methylallene	−154.0414787	−154.8993827	−155.4036729
1-Butyne	−154.0448246	−154.8990736	−155.4074396
Cyclobutene	−154.0307180	−154.8996225	−155.4101836
2-Butyne	−154.0536453	−154.9092384	−155.4156988
Trans-butadiene	−154.0594563	−154.9196540	−155.4226556
Cyclobutene-opening TS	−153.9643540	−154.8248250	−155.3505590

quality of model chemistry. We use GAUSSIAN for the following Studies and begin with the HF model and the simplest correlation correction, provided by MP2 (Table 9.4).

The relative energies are much easier to judge, and to compare with thermodynamic quantities. Of course, we should keep in mind that to make a just comparison between computed relative energies and thermo-dynamic values, we should allow for the zero-point vibrational energy corrections and the thermal corrections needed to bring values to the thermodynamic standard temperature (Table 9.5).

Even in the simplest model, HF/3-21G, *trans*-butadiene is recognized as the most stable isomer and the cyclopropene systems are estimated to be the least stable species. Enhancing the basis set reduces the relative instability of these very strained rings, and adding MP2 correlation correction further reduces their energy relative to unstrained *trans*-butadiene. Standard heats of formation are available for all but one of these systems. Surprisingly, the simple differences in electronic energy at the RHF/6-31G(d) match the thermodynamic values very closely. We may reasonably suspect that this

TABLE 9.5

Relative Energies of C_4H_6 Systems (kcal/mol: no ZPE or Thermal Corrections)

Species: Model	REL3-21G	REL6-31G(d)	RELMP2	ΔH Formation
2-Methylcyclopropene	47.20	37.02	29.86	Abs (rel)
1-Methylcyclopropene	43.47	32.38	25.95	58.2 (32.2)
Bicyclo[1.1.0]butane	45.69	30.05	20.51	51.9 (25.9)
Methylenecyclopropane	25.56	20.27	16.29	48.0 (22.0)
Methylallene	11.28	12.72	11.91	38.8 (12.8)
1-Butyne	9.18	12.91	9.55	39.5 (13.5)
Cyclobutene	18.03	12.57	7.83	37.5 (11.5)
2-Butyne	3.65	6.54	4.37	34.6 (8.6)
Trans-butadiene	0.00	0.00	0.00	26.0 (-0-)
MAD (thermo: kcal/mol)	7.42	1.41	4.30	
Cyclobutene-opening TS	41.64	46.94	37.41	

TABLE 9.6

Energies of C_4H_6 Systems (kcal/mol: no ZPE or Thermal Corrections)

Species: Model	MP2/ 6-31G(d)	B3LYP/ 6-31G(d)	CCSD/ 6-31G(d)	CCSD(T)/ 6-31G(d)
2-Methylcyclopropene	−155.3750712	−155.9363950	−155.4129375	−155.4336019
1-Methylcyclopropene	−155.3812960	−155.9445960	−155.4187989	−155.4393418
Bicyclo[1.1.0]butane	−155.3899740	−155.9480540	−155.4246145	−155.4446695
Methylenecyclopropane	−155.3966988	−155.9626250	−155.4368725	−155.4569883
Methylallene	−155.4036729	−155.9758370	−155.4471030	−115.4678631
1-Butyne	−155.4074396	−155.9669650	−155.4453431	−155.4654148
Cyclobutene	−155.4101836	−155.9733010	−155.4499584	−155.4703495
2-Butyne	−155.4156988	−155.9789740	−155.4533356	−155.4732077
Trans-butadiene	−155.4226556	−155.9921340	−155.4659043	−155.4871405
Cyclobutene-opening TS	−155.3505590	−155.9165610	−155.3869259	−155.3869259

close match is coincidental, since far more general studies presented by Foresman and Frisch [7] and Hehre [62] show much poorer performance of this level of theory.

For the same C_4H_6 set, we find the energies in correlated models reported below. There is continued improvement in the match of relative computed energies to relative heats of formation (Tables 9.6 and 9.7).

We will consider the thermal corrections as we discuss schemes devised expressly to achieve chemical accuracy—i.e., differences in relative energy near 1 kcal/mol.

As an example of appeal to coupled cluster calculations—even in a small basis—as authoritative, consider the brief analysis of the relative energy of cumulene double bonds relative to adjacent single and triple bonds. This was intended to bear on the question of diradical and carbonic nature of large triplet systems [26] (Table 9.8).

TABLE 9.7

Relative Energies of C_4H_6 Systems (kcal/mol: no ZPE or Thermal Corrections)

Species: Model	REL-DFT	REL-CCSD	REL-CCSD(T)	CBS-QB3	ΔH Formation
2-Methylcyclopropene	34.98	33.24	33.59	33.90	(rel)
1-Methylcyclopropene	29.83	29.56	29.99	30.61	(32.2)
Bicyclo[1.1.0]butane	27.66	25.91	26.65	26.78	(25.9)
Methylenecyclopropane	18.52	18.22	18.92	19.27	(22.0)
Methylallene	10.23	11.80	12.10	11.92	(12.8)
1-Butyne	15.79	12.90	13.63	13.23	(13.5)
Cyclobutene	11.82	10.01	10.54	12.13	(11.5)
2-Butyne	8.26	7.89	8.74	8.40	(8.6)
Trans-butadiene	0.00	0.00	0	0	(-0-)
MAD (relative to thermo)	1.88	1.46	1.14	1.02	
Cyclobutene-opening TS	47.42	38.81	52.3		

TABLE 9.8

Judgment of Bias in B3LYP

Species: Model	B3LYP/6-31G(d)	CCSD/6-31G(d)	QCISD(T)/6-31G(d)
Methyl acetylene	−116.6532701	−116.2684833	−116.359666
Allene	−116.6576762	−116.2668647	−116.356558
Preference	Allene	Methylacetylene	Methylacetylene
	(3.0 kcal/mol)	(1.0 kcal/mol)	(0.8 kcal/mol)
Methyl nitrile	−132.7549284	−132.3567318	−132.445283
Ethylideneimine	−132.7111412	−132.3016999	−132.395368
Preference	Methyl nitrile	Methyl nitrile	Methyl nitrile
	27.5 kcal/mol	36.4 kcal/mol	31.3 kcal/mol
Methylisonitrile	−132.7165728	−132.3165899	−132.406690
Unnamed ylide	−132.6696751	−132.2544591	−132.348210
Preference	Methylisonitrile 29.4	Methylisonitrile 39.0	Methylisonitrile 36.7

Note: Energies in hartrees; relative energies in kilocalories per mole. QCISD/6-311G(d,p)//
MP2/6-31G(d) as computed in the G2(MP2) sequence.

The B3LYP/6-31G(d) method places the structures with adjacent single and triple bonds relatively low in energy relative to the isomers with cumulenic double bonds; the bias in favor of the cumulene is 5–10 kcal/mol, which is decisive when the energy balance is very close.

Another, more characteristic, use of coupled cluster calculations is to support density functional descriptions of chemical reactions. Su et al. [27] studied the 1,3-dipolar cycloadditions of nitrile ylide, nitrile imine, nitrile oxide, diazomethane, azine, and nitrous oxide to ethylene. They located optimum geometries for the reactants, transition states, and products for each cycloaddition with the far less expensive B3LYP/6-31G(d) method. Using those geometries, they performed single-point coupled cluster calculations in an extended basis, CCSD(T)/6-311G(d,p). The correlations between CC and DFT results are shown in Figure 9.3.

B3LYP, being calibrated to the atomization energies of molecules at equilibrium geometries, tracks the CCSD(T) values of reaction energies faithfully; the correlation coefficient $R = 0.984$ (see Figure 9.4). The maximum errors are about 5 kcal/mol.

Activation energies are outside the training set of atomization energies, ionization energies, and electron affinities, so it is not a great surprise that the representation of the barriers is not so consistent with the CCSD(T) results (correlation coefficient $R = 0.904$).

Thermochemical Standards by Quantum Chemistry

We have made passing references to schemes which have the purpose of emulating thermochemical data, including heats of formation, ionization potentials, and electron affinities. The first of these—the G1 series [1–3]

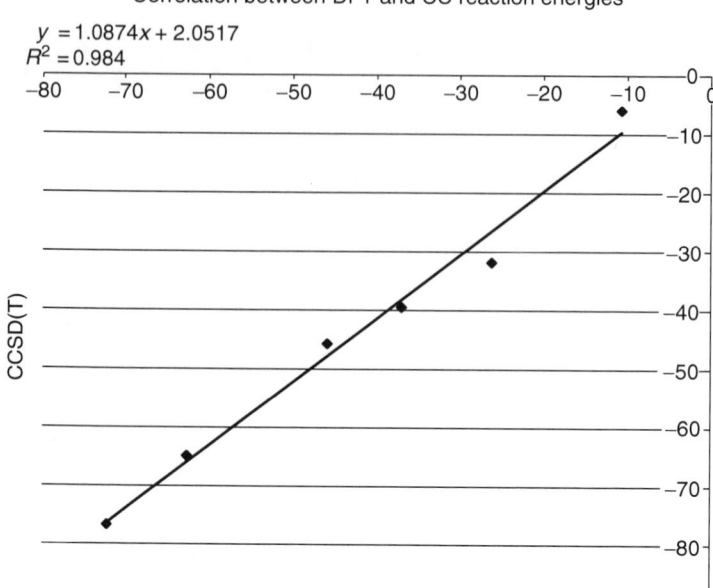

FIGURE 9.3

Correlation between CCSD(T)/6-31G(d) and DFT [B3LYP/6-31G(d)] reaction energies for a set of 1,3-dipolar cycloadditions. The vertical axis is the exothermicity in kilocalories per mole as estimated by CCSD(T); the horizontal axis is the corresponding DFT quantity. The regression coefficient of 0.984 shows that DFT calculations constitute an economical and reliable alternative to coupled cluster calculations for these reaction energies.

from the Pople group—set the goal of accuracy of 1 kcal/mol in atomization energies. It is interesting to note that this project required not only very detailed studies of computational methods but also critical reexamination of experimental data. The succeeding variants G(n), the CBS series by Petersson et al. [28], and most recently the W(n) series, from Martin and De Oliveira [29,30], have met this goal. These methods are applicable by means of workstations of modest cost for small systems of first- and second-row atoms. Jensen [8] has provided a clear account of the G and CBS methods, so we will focus primarily on the DFT modification of CBS-Q [31–33] and then the newer W1 process.

Every thermochemical scheme begins with geometry optimization and an estimate of the zero-point vibrational energy. CBS-Q used (U)HF in a compromise basis, 6-31G† (Table 9.9).

The basis set symbol (†) refers to a combination of the 6-31G sp functions with 6-311G(d,p) polarization exponents. The notation G(3d2f;2df;2p) means that the polarization set 3d2f is used on second-row atoms (Na–Cl), 2df is used for first-row (Li = F) atoms, and the set 2p is used for H atom.

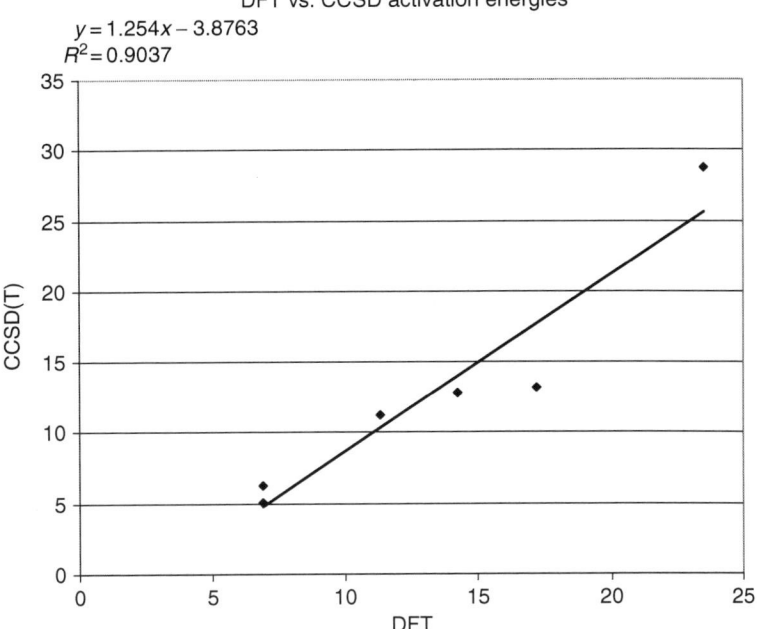

FIGURE 9.4
Correlation between CCSD(T)/6-31G(d) and DFT [B3LYP/6-31G(d)] activation energies for a set of 1,3-dipolar cycloadditions. The vertical axis is the exothermicity in kilocalories per mole as estimated by CCSD(T); the horizontal axis is the corresponding DFT quantity. The regression coefficient of 0.904 shows that DFT calculations are a less reliable alternative to coupled cluster calculations for activation energies.

The notation G(d/f;p) means a d set on both first- and second-row atoms, and an f set on second-row atoms, and p on H atom.

CBS extrapolation estimates energy increments from

$$\Delta E_{MP4} = E[\text{MP4(SDQ)}/6\text{-}31\text{+}G(d/f;p)] - E[\text{MP2}/6\text{-}31\text{+}G(d/f;p)]$$
$$\Delta E_{QCI} = E[\text{QCISD}(T)/6\text{-}31\text{+}G\dagger\,] - E[\text{MP4(SDQ)}/6\text{-}31\text{+}G\dagger\,)]$$

W1 uses density functional theory, B3LYP, in a valence triple-zeta (cc-pVTZ) basis for geometry optimization and frequencies. For second-row atoms this basis is supplemented by a high-exponent d set borrowed from Dunning's cc-pV5Z basis, which serves as an "inner polarization function." (The W-nomenclature calls this "VTZ + 1." Further additions to the basis are introduced below.) The ZPE is scaled from the density functional results by the factor 0.985 to account simultaneously for anharmonicity and the innate errors in the model chemistry.

The total atomization energy is considered to have an SCF part and several correlation parts. The SCF part is extrapolated by the form

TABLE 9.9

Sequence of Steps in Extrapolation Schemes

Step	CBS-Q	CBS-QB3	W1
Frequencies	(U)HF/6-31G†	B3LYP/6-31G†	B3LYP/cc-pVTZ
Scaling	0.91844	0.99	0.985
Optimized geometry	MP2(FC)/6-31G†	B3LYP/6-31G†	B3LYP/cc-pVTZ
Extrapolation	(U)MP2/6-311+G (3d2f;2df;2p)	(U)MP2/6-311+G (3d2f;2df;2p)	Based on CCSD(T) in Dunning basis sets (see below)
Further correlation estimates	MP4(SDQ)/6-31+G (d/f;p)	MP4(SDQ)/6-31+G (d/f;p)	
Final correlation estimate	QCISD(T)/6-31G†	CCSD(T)/6-31G†	
Empirical adjustment	Yes	Yes	No
Spin contamination adjustment	Yes	Yes	No
Spin–orbit coupling		Experimental atomic	Computed (CASSCF)
Inner core correction		Approximate	CCSD (full) – CCSD (valence)
Relativistic correction	No	No	Yes (Watson)
G2 MAD (MAX)	0.98 (3.8) kcal/mol	0.87 (2.8) kcal/mol	<0.5 (<2) kcal/mol

$A + B/C^n$ using values from a sequence of calculations in basis sets of level n (see Table 9.10).* Estimates of high-level valence correlation effects are obtained by a sequence of single coupled cluster calculations. The valence

TABLE 9.10

Extrapolation Schemes for the Valence Shell Correlation Energy in W1 Theory

Component	SCF	Valence CCSD	Valence (T)
Extrapolation	$A + B/C^n$	$A + B/l^\beta$	$A + B/l^\beta$
SP calculations	SCF/aug-cc-pVDZ+2d ($n = 2$)	CCSD/aug-cc-pVTZ+2d1f	CCSD(T)/aug-cc-pVDZ+2d
	SCF/aug-cc-pVTZ+2d1f ($n = 3$)	CCSD/aug-cc-pVQZ+2d1f	CCSD(T)/aug-cc-pVTZ+2d
	SCF/aug-cc-AVQZ+2d1f ($n = 4$)		

---

* Later Parthiban and Martin [35] recommended a two-parameter extrapolator of form $A + B/l^5$ defined without reference to the SCF energy value obtained in the smaller basis. This served to enhance the stability of the extrapolation.

correlation part of the total atomization energy is separately extrapolated with the form $A + B/l^\beta$ where the exponent $\beta = 3.22$ is empirically fit to produce best-possible total atomization energies.

Core correlations, scalar relativistic (Watson and mass–velocity) contributions, and spin-orbital coupling as needed are obtained in single-point calculations using the specially decontracted and enhanced Martin–Taylor basis [34]. The core correlation is the difference between the results for CCSD(T) calculations exciting all electrons and only valence electrons. For second-row atoms this calculation requires a substantial investment of time.

Parthiban and Martin [35] have subjected W1 and W2—for which fewer compromises are made—to thorough evaluation by comparison with various subsets of the databases G2-1 and G2-2. For heats of formation, the MAD may be less useful an index of quality since the atomization energies themselves carry experimental errors of about half a kilocalorie per mole (0.6 kcal/mol). The figure of merit for CBS-QB3 generally lies between G2 and G3 values (Table 9.11).

For a case study of the W1 scheme we consider the proton affinities for the very small systems CO and CN anion. These are, in principle, ambidentate; thus, we can evaluate the differences between the HCN and HNC isomers, as well as the HOC and HCO cations (Table 9.12).

The enthalpy change for the isomerization $HCO(+) \rightarrow HOC(+)$ is 38.4 kcal/mol according to the CBS-QB3 scheme and 38.9 kcal/mol in the W1 scheme. The enthalpy change for the reaction $H(+) + CO \rightarrow HCO(+)$ (i.e., the proton affinity for CO) is 139.93 kcal/mol according to CBS-QB3 and 140.39 kcal/mol according to W1. The two available experimental values are 141.5 and 141.9 kcal/mol [36]. The isomerization enthalpy for the conversion $HCN \rightarrow HNC$ is 14.26 kcal/mol according to the CBS-QB3 method and 14.76 kcal/mol according to W1. The experimental value is 14.9 kcal/mol [37].

$CHO(+)$ protonates water, and it appears that it is a stronger acid than $H_3O(+)$ by about 22 kcal/mol. Our W1 value is 22.30 kcal/mol, while CBS-QB3 estimates the value as 22.00 kcal/mol. This is the difference in proton affinities, which for water is evaluated as 161.93 kcal/mol by CBS-QB3 and 162.69 kcal/mol by W1.

It might be of interest to anyone making choices between these methods that all six species could be described within the CBS scheme in the time

TABLE 9.11

Mean Absolute Deviations and Maximum Errors() in a Subset of the G2 Database

Property	Exp. Error	W1	G2	G3
EA (eV)	0.010 (0.037)	0.019 (0.075)	0.065 (0.173)	0.039 (0.159)
IP (eV)	0.011 (0.100)	0.025 (0.129)	0.062 (0.161)	0.044 (0.144)
ΔH_f (kcal/mol)	0.4 (1.2)	0.7 (1.7)	1.5 (5.5)	0.7 (3.2)
PA (kcal/mol)	—	0.44 (0.8)	—	1.2 (2.3)

TABLE 9.12

Thermodynamic Quantities Estimated by Extrapolation Schemes

Species	W1:	ΔH	ΔG	CBS-QB3:	ΔH	ΔG
HCO(+)		−113.593421	−113.616298		−113.401695	−113.424569
CO		−113.369696	−113.392119		−113.178703	−113.201128
HOC(+)		−113.531431	−113.556219		−113.340540	−113.364978
CN(−)		−92.886309	−92.908645		−92.727762	−92.750102
HCN		−93.443440	−93.466294		−93.284059	−93.306892
HNC		−93.419926	−93.443240		−93.261327	−93.284617
H_2O		−76.333703	−76.355129		−76.458282	−76.479704
$H_3O(+)$		−76.717548	−76.738732		−76.717548	−76.738732

required for a single W1 estimation. Whether the increment in accuracy justifies the investment of resources is a question which can be addressed in specific cases.

Aspiration to Exact Description: Quantum Monte Carlo Calculations

In this section we depart from our practice of establishing the theoretical and algorithmic background of methods now widely used in electronic structure modeling. A quantum Monte Carlo (QMC) calculation is not an option in most commercial or shared electronic structure codes. Yet, simply because it is capable, in principle, of exact solutions to the Schroedinger equation, it deserves our attention.

QMC was introduced to chemistry by James Anderson [38,39]. The method itself is venerable; Anderson provides a timeline extending to Wigner's work [40] in the early 1930s. The method was automated during the Manhattan Project at Los Alamos, being used to describe the diffusion of neutrons in simulated chain reactions. One early landmark in the open literature was the famous $M(RT)^2$ paper by Metropolis et al. [41]. The general technique was reviewed by Meyer in 1956 [42].

In compiling this basic survey of the QMC approach to electronic structure modeling, we are guided primarily by the work of Anderson and the 1994 text by Hammond, Lester, and Reynolds (HLR) [43]. Anderson's 1995 review [44] provides an overview of his later work. Lester's edited volumes in 1997 and 2002 contain summaries of more recent advances [45]. Here our purpose is to give a broad impression of the assumptions of the method, and to summarize some of its achievements. It appears to us that for some problems this approach may have unique advantages.

QMC software is available; the most recent public-access code is the QMC@HOME by Grimme and Luchow [46] which is based on Anderson's ideas. Lester's Zori code is also available from his research group [47].

Random (?) Numbers

The heart of any Monte Carlo method is the use of random numbers. We lose no generality if we constrain the range of a set of numbers which may be somehow considered random to a uniform distribution in the closed interval (0,1). A sequence of random numbers should be unpredictable in detail; yet, methods to generate values for Monte Carlo calculations have generally been deterministic. For example, Metropolis et al. [41] used the technique of the "middle square process." A Maple procedure for this method is shown

```
> restart;ndigits:=6;
```
$$ndigits := 6$$
```
> seed:=423456;
```
$$seed := 423456$$
```
> sq:=seed^2;
```
$$sq := 179314983936$$
```
> C1:=round(sq/10^9);
```
$$C1 := 179$$
```
> Val1:=(sq-10^9*C1);
```
$$Val1 := 314983936$$
```
> V1:=Val1-trunc(Val1/10^3);
```
$$V1 := 314668953$$
```
> V1e:=value(Val1-V1);
```
$$V1e := 314983$$

Beginning the middle square process requires a seed. There is a paradox here. The sequence may share some of the properties of a "truly" random set including a uniform distribution over its interval and essentially zero correlation between the jth and $j + n$th member of a sequence; but the sequence is, by its very construction, perfectly predictable. A given seed will always produce the same sequence of descendents. The sequence will repeat itself if the seed is regenerated somewhere in the process. A thorough discussion of the subtleties of random number mathematics is given by Haile [48].

Pseudorandom numbers (PRNs) can be useful in exploratory work where it is desirable to repeat the sequence of variable values. Recurrence may not be a serious drawback if the cycle is long. Then, the steps dictated by the PRNs in that cycle can lead to essentially all possible arrangements of the system—the coverage will be close to "ergodic." Hammond et al. [43] recommend a "linear congruential generator" based on three constants a, b, and m used as

$$I_{j+1} = (aI_j + b)\bmod(m)$$

A seed is required for this generator. One divides the number produced by the expression by $(m-1)$ to normalize the range to the interval (0,1). HLR comment that some real numbers cannot be reached by this process; using the modulus (m) requires that only integer multiples of $(m-1)$ can be produced. The "grain size" is $(m-1)^{-1}$. If m is chosen to equal 2^b, the PRN can distinguish binary choices with precision defined by b. Then, if b is large the set of PRNs in a cycle can approach ergodicity.

Anderson's Intuitive Application of Randomness

Anderson [38] applied a very simple algorithm to the study of H_3^+. His entire program consisted of only about 75 FORTRAN-H statements. (More recently [45] Anderson mentioned his program is still less than 600 lines of FORTRAN.) We will follow Anderson's first development of QMC, for its conceptual clarity. Consider the time-dependent Schroedinger equation as it is usually written

$$i\hbar \frac{\partial \psi}{\partial t} = -\frac{\hbar^2}{2m}\frac{\partial^2 \psi}{\partial x^2} + V\psi$$

Anderson chose a fundamental unit of time

$$\tau = it/\hbar$$

This transforms the Schroedinger equation to

$$\frac{\partial \psi}{\partial \tau} = \frac{\hbar^2}{2m}\frac{\partial^2 \psi}{\partial x^2} - V\psi$$

which can be compared to the standard form of a diffusion equation with a first-order rate process:

$$\frac{\partial C}{\partial t} = D\frac{\partial^2 C}{\partial x^2} - kC$$

Here D is the diffusion constant and k is the first-order rate constant for the decline of the concentration C; this one-dimensional equation is easily generalized to three dimensions. One can assume that the solution is a product of spatial and time-dependent factors. Then, it is easy to separate the equation into space and time parts. Then,

$$H\psi = E\psi \quad \text{where} \quad H_{op} = -\frac{\hbar^2}{2m}\frac{\partial^2}{\partial x^2} + V \quad \text{and} \quad \frac{\partial \psi}{\partial \tau} = -E\psi$$

The total wave function assumes the form

$$\psi(x,\tau) = \psi(x)e^{-E\tau}$$

For large values of the scaled time the wave function will approach the time-independent factor. At any position in space, the wave function amplitude (and the probability distribution's value at that point) will vary in time. Anderson postulated "psips," nonphysical entities which help define the probability distribution. As the probability increased or decreased such psips could appear, diffuse, or disappear. The motion of these entities (which in more recent work are termed "walkers") is governed by the diffusion form of the Schroedinger equation.

An object diffuses into the neighborhood of a point readily if it is more stable there, but reluctantly and less frequently if it is less stable there. Diffusion is typically Gaussian. Anderson generated six coordinates for the two electrons. The probability of a psip appearing at a separation Δx from an origin is the Gaussian W. A random number generator in the interval $(0,1)$ was used for each psip with a weight function of

$$W(\Delta x) = \frac{1}{\sqrt{2\pi}\sigma} \exp\left(-\frac{(\Delta x)^2}{2\sigma^2}\right) \quad (-\infty < x < \infty) \text{ with } \sigma = [2D(\Delta \tau)]^{1/2}$$

Here the "diffusion constant" D has the units defined in the Schroedinger equation. The analogy of the rate constant in the diffusion equation is the potential. Anderson used two rules as for the probability of "birth" or "death" of a psip:

1. $P_b = -k\Delta t = -(V - V_{ref})(\Delta \tau)$ if V is less than V_{ref} or $= 0$ if $V \geq V_{ref}$
2. $P_d = +k\Delta t = (V - V_{ref})(\Delta \tau)$ if V is greater than V_{ref} or $= 0$ if $V \leq V_{ref}$

To maintain normalization, the number of psips must stay near 1000 psips. This is assured by requiring

$$V_{ref} = \bar{V} - \frac{N - 1000}{1000(\Delta \tau)}$$

where $\bar{V}$ is the average of the potential energy for all the psips at a given iteration.

Recall that the original scaled time-dependent Schroedinger equation was

$$\frac{\partial \psi}{\partial \tau} = \frac{\nabla_1^2}{2} + \frac{\nabla_2^2}{2} - (V - V_{ref})\psi = -(E - V_{ref})\psi$$

Rearranging, we can isolate the fractional change of the wave function in the unit of scaled time:

$$(E - V_{ref}) = -\frac{1}{\psi}\frac{\partial \psi}{\partial \tau} = -\frac{1}{N}\frac{\partial N}{\partial \tau}$$

This equation states that the fractional change in the wave function is also the fractional change in psips. Diffusion does not change the number of psips so that the only change in that number is defined by "birth" and "death" as defined above.

$$\frac{\partial N}{\partial \tau} = -\sum_{i=1}^{N}(V_i - V_{ref})$$

Thus, the energy can be defined (to 1 part in 1000 psips) by the average potential energy relative to the value of V_{ref}

$$E - V_{ref} = -\frac{1}{N}\frac{\partial N}{\partial \tau} = \frac{1}{N}\sum_{i-1}^{N}(V_i - V_{ref}) = \bar{V} - V_{ref}$$

Using this process with scaled time-intervals varying between 0.001 and 0.010 and up to 400 steps, the mean energy for the H_3 cation was obtained as -1.344 hartree. This value is comparable to the best values calculated by a variety of other methods [38].

Anderson's ideas illustrate some persistent themes in the development of QMC. First, the random process is "guided" by the known properties of the diffusion process. Second, the eventual value of the energy is determined by an average of "local energy" values. The QMC energy is a statistical property, which means a standard error can be defined. Third, the efficiency of the calculation is sensitive to the size of the time step.

This example is deceptively simple. For a system with more than two electrons, the Pauli principle will require a nodal surface to assure antisymmetry. While this first application would require considerable development to be applicable to many-electron systems, it was a bold step into a radically novel mode of electronic structure modeling.

An Early Example of Diffusion Quantum Monte Carlo Calculation

In the second of Anderson's pioneering papers [39], QMC was extended to many-electron systems and unpaired electron spins. This represented a major development of the diffusion quantum Monte Carlo (DQMC) approach, which has been extended by Caffarel and Claverie [49].*

---

*The major alternative to the diffusion-based quantum Monte Carlo method is a hybrid method with variational aspects. The mathematics was developed by Reynolds et al. [50]; the method has been widely applied by Lester [51].

Anderson defined $\tau = it$, obtaining a slightly revised form of the Schroedinger equation:

$$\frac{\partial \psi(x,\tau)}{\partial \tau} = \frac{1}{2} \frac{\partial^2 \psi(x,\tau)}{\partial x^2} - (V - V_{ref})\psi(x,\tau)$$

Again the value of V_{ref} is chosen to maintain a constant number of psips. The total time-dependent wave function can be factored into a time-independent part multiplied by an energy-dependent exponential:

$$\psi(x,t) = \psi(x)e^{-(E-V_{ref})\tau}$$

Once again we see that for large values of τ the total wave function will tend to the time-independent solution $\psi(x)$. For short (imaginary) times the diffusion equation is dominant but longer (imaginary) times are dominated by the exponential terms. For numerical accuracy, short time steps are desirable but then several steps are needed to reach a steady state where the average energy can be calculated with a small variance arising from small fluctuations in the value calculated as the "local energy." Thus, Anderson previously used the differential expressions

$$P_b = -k\Delta t = -(V - V_{ref})(\Delta \tau)$$
$$P_d = +k\Delta t = (V - V_{ref})(\Delta \tau)$$

for the birth or death of a psip. Here Anderson modified the birth/death rule for a psip to the integrated forms

1. $P_b = \exp[-(V - V_{ref}) \Delta \tau] - 1$ if V is less than V_{ref} or $= 0$ if $V \geq V_{ref}$
2. $P_d = 1 - \exp[-(V - V_{ref}) \Delta \tau]$ if V is greater than V_{ref} or $= 0$ if $V \leq V_{ref}$

This revision allowed longer time steps without loss of accuracy and reduced the total computer run times by a factor of about 10. Our understanding of this innovation is that the exponential term of the diffusion process can be integrated independent of the diffusion Gaussian term for short times and the integrated form is more accurate and efficient than the simple finite difference used in the previous paper.

Several other improvements and modifications of the first primitive algorithm were made in Anderson's second paper [39]. For example, it is beneficial to place initial psips near nuclei. Further, one may use an average of the nuclear potential at a radius r_c from a nucleus of charge Z as the constant value for all points within the volume enclosed by a shell of that radius, i.e., the "crossing/recrossing" situation.

$$\langle V \rangle = -\frac{3}{2} \frac{Z}{r_c}$$

Anderson estimated that the probability of moving a psip from $x' \rightarrow x''$ is given by the Gaussian diffusion process as

$$P_{AB} = \frac{1}{\sqrt{2\pi}\sigma} \exp\left[-\frac{(x'' - x')^2}{2\sigma^2} \right]$$

or

$$P_{BA} = \frac{1}{\sqrt{2\pi}\sigma} \exp\left[-\frac{(x'' + x')^2}{2\sigma^2} \right]$$

Anderson reasoned that the step into the negative region would be likely to be followed by a step crossing back to a position on the original side of the node. Thus, he stored the values of x' and x'' until the end of the step and estimated the probability of the crossing–recrossing sequence as the ratio of the two probabilities. This is easily shown to be

$$P_X = \exp\left(-\frac{4x'x''}{2\sigma^2} \right)$$

If P_X for a particular psip were greater than a random number in the interval (0,1), that psip was erased. Conversely, the probability of a psip moving without crossing and recrossing the node plane is given by $(1 - P_X)$. The product $P_X(1 - P_X)$ of these two probabilities was used to create or delete a given psip.

However, the most significant problem addressed in the second paper (and still significant today) is how to treat the problem of nodes (i.e., sign changes) in the wave function. This occurs as a consequence either of angular momentum (the form of p, d, orbitals) or of the antisymmetry of the many-electron wave function. When the sign of a wave function changes, the birth and death psip rules are reversed. The way Anderson avoided the problem was to restrict the movement of the psips, forcing them to stay on the same side of the node. In most cases, nodes due to p orbitals, d orbitals, etc. will have curved shapes so that a movement of a psip near a nodal plane may pass through a curved part of the nodal surface and actually come back on to the same side of the node as it left. This is a persistent problem in Monte Carlo technology; a number of ways have been proposed to cope with the puzzle.

Anderson [39] noted that excited states would introduce additional nodes. His simple analysis based on the nodal patterns of the particle-in-a-box problem led to the notion of the "fixed node." Such nodes also arise from indistinguishability of electrons. Anderson divided the configuration space for each pair of indistinguishable pairs with a symmetric surface on which $\psi = 0$ and moved the psips only in one part (usually the positive part) of the space. He applied this idea to the description of the triplet state of the hydrogen molecule and the Be atom. For Be he defined a radial node between the $1s^2$ and $2s^2$ shells.

This idea of a fixed-node surface comes up again and again in molecular QMC calculations. On the one hand restriction of random movement to certain regions of space is a simplification in the dynamics, but on the other hand it is not always clear where the nodes are located. In Anderson's examples symmetry was a considerable advantage, but if we consider any partially ionic bond between different atoms it is clear that the nodal surfaces will not be known until the problem is solved. Approximate wave functions are used as guiding functions, defining the location of nodes.

Recently Aspuru-Guzik and Lester have developed the software suite "Zori" [47] which offers a number of variations and paths through forms of QMC in one program based primarily on the fixed-node approximation. Here is a sketch of the structure of a fixed-node QMC calculation:

1. Obtain an RHF or UHF solution for a given molecule to obtain the SCF energy ($E_T = V_{ref}$) and the "guiding wave function." Any geometry optimization of the molecule should be done here. (Zori can use input from GAMESS or ADF.) An MCSCF wave function is highly desirable at this point.

2. Set up $\Psi = \Phi e^U$, a correlated guide function. Use initial guess values of the Boys–Handy parameters in the correlated part of the wave function e^U and use the RHF/UHF wave function as Φ. The parameters in U may/may not be optimized with further runs.

3. Set up a number of walkers with randomized initial positions.

4. Calculate the values of $F(x)$ which weights the probability of approving a step and cycle through the list of walkers. Save the values of $\Psi(x)$ and the individual values of the local energy $E_L(x)$ for each value of x. Use an algorithm to avoid crossing a node in the multielectron wave function.

5. Compute the average value and the variance of the local energy for this set of walkers. Print out the energy and variance to evaluate convergence.

6. If the variance meets a desired minimum threshold, proceed to step 7; if not go back to step 4 and move the set of walkers again from their present positions depending on the Metropolis yes/no criteria.

7. Use all the accumulated values of $\Psi(x)$ and the values of $E_L(x)$ to compute a final energy and variance as well as to calculate any desired property from the set of values of $\Psi(x)$ at the values of x.

The key words used to run Zori for a variety of options are given by Aspuru-Guzik et al. [47]. With use of a CI guide function in a 15-member basis of Gaussian spheres to define the nodes for He atom, after 10,000 steps of length 0.001 scaled time-units the energy converges to -2.903573 hartree. This is exact to the reported number of decimal places.

Shortcuts: Pseudopotentials and the Sparse Algorithm

The QMC approach to electronic structure calculations is computationally intensive; thus, enhancing its efficiency is urgent. In chemical applications for which energy differences are most significant, the use of pseudopotentials for inner core electrons is appealing. Sokolova and Luchow [52] applied diffusion-QMC to study TiC using pseudopotentials and Hammond et al. [53] have developed core pseudopotentials for inner orbitals. They investigated Na_2 and NaH and estimated the ionization potential and electron affinitiy of Li, Na, and Mg. The calculated ionization potentials were comparable with the best available computational results. The binding energies for Na_2 and NaH were competitive with all-electron MCSCF results. The values obtained with pseudopotentials were as accurate as the results of all-electron QMC calculations, and the calculations with pseudopotentials were almost a factor of 1000 faster. When pseudopotentials are used, the energy obtained is the valence shell energy and so cannot be compared to a total energy value.

Formaldehyde (Again) by QMC

Galek et al. [54] have described formaldehyde by both variational and fixed-node QMC. Their geometry was not optimized within QMC and their guide function was an HF single determinant. They employed the ADF representations of STOs, specifically the TZP basis, which satisfied the limiting condition on the wave function's derivative at the nuclear positions (the "cusp condition"). The motion of electrons was governed in part by the Boys–Handy Jastrow correlation function. As Table 9.13 shows, the fixed-node method is remarkably accurate.

Schuetz, Frenklach, Kollias, and Lester (SFKL) [55] studied formaldehyde by QMC, using core pseudopotentials. This is one of the first polyatomic vibrational analysis treatments using QMC; it was a major advance in QMC applications. Their exploration of the potential energy surface (PES) consisted of 15 QMC calculations of the energy at selected geometries, and a fit to those points by a polynomial with 10 parameters. This PES was then used

TABLE 9.13

Formaldehyde Total Energies (Hartrees)

E (HF)	E0	E (VMC)	% E corr	E (FN)	% E corr
-113.8958	-114.5100	-114.1510 (10)	41.5	-114.4689 (14)	93.3

Note: The uncertainties are listed in parentheses, as entries in following decimal places.

TABLE 9.14

Optimized Bond Lengths and Angles for Formaldehyde

Method	Valence-E (au)	R(C=O), A	R(C—H), A	∠ (COH), deg.
VMC				
HF	−22.7868295	1.197(15)	1.101(15)	121.3(2.25)
MCSCF	−22.7984708	1.204(18)	1.08(3)	120.8(1.25)
DMC				
HF	−22.8445736	1.20(1)	1.11(1)	121.4(1.0)
MCSCF	−22.8515749	1.212(15)	1.08(15)	120.2(1.15)
Experiment		1.21(1)	1.12(1)	121(1)

Note: The uncertainties are listed in parentheses, as entries in following decimal places. VMC = variational; DMC = diffusional; HF = Hartree–Fock guide functions; MCSCF = multiconfigurational self-consistent-field guide functions.

to optimize the C—H bond length, the C=O bond length, and the HCH bond angle. All three values were found to fall close to experimental values; the most serious discrepancy was in the C–H bond length, which was found to be a bit short (Table 9.14).

The fitted PES polynomial determined three of the six vibrational frequencies. Table 9.15 shows the DMC results to be better than the VMC results using the same guide function, whether HF or MCSCF. The quality of these estimates is somewhat disappointing in view of the substantial computational investment in the method; DFT will typically reproduce vibrational frequencies within 50 cm^{-1} or better.

Aspuru-Guzik et al. [56] developed methods which permit QMC calculations on large systems, up to 400 electrons. Their breakthrough idea can be illustrated by a simple example. Imagine a collection of five hydrogen diatomics. In your guide function, use only a 1s orbital on each atom and transform the set of atomic orbitals into localized sigma orbitals. The interaction of electrons in sigma orbitals removed by a considerable distance will be small. Any aspect of the calculation—such as the reestablishment of

TABLE 9.15

Selected Formaldehyde Vibrational Frequencies

Method	ν(cm^{-1}, C=O)	ν(cm^{-1}, C—H)	ν(cm^{-1}, CH—bend)
VMC			
HF	1820 (194)	3311 (43)	1275 (88)
MCSCF	1639 (59)	2759 (85)	2327 (173)
DMC			
HF	1785 (56)	3109 (60)	2087 (55)
MCSCF	1732 (106)	2446 (75)	1975 (146)
Experiment	1764 (17)	2944 (100)	1563 (15)

Source: Schuetz, C.A., Frenklach, M., Kollias, A.C., et al., *J. Chem. Phys.*, 119, 9386, 2003.

antisymmetry with respect to electron exchange—that depends on that interaction can then be approximated. In the sparse algorithm of the Zori code the orbitals of the guiding function are localized. The molecule is enclosed by a volume of test grid points. Zori evaluates the overlap product of the localized orbitals at such points. If the overlap product is very small (less than 10^{-11}) the task of antisymmetrizing is abandoned. Owing to the localization, a coarse grid can give accurate results. Calculations on a series of large molecules with up to 400 electrons maintained the same accuracy as calculations without the screening while reducing the total run time by roughly 90% in all cases. The most spectacular result is that for more than about 80 electrons the run times only increase linearly in the number of electrons [56,57]. This innovation brings QMC to a point of routine use for large molecules, although one still needs a large computer cluster.

There is so far only one study comparing QMC results for the G1 set of molecules [58,59]. That test [60] yielded an average maximum deviation from experiment (atomization energies) of 2.9 kcal/mol with a maximum deviation of 14 kcal/mol for diatomic phosphorous. This is hardly any better than using DFT with the B3LYP functional and the QMC calculation is clearly a much more computationally demanding process. Grossman [60] compares his results to the accuracy available from CCSD(T)/cc-pVQZ but DFT and the G1 theory yield similar uncertainty levels. Part of that problem might be due to the use of pseudopotentials for first and second inner cores instead of all electron treatments. Grossman [60] estimates that ~2 kcal/mol error in the P_2 calculation is due to the use of pseudopotentials and 3.5 kcal/mol error is due to using a single-determinant reference.

Conclusion

We have seen that modern methods of electronic structure modeling can aspire to experimental accuracy for thermodynamic quantities for gas phase species of modest size. This is a considerable achievement—the result of major advances in theory, algorithms, and computer power, both in software and hardware. Yet this triumph—that is what it surely must be considered—is so limited in scope, considering the vast realms of chemical species which lie beyond reach of the extrapolation and post-HF methods. The quest to get the right answer for the right reasons seems to have confined us to a small corner of the chemical universe. If we wish to characterize larger systems of great chemical importance, are we not in the unfortunate position of Winston Churchill's golfer, who tries so hard to put a little ball into a little hole, with tools singularly ill-suited to the job? Yet we are undeterred, and computational studies of extensive systems are commonplace, using simplified and approximate versions of the methods we now know capture a broad range of chemical phenomena when they can be applied. Use of approximate methods requires not just access to the most powerful computers and most accurate methods, but experience to help us develop the judgment allowing us to frame questions which can be settled by the methods we can bring to bear. Asking answerable questions is one definition of the scientific enterprise.

Communications with Prof. Bill Lester and Prof. Alan Aspuru-Guzik were extremely helpful. Prof. Jim Harrison was helpful in interpreting Anderson's H_2 calculation and a seminar by Prof. Jim Anderson at Virginia Commonwealth University several years ago inspired our attempt to try and explain QMC to graduate students and postdocs who now have access to parallel computing and so can embark on a new wave of high accuracy calculations in quantum chemistry.

References

1. Pople, J.A., Head-Gordon, M., Fox, D.J., et al., *J. Chem. Phys.* 90:5622, 1989.
2. Curtiss, L.A., Jones, C., Trucks, G.W., et al., *J. Chem. Phys.* 93:2537, 1990.
3. Curtiss, L.A., Raghavachari, K., Trucks, G.W., et al., *J. Chem. Phys.* 94:7221, 1991.
4. Curtiss, L.A., Raghavachari, K., Redfern, P.C., et al., *J. Chem. Phys.* 93:1063, 1997.
5. Curtiss, L.A., Redfern, P.C., Raghavichari, K., et al., *J. Chem. Phys.* 109:42, 1998.
6. Curtiss, L.A., Raghavachari, L., Redfern, P.C., et al., *J. Chem. Phys.* 109:7764, 1998.
7. Foresman, J.B. and Frisch, A., *Exploring Chemistry with Electronic Structure Methods*, 2nd Ed., Gaussian Inc., Pittsburg, PA, 1996.
8. Jensen, F., *Introduction to Computational Chemistry*, Wiley, New York, 1999.
9. http://simons.hec.utah.edu/school/index.html
10. Szabo, A. and Ostlund, N.S., *Modern Quantum Chemistry: Introduction to Advanced Electronic Structure Theory*, Macmillan, New York, NY, 1982.
11. Crawford, T.D. and Schaefer, H.F. III, *Rev. Comp. Chem.* 14:33, 2000.
12. (a) Bartlett, R.J., *Ann. Rev. Phys. Chem.* 32:359, 1981; (b) Bartlett, R.J. and Stanton, J.F., *Rev. Comp. Chem.* 5:65, 1994; (c) Bartlett, R.J. and Musial, M., *Rev. Mod. Phys.* 79:291, 2007.
13. Čížek, J. and Paldus, J., *Int. J. Quantum Chem.* 5:359, 1971.
14. Brillouin, L., *J. Phys.* 3:373, 1932. (Brillouin's theorem shows that the SCF procedure of variational orbital optimization is equivalent to guaranteeing that the SCF wave function does not mix with any singly excited configurations.)
15. Bartlett, R.J. and Watts, J.D., *ACES II*, in *Encyclopedia of Computational Chemistry*, Eds. Schleyer, P.V.R., et al., Wiley, New York, NY, 1998.
16. The requirement for Brueckner orbitals is that in the configuration-interaction expansion of the exact wavefunctions in the intermediate normalization there are no singly-excited configurations.
17. Raghavachari, R., Trucks, G.W., Pople, J.A., et al., *Chem. Phys. Lett.* 157:479, 1989.
18. Lee, T.J. and Scuseria, G.E., Achieving chemical accuracy with coupled-cluster theory, in *Quantum Mechanical Electronic Structure Calculations with Chemical Accuracy*, Ed. Langhoff, S.R., Kluwer Academic Publishers, Dordrecht, The Netherlands, 1995.
19. Feller, D. and Peterson, K.A., *J. Chem. Phys.* 108:154, 1998.
20. Wilson, A., van Mourik, T., and Dunning, Jr. T.H., *J. Mol. Struct.* (*Theochem*) 388:339, 1997 and citations therein.

21. www.ccs.uky.edu/workshop/SURA2003/presentation/ThomDunning.ppt
22. Dunning, Jr. T.H., *J. Phys. Chem. A* 104:9062, 2000; Bak, K.L., Jørgensen, P., Olsen, J., Helgaker, T., Klopper, W., *J. Chem. Phys.* 112:9229, 2000.
23. Frisch, M.J., Trucks, G.W., Schlegel, H.B., et al., Gaussian 03, Revision C.02, Gaussian Inc., Wallingford, CT, 2004.
24. Crawford, T.D., Virginia Polytechnic Institute and State University Department of Chemistry (private communication).
25. Cole, S.J., Purvis, III G.D., and Bartlett, R.J., *Chem. Phys. Lett.* 113:271, 1985.
26. Trindle, C., *J. Org. Chem.* 68:9669, 2003.
27. Su, M.-D., Liao, H.-Y., Chung, W.-S., et al., *J. Org. Chem.* 64:6710, 1999.
28. Montgomery, Jr. J.A., Ochterski, J.W., and Petersson, G.A., *J. Chem. Phys.* 101:5900, 1994.
29. Martin, J.M.L. and De Oliveira, G., *J. Chem. Phys.* 111:1843, 1999.
30. Boese, A.D., Oren, M., Atasoylu, O., et al., *J. Chem. Phys.* 120:4129, 2004.
31. Ochterski, J.W., Petersson, G.A., and Montgomery, Jr. J.A., *J. Chem. Phys.* 104:2598, 1996.
32. Montgomery, Jr. J.A., Frisch, M.J., Ochterski, J.W., et al., *J. Chem. Phys.* 109:6505, 1998.
33. Montgomery, Jr. J.A., Frisch, M.J., Ochterski, J.W., et al., *J. Chem. Phys.* 110:2822, 1998.
34. Martin, J.M.L. and Taylor, P.R., *Chem. Phys. Lett.* 225:473, 1994.
35. Parthiban, S. and Martin, J.M.L., *J. Chem. Phys.* 114:6014, 2001.
36. Szulejko, J.E. and McMahon, T.B., *J. Am. Chem. Soc.* 115:7839, 1993; Lias, S.C., Bartimess, J.E., Liebman, J.F., et al., *J. Phys. Chem.* Ref data Suppl 1:17, 1988.
37. Gurvich, L.V., Veyts, I.V., and Alcock, C.B., *Thermodynamic Properties of Individual Substances*, 4th Ed., Hemisphere Publishing Co., New York, 1989; *NIST Chemistry Webbook* (http://webbook.nist.gov/chemistry).
38. Anderson, J.B., *J. Chem. Phys.* 63:1499, 1975.
39. Anderson, J.B., *J. Chem. Phys.* 65:4121, 1976.
40. Wigner, E.P., *Phys. Rev.* 40:749, 1932.
41. Metropolis, N., Rosenbluth, A.W., Rosenbluth, M.N., et al., *J. Chem. Phys.* 21:1087, 1953.
42. Meyer, H.A., *Symposium on Monte Carlo Methods*, Wiley, New York, 1956.
43. Hammond, B.L., Lester, W.A., and Reynolds, P.J., *Monte Carlo Methods in Ab Initio Quantum Chemistry*, World Scientific, Singapore, 1994.
44. Anderson, J.B., *Quantum Mechanical Electronic Structure Calculations with Chemical Accuracy*, Ed. Langhoff, S.R., Kluwer Academic Publishers, Dordrecht, The Netherlands, 1995.
45. Lester, W.A., Ed. *Recent Advances in Quantum Monte Carlo Methods (I)*, World Scientific, Singapore, 2002; ibid. 1997.
46. Grimme, S. and Luchow, A., QMC@HOME, http://qah.uni-muenster.de/
47. Aspuru-Guzik, A., Salomon-Ferrer, R., Austin, B., et al., *J. Comput. Chem.* 26:856, 2005, Zori v1.0.
48. Haile, J.M., *Molecular Dynamics Simulation*, Wiley, New York, Appendix H, 1992.
49. Caffarel, M. and Claverie, P., *J. Chem. Phys.* 88:1088, 1988; ibid. 88:1100.
50. Reynolds, P.J., Ceperley, D.M., Alder, B.J., et al., *J. Chem. Phys.* 77:5593, 1982.
51. Lester, W.A., Jr. and Hammond, B.L., *Ann. Rev. Phys. Chem.* 41:283–311, 1990.
52. Sokolova, S. and Luchow, A., *Chem. Phys. Lett.* 320:421, 2000.
53. Hammond, B.L., Reynolds, P.J., and Lester, W.A., *J. Chem. Phys.* 87:1130, 1987.
54. Galek, P.T.A., Handy, N.C., and Lester, W.A., *Mol. Phys.* 104:3069, 2006.

55. Schuetz, C.A., Frenklach, M., Kollias, A.C., et al., *J. Chem. Phys.* 119:9386, 2003.
56. Aspuru-Guzik, A., Salomon-Ferrer, R., Austin, B., et al., *J. Comput. Chem.* 26:708, 2005.
57. Lester, W.A. and Salomon-Ferrer, R., *J. Mol. Struct.* (*Theochem*) 771:51, 2006.
58. Pople, J.A., Head-Gordon, M., Fox, D.J., et al., *J. Chem. Phys.* 90:5622, 1989.
59. Curtis, L.A., Jones, C., Trucks, G.W., et al., *J. Chem. Phys.* 93:2537, 1990.
60. Grossman, J.C., *J. Chem. Phys.* 117:1434, 2002.
61. Merzbacher, E., *Quantum Mechanics*, 2nd Edition, John Wiley and Sons, New York, 1970.
62. Hehre Warren, J., A guide to molecular mechanics and quantum chemical calculations, Wavefunction, Irvine, CA, 2003.

10

Modeling the Coulomb Hole

So far our studies have followed the development of the wave function, certainly the most influential idea in quantum chemistry's first half century. It seems safe to say that its importance will continue. However, an increasingly prominent position is being taken by methods that focus on the density. To effect a transition in this direction, we pause to consider an unusual way to represent the properties of an electron distribution, beginning with an emphasis on the interelectronic interaction rather than adding electron correlation as an afterthought. The central concept here is the "hole" surrounding each electron which can arise from the fundamental statistics of fermions or their mutual repulsion according to Coulomb's law.

If we define a spatial form for an electron in a field of one nucleus, such as that given by the exact solution for He cation, we have a very clear idea of the probability of an electron to be located in the neighborhood of any point. When a second electron is added to the system, we are faced with the complications of electron correlation. We will introduce the pair probability distribution or correlation function as a means for discussion of this problem.

$$\Pr(r_1, r_2) = \Pr(r, u); \quad u = r_2 - r_1$$

This function defines the probability of electron 1 being at r and electron 2 being separated from electron 1 by the distance u. The probabilities will be described by the wave function, according to the Copenhagen interpretation.

$$\int d\mathbf{r}_1 \, d\Omega |\psi(\mathbf{r}_1, \mathbf{r}_1 + \mathbf{u})|^2 = f(u)$$

Here $d\Omega$ is the element of solid angle; integration yields the spherical average of the three-dimensional hole function.

The simplest function describing two electrons is the Hartree product

$$\Psi_{\text{Hartree}} = 1s^\alpha(1)2s^\beta(2) \quad \text{or} \quad 1s^\alpha(1)2s^\alpha(2)$$

The Hartree product, which labels each electron uniquely, violates the fundamental indistinguishability and antisymmetry of electrons. By definition, it represents no correlation at all, neither the influence on electrons' relative distribution in space arising from antisymmetry nor the influence of interelectronic repulsion.

The Fermi Hole and Exchange–Correlation

The Fermi hole (or exchange–correlation hole) was defined by Boyd and Coulson [1] as the difference in pair probability distribution functions for antisymmetrized product and the simple Hartree product. The variable u is the distance between electrons

$$f_C(u) - f_N(u) = \left\langle \psi_C \left| \sum_{i<j} \delta(r_{12} - u) \right| \psi_C \right\rangle - \left\langle \psi_N \left| \sum_{i<j} \delta(r_{12} - u) \right| \psi_N \right\rangle$$

Here C denotes a correlated wave function; that is, a Slater determinant [2], of Hartree orbitals, and N denotes a noncorrelated wave function, the Hartree product. Alternatively, one might reoptimize the Hartree orbitals once they are placed in the determinant, and consider the difference between the pair distribution functions defined by a product of Hartree orbitals and by the determinant of the best HF orbitals. Moiseyev et al. [3] present just such graphs for He ^{3}S and isoelectronic ions. The Fermi correlation operates at short range, less than an (Z-scaled) interelectronic distance of about four units—for He, about 1 bohr. The charge excluded from this sphere is not uniformly dispersed but moves into a radial shell just outside the exclusion zone, and within ten units (Figure 10.1).

Coulomb Correlation Hole

The Hartree–Fock (HF) method represents the influence of the repulsion of electrons by a mean field, incorporating the exchange but no details of individual electron–electron repulsion. That is, there is an intrinsic coulombic correlation error which may account for 1%–2% of the total energy. Consider the case of He, in which the SCF function is pretty well represented by an antisymmetrized product of one-electron functions. If one uses correlation-corrected wave functions such as Hylleraas functions which introduce the interelectronic distance into the variational optimization [4] or large-scale configuration interaction (CI), which in principle is perfectly flexible, the flaw is corrected and the energy is lowered. The Coulomb hole function can be envisioned by the difference of the pair distribution

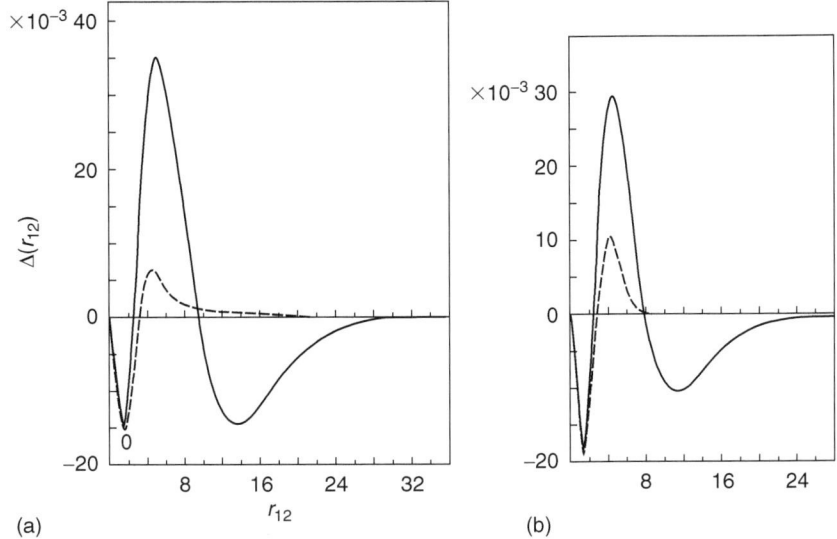

FIGURE 10.1
Fermi holes from Moiseyev et al. [3] displaying the short-range displacement of electrons from the neighborhood of an electron of like spin. The horizontal axis is the distance between electrons in bohrs; the vertical axis is the density difference or hole function defined by Boyd and Coulson [1]. Dashed lines display differences in density between that formed with Hartree orbitals defining a simple product wave function and those orbitals properly antisymmetrized. Reoptimizing the orbitals in the determinantal form alters the Fermi hole at long range. Graph(a) refers to the hole defined with Hartree orbitals; (b) refers to the hole obtained after reoptimizing the Hartree orbitals within the antisymmetrized form.

functions computed with a correlation-corrected wave function and the pair distribution function defined by the Hartree–Fock wave function.

Once again correlation forces a decrease in density in the neighborhood of an electron at position r_0. The "Coulomb hole graph" illustrates the effect of the more flexible wave function's description of the charge density in the correlation region around each electron. Plots of the Coulomb hole have been obtained in a number of studies [5–9]. Figure 10.2 shows the general form of the Coulomb hole. In Banyard and Mobbs' more detailed diagrams [7], the departure of the radially averaged density calculated with correlated wave functions from density calculated with HF wave functions is shown for several choices of correlated function. Regardless of the function chosen, the holes are larger for the 2s (L-shell) region; this reflects the lesser effective nuclear charge. But while the holes for the K shell are smaller than the holes for the L shell, the shapes of the two kinds of hole are similar. This suggests that the hole size (i.e., the parameter ω) might be scaled according to the effective nuclear charges.

In the DFT chapter, we discuss the exchange and correlation holes in more detail. We quote the key point here, that in this context the hole is linked to both the exchange and correlation energy.

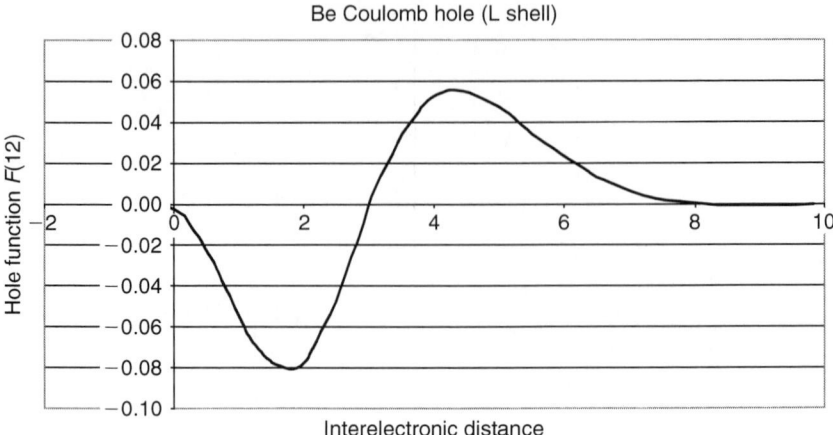

FIGURE 10.2

Representation of the Coulomb hole for the 2s shell for Be atom described by Banyard and Mobbs [7]. The horizontal axis is interelectronic separation in bohrs; the vertical axis is the hole function, defined by a difference in density between RHF and correlated wave functions. A variety of well-correlated functions agree that charge is displaced from the immediate vicinity of a specific 2s electron, but is redistributed to a shell with slightly larger radius. This permits avoidance of direct contact of one electron with another, but permits clustering of electrons around a nuclear center.

$$E_{XC} = \int dr_1 \, dr_2 \frac{\rho(r_1)h(r_1,r_2)}{2|r_1 - r_2|} = \int dr \, du \frac{\rho(r)h(r+u)}{2u}$$

Here u is the interelectronic distance. To emphasize the relation between the hole and the two-electron term in the perfect pairing single-determinant energy we write

$$E_{XC} = \int dr \, du \sum_{I\mu\sigma} C_{I\mu} C_{I\sigma} \chi_\mu(r)\chi_\sigma(r) \frac{h(r,u)}{u}$$

For the HF method there is by definition no Coulomb correlation, but there is exact exchange–correlation.

$$E_{XC}^{HF} = E_X = -\sum_{IJ} K_{IJ} = -\sum_{IJ} \langle IJ|JI \rangle$$

We will find convenient an expression of the exchange in the AO basis.

$$E_X^{HF} = -\sum_{IJ,\mu\nu\rho\sigma} C_{I\mu} C_{J\nu} C_{Jp} C_{I\sigma} \langle \mu\nu|\rho\sigma \rangle$$

This helps to make clear that any change in the two-electron integrals is actually an assumption on the form of the correlation hole.

In 1989, Chakravorty and Clementi [10] published a way to introduce a "soft Coulomb hole" directly into the two-electron integrals of the Hartree–Fock–Roothaan SCF equation. The two-electron integrals are to be modified from

$$\left\langle \mu\nu \left| \frac{1}{r_{12}} \right| \rho\sigma \right\rangle = R$$

to

$$\left\langle \mu\nu \left| [1 - \exp(-\omega r_{12}^2)/r_{12}] \right| \rho\sigma \right\rangle = R^{CC} = R - W^{CC}$$

Define the operator h

$$h = \frac{-\exp(-\omega r_{12}^2)}{r_{12}}$$

At short interelectron distance the operator h is linear in r_{12} and the operator approaches zero at large interelectronic separations. The parameter ω defines the range of the effect. The single-determinant energy becomes

$$E = E_{HF} + \sum_{IJ,\mu\nu\rho\sigma} 2C_{I\mu}C_{I\nu}C_{J\rho}C_{J\sigma}\langle \mu\rho|h|\nu\sigma \rangle - \sum_{IJ,\mu\nu\rho\sigma} C_{I\mu}C_{J\nu}C_{J\rho}C_{I\sigma}\langle \mu\rho|h|\nu\sigma \rangle$$

The post-Hartree–Fock energy correction is made explicit. It appears that the correction must be negative. Note that the CC device defines a spherical hole. The work of Boyd and Yee [8] indicates that the Coulomb hole is not spherical, but analysis by Perdew et al. [11] shows that only the spherical average of the exchange–correlation hole is required to define an effective energy functional for density functional calculations.

Take the simplest wave function for helium

$$\psi(r_1,r_2) = N\exp(-\alpha[r_1 + r_2]) = N\exp(-2\alpha r - \alpha u); \quad u = r_1 - r_2$$

and form the probability density of the first electron being near **r** and the second electron being **u** removed from it.

$$\psi^2\,d\tau_1\,d\tau_2 = N^2\exp(-4\alpha r)r^2\,dr\,d\Omega_r\exp(-2\alpha u)u^2\,du\,d\Omega_u$$

This can be written as

$$F(r)\,d\tau_r\,G(u)u^2\,du\,d\Omega_u$$

The u^2G factor defines the distribution of relative distances after spherical averaging. For the uncorrelated wave function, it is defined by one simple

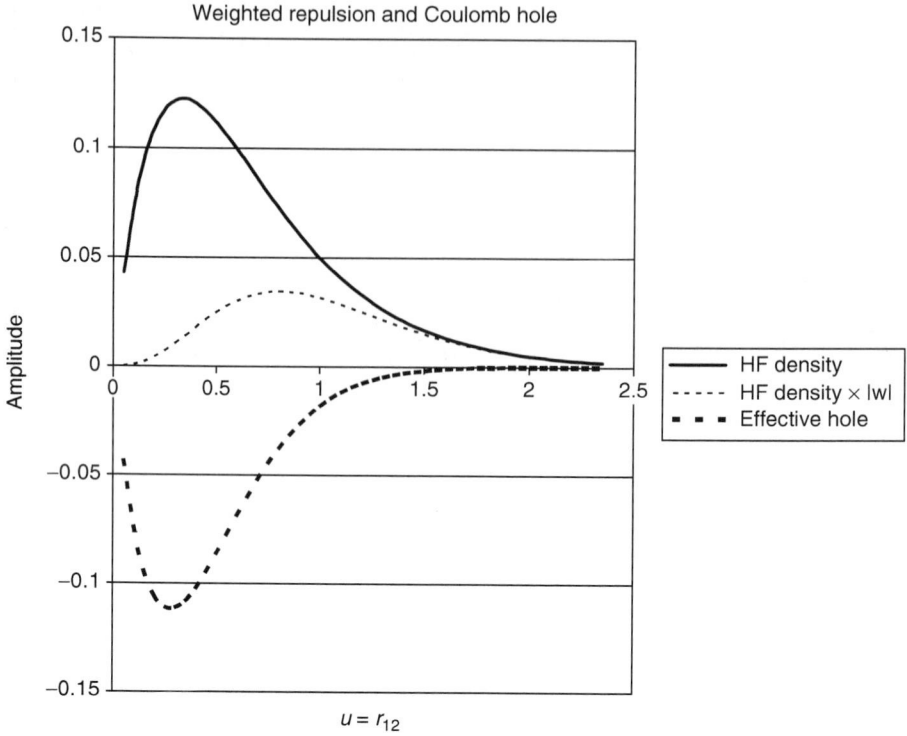

FIGURE 10.3

Qualitative representation of the Clementi hole definition. The horizontal axis is the distance between two electrons in bohrs; the vertical axis is the radially averaged density distribution around an electron. The dotted line displays the density as attenuated by the factor $[1 - \exp(-\omega r_{12}^2)]$. The difference is an estimate of how charge must be displaced so that the interelectronic repulsion is reduced by an amount equivalent to that produced by the factor. The displacement can be interpreted as the effective Coulomb hole in Clementi's approximation.

exponential, and is displayed by the solid line in Figure 10.3. Correlation will shift charge away from small interelectronic distances, and the density-weighted repulsion at short distance will decline. We have written the effective repulsion remaining after the density shift as

$$R^* \approx \frac{1 - \exp\left(-\omega u^2\right)}{u} G(u)$$

This is the dotted line in Figure 10.3. The difference between the two, represented in the dashed line in Figure 10.3, expresses the effect of the density shift on the Coulomb repulsion. As must happen, the device moderates repulsion at short range.

The overall energy must be lowered by an amount which is an estimate of the correlation energy. This qualitative picture of the impact of correlation as modeled by the revised Coulomb operator is consistent with what we expect for elaborate correlation-corrected densities, but does not require any wave function more complicated than a single determinant.

Chakravorty and Clementi [10] evaluated the necessary one-center integrals of the operator

$$O = \frac{\exp(-\eta r_{12})}{r_{12}}$$

for the atomic case, and developed parameters for s, p, d, and f orbitals. Their semiempirical estimates of correlation energies and ionization potentials for atoms up to no. 55 are excellent. They did not, however, investigate applications to molecules. Otto et al. [12] presented integrals of the operator for Gaussian lobes, as we have shown in the discussion of the lobe basis. Shillady et al. [13] adapted the soft hole formulation to the molecular problem. The curves given by Banyard and Mobbs [7] for the Coulomb hole of the Be atom provided guidance.

PCLOBE assumes that a single Gaussian sphere can be used to model a universal form for the Coulomb hole, and assumes that the size of the sphere depends on the effective potential field defined by the effective nuclear charge. The effective potential depends on position. As an electron moves in close to a nucleus the effective nuclear charge is large, whereas if the electron is far from the nucleus screened by other electrons the effective nuclear charge is smaller. This nonuniform screening effect is to some extent reflected in values of the various Gaussian exponents of the primitive components of the lobe basis.

The PCLOBE code computes an integral-specific ω scaling value in terms of the scaling factors α_i for the lobes taking part in the integral. Following previous work [10], the Coulomb hole exponent ω is taken as the geometric mean of the effective scaling of the four lobes in the integral

$$\omega(A,b) = A(Z_1 Z_2 Z_3 Z_4)^{1/2}$$

The Z_i values are proportional to the square root of the α_i values. The single parameter ω was first evaluated [12] by fitting the energy of H_2 to within 0.000001 hartree of the exact value of -1.174474 hartree found by Kolos and Wolniewicz [14]. The value $A = 6.4148$ was the result of this fit.

For improved flexibility, a two-parameter fit seemed desirable. First the scaling ratio z_i is computed for each lobe, with the H atom scale factor as reference. The need for additional flexibility to represent first-row atoms [12] suggested that a quadratic form $Z_i(b)$ be used.

$$z_i = (\alpha_i/\alpha_H)^{1/2} \quad \text{and} \quad Z_i(b) = z_i(1 + bz_i)$$

The only parameters now to be chosen are A and b. Neither depends on the scale of the basis function nor the nuclear charge. Forcing a match to

TABLE 10.1

Results of Training Set for the 6-31GL(d,p) Basis Set

Molecule	E (SCF, a.u.)	Dipole (SCF)	E (CSCF, a.u.)	Dipole (CSCF)
H_2O	−76.019218	2.182 D	−76.420818	2.368 D
HF	−100.008237	1.990 D	−100.455523	2.137 D
LiH	−7.980897	5.884 D	−8.077331*	5.982 D
BH	−25.118950	1.468 D	−25.282657	1.497 D
H_2	−1.131277	0.000 D	−1.168952	0.000 D

quantum Monte Carlo energies for HF and LiH produced values $A = 6.31380$ and $b = 0.284126$ [15]; the A value is not very different from the single-parameter fit, but the quadratic term is significant.

The Coulomb hole method implemented in PCLOBE as the "correlated-SCF method" or CSCF produced results for the static potential energy obtained for the hydrogen bond in the water dimer [14] very similar to those of the BPW91-DFT method in the GAUSSIAN-03 program [16]. The parameters have been further refined by fits to the quantum Monte Carlo energies of a larger training set of molecules including H_2, LiH, BH, H_2O, and HF, and using both the 6-31GL(d,p) and 6-311GL(d,p) basis sets. A nonlinear least-squares method, due to Ransil [17], produced new minimum-error parameters. The new values $A(new) = 1.843076$ and $b(new) = 0.400525$ were considerably different from the original values fitted to only LiH and HF. This must reflect the greater variety of electronic environments provided by the new species.

The lobe basis shows its advantages of simplicity and efficiency in this application. Many of the algebraic expressions in the Coulomb hole integrals use forms already defined for conventional two-electron integrals so that the time for integral evaluation is increased by only about 40%; the SCF process is unchanged. Overall, the CSCF total run times are only about 20% longer than the standard RHF–SCF runs. This is similar to the time requirements of DFT methods relative to conventional RHF calculations (Tables 10.1 through 10.6).

TABLE 10.2

Results of Training Set for the 6-311GL(d,p) Basis Set

Molecule	E (SCF, a.u.)	Dipole (SCF)	E (CSCF, a.u.)	Dipole (CSCF)
H_2O	−76.044256	2.146 D	−76.414234	2.315 D
HF	−100.046358	1.990 D	−100.469876*	2.119 D
LiH	−7.985777	6.009 D	−8.072661*	6.056 D
BH	−25.127502	1.675 D	−25.275044	1.667 D
H_2	−1.131277	0.000 D	−1.168952	0.000 D

TABLE 10.3

Results of Training Set for the 6-311GL(d,p)++ Basis Set

Molecule	E (SCF, a.u.)	Dipole (SCF)	E (CSCF, a.u.)	Dipole (CSCF)
H_2O	−76.050493	2.219 D	−76.417824	2.375 D
HF	−100.052784	2.035 D	−100.462880*	2.156 D
LiH	−7.985809	6.023 D	−8.072304*	6.037 D
BH	−25.128148	1.754 D	−25.274975	1.720 D
H_2	−1.132482	0.000 D	−1.165846	0.000 D

TABLE 10.4

Results of Training Set for the DH11s/7p1d/H3,1s/1p Lobe-mimic Basis Set

Molecule	E (SCF, a.u.)	Dipole (SCF)	E (CSCF, a.u.)	Dipole (CSCF)
H_2O	−76.047975	2.163 D	−76.409660	2.318 D
HF	−100.055942	2.049 D	−100.459500	2.174 D
LiH	−7.984998	5.892 D	−8.070089	5.884 D
BH	−25.127612	1.781 D	−25.271839	1.798 D
H_2	−1.122962	0.000 D	−1.154871	0.000 D

TABLE 10.5

Energy Standards Used for Training Set

Molecule	E (a.u.)	Source	Dipole (D)	Source
H_2O	−76.4383	Exp. [18]	1.8546	87th CRC [19]
HF	−100.4595	Nonrel. [20]	1.826178	87th CRC [19]
LiH	−8.070215	Nonrel. [21]	5.88	[23]
BH	−25.2879	Nonrel. [20]	1.784	Calc. [22]
H_2	−1.174475	Exact [18]	0.0000	Symmetry

TABLE 10.6

Root-Mean-Square Energy Error for Test Set

Basis Set	RMS-Error (a.u.)[a]	RMS-Error (a.u.)[b]
6-31GL(d,p)	0.019989	0.020738
6-311GL(d,p)	0.028679	0.027397
6-311GL(d,p)++	0.026011	0.024538
DH11s/7p1dH3,1s/1p	0.038243	0.032836

[a] H_2 is not in the training set.
[b] H_2 is included in the training set.

As is often the case in density functional theory, modest basis sets will sometimes yield better results than larger basis sets. The (A, b) parameters were carefully optimized for the 6-31GL(d,p) and 6-311GL(d,p) basis sets. Considerable effort with the latter basis showed it to be less reliable than the simpler 6-31GL(d,p) basis. The Pople triple-zeta polarized basis set, which 6-311GL ressembles, was designed [24] to perform well with the MP2 correlation correction, which requires a flexible set of virtual MOs. The hole method is of very different nature from MP2—it does not rely on admixture of excited states but deals with the ground state of a model Hamiltonian—so perhaps it is not amazing that the basis tuned to one purpose is not ideal for the other. The two larger basis sets, 6-311GL(d,p)++ and the Dunning–Huzinaga (11s/7p/1d), were treated less extensively.

RHF functions generally overestimate dipole moments. CI typically reduces the dipole moment as it lowers the energy by mixing in excited states that often reverse the direction of the dipole moment. The CSCF method does not correct that flaw of single-determinant wave functions. The Coulomb hole correction increases the dipole moment beyond the RHF overestimate by a small amount.

By parametrizing the exchange–correlation contribution to the energy, the CSCF method forsakes the variational principle. Some energies go below the accepted variational limit. These are marked with a trailing "*" in the tables above. This can happen in other types of DFT modeling, which also sacrifice strict variational rigor. Yet it does appear that we are appealing to a variational principle by solving SCF equations. This is a paradox. But it is a consequence of our revision of the Hamiltonian; we are seeking the minimum of a model, a redefined energy-operator-in-a-basis. Disobeying the variational limit is alarming, but perhaps not fatal to the practical use of correlation correction.

Overall, it would appear that the present values of (A,b) give a better account of the heavier atoms O and F than H and Li. Moiseyev et al. [3] noticed that the form of their Fermi holes in isoelectronic 1s2s triplet ions approached a limiting form as the nuclear charge increased; thus, it may be that the Otto hole resembles the high charge limit. Perhaps when the nuclear charge is low there is a need for a more accurate rim shape whereas in the presence of a higher nuclear charge a small sphere describes the main effect of the Coulomb hole. Further research could be carried out using a more complicated form of the hole function. The first improvement might be something like a "double-zeta" hole, with terms representing long- and short-range features of the hole.

$$\left(\frac{1 - e^{-\omega_1 r_{12}^2} + e^{-\omega_2 r_{12}^2}}{r_{12}} \right), \quad \omega_1 \neq \omega_2$$

No new integrals would be required. Alternatively, one may introduce a nonspherical form to the holes, at the expense of greater complexity in the parameter set and a need to evaluate new integrals.

Applying the Correlated-SCF Method to Hydroxylamine

Table 10.7 shows that for H_3NO the CSCF method results in an energy lower than the RHF–SCF value—by over 0.7 hartree. The dipole moment is higher by about 0.1 debye as was the pattern in the test set of molecules. The CSCF bond lengths are shorter than found by RHF calculations using the same basis, but the CSCF bond angles are similar to RHF values.

The same trends are evident in the correlated calculation of hydroxylamine, as shown in Table 10.8. The hydroxylamine conformation is that found by Fink et al. [25] to be lowest in energy. Again we see that the dipole moment is slightly higher and the bond lengths are shorter with the correlated method, but the energy is lower than the RHF–SCF energy by over 0.7 hartree.

In PCLOBE, we report the virial relationship as $-\langle E \rangle / \langle T \rangle$. This ratio of $-\langle E \rangle / \langle T \rangle$ should be precisely unity for an exact wave function according to the virial theorem. Usually the theorem is expressed as $2\langle T \rangle = -\langle V \rangle$ but with the use of the screening it is not totally clear what is meant by the quantity $\langle V \rangle$ (Table 10.8).

In Table 10.9, we present a comparison of the RHF–SCF and the CSCF for linear HNO. Once again CSCF produces a slightly higher dipole moment, a much lower energy, and shorter bond lengths relative to RHF.

In Table 10.10, we see similar results for formaldehyde. One way to judge the accuracy of a calculation is by its estimate of energies of reactions. As shown in Table 10.11, the CSCF agrees with much more demanding calculations that hydroxylamine is much more stable than the fragments HNO and H_2, while formaldehyde is not very much more stable than fragments CO and H_2 [26]. However, the discrepancy between the calculations is substantial, more than 0.01 hartree or about 5–10 kcal/mol. One might note that early formulations of DFT for chemical systems achieved roughly comparable accuracy.

TABLE 10.7

Comparison of SCF and CSCF Results for H_3NO

Property	Hartee–Fock–Roothaan SCF	Correlated-SCF
E (a.u.)	−130.940400	−131.657996
$-\langle E \rangle / \langle T \rangle$	1.000279	1.003822
Dipole (D)	5.591	5.700
$R(N–H)$, $A°$	1.0087	0.9960
$\angle$ HNH°	107.20	107.91
$R(N=O)$, $A°$	1.3771	1.3383
$\angle$ HNO°	111.65	110.99
Grad. (a.u.)	0.0006173	0.0008528

Note: Basis 6-31GL(d,p).

TABLE 10.8

Comparison of SCF and CSCF Results for H_2NOH

	Hartee–Fock–Roothaan SCF	Correlated-SCF
E (a.u.)	−130.989572	−131.706239
$-\langle E \rangle / \langle T \rangle$	1.000646	1.004143
Dipole (D)	0.667	0.723
R(N–H), $A°$	1.0014	0.9882
∠HNH°	106.74	106.95
R(N–O), $A°$	1.3990	1.3685
∠NOH°	104.21	103.74
R(O–H), $A°$	0.9430	0.9332
Grad. (a.u.)	0.0006101	0.0009358

Note: Basis 6-31GL(d,p).

TABLE 10.9

Comparison of SCF and CSCF Results for HNO, $C_{\infty v}$

	Hartee–Fock–Roothaan SCF	Correlated-SCF
E (a.u.)	−129.669755	−130.310781
$-\langle E \rangle / \langle T \rangle$	1.001068	1.004185
Dipole (D)	2.908	3.081
R(N–H), $A°$	0.9831	0.9737
R(N–O), $A°$	1.1899	1.1667
Grad. (a.u.)	0.0006151	0.0007517

Note: Basis 6-31GL(d,p).

TABLE 10.10

Comparison of SCF and CCF Results for H_2CO

	Hartree–Fock–Roothaan SCF	Correlated-SCF
E (a.u.)	−113.867715	−114.480221
$-E/T$	1.001939	1.005093
Dipole (D)	2.620	2.959
R(C–H), $A°$	1.0938	1.0717
∠HCH°	115.68	116.58
R(C–O), $A°$	1.1855	1.1619
∠HCO°	122.16	121.71
Grad. (a.u.)	0.0006151	0.0009855

Note: Basis 6-31GL(d,p).

TABLE 10.11

Energies of Reaction by CSCF and CCSD(T) Calculations

Thermochemistry for Hydroxylamine		
Species: Model	**CCSD(T)/6-311+G(d,p)**	**CSCF/6-31GL(d,p)**
HNO	−130.1119407	−130.310781
H_2	−1.1683225	−1.1693554
H_2NOH	−131.4569645	−131.706239
ΔE (assembly)	0.182018	0.226103

Thermochemistry for Formaldehyde		
CCSD(T)/6-311+G(d,p)		**CSCF/6-31GL(d,p)**
CO	−113.0964674	−113.295500
H_2	−1.1683225	−1.1693554
H_2CO	−114.2666864	−114.480221
ΔE (assembly)	0.003168	0.015366

Note: Energy in hartrees; geometries for the CCSD calculation obtained by MP2/6-311+G(d,p) optimization.

Summary

The Coulomb hole simulation presented here is a modification of the methods developed by Chakravorty and Clementi [10] and adapted to the lobe basis by Otto et al. [12]. We are grateful to professor Enrico Clementi for encouragement of our research. The main postulate here is that the Coulomb hole is a scalable universal shape and can be modeled by a single Gaussian sphere related to the local orbital screening at a given region in the space around positive nuclei. This model is crude but yields drastically lower total energies, probably within ±5% of the quantum Monte Carlo energy for a molecule if the molecule has few H atoms in the structure.

References

1. Boyd, R.J. and Coulson, C.A., *J. Phys. B* 6:782, 1973.
2. Slater, J.C., *Phys. Rev.* 34:1293, 1929.
3. Moiseyev, N., Katriel, J., and Boyd, R.J., *J. Phys. B: Atom. Mol. Phys.* 8:L130, 1975.
4. Hylleraas, E.A., *Z. fur Physik* 65:209, 1930.
5. Coulson, C.A. and Neilson, A.H., *Proc. Phys. Soc.* 78:831, 1961.
6. Boyd, R.J., *Can. J. Phys.* 53:592, 1976.
7. Banyard, K.E. and Mobbs, R.J., *J. Chem. Phys.* 75:3433, 1981.
8. Boyd, R.J. and Yee, M.C., *J. Chem. Phys.* 77:3578, 1982.
9. Baratan, D.N. and Luken, W.D., *Theor. Chem. Acta* 61:265, 1982.

10. Chakravorty, S.J. and Clementi, E., *Phys. Rev. A* 39:2290, 1989.
11. Perdew, J.P., Burke, K., and Wang, Y., *Phys. Rev. B* 54:16533, 1996.
12. Otto, P., Reif, H., and Hernandez-Laguna, A., *J. Mol. Struct.* 340:51, 1995.
13. Shillady, D.D., Craig, J., and Rutan, S., *Int. J. Quantum Chem.* 85:520, 2001.
14. Kolos, W. and Wolniewicz, L., *J. Chem. Phys.* 49:404, 1968.
15. Shillady, D.D., Craig, J., Rutan, S., et al., *Int. J. Quantum Chem.* 90:1414, 2002.
16. Gaussian 03 Revision C02 Frisch, M.J., Trucks, G.W., Schlegel, H.B., et al., Gaussian Inc., Wallingford, CT, 2004.
17. Ransil, B.J., *Rev. Mod. Phys.* 32:239, 1960.
18. Luchow, A., Anderson, J.B., and Feller, D., *J. Chem. Phys.* 106:7706, 1997.
19. Lide, D.R., Ed. *CRC Handbook of Chemistry and Physics*, 87th Ed., Taylor & Francis, Boca Raton, 10:242, 2006–2007.
20. Luchow, A. and Anderson, J.B., *J. Chem. Phys.* 105:7573, 1996.
21. Chen, B. and Anderson, J.B., *J. Chem. Phys.* 102:4491, 1995.
22. Harrison, J.F. and Allen, L.C., *J. Mol. Spectr.* 29:432, 1969.
23. Garmer, D.R. and Anderson, J.B., *J. Chem. Phys.* 86:4025, 1987.
24. Krishnan, R., Binkley, J.S., Seeger, R., et al., *J. Chem. Phys.* 72:650, 1980.
25. Fink, W.H., Pan, D.C., and Allen, L.C., *J. Chem. Phys.* 47:895, 1967.
26. Trindle, C. and Shillady, D.D., *J. Am. Chem. Soc.* 95:703, 1973.

11

Density Functional Theory

Introduction

The introduction of efficient and capable versions of density functional theory (DFT) in the 1990s has altered computational chemistry profoundly. It is fair to say that computational study of medium-sized organometallic systems and rather large organic systems is now dominated by density functional methods. The popular B3LYP method has usually proved comparable to MP2 calculations in quality of characterization of structures and energetics, and by good fortune produces harmonic vibrational frequencies closely comparable with observed vibrational absorptions. DFT often is competitive with far more demanding coupled cluster methods.

Density functional methods rest on the assertion that the electron density distribution is sufficient to define exactly the ground state of a system of electrons and nuclei. Wave functions, our focus to this point, play no necessary role in the description of molecular systems according to this point of view. The object of investigation is the "functional" of the density which actually accomplished the specification of the exact energy.

Density functional theory has a sound basis in the Hohenberg–Kohn theorem [1,2] and its refinement by Levy [3] which guarantees that there is some definite link between the density and the exact ground state energy. The development of DFT is the story of the invention of functionals which contain the physics of electron exchange–correlation. There is no systematic way to improve exchange–correlation functionals such as is provided in wave function–based electronic stricture theory by the variation method. Functionals can be chosen to obey certain proved constraints on the exact functionals, to capture the behavior of fully understood limiting cases such as the homogeneous electron gas, and (by the incorporation of disposable parameters) to represent faithfully collections of well-established experimental data. This last requirement, so central to the success of the methods in applications, is not generally compatible with such desirable features as simplicity of mathematical form or computational efficiency.

In this chapter, we wish to provide some of the historical background which laid the groundwork to the modern implementations of DFT, to describe some of the strategems which lead to functionals capable of very accurate estimates of energy, to provide some guidance on which functionals are effective choices for particular tasks, and to describe what we might expect as the field advances. We are indebted to Peter Gill's [4] survey of the early days of density functional theory and the unified view he gives of electronic structure theories which are orbital-based, density-based, and mixtures of the two. A useful and compact guide to the past 15 years of development is provided by Koch and Holthausen's book [5], a very accessible guide to the performance of some of the most popular functionals. We will draw on these sources and augment them. Additional information is found, in the guide to GAUSSIAN-by Foresman and Frisch [6], and in the review by Perdew and Schmidt [7]. More recent advances are discussed by Jensen [8].

A second purpose of this chapter is to survey the conceptual and practical challenges and current preoccupations in this rapidly developing field of research. One way to organize this survey has been provided by Perdew's fanciful image of a hierarchy of DFT formulations leading from earth (Hartree–Fock theory) to heaven (chemical accuracy).

John Perdew's Ladder*

In June 2000, at the DFT2000 symposium in Menton, France, John Perdew presented his vision of five levels of functionals in the form of "Jacob's Ladder,"[†] more or less as follows (read from bottom to top):

Heaven (Chemical Accuracy)		
+ Explicit dependence on unoccupied orbitals	Rung 5	Fully nonlocal Grimme's dispersion corrected DFT
+ Explicit dependence on occupied orbitals	Rung 4	Hybrid functionals: SAOP
+ Explicit dependence on kinetic energy density	Rung 3	Meta-GGAs: TPSS
+ Explicit dependence on gradients of the density	Rung 2	GGAs: PW86, PW92, PBE
Local density only	Rung 1	Thomas–Fermi–Dirac, LDA: VWN
Earth (Hartree Theory)		

---

*John P. Perdew and Karla Schmidt, in *Density Functional Theory and Its Applications to Materials*, edited by V.E. Van Doren, K. Van Alseoy, and P. Geerlings (American Institute of Physics, 2001).

[†] According to the Bible, Jacob had a dream in which he saw a ladder descending from heaven to earth and angels climbing and descending the ladder. In John Perdew's dream, the angels are users of DFT who climb the ladder to gain greater precision (at greater cost), but who may also descend the ladder depending upon their needs.

TABLE 11.1

Some Important Quantities in Density Functional Theory

Variable	Symbol/Formula	Comments
Density	ρ	Sometimes n, esp. in physics literature
Wigner–Seitz radius	$r_s = [3/4\pi\rho]^{1/3}$	Radius of sphere holding on average one electron at that density
Spin polarization	$\zeta = \dfrac{\rho^\alpha - \rho^\beta}{\rho^\alpha + \rho^\beta}$	Ranges from -1 to $+1$
Fermi wave number	$k_F = (3\pi^2 \rho)^{1/3}$	Note $k_F r_s = (9\pi/4)^{1/3}$
Reduced gradient	$x = |\nabla\rho|/\rho^{4/3}$	Key quantity in GGAs
Thomas–Fermi kinetic energy density	$t_s = k_F^5/10\pi^2 \sim \rho^{5/3}$	Kinetic energy density of homogeneous gas of noninteracting electrons
Local kinetic energy density	$\tau = \sum_i |\nabla\psi_i|^2$	Kinetic energy of interacting system is half the integral of τ
Dirac exchange energy density (spin σ)	$\varepsilon_X^\sigma = -(3/4)(6/\pi)^{1/3}\rho_\sigma^{1/3}$	Spin-unpolarized form $\varepsilon_X = -(3/4)(3/\pi)^{1/3}\rho^{1/3}$

We have made some effort to unify notation, but cannot promise a simple tour of density functional theory, which has become increasingly complex. The guide to key variables in DFT will help, since these recur in almost every formulation of DFT (Table 11.1).

Early Forms of Density Functional Theory: Gill's History

The idea that the electron density should be sufficient to define a system's energy was current long before the proof of the existence theorem of DFT put this hope on firm footing. Gill [4] gives an excellent history of the development of DFT from 1927 to 1998.

Thomas–Fermi–Dirac Theory

The earliest mathematical expression of a link between density and energy was the Thomas–Fermi–Dirac [9–11] model. According to Gill's sketch [4], Thomas and Fermi represented the total energy by terms defining the kinetic energy, electron–nuclear attraction, and electron–electron repulsion.

$$E_{TF}[\rho] = T_{TF}[\rho] + E_{ne}[\rho] + J[\rho]$$

Thomas and Fermi developed the kinetic energy with reference to the particle-in-a-box, while the electrostatic terms followed classical forms.

$$T_{TF}[\rho] = \frac{3(3\pi^2)^{2/3}}{10}\int \rho^{5/3}(\mathbf{r})d\mathbf{r} \quad E_{ne}[\rho] = \sum_a \int \frac{Z_a\rho(\mathbf{r})d\mathbf{r}}{|\mathbf{R}_a - \mathbf{r}|} \quad J[\rho] = \frac{1}{2}\int\int \frac{\rho(\mathbf{r})\rho(\mathbf{r}')d\mathbf{r}\,d\mathbf{r}'}{|\mathbf{r} - \mathbf{r}'|}$$

The Thomas–Fermi formulation is entirely local, meaning that only the density at one or two points is required to evaluate the integrands in the expressions quoted. This makes for a very efficient calculation. The original Thomas–Fermi expression omits reference to the need for antisymmetry of the wave function with respect to permutation of any pair of electrons, which in Hartree–Fock theory produces an exchange contribution to the total energy [12,13]. Dirac [11] added an exchange term, which defines the Thomas–Fermi–Dirac equation

$$E_{\text{TFD}} = E_{\text{TF}} - K_D[\rho], \quad K_D[\rho] = \frac{3}{4}\left(\frac{3}{\pi}\right)^{1/3}\int \rho^{4/3}(\mathbf{r})\,d\mathbf{r}$$

It might be convenient now to distinguish the energy terms—which are numerical values of integrals—from the energy functionals, which are the integrands dependent on the density. Here E is the functional for the energy, ε is the energy density functional, and the integral defines the contribution to the total energy.

$$E[\rho] = \int \rho\varepsilon[\rho]\,d\mathbf{r}$$

Unfortunately, the Thomas–Fermi–Dirac method, so simple and in its way beautiful, was unable to predict that atoms would be bound into molecules. Part of its weakness lay in the initial particle-in-a-box treatment of the kinetic energy. An important improvement was given by von Weizsäcker [14] in 1935, who defined the kinetic energy in his work on the (slightly) nonuniform electron gas:

$$T = A\rho^{5/3} + B\frac{(\nabla\rho)^2}{\rho}$$

Here A and B are collections of fundamental constants, not adjustable parameters. We will see this gradient term again, since it plays a role in recent DFT formulations as τ_W.

$$\tau_W(\mathbf{r}) = \frac{1}{8}\frac{|\nabla\rho(\mathbf{r})|^2}{\rho(\mathbf{r})}$$

Another shortfall of Thomas–Fermi–Dirac theory is that there is no explicit recognition of the electron correlation; J and K already defined constitute a kind of mean-field appreciation of the influence of the charge distribution on any particular electron. Wigner [15] introduced an expression for a "correlation" correction of the form

$$E_c = -4a\int \frac{\rho_\alpha\rho_\beta}{\rho}\left(\frac{1}{1+d\rho^{-1/3}}\right)d\mathbf{r} \text{ with } a = 0.04918 \text{ and } d = 0.349$$

This expression was not justified by any kind of derivation; the presence of parameters suggests that it is intended to be used empirically.

In chemistry the valence-bond (Pauling) [16] and molecular orbital method (Mulliken) [17] dominated electronic structure study at mid-century, especially after the Roothaan–Hall [18,19] self-consistent field (SCF) method transformed the molecular project from what had been an impossible numerical problem into a merely difficult algebraic exercise. In physics, DFT still was valuable, especially in the solid state community since it dealt effectively with the electrons in metals. In an anticipation of the Kohn–Sham procedure which merges some ideas from both approaches, Slater [20] introduced a DFT version of the exchange as an alternative to the nonlocal HF exchange which occurs in the SCF equations. His local density approximation (LDA) form for the exchange is

$$E_x^{LDA} = -\frac{3}{2}\left(\frac{3}{4\pi}\right)^{1/3}\sum_\sigma\int\rho_\sigma^{4/3}\,d\mathbf{r}$$

The index σ refers to spin. Slater retained the HF expression of the kinetic energy as a sum of integrals over one-electron functions (molecular orbitals). This HF–Slater method might be placed on the first (LDA) rung of Perdew's ladder.

The Hohenberg–Kohn Existence Theorem

Density functional theory was an intuitively appealing and to a degree even successful, but only with the existence theorem of Hohenberg and Kohn [1,2] was it established that one could hope for rigorously accurate results. Here we sketch the remarkably compact proof of the proposal that the ground state density in fact defines the ground state energy exactly.

In a molecular system, electrons interact with one another and the set of fixed positively charged nuclei. The field exerted by the nuclei is called an "external" field and can be considered to define the Hamiltonian. Now consider that we have in hand a ground state electron density ρ_0. Is the external potential, i.e., the arrangement of nuclei, uniquely determined? If so, the Hamiltonian is determined as well, and thus the ground state energy is also fixed. Assume the contrary that ρ_0 is consistent with two potentials V_A and V_B. Then, there must be two Hamiltonians H_A and H_B, as well as two ground state wave functions ψ_A and ψ_B and associated eigenvalues E_A and E_B.

By the variational theorem

$$E_A < [\psi_B|H_A|\psi_B\rangle = \langle\psi_B|H_A - H_B + H_B|\psi_B\rangle]$$

so

$$E_A < [E_B + \langle\psi_B|H_A - H_B|\psi_B\rangle = E_B + \int(V_A - V_B)\rho_0\,d\mathbf{r}]$$

We could begin again and produce

$$E_B < E_A + \int (V_B - V_A)\rho_0 \, d\mathbf{r}$$

Adding these together we find

$$E_A + E_B < E_A + E_B$$

This is a contradiction, so it cannot be the case that a single density is consistent with two potentials V_A and V_B. Thus, the density must determine the Hamiltonian, and hence the ground state energy. While this gives reassurance regarding the existence of a solution, we should note that it does not tell us how to construct a solution.

Kohn–Sham Procedure for Finding the Density

Kohn and Sham [1,2] introduce the notion that the true density for the system of interacting electrons may be identical with the density for a system of noninteracting electrons. If so, the Schröedinger equation is separable and a Slater determinant of orbitals for the noninteracting electrons is an exact solution, defining the exact density.

$$\rho = \sum_{i}^{n} N_i \chi_i^* \chi_i; \text{ and since } E[\rho] = T_{non} + V_{nuc-el} + V_{el-el} + \Delta T$$

$$E[\rho] = \sum_{i}^{n} <\chi_i | t(i) | \chi_i> - \sum_{i,A} \left\langle \chi_i \left| \frac{Z_A}{|r_i - R_A|} \right| \chi_i \right\rangle$$

$$+ \sum \left\langle \chi_i \left| \frac{1}{2} \int \frac{\rho(r')}{|r_i - r'|} dr' \right| \chi_i \right\rangle + E_{XC}$$

This is not quite exact, since the kinetic energy defined by the noninteracting orbitals must be altered by electron interactions by some amount ΔT. The total energy includes the electrostatic repulsion between the bare nuclear charges.

$$E_{tot} = E[\rho] + \sum_{A>B}^{atoms} \frac{Z_A Z_B}{|\mathbf{R}_A - \mathbf{R}_B|}$$

The usual SCF manipulations—minimizing the energy with respect to the forms of the orbitals for the noninteracting electrons while maintaining their orthonormality—produce the Kohn–Sham equations, and introduce the central problem of DFT, the unknown form of the exchange–correlation functional V_{XC}.

$$h_i^{KS} \chi_i = \varepsilon_i \chi_i$$

$$h_i^{KS} = -\frac{1}{2} \nabla_i^2 - \sum \frac{Z_A}{|r_i - R_A|} + \int \frac{\rho(r')}{|r_i - r_i'|} dr' + V_{XC}[\rho]$$

If we knew V_{XC}, we would have a way to compute the exact energy. But even an approximate V_{XC} might do better than SCF theory. The Kohn–Sham formulation seeks an accurate method within DFT without abandoning the concepts of one-electron orbitals.

DFT and SCF Calculations Have Common Features

SCF and DFT calculations in the Kohn–Sham formulation share many features. We are indebted to Köster et al. [21] for the following outline.

Flowchart for Generic-DFT Program

1. Specify the location of atoms and the total charge of the molecule.
2. Choose a basis set; this permits use of much of the machinery of the SCF calculation.
3. Define grid points and weights for numerical integration of the functionals.
4. Define auxiliary basis sets for expansion of ρ and $\rho^{1/3}$.
5. Construct one-electron matrix elements for the core Hamiltonian.
6. Diagonalize the core-Hamiltonian matrix to obtain an initial guess of the orbitals; construct the first trial density.
7. Form the density matrices, the density functions, and the coulomb and exchange term.
8. Construct the effective mean-field (Fock) operator, including the exchange–correlation term (in the SCF case, no correlation correction is available).
9. Solve the secular equation for the matrix of LCAO coefficents **C**.
10. Form a new density matrix and density function.
11. Integrate V_{XC} over basis functions (SCF) or atomic grid points (DFT) to find $E_{xc}(\rho, \nabla\rho, \rho^{1/3})$.
12. Calculate the electronic energy

$$E = \sum_{\mu\nu} P_{\mu\nu} H_{\mu\nu} + \sum_k J_k + (K \text{ or } E_{xc}(\rho, \nabla\rho, \rho^{1/3}))$$

13. Check for convergence of the energy; if not converged go back to step 7. If the energy converges to within some desired threshold proceed to step 14.

14. Add the nuclear repulsion to the electronic energy to get the total energy

$$E_{\text{tot}} = E[\rho] + \sum_{A > B}^{\text{atoms}} \frac{Z_A Z_B}{|\mathbf{R}_A - \mathbf{R}_B|}$$

15. Print final energy and use the final orbital matrix C to evaluate one-electron properties such as dipole moments and bond orders.

This sketch shows that the well-established SCF methods and associated integral evaluation techniques are readily adapted to the construction of Kohn–Sham orbitals. The significant difference is the extensive use of numerical quadrature in DFT.

Two main features of DFT distinguish it from the Hartree–Fock method.

1. The "true" V_{XC} is still unknown, but modern exchange–correlation functionals capture 95%–105% of the correlation energy missing in the Hartree–Fock method. DFT has its main justification as a way to defeat the problem of the correlation error in the Hartree–Fock method.

2. During the various implementations of DFT technology, several ways were found to reduce the computational burden of the N^4 list of four-center two-electron interactions by using fitted auxiliary basis sets to charge densities and restrict integrations over one-center cases to reduce the N^4 list to a much smaller list of three-center integrals over each atom resulting in an overall N^2 dependence for a problem with N orbitals.

We have already discussed the auxiliary basis representation of the density and the reduction of troublesome integrals it permits. Here we should remark that maintaining normalization of the charge density is essential.

Search for the Functionals

The invention of useful approximations to the functional V_{XC} has made DFT powerful and popular. In DFT methods, unlike wave function techniques for which we can extend the CI or perturbation series, we have no systematic means to improve the calculations. We see three tracks in the research: fitting a functional to experimental data, fitting a functional to mathematical and physical constraints, and for now most effective, hybrids of both methods. There are six features of the search: (1) use of the LSDA approximation as a starting point (rung 1 of Perdew's ladder); (2) introduction of the density gradient (rung 2); (3) use of models of the exchange–correlation hole; (4) mixing Hartree–Fock and DFT exchange (rung 4); (5) imposition of constraints guaranteeing conformity with known properties of the exact functionals; and/or (6) empirical fits to thermochemical data. Successes come by combinations of many or all these strategems.

In one major strand of research, empirical parameters are introduced into functionals, to be estimated by minimizing the error in representation of well-established experimental data. In another, forms of functionals are chosen to obey conditions known to apply to the exact functionals. Other schemes impose theoretically required properties on the functional form but choose parameters by reference to experimental data.

The first well-established databases include thermodynamic values, most frequently heats of atomization, electron affinities, and ionization energies. Databases have become more extensive and have come to include kinetic barriers, bond lengths, and vibrational frequencies. We will see that it is possible to design functionals to represent certain types of data very accurately, although the full promise of DFT remains to be fulfilled.

The Idea of the Hole

The exchange–correlation functional is defined by the relevant "hole" $h(\mathbf{r},\mathbf{u})$; $\mathbf{r}$ is a position in space and $\mathbf{u}$ is a displacement from that location.

$$E_{XC} = \int dr_1\, dr_2 \frac{\rho(r_1)h(r_1,r_2)}{2|r_1 - r_2|} = \int dr\, du \frac{\rho(\mathbf{r})h(\mathbf{r}+\mathbf{u})}{2\mathbf{u}}$$

The hole derives from the fact that the presence of an electron at point $\mathbf{r}$ decreases the probability of an electron's presence in the immediate neighborhood of $\mathbf{r}$, i.e., $\mathbf{u}$. For the unpolarized homogeneous electron gas Wang and Perdew [22] offer an analytic fit to a quantum Monte Carlo description. The hole must contain one electron:

$$\int dr'h(\mathbf{r},\mathbf{r}'-\mathbf{r}) = -1 \quad \text{all } \mathbf{r}$$

Its mean radius can be defined by the spherical average

$$\langle R^{-1}\rangle_r = -\int d\mathbf{u}\,\frac{h(\mathbf{r},\mathbf{u})}{\mathbf{u}}$$

The exchange–correlation contribution to the energy is

$$E_{XC} = -\frac{1}{2}\int d^3r\,\rho(r)\langle R^{-1}\rangle_r$$

This suggests that the role of the energy density functional $\varepsilon(\rho)$ is mainly to define the radius of the hole. Further, the final exchange–correlation energy requires only the spherical average of the hole. Perhaps even an approximation to the hole which is wrong in angular form can be useful. The simplest assignment would be something like the Seitz radius, proportional to $\rho^{-1/3}$ which suggests that the integrand should be $\rho^{4/3}$. This is indeed precisely the form employed by Slater and many others.

Evaluating the DFT Energy Using Functionals (Slater)

We introduce the energy density functional $\varepsilon(\rho)$. We divide the energy density functional into exchange and correlation parts; then

$$E_{XC}[\rho] = E_X[\rho] + E_C[\rho] = \int \rho \varepsilon_X \, d\mathbf{r} + \int \rho \varepsilon_C(\rho) \, d\mathbf{r}$$

We first seek the exchange functional. As already mentioned, Slater introduced an LDA form for the exchange with the hole

$$E_X^{LDA} = -\frac{3}{2} \left(\frac{3}{4\pi} \right)^{1/3} \sum_\sigma \int \rho^{4/3} \, d\mathbf{r}$$

The terms in the sum (indexed by σ) refer to the α and β spins. It is often the case that the spin densities are equal, but the expression can be generalized to nonequivalent spin densities.

$$E_x^{LSDA}[\rho] = -2^{1/3} C_x \int \left[\rho_\alpha^{4/3} + \rho_\beta^{4/3} \right] d\mathbf{r} \quad \text{and} \quad \varepsilon_x^{LSDA}[\rho] = -2^{1/3} C_x \left[\rho_\alpha^{1/3} + \rho_\beta^{1/3} \right]$$

This is called the local spin density approximation (LSDA), and is part of the family of density functional formulations which occupy the first rung of Perdew's ladder.

A very important early correlation-functional was an analytical fit to the results of Ceperley and Alder's [23,24] numerical evaluation of the energy for the uniform electron gas for a range of densities. Define

$$\zeta = \frac{\rho^\alpha - \rho^\beta}{\rho^\alpha + \rho^\beta}$$

Vosko, Wilk, and Nusair's (VWN) [25] formula interpolates between the unpolarized case of $\zeta = 0$ to the spin-polarized limit of $\zeta = 1$.

The Seitz radius r_s (of a sphere which contains one electron at the given density) is defined by

$$\frac{4}{3} \pi r_s^3 = \rho^{-1} = \frac{1}{\rho} \quad \text{or} \quad r_s = \left(\frac{3}{4\pi\rho} \right)^{1/3}$$

$$\varepsilon_c^{VWN}(r_s, \zeta) = \varepsilon_c(r_s, 0) + \varepsilon_c(r_s) \left[\frac{f(\zeta)}{f''(0)} \right] [1 - \zeta^4] + [\varepsilon_c(r_s, 1) - \varepsilon_c(r_s, 0)] f(\zeta) \zeta^4$$

where

$$f(\zeta) = \frac{(1+\zeta)^{4/3} + (1-\zeta)^{4/3} - 2}{2(2^{1/3} - 1)}$$

In the limit that $\zeta \to 0$, the function f also becomes zero and only the leading term survives. The $\varepsilon_c(r_s, \zeta)$ and $\varepsilon_a(r_s)$ functionals are dependent on parameters x_0, b, and c.

$$\varepsilon_{c/a}(x) = A \left\{ \ln \frac{x^2}{X(x)} + \frac{2b}{Q} \tan^{-1}\left(\frac{Q}{2x+b}\right) - \frac{bx_0}{X(x_0)} \left[\ln \frac{(x-x_0)^2}{X(x)} \right. \right.$$
$$\left. \left. + \frac{2(b+2x_0)}{Q} \tan^{-1}\left(\frac{Q}{2x+b}\right) \right] \right\}$$

The variable $x^2 = r_s$ while $X(x) = x^2 + bx + c$ and $Q = (4c - b^2)^{1/2}$. Here $A(\text{paramagnetic}) = 2A(\text{ferromagnetic}) = 0.0621814$. The parameters x_0, b, and c were fitted to match the numerical density functional over a range of r_s values of 10, 20, 50, and 100 bohr. The VWN parametrized analytic forms (plural; Vosko et al. provide two expressions in wide use called VWN3 and VWN5) are used in essentially all later refinements of density functional theory. Because the VWN forms contain parameters, it is not quite correct to say that any later DFT method using the electron gas results is parameter-free. However, these parameters are chosen to fit to an exact (numerical) theoretical result rather than chosen to reduce errors in modeling experimental data, so they are not empirical in the usual sense. Although these expressions for the electron gas density functionals seem so complicated that they defy intuition, it is easy to envision how one can write computer routines to evaluate them. One might take comfort that once the computer code is correct, the computer will do the numerical work; such is the wonder of modern computation.

More on the Local Density Approximation's Hole

The LDA hole was modeled by Ernzerhof and Perdew [26] as

$$h(r, r+u) = \rho J(k_F u)$$

with $k_F = (3\pi^2 \rho)^{1/3}$ defined as the Fermi wave number. The function J^{PW92} is modified so to be nonoscillatory, although the exact LDA factor does oscillate. Let the dimensionless distance variable $k_F u = y$. Then, Ernzerhof and Perdew write

$$J^{\text{PW92}}(y) = -\frac{A}{y^2} \frac{1}{1 + 4Ay^2/9} + \left(\frac{A}{y^2} + B + Cy^2\right) \exp\left(-Dy^2\right)$$

A, B, C, and D are parameters. Their values are chosen so that this model of the hole h satisfies the normalization condition

$$\int\limits_0^\infty du 4\pi u^2 h(r,u) = 4\pi\rho(r) \int\limits_0^\infty du\, u^2 J^{PW92}(k_F u)$$

$$= \frac{4}{3\pi} \int\limits_0^\infty dy\, y^2 J^{PW92}(y) = -1$$

This reproduces the uniform electron gas expression for the exchange energy, and as u approaches zero, h contains the same second derivative as the exact LDA hole density.

One might wonder how one can justify using the uniform electron gas model as a starting point for description of the very nonuniform density distributions in molecules. One pragmatic response to this question might be to judge the performance of the method. First, the VWN-based calculations do predict molecular binding, in contrast to the Thomas–Fermi–Dirac formulation of DFT. In fact VWN DFT systematically overbinds molecules. More remarkably, the VWN functionals produce useful estimates of bond lengths and angles in molecules. One could say that molecular geometries are determined primarily by one-electron features; wave function methods with very approximate treatments of the two-electron effects produce reasonable molecular geometries. Alternatively, one might conjecture that the LSDA hole, although far from an accurate rendering of the exact exchange–correlation hole, is a reasonable approximation to the spherically averaged hole which actually defines the exchange–correlation energy. Still it seems inescapable that some recognition must be taken that densities in molecules are far from uniform.

Refinements to the VWN Exchange–Correlation Functionals

Enhancements to the electron gas model typically refer to the gradient through a form of the dimensionless von Weizsäcker [14] kinetic variable

$$x = \frac{|\nabla\rho|}{\rho^{4/3}}$$

Formulations including this variable are called gradient-corrected (GC) theories and belong to the second rung of Perdew's ladder. There are two pioneering lines of investigation: one by Perdew and another by Becke. Here we sketch some of their early work.

Perdew Functionals

As we will see often in later work, the LDA electron gas exchange functional is improved by multiplication by an "enhancement" function. Perdew and Wang [27] proposed the exchange functional form

$$\varepsilon_x^{PW86} = \varepsilon_x^{LDA}(1 + ax^2 + bx^4 + cx^6)^{1/15}$$

Here a, b, and c are constants and x is a dimensionless gradient variable related to Weizsäcker's scaled quantity. Only even powers of x appear in the enhancement factor. Perdew and Wang later refined the correction factor to

$$\varepsilon_x^{PW91} = \varepsilon_x^{LDA}\left(\frac{1 + a_1 x \sinh^{-1}(a_2 x) + (a_3 + a_4 \exp(-bx^2))x^2}{1 + a_1 \sinh^{-1}(a_2 x) + a_5 x^2}\right)$$

The quantities a_i and b are again disposable constants. The form reduces to the LDA exchange when the kinetic variable approaches zero. This pioneering GGA (generalized gradient approximation) functional provided estimates of the exchange energies in first-row atoms and of the exchange contribution to their ionization energies which were substantial improvements over LSDA values. (The reference was the HF exchange evaluated with Kohn–Sham orbitals.) In their development of the correlation corrections, Perdew and Wang first introduced the Seitz radius r_s

$$r_s = \left[3/4\pi(\rho_\alpha + \rho_\beta)\right]^{1/3}$$

The spin polarization is again defined by

$$\zeta = (\rho_\alpha - \rho_\beta)/(\rho_\alpha + \rho_\beta)$$

The spin interpolation form is borrowed from Ceperley and Alder [23,24].

$$\varepsilon_c^{PW91}(r_s, \zeta) = \varepsilon_c^{PW91}(r_s, 0) + [\zeta\text{-dependent terms}]$$

The ζ-dependent terms include one proportional to the difference

$$\varepsilon_c(r_s, 1) - \varepsilon_c(r_s, 0)$$

Perdew and Wang define a flexible form to replace ε_c

$$G(r_s : A, \alpha, \beta_1, \ldots, \beta_4, p) = -2A(1 + \alpha r_s)\ln\left(1 + \frac{1}{2A\left(\sum_{k=1}^{3}\beta_k r_s^{k/2} + \beta_4 r_s^{p+1}\right)}\right)$$

They relate certain of the parameters to the well-defined high density expansion of the correlation energy density functional for the electron gas, and fit to the results of the VWN numerical calculation. We will provide a more thorough discussion of the performance of the PW86 formulation of density functional theory in later sections.

Becke Functionals

Axel Becke was one of the pioneers in bringing DFT to a useful form. Just as the numerical solutions to the electron gas at various density served to define the VWN functionals and the Perdew–Wang refinements, the numerical results obtained by Becke and Dickson [28] by their basis-free DFT program allowed Becke to test and refine a series of forms for exchange–correlation functionals [29–31]. Becke proposed an exchange functional (the "B," "Becke88," or "B88" [32] form) which had correct asymptotic behavior

$$E_{xc}(r \to \infty) = -\frac{1}{2} \int \frac{\rho}{r} d\mathbf{r}$$

and a proper hole normalization constraint. The Becke correction appears as an additional term in the exchange energy density functional (not an enhancement factor as in Perdew's approach)

$$\varepsilon_x^{B88} = \varepsilon_x^{LDA} + \Delta\varepsilon_x^{B88}$$

$$\Delta\varepsilon_x^{B88} = -\beta \left(\frac{x^2}{1 + 6\beta x \sinh^{-1}(x)} \right)$$

Here x is again a scaled (dimensionless) von Weizsäcker kinetic variable [14] defined above. This form reproduced values of the Hartree–Fock exchange for rare gas atoms. Inclusion of the gradient variable makes this a second-rung GC or GGA formula. The $\sinh^{-1}(x_\sigma)$ in the exchange formula is a fairly simple and conveniently smooth function which allows the proper long-range asymptotic behavior of the spin density:

$$\rho_\sigma(r \to \infty) = e^{-a_\sigma r}$$

According to Becke, a_σ is related to the ionization potential of the system. (In orbital language, we would recognize that the long tail of the density would be proportional to $\exp(-\varepsilon_{HOMO} r)$, and the energy of the highest occupied MO is the ionization energy according to Koopmans' theorem.

The exchange density functional is related to the exchange energy according to

$$E_X^{SE} = E_X^{LDA} - \beta \sum_\sigma \int \rho_\sigma^{4/3} \frac{x_\sigma^2}{(1 + \gamma x_\sigma^2)} d\mathbf{r} \quad \text{where } x_\sigma = \frac{|\nabla\rho_\sigma(\mathbf{r})|}{\rho_\sigma^{4/3}(\mathbf{r})}$$

The superscript SE refers to the possibility of adjusting the value of parameters β and γ, which would make this a "semiempirical" fit. However, enforcing the condition that the Fermi hole should contain precisely one electron and imposing the proper limits on the spin density (zero at infinite removal from the molecule) and the asymptotic form of the exchange energy

$$E_{xc}(r \to \infty) = -\frac{1}{2} \int \frac{\rho}{r} d^3r$$

required the form

$$E_X = E_X^{\text{LDA}} - \beta \sum_\sigma \int \rho_\sigma^{4/3} \frac{x_\sigma^2}{(1 + 6\beta x_\sigma \sinh^{-1} x_\sigma)} d^3\mathbf{r} \quad \text{where } x_\sigma = \frac{|\nabla \rho_\sigma|}{\rho^{4/3}}$$

The constraints are not met by the Perdew–Wang [27] exchange functional quoted above.

To judge the quality of the gradient-correction to the VWN-LSDA formulation of density functional theory for molecules, Becke [33,34] characterized the molcules in the G1 set [35–37], for which heats of atomization were available. Each molecule was geometry optimized in VWN-LSDA. The VWN-LSDA tends to give good optimized geometries and vibrational frequencies but predicts "overbinding"; its heat of atomization is systematically high. When the new GC exchange correction term was subtracted from the VWN-LSDA energy, the mean absolute error in heats of atomization as estimated in this GC-VWN scheme was considerably improved, from 36.2 to ca. 3.7 kcal/mol. The best means of estimating thermochemical values available then was "G1 theory," a composite method with accuracy of roughly 1.6 kcal/mol.

Becke used the Perdew–Wang correlation functional in combination with his own exchange functional on the molecular data for the G1 set [35–37] and found that the correlation correction improved ionization energies considerably, but had lesser effects on heats of atomization.

Lee–Yang–Parr Correlation

Colle and Salvetti [38–40] developed a formula to calculate correlation energies from Hartree–Fock wave functions using a spherical Gaussian hole and the diagonal elements of the second-order density matrix. Lee, Yang, and Parr [41] (LYP) recast Colle and Salvetti's equations introducing the electron density and treating both open- and closed-shell cases so that the functional could be used in a DFT calculation. Recall that

$$E_{xc}[\rho] = E_x[\rho] + E_c[\rho] = \int \rho(\mathbf{r}) \varepsilon_x[\rho(\mathbf{r})] \, d\mathbf{r} + \int \rho(\mathbf{r}) \varepsilon_c[\rho(\mathbf{r})] \, d\mathbf{r}$$

Jensen [6] quotes the ε_c formulas shown below.

$$\varepsilon_c^{\text{LYP}} = -a\frac{\gamma}{(1 + d\rho^{-1/3})} - ab\frac{\gamma \exp(-c\rho^{-1/3})}{9(1 + d\rho^{-1/3})\rho^{8/3}}$$
$$\times \left[18(2^{2/3})C_F(\rho_\alpha^{8/3} + \rho_\beta^{8/3}) - 18\rho t_W + \rho_\alpha(2t_W^\alpha + \nabla^2 \rho_\alpha) + \rho_\beta(2t_W^\beta + \nabla^2 \rho_\beta) \right]$$

where

$$\gamma = 2\left[1 - \frac{\rho_\alpha^2 + \rho_\beta^2}{\rho^2}\right] \quad \text{and} \quad t_W^\sigma = \frac{1}{8}\left(\frac{|\nabla\rho_\sigma|^2}{\rho_\sigma} - \nabla^2\rho_\sigma\right)$$

The quantities a, b, c, and d are parameters fitted to match results for He. The ε form of the LYP functional (see also [42]) given above contains second-derivative (∇^2) terms. In their investigation of the performance of this functional, Miehlich et al. [43] recast the expression by partial integration to produce a more elaborate but more easily computed form with at worst $|\nabla\rho|^2$ to evaluate. This is

$$E_c = -a\int \frac{4}{(1 + d\rho^{-1/3})} \frac{\rho_\alpha\rho_\beta}{\rho} d^3r$$

$$- ab\int \omega\left\{\rho_\alpha\rho_\beta\left[2^{11/3}C_F\left(\rho_\alpha^{8/3} + \rho_\beta^{8/3}\right) + \left(\frac{47}{18} - \frac{7}{18}\delta\right)|\nabla\rho|^2\right.\right.$$

$$\left.- \left(\frac{5}{2} - \frac{1}{18}\delta\right)(|\nabla\rho_\alpha|^2 + |\nabla\rho_\beta|^2) - \left(\frac{\delta - 11}{9}\right)\left(\frac{\rho_\alpha}{\rho}|\nabla\rho_\alpha|^2 + \frac{\rho_\beta}{\rho}|\nabla\rho_\beta|^2\right)\right]$$

$$\left.- \frac{2}{3}\rho^2|\nabla\rho|^2 + \left(\frac{2}{3}\rho^2 - \rho_\alpha^2\right)|\nabla\rho_\beta|^2 + \left(\frac{2}{3}\rho^2 - \rho_\beta^2\right)|\nabla\rho_\alpha|^2\right\} d^3r$$

Here $\omega = \dfrac{\exp(-c\rho^{1/3})}{(1 + d\rho^{-1/3})\rho^{14/3}}$, $\delta = c\rho^{1/3} + \dfrac{d\rho^{-1/3}}{(1 + d\rho^{-1/3})}$, and $C_F = \dfrac{3}{10}(3\pi^2)^{2/3}$ and the parameters are adopted from the Colle–Salvetti fit to He. This correlation functional can then be used with the B88 exchange functional to form the "BLYP" exchange–correlation functional. The mean error for ionization energies, electron affinities, and dissociation energies is on the order of tens of millihartrees (6–12 kcal/mol).

The Adiabatic Connection and Hybrid Functionals

Kohn and Sham [1,2] introduced the notion that the true density for the system of interacting electrons may be identical with the density for a system of noninteracting electrons. Then, as we have seen, the Schroedinger equation is separable into equations for one-electron Kohn–Sham orbitals

$$-\frac{1}{2}\nabla^2\psi_i + V_{KS}\psi_i = e_i\psi_i$$

A Slater determinant of orbitals ψ_i for the noninteracting electrons is an exact solution. Its density is

$$\rho = \sum_i^n N_i \psi_i^* \psi_i; \quad \text{and since } E[\rho] = T_{\text{non}} + V_{\text{nuc-el}} + V_{\text{el-el}} + \Delta T$$

$$E[\rho] = \sum_i^n <\psi_i|t(i)|\psi_i> - \sum_{i,A} \left\langle \psi_i \left| \frac{Z_A}{|r_i - R_A|} \right| \psi_i \right\rangle$$

$$+ \sum \left\langle \psi_i \left| \frac{1}{2} \int \frac{\rho(r')}{|r_i - r'|} dr' \right| \psi_i \right\rangle + E_{XC}$$

Then, we express the total electronic energy of the real fully interacting system as

$$E_{\text{tot}} = T_0 + \int \rho V_{\text{nuc}} \, d\mathbf{r} + \frac{1}{2} \int \int \frac{\rho(r_1)\rho(r_2)}{|\mathbf{r}_1 - \mathbf{r}_2|} d\mathbf{r}_1 \, d\mathbf{r}_2 + E_{XC}$$

Here T_0 is the kinetic energy of the noninteracting reference system, the second term is the nuclear interaction term (note electrostatic attractions are "turned on"), the third term is only the classical coulomb self-energy and the last term, E_{XC}, is the density-functional "exchange–correlation" energy (of as yet noninteracting electrons, distributed by the exact density).

How can the transformation to the system of interacting charges be effected? Let there be some unknown functional $U_{XC}(\lambda)$ which depends on a parameter λ

$$E_{XC} = \int_0^1 U_{XC}(\lambda) \, d\lambda$$

which "switches on" the coulomb repulsion between the electrons. The density, defined by noninteracting electrons in the Kohn–Sham orbitals, is unchanged throughout the integration. Then, $U_{XC}(\lambda)$ is the potential energy of the exchange–correlation at some intermediate coupling strength λ, which connects some noninteracting Kohn–Sham reference system ($\lambda = 0$) to the fully interacting real system ($\lambda = 1$).

Becke identifies the limiting cases as the exact exchange for Kohn–Sham orbitals ($\lambda = 0$) and (approximately) the LSDA exchange–correlation ($\lambda = 1$). The simplest guess at the integral value is $E_{XC}^{HH} = \frac{1}{2}(E_X + U_{XC})$ [44]. Here the first term is the exchange in the Kohn–Sham set of orbitals (which Becke approximates with HF orbitals) and the second term is estimated as an electron gas exchange–correlation potential energy, making no reference to the kinetic energy correction, which nonetheless must accompany the turning on of the electron repulsion. Tests with the G1 set of data on small molecules show a mean absolute deviation of about 6 kcal/mol. Of course, the idea of mixing HF exchange and electron gas exchange–correlation immediately emerges as a promising semiempirical procedure. Becke's first try improved the fit to atomization energies, but at the expense of accuracy in ionization potentials and proton affinities.

The "half and half" [44] combination points the way to a less approximate treatment of the adiabatic connection function. Unfortunately, we have no deep insight into the form for $U_{XC}(\lambda)$. It is not even clear that one should introduce the so-called exact (Hartree–Fock) exchange at $\lambda = 0$.

In Paper III [45] of Becke's thermochemistry series, he used this adiabatic connection idea to merge three functionals beyond the LSDA functional:

$$E_{XC} = E_{XC}^{\text{LSDA}} + a_0(E_X^{\text{exact}} - E_X^{\text{LSDA}}) + a_X \Delta E_X^{\text{B88}} + a_c \Delta E_c^{\text{PW91}}$$

The first term sets the reference point, the LSDA term from the electron gas. The second term introduces a portion of exact (Hartree–Fock) exchange, and the third term provides a contribution from Becke's GC exchange functional. There is still a need for the Hartree–Fock form of "exact" and "nonlocal" exchange. Regardless of choice of PW91 or LYP correlation, the Hartree–Fock exchange component is on average near 25%. Becke replaced the PW91 [45] correlation functional by the LYP [41,42] correlation functional. A version of this formulation became B3LYP. An alternative way to write the B3LYP exchange–correlation energy is

$$E_{\text{B3LYP}}^{XC} = E_{\text{LDA}}^X + c_0(E_{\text{HF}}^X - E_{\text{LDA}}^X) + c_X(\Delta E_{\text{B88}}^X) + E_{\text{VWN3}}^C + c_C(E_{\text{LYP}}^C - E_{\text{VWN3}}^C)$$

For the fit to the G1 set of molecular data $c_0 = 0.2$, $c_X = 0.72$, and $c_C = 0.81$ [45]. The B3LYP functional gives good results for atomization energies, ionization potentials, proton affinities, and total energies. Although there are many newer pure and hybrid density functionals, performance of newer methods often falls short of B3LYP's quality.

Good performance in fit of computed quantities to experimental data, as desirable as that must be, does not by itself convince us that our parametrized expressions help us understand better the physics and chemistry of molecules. Nor are we confident that we have approached the ultimate object of DFT, finding the elusive exact exchange–correlation energy via the "universal functional." The next step for improvement is not obvious. In response to all these felt needs, Becke sought simpler formulas with fewer constants and perhaps enhanced performance.

Becke 1995—Impact of Imposing Constraints

In Paper IV [46] of his studies on thermochemistry, Becke set out conditions to be met by the correlation functional:

1. It should reduce to the exact uniform electron gas limit.
2. It should allow distinct treatment of α and β electron spins.

3. It should be free of self-interaction and produce zero correlation for one-electron systems.

4. It should achieve a good fit to exact correlation energies of atoms.

Becke asserts that fitting a correlation functional to He is not sufficiently general to capture all possible spin interactions. Further, dynamical correlations are intrinsically short range and should be calibrated to atoms rather than to molecules where long-range exchange can cancel out some of the dynamical correlation. Thus, Becke fit his new correlation functional, called Bc95, to Ne data.

One of the first priorities was to force the correlation energy of H to be zero. Define

$$x = \frac{|\nabla \rho|}{\rho^{4/3}} \quad \text{and} \quad D = \sum_i^N |\nabla \phi_i|^2 - \frac{(\nabla \rho)^2}{4\rho} \quad \text{with} \quad D_\sigma^{\text{UEG}} = \frac{3}{5}(6\pi^2)\rho_\sigma^{5/3} \quad \text{so that} \quad E_C^{\sigma\sigma}$$

$$= \frac{D_\sigma}{D_\sigma^{\text{UEG}}} E_{C\sigma\sigma}^{\text{UEG}}$$

The second term defines a corrected kinetic energy. Here "UEG" refers to the uniform electron gas. Multiplying $E_C^{\sigma\sigma}$ by the ratio of the D_σ values leads to a zero correlation energy for H and any one-electron system as the limiting case for any interaction of like spins. Then, relatively simple formulas were used for unlike spin ($\alpha\beta$) and like spin ($\sigma\sigma$, $\sigma = \alpha$ or β) interactions.

$$E_C^{\alpha\beta} = \frac{E_{C\alpha\beta}^{\text{UEG}}}{\left[1 + a(x_\alpha^2 + x_\beta^2)\right]}, \quad a = 0.0031 \quad E_C^{\sigma\sigma} = \frac{E_{C\sigma\sigma}^{\text{UEG}}}{(1 + bx_\sigma^2)} \frac{D_\sigma}{D_\sigma^{\text{UEG}}}, \quad b = 0.038$$

This correlation functional combined with the Becke88 exchange was tested on the G2 data set [47–49] for 56 atomization energies, 42 ionization potentials, and 8 electron affinities. For accurate results, it still proved necessary to add exact Hartree–Fock exchange. The final result

$$E_{XC} = E_{XC}^{\text{Bx88/Bc95}} + 0.28(E_X^{\text{exact}} - E_X^{\text{Bx88}})$$

is a best fit to atomization energies, ionization potentials, and proton affinities. Use of the B86 [29] exchange potential reduced the error slightly.

In Paper V [50], Becke abandoned the requirements 2 and 3 cited above and achieved what might be the limit of the strategy of fitting experimental data within the generalized gradient approximation with added exact exchange. His 10-parameter fit produces a mean absolute deviation (MAD) of 1.8 kcal/mol in the G2 atomization energies, and errors in ionization energies and electron affinities of about 0.1 eV. This is an improvement over B3LYP which has MAD values of 2.4 kcal/mol for energies of atomization and 0.14 for ionization potentials. This formulation, however, has not been so widely tested as B3LYP in other experimental contexts.

PBE Exchange–Correlation Functional

The philosophy of the Perdew endeavor has been to avoid empirical parameters as much as possible in corrections to the LSDA density functionals. Perdew relies on improvement of the enhancement functions f and F, the latter being a hole function.

$$E_{XC}^{GGA}(\rho_\alpha, \rho_\beta) = \int d\mathbf{r} f(\rho_\alpha, \rho_\beta, \nabla\rho_\alpha, \nabla\rho_\beta) = \int d\mathbf{r} \rho F(\rho_\alpha, \rho_\beta, \nabla\rho_\alpha, \nabla\rho_\beta)$$

The PW91* GGA F was an analytical form fit to a numerical evaluation of the exchange–correlation hole attending an electron in a system of nearly uniform density.

In his 1996 paper with Burke and Ernzerhof [51] several deficiencies of PW91 were addressed:

1. Developing the function f involves a long, difficult derivation.
2. The function f is itself complicated and uninformative.
3. The function f has too many parameters.
4. These parameters are not "seamlessly joined," leading to anomalies in the high and low gradient limit.
5. PW91's analytical form does not obey scaling to the high density limit.
6. PW91 is less satisfactory than LSDA in the limit of a uniform electron gas.

Thus motivated, Perdew, Burke, and Wang published the mordantly named article "Generalized gradient approximation made simple." They dealt first with the correlation functional, which they adapted from the PW91 form:

$$E_c^{PBE}[\rho_\alpha, \rho_\beta] = \int \rho \{\varepsilon_C^{UEG}(r_s, \zeta) + H^{PBE}(r_s, \zeta, t)\} \, d^3r$$

H is a function of t, the scaled gradient variable:

$$t = \frac{|\nabla\rho|}{2\phi k_s \rho}$$

---

*Usage varies; Perdew himself refers to the PW92 model, which makes some sense since the details appeared in 1992.

as well as r_s which is the Seitz radius of a sphere which contains one electron at the prescribed density (small for high density and vice versa):

$$r_s = (3/4\pi\rho)^{1/3}$$

The function $\varphi(\zeta)$ of the spin polarization ζ becomes unity for systems with zero spin polarization:

$$\zeta = \frac{\rho_\alpha - \rho_\beta}{\rho_\alpha + \rho_\beta} \quad \text{and} \quad \varphi(\zeta) = \frac{1}{2}\left[(1+\zeta)^{2/3} + (1-\zeta)^{2/3}\right]; \quad \text{at} \quad \zeta = 0, \varphi(0) = 1$$

The Fermi level of the system appears as k_F.

$$k_s = (4k_F/\pi)^{1/2}$$

The following conditions constrain the function H:

1. As $t \to 0$, H approaches its second-order series expansion form $\beta\varphi^3 t^2$; β is an empirical constant.
2. As $t \to \infty$, H approaches zero; the correlation hole must obey $\int du h_C(\mathbf{r}, \mathbf{r} + \mathbf{u}) = 0$ which requires the correlation hole density, h_C, to vanish in the high—t limit.
3. Upon scaling to high density the correlation energy must approach a constant value; then the function H must cancel the logarithmic singularity of the uniform electron gas correlation.

$$\varepsilon_C^{UEG}(r_s, \zeta) \to \varphi^3[\gamma \ln(r_s) - \omega]$$

PBE replace the functions γ and ω by their values at $\zeta = 0$.
The detailed PBE form is

$$H^{PBE} = \gamma\phi^3 \ln\left\{1 + \frac{\beta}{\gamma}t^2\left[\frac{1+At^2}{1+At^2+A^2t^4}\right]\right\} \quad A = \frac{(\beta/\gamma)}{[\exp(-\varepsilon_c^{UEG}/\gamma\phi^3) - 1]}$$

At high density

$$E_C^{GGA} \to -\int d^3r\,\rho(r)\gamma\varphi^3 \ln[1 + g(t, \varphi)]$$

The quantity β is the only empirical parameter, with the value 0.066725.
The PBE exchange functional is based on the Uniform Electron Gas (UEG)

$$E_X^{GGA}[\rho] = \int \rho\varepsilon_X^{GGA}(\rho)\,d\mathbf{r} = \int \rho\varepsilon_X^{LDA}(\rho)F_X^{GGA}(s)\,d\mathbf{r}$$

The dimensionless gradient variable s is

$$s = |\nabla\rho|/2k_F\rho$$

The uniform electron gas exchange is

$$\varepsilon_X^{\text{UEG}}(\rho) = -3k_F/4\pi$$

The characteristic (Fermi) wave number is

$$k_F = (3\pi^2\rho)^{1/3}$$

F_X is the enhancement function, which reduces to -1 for $s=0$, and is related to the hole function.

$$\frac{8}{9}\int dy\, y J^{\text{GGA}}(s,y) = -F_X^{\text{GGA}}(s)$$

The assumed form for J^{PBE} is

$$J^{\text{PBE}} = \exp\left(-K(s)s^2y^2\right)\left[T_1 + (T_2 + T_3)\exp\left(-D'y^2\right)\right]$$

$$T_1 = -\frac{A'}{y^2}\frac{1}{1 + (4A'y^2/9)}$$

$$T_2 = \frac{A'}{y^2} + B' + C'y^2\left[1 + s^2F(s)\right]$$

$$T_3 = Hy^4\left[1 + s^2G(s)\right]$$

The primed parameter values are distinct from the parameters of the LDA hole. The integral defining F can be evaluated analytically. Details of the functions E, G, and K are given in the original report (Figure 11.1).

The function $F_X^{\text{PBE}}(s)$ was designed to meet several criteria. It is required to

1. Obey the limit of uniform electron gas density; $F_X(0) = 1$
2. Recover proper linear response, i.e.,

$$F_X(s) \rightarrow 1 + \mu s^2, \ \mu = \beta(\pi^2/3)$$

3. Meet the Lieb–Oxford bound [54]

$$E_X[\rho_\alpha, \rho_\beta] \geq E_{XC}[\rho_\alpha, \rho_\beta] \geq -1.679\int\rho^{4/3}\, d^3r$$

The spin-scaling condition requires

$$E_X^{\text{GGA}}[\rho_\alpha, \rho_\beta] = \frac{1}{2}E_X^{\text{GGA}}[2\rho_\alpha] + \frac{1}{2}E_X^{\text{GGA}}[2\rho_\beta]$$

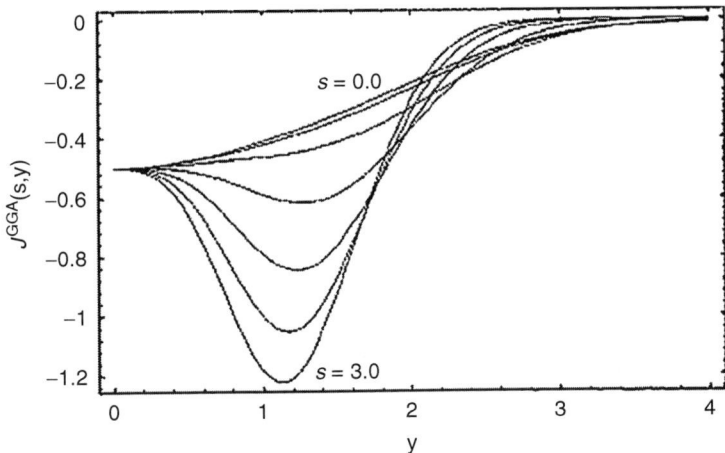

FIGURE 11.1
The shape of Ernzerhof and Perdew's hole function J^{GGA} in a generalized gradient approxima-
tion (GGA) is displayed for several values of the dimensionless Weizsäcker kinetic variable s.
The horizontal axis is the interelectronic distance scaled by the Fermi wavelength. (From
Enzerhof, M. and Perdew, M., *J. Chem. Phys.*, 109, 331, 1998. With permission.)

The Lieb–Oxford bound requires the scaling

$$F_X(\zeta = 1, s) = 2^{1/3} F_X(s/2^{1/3})$$

PBE recovered the form

$$F_X(s) = 1 + \kappa - \frac{\kappa}{(1 + (\mu s^2/\kappa))}$$

The parameters $\kappa = 0.804$ and $\mu = 0.21951$ assure that the bound is met
locally.

In 1986, Becke [53] had already proposed a function of similar form but
with empirically determined constants $\kappa = 0.967$ and $\mu = 0.235$; this can be
shown to violate the Lieb–Oxford local bound [52]. In 1998, Zhang and
Wang [54] showed that for systems without spin polarization, the reparametrized PBE still obeys the local bound. For all spin polarization the
reparametrized PBE form satisfies an integrated version of the Lieb–Oxford
bound. The value $\kappa = 1.245$ best matches exchange energy in first-row atoms
He–Ne (MAD = 4 millihartree). Their "revPBE" formulation reduced the
PBE MAD for a set of 24 small molecules from 8.1 to 4.9 kcal/mol—not yet
competitive with hybrid methods with optimized admixture of exact
exchange, but a substantial step.

In 1999, Hammer, Hansen, and Norskov [55] defined

$$F_X^{HHN}(s) = 1 + \kappa \left[1 - \exp\left(-\mu s^2/\kappa\right)\right]$$

which retains the general form of $F_X(s)$ but approaches the Lieb–Oxford bound as an asymptote at high values of s. That is,

$$E_X[\rho_\alpha, \rho_\beta] \geq E_{XC}[\rho_\alpha, \rho_\beta] \geq -1.679 \int \rho^{4/3} \, d^3r$$

$$E_X^{GGA}[\rho] = \int \rho \varepsilon_X^{UEG}(\rho) F_X(s) \, d\mathbf{r} = -(3/4\pi)(3\pi^2)^{1/3}$$

$$\int \rho^{4/3} \{1 + \kappa[1 - \exp(-\mu s^2/\kappa)]\} \, d\mathbf{r}$$

This new formula "RPBE" was intended for applications of gas absorption on metal surfaces. In this context, it improved on the performance of the PBE functional while avoiding violation of the strict local version of the Lieb–Oxford bound (Figure 11.2).

Ernzerhof and Scuseria [56] showed that the PBE exchange–correlation functional match to heats of atomization reported for the G2 [47–49] set of molecules was considerably improved by addition of a fractional amount of Hartree–Fock exchange. The addition was simply one-fourth of Hartree–Fock exchange, as suggested by MP4 perturbation theory

$$E_{XC}^{hyb} = a(E_X - E_X^{DFT}) + E_{XC}^{DFT}, \quad a = (1/4)$$

and the resulting functional was near to B3LYP in performance. The introduction of this Hartree–Fock exchange, a nonlocal quantity, seems essential to high accuracy estimates of heats of atomization; purely local functionals are so far not competitive with hybrid functionals. It does not seem to be critical just what local functionals are employed. Gill [57] remarks that this

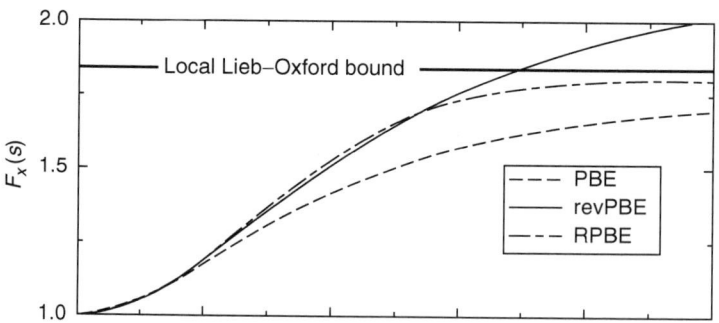

FIGURE 11.2

Three enhancement factors and the local Lieb–Oxford bound [54]. The horizontal axis is the reduced density gradient as in Figure 11.1; the enhancement factor requires a parameter κ which is 0.804 in the PBE functional [51] and 1.245 in the reformulation revPBE by Zhang and Wang [54]. Hammer et al. [55] define RPBE to fulfill the bound locally. (From Hammer, B., Hansen, L.B., and Norskov, J.K., *Phys. Rev. B*, 59, 7413, 1999. With permission.)

abandonment of pure density functional concept to embrace semiempirical combinations of functionals is a major departure from the hopes and intentions of Kohn and Sham. In practice, the errors the pure (and local) functionals suffer and whatever errors the Hartree–Fock exchange contains seem to compensate in the hybrid forms. We will return to this point in our discussion of Truhlar's contribution.

Correlation Effects on Kinetic Energy

It will have occurred to the reader that the definition of the kinetic energy using Kohn–Sham orbitals cannot be exactly correct. Third-rung (meta-GGA) theories address this flaw. This kinetic energy is a property of the orbitals which make up a single determinant and refer to noninteracting electrons (but which do nonetheless compose the exact density). It is closely analogous to the kinetic energy defined by HF orbitals which incorporate a mean-field estimate of the electron–electron interaction.

$$E_T = \langle T \rangle = -\frac{1}{2} \sum_i^{\text{occ}} \int \psi_i^* \nabla^2 \psi_i \, d\mathbf{r}$$

This simple sum is a consequence of the single determinant form. But the kinetic energy in density functional theory should accommodate all correlation effects. This must mean that there must be some correction of the kinetic energy incorporated into the correlation functional. Let $T_S[\rho]$ be the kinetic energy of the Slater determinant of Kohn–Sham orbitals and $T[\rho]$ be the actual kinetic energy after the interaction is turned on. Then,

$$E[\rho] = V_{\text{ne}}[\rho] + (T_S[\rho] + \{T[\rho] - T_S[\rho]\}) + (J[\rho] + V_{\text{ee}}[\rho] - J[\rho])$$

If we group together the terms which are responsive when the interaction is turned on we can write them as

$$E_{XC}[\rho] = (T[\rho] - T_S[\rho]) + (V_{\text{ee}}[\rho] - J[\rho])$$

so that we see that the exchange–correlation term will include a correction to the kinetic energy. The "τ-dependent" construct of "meta-GGA" functionals—the next step up Jacob's ladder—is an attempt to incorporate such corrections to the exchange–correlation functional.

Perdew, Kurth, Zupan, and Blaha (PKZB) [58] proposed what they called a meta-GGA functional, often called the PKZB functional. The improvement is based on the idea of introducing the kinetic energy density of the occupied Kohn–Sham orbitals as

$$\tau_\sigma = \frac{1}{2} \sum_\alpha^{\text{occ}} |\nabla \psi_{\alpha\sigma}|^2 \text{ where } \sigma = \alpha \text{ or } \beta \text{ and } \rho_\sigma = \sum_\alpha |\psi_{\alpha\sigma}|^2$$

Instead of using the actual orbital expansion suggested by these formulas, PKZB introduced a second-order gradient expansion of the kinetic energy density:

$$\tau^{\text{GGA}} = \frac{3}{10}(3\pi^2)^{2/3}\rho^{5/3} + \frac{1}{72}\frac{|\nabla\rho|^2}{\rho} + \frac{1}{6}\nabla^2\rho$$

adopted from Brack et al. [59]. Then, three additional constraints were applied:

1. The spin-scaling condition:

$$E_X[\rho_\alpha, \rho_\beta] = \tfrac{1}{2}E_X[2\rho_\alpha] + \tfrac{1}{2}E_X[2\rho_\beta]$$

2. The uniform density-scaling requirement:

$$E_X[\lambda\rho] = \lambda E_X[\rho]$$

3. The Lieb–Oxford bound:

$$E_X[\rho_\alpha, \rho_\beta] \geq E_{XC}[\rho_\alpha, \rho_\beta] \geq -1.679 \int d^3\mathbf{r}\,\rho^{4/3}$$

PKZB employed the expansion of the enhancement factor F_X from Svendsen and von Barth [60].

$$F_X = 1 + \frac{10}{81}p + \frac{146}{2025}q^2 - \frac{73}{405}qp + Dp^2 + O(\nabla^6\rho)$$

This incorporates both the density gradient and curvature through the definitions of p and q.

$$p = \frac{|\nabla\rho|^2}{\left[4(3\pi^2)^{2/3}\rho^{5/3}\right]}$$

$$q = \frac{\nabla^2\rho}{\left[4(3\pi^2)^{2/3}\rho^{5/3}\right]}$$

Defining the quantity

$$\tilde{q} = \frac{3\tau}{\left[2(3\pi^2)^{2/3}\rho^{5/3}\right]} - \frac{9}{20} - \frac{p}{12}$$

allows the recovery (against all expectation) of the functional form used in PBE96

$$F_X(p, \tilde{q}) = 1 + \kappa - \frac{\kappa}{(1 + (x/\kappa))}$$

where now

$$x = \frac{10}{81}p + \frac{146}{2025}\tilde{q}^2 - \frac{73}{405}\tilde{q}p + \left[D + \frac{1}{\kappa}\left(\frac{10}{81}\right)^2\right]p^2$$

instead of simply $0.21951p$ as in PBE96. The parameter $\kappa = 0.804$ remains as before. The PKZB functional was shown to still maintain the Lieb–Oxford [52] bound when p and q are substituted, a process which challenges our algebra. D was estimated by minimizing predicted error in the atomization energies of 20 small molecules. The value $D = 0.113$ produced the best mean absolute deviation, 3.06 kcal/mol.

The PKZB functionals have properties which are desirable on theoretical grounds; for example the PKZB correlation functional is constrained to be free of self-interaction and hence zero for a spin-polarized one-electron density. But the scheme falls short of useful accuracy in chemical applications. Partly because the PKZB functional is not a great improvement in energetics over methods lower on Perdew's ladder and also tends to predict bond lengths which are too long, an overcorrection to the usual overbinding in LSDA, further improvement in meta-GGA formulations was needed.

TPSS Functional*

We saw that the form of $F(s)$ obtained by PKZB obeys the Lieb–Oxford [52] bound at high densities (high s). Tao, Perdew, Saroverov and Scuseria (TPSS) [61,62] turned their attention to the behavior of $F(s)$ in the limit of low s. TPSS consider two extreme cases of electron density. Case A is the situation in conducting solids where the electron density tends to be slowly varying within a crystal lattice while Case B is characteristic of small low symmetry molecules where the density varies wildly from extremely large values near a nucleus to intermediate values in covalent bonds, and very low density in the outer reaches of a molecule. These are all important to

---

*Some scientists think that nature is described by simple powerful equations while others consider nature beyond such compact summary. What follows in the TPSS derivation belongs to the latter philosophy.

chemical phenomena. The gradient is as usual expressed with reference to a version of the Weizsäcker kinetic variable

$$\tau_\sigma^W(\mathbf{r}) = |\nabla \rho_\sigma|^2 / 8\rho_\sigma$$

For slowly changing spin-unpolarized density, TPSS uses the same gradient expansion as PKZB from Brack et al. [59].

$$\tau^{GEA} = \tau^{unif} + \frac{1}{72}\frac{|\nabla\rho|^2}{\rho} + \frac{1}{6}\nabla^2\rho + O(\nabla^4\rho) \quad \text{where} \quad \tau^{unif}$$

$$= \frac{3}{10}(3\pi^2)^{2/3}\rho^{5/3} \quad \text{and} \quad \tau = \sum_\sigma \tau_\sigma$$

The GEA notation refers to a "gradient expansion approximation" which is to be distinguished from the generalized gradient approximation (GGA). As usual, TPSS begins with the general formula for an exchange–correlation functional:

$$E_{xc}[\rho_\alpha, \rho_\beta] = \int d\mathbf{r}\, \rho\varepsilon_{XC}(\rho_\alpha, \rho_\beta, \nabla\rho_\alpha, \nabla\rho_\beta)$$

The heart of this method is the function F_{XC} which serves as an "enhancement factor" for the exchange relative to the uniform electron gas.

$$\varepsilon_{XC} = \varepsilon_X^{UEG}(\rho)F_{XC}; \quad \varepsilon_X^{UEG}(\rho) = -(3/4\pi)(3\pi^2\rho)^{1/3}$$

The TPSS functional, following its PKZB predecessor, enforces proper spin scaling and uniform coordinate scaling. The TPSS functional introduces two dimensionless inhomogeneity measures p (referring to the square of the gradient of the density) and z (referring to the departure of the kinetic variable from the Weizsäcker value).

$$p = |\nabla\rho|^2 \Big/ \left[4(3\pi^2)^{2/3}\rho^{8/3}\right] = s^2$$
$$z = (\tau^W/\tau) \leq 1$$
$$\tau = \sum_\sigma \tau_\sigma \quad \text{and} \quad \tau^W = |\nabla\rho|^2/8\rho$$

As with PKZB, the fourth-order expansion of Svendsen and von Barth [62] is used for F_X. One of the constraints on the TPSS is that the enhancement factor should approach the original PBE96 GGA form [64] for large values of p.

$$F_X \sim 1 + \kappa - \kappa^2/(\mu p) \quad \text{as} \quad (p \to \infty)$$

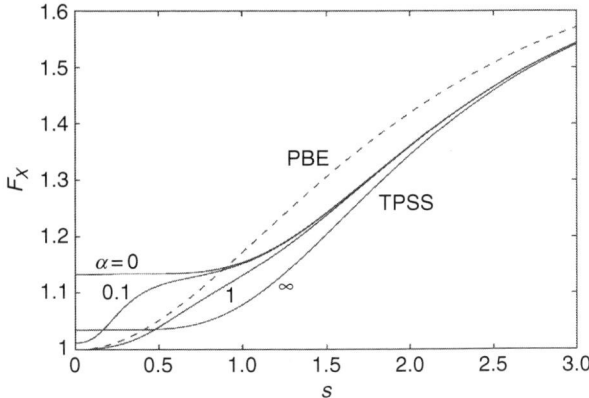

FIGURE 11.3
Enhancement functions from TPSS [61] and PBE [51]. The horizontal axis is the reduced kinetic variable. The parameter α is a measure of the departure of the kinetic variable τ from the Weizsäcker value. In principle, it is dependent on location (see text). (From Perdew, J.P., Tao, J., Staroverov, V.N., et al., *J. Chem. Phys.*, 120, 6898, 2004. With permission.)

The PKZB derivation modified this to fulfill the limit of the Lieb–Oxford boundary (Figure 11.3) using

$$F_X = 1 + \kappa - \kappa/(1 + (x/\kappa))$$

In the TPSS modification to F_X, the quantity q is replaced by $\tilde{q}$ defined as

$$\tilde{q} = (9/20)(\alpha - 1) + (2p/3)$$

Here $\alpha = (\tau - \tau^W)/\tau^{UEG} = (5p/3)(z^{-1} - 1) \geq 0$. The quantity α is a measure of the departure of the kinetic variable τ from the Weizsäcker value. It tends to 0 for a one- or two-electron density for which $\tau = \tau^w$, and tends to 1 for a slowly varying density where $\tau \approx \tau^{UEG}$. The departure can be large in regions between shells (i.e., where density is small but rapidly changing) and can diverge as the density approaches zero, in the exponential tail of a finite atom or molecule.

TPSS point out that if F_X is constructed from p and $\tilde{q}$ there is no problem for large arguments; the functional is guaranteed to satisfy the Lieb–Oxford [52] bound. However, when p is small and α is large, which makes $\tilde{q}$ large, then $F_X \rightarrow 1.804$ so that the combined F_{XC} can exceed the Lieb–Oxford bound. This problem is (amazingly) solved in the TPSS derivation by adding a parameter b in $\tilde{q}$; thus,

$$\tilde{q}_b = (9/20)(\alpha - 1)/[1 + b\alpha(\alpha - 1)]^{1/2} + (2p/3)$$

When $\alpha = 0$, $\tilde{q}_b \rightarrow \tilde{q}$. When $\alpha = 1$, $\tilde{q}_b$ is also independent of b. For very large α, $\tilde{q}_b$ becomes $\tilde{q}_b \rightarrow (9/20)\sqrt{b} + (2p/3)$ which is linear in p. Going back to the x of

$$F_X = 1 + \kappa - \kappa/[1 + (x/\kappa)]$$

$$x = \left\{ \left[\frac{10}{81} + c\frac{z^2}{(1+z^2)^2} \right] p + \frac{146}{2025}\tilde{q}_b^2 - \frac{73}{405}\tilde{q}_b\sqrt{\frac{1}{2}\left(\frac{3}{5}z\right)^2 + \frac{1}{2}p^2} + \frac{1}{\kappa}\left(\frac{10}{81}\right)^2 p^2 \right. $$

$$\left. + 2\sqrt{e}\frac{10}{81}\left(\frac{3}{5}z\right)^2 + e\mu p^3 \right\} \Big/ (1 + \sqrt{e}p)^2 \quad \text{where} \quad c = 1.59096 \quad \text{and} \quad e = 1.537$$

The parameter b is assigned the value 0.40, the smallest positive number that makes F_X a nondecreasing function of s for any given value of α. Fortunately, when tests were carried out it was found that all the values of $0 \leq b \leq 0.40$ give essentially the same results for molecules and the surfaces of solids.

The TPSS functional also modifies the PKZB correlation part, defining the "revPKZB" correlation functional.

$$E_C^{MGGA}[\rho_\alpha, \rho_\beta] = \int d\mathbf{r}\, \rho\varepsilon_C^{revPKZB}\left[1 + d\varepsilon_C^{revPKZB}(\tau^W/\tau)^3\right]$$

$$\varepsilon_C^{revPKZB} = \varepsilon_C^{GGA}(\rho_\alpha, \rho_\beta, \nabla\rho_\alpha, \nabla\rho_\beta)\left[1 + C(\zeta, \xi)(\tau^w/\tau)^2\right]$$

$$- [1 + C(\zeta, \xi)](\tau^W/\tau)^2 \sum_\sigma \left(\frac{\rho_\sigma}{\rho}\right)\tilde{\varepsilon}_C^\sigma$$

$$\text{and} \quad \tilde{\varepsilon}_C^\sigma = \max\left[\varepsilon_C^{GGA}(\rho_\sigma, 0, \nabla\rho_\sigma, 0), \varepsilon_C^{GGA}(\rho_\alpha, \rho_\beta, \nabla\rho_\alpha, \nabla\rho_\beta)\right]$$

In these equations we have $\zeta = (\rho_\alpha - \rho_\beta)/\rho$ and $\xi = |\nabla\zeta|/2(3\pi^2\rho)^{1/3}$. Choosing $C(0,0) = 0.53$ allows the form to reduce to the PKZB result for a spin-unpolarized system in the low density limit. To make things even more complicated there are several special cases depending on the local Seitz radius r_s. We do not pursue the details of these elaborations. We should mention that Zhang and Salahub [63] have reoptimized the TPSS meta-GGA correlation functional to match the exchange functional in a more balanced way and included this new functional in the deMon2k [21] program. This modification is called TPSSτ3. It gives greatly improved estimates of H-bonding energies. The treatment of weak interactions remains a challenge for DFT functionals.

It will have occurred to the reader that this version of applied mathematics requires that forms be shaped to conform to bounds and constraints, and can otherwise be chosen as pleases the investigator. Simplicity of form is not required so long as one deploys the sophisticated grids and weights developed for the evaluation of integrals of the exchange–correlation functionals. This mode of research might be called "meta-nonempirical" research as compared to "ab initio" developmental work typical of wave function theory.

We will now leave Perdew's train of research which tries to avoid the temptation to introduce extensive parametrization to match collections of experimental data. This is a serious effort to avoid the trap of trusting numerical predictions made by methods which do not capture the essential physics.

The Empirical Thread

The other major theme in density functional design has been the attempt to match experimental data by parametrization of a mathematically defensible and computationally feasible formulation of density functionals. Our experience with the analogous semiempirical SCF methods suggests that distinct parametrizations would be necessary to represent faithfully specific types of experimental data.

The latest density functional developments along this line are by Zhao and Truhlar [64–68]. In 2004, Zhao et al. [64] surveyed the performance of a number of formulations including "first-generation," "second-generation," and "third-generation" functionals. These categories do not correspond perfectly to rungs on Perdew's ladder, but are roughly chronological designations. This team chose small representative data sets for atomization energies and kinetic barrier heights, and reported mean absolute deviations between these experimental values and energy values computed in a modest 6-31+G(d,p) basis set and a large basis set 6-311++G(3d2f,2df,2p) for H-Si. Table 11.2 selects some examples of each type, obtained with the smaller basis.

The hybrid methods, which add some nonlocal exact exchange of Hartree–Fock form, perform very well. For a decade or more the B3LYP hybrid set the standard and is still the method of choice for most applications. As we will see, its period of ascendancy may slowly be passing.

Density functional theory systematically underestimates barrier heights, and also does not describe weak interactions well. Zhao and Truhlar [65] provide a noteworthy history of DFT up to 2005 and develop useful databases for nonbonded interactions, including (1) hydrogen bonding, (2) charge transfer, (3) dipole interactions, and (4) weak interactions such as van der Waals and dispersion interactions. Their very extensive data show that as one ascends Perdew's ladder from LSDA to PBE to TPSS, a composite figure of merit improves from 3.83 to 1.05 to 0.95 kcal/mol. The popular hybrid method B3LYP lies at 0.74, PBE1PBE at 0.55, and MP2 (!) at 0.49 kcal/mol. Zhao and Truhlar's hybrid meta-GGA methods, called MPW1B95 and MPWB1K, are superior to other tested hybrid DFT methods, with composite error measures

TABLE 11.2

Development of Generations of Density Functionals

Generation	Examples	MAD (AE)	MAD (BH)	Composite
First (1988–1994)	BP86	2.7	9.4	6.2
	BLYP	1.4	8.3	4.7
	PW91	2.5	9.8	6.3
	B3LYP (hybrid)	1.5	5.0	3.1
Second (1996–2001)	mPWPW91	1.5	8.6	5.2
	mPW1PW91 (hybrid)	1.7	3.9	2.6
	MPW1K (hybrid)	3.1	1.4	2.1
	O3LYP(hybrid)	1.2	4.7	2.7
	PBE	2.5	9.5	6.1
	PBE0 (hybrid)	1.5	4.6	3.0
Third (1996 and 2001)	B1B95 (hybrid)	0.9	3.2	2.0
	VSXC	0.7	5.1	2.8

Note: Mean absolute errors in kcal/mol. Second-generation methods are mainly of type GGA (rung 2 on Perdew's ladder) and include hybrid methods; third-generation methods are of type meta-GGA (rung 3 on Perdew's ladder). The MAD values hide the fact that all barrier heights (BH) are systematically low. The composite measure is an average of BH and AE errors in each basis.

of 0.51 and 0.46 kcal/mol, respectively. It is noteworthy that after all the accumulated effort to develop some 45 different methods, the MP2 results are more reliable for the 4 categories listed than 43 of the 45 methods; only Truhlar's MPWB1K [66] seems superior to MP2 in this comparison. Truhlar's recent work on a local M06-L [67] functional and the nonlocal M05-2X [68] is not included in this list so we discuss them in detail below.

The Zhao–Truhlar M06-L DFT Functional

Owing to the considerable computational speed advantage of entirely local DFT methods,* Zhao and Truhlar [67] sought the best possible formulation of this type. The designation "L" in M06-L indicates that the formulation uses only totally local exchange–correlation functionals. "Local" means that the exchange–correlation is defined only by properties of the electron density in grid-point space. Hartree–Fock exchange, in contrast, refers to the entire density distribution. The M06-L functional excludes Hartree–Fock exchange, but alternative forms called M06-2X and M06-HF contain various amounts of Hartree–Fock exchange mixed with the M06-L functional. We will return to these after an outline of the structure of the local theory.

---

* Prof. Truhlar tells us that including any Hartree–Fock exchange increases the run time by an order of magnitude; the local DFT E_{XC} requires significantly less computation time than methods including any HF exchange.

Zhao and Truhlar adopt some familiar quantities, including a reduced gradient variable $x_\sigma = \dfrac{|\nabla \rho_x|}{\rho_\sigma^{4/3}}$, $\sigma = \alpha, \beta$ and the kinetic energy $\tau_\sigma = \frac{1}{2}\sum_i^{occ}|\nabla \psi_{i\sigma}|^2$. They define some special-purpose objects

$$z_\sigma = \frac{\tau_\sigma}{\rho_\sigma^{5/3}} - C_F, \quad C_F = \frac{3}{5}(6\pi^2)^{2/3} \quad \text{and} \quad \gamma(x_\sigma, z_\sigma) = 1 + \alpha(x_\sigma^2 + z_\sigma)$$

These allow the expression of the M06-L meta-GGA exchange functional as

$$E_X^{\text{M06-L}} = \sum_\sigma \int dr \left[F_{X\sigma}^{\text{PBE}}(\rho_\sigma, \nabla \rho_\sigma) f(w_\sigma) + \varepsilon_{X\sigma}^{\text{LSDA}} h_X(x_\sigma, z_\sigma) \right]$$

Zhao and Truhlar alter the LSDA exchange and the PBE exchange by enhancement functions f and h. The LSDA exhange functional is

$$\varepsilon_{X\sigma}^{\text{LSDA}} = \frac{3}{2}\left(\frac{3}{4\pi}\right)^{1/3} \rho_\sigma^{4/3}$$

Its enhancement function is

$$h(x_\sigma, z_\sigma) = \left(\frac{d_0}{\gamma(x_\sigma, z_\sigma)} + \frac{d_1 x_\sigma^2 + d_2 z_\sigma}{\gamma_\sigma^2(x_\sigma, z_\sigma)} + \frac{d_3 x_\sigma^4 + d_4 x_\sigma^2 z_\sigma + d_5 z_\sigma^2}{\gamma_\sigma^3(x_\sigma, z_\sigma)} \right), \quad \sigma = \alpha, \beta$$

The parameters $\{d_j\}$ are given by Voorhis and Scuseria [69]. The PBE exchange functional is altered by $f(w_\sigma)$, the spin kinetic energy density enhancement factor. This is expressed in a power series in w_σ

$$f(w_\sigma) = \sum_{i=0}^{m} a_i w_\sigma^i$$

Here w_σ is a function of t_σ and spin density ρ_σ, specifically

$$w_\sigma = (t_\sigma - 1)/(t_\sigma + 1); \quad t_\sigma = \tau_\sigma^{\text{LSDA}}/\tau_\sigma; \quad \text{and} \quad \tau_\sigma^{\text{LSDA}} = \frac{3}{10}(6\pi^2)^{2/3}\rho_\sigma^{5/3}$$

M06-L Meta-GGA Correlation Functional

The M06-L correlation functional derives from PBE functionals.

$$E_C^{\alpha\beta} = \int \rho \varepsilon_{\alpha\beta}^{\text{UEG}}\left[g_{\alpha\beta}(x_\alpha, x_\beta) + h_{\alpha\beta}(x_{\alpha\beta}, z_{\alpha\beta})\right] dr$$

Here the enhancement function g is defined by

$$g_{\alpha\beta}(x_\alpha, x_\beta) = \sum_{i=0}^{n} c_{C\alpha\beta, i} \left(\frac{\gamma_{C\alpha\beta}(x_\alpha^2 + x_\beta^2)}{1 + \gamma_{C\alpha\beta}(x_\alpha^2 + x_\beta^2)} \right)^i$$

and $h_{\alpha\beta}(x_{\alpha\beta}, z_{\alpha\beta})$ is given above. The variables x and z are

$$x_{\alpha\beta}^2 = x_\alpha^2 + x_\beta^2 \text{ and } z_{\alpha\beta} = z_\alpha + z_\beta$$

For parallel spins

$$E_C^{\sigma\sigma} = \int e_{\sigma\sigma}^{\text{UEG}} [g_{\sigma\sigma}(x_\alpha) + h_{\sigma\sigma}(x_\sigma, z_\sigma)] D_\sigma \, dr$$

The new enhancement function g is expressed in a series expansion

$$g_{\sigma\sigma}(x_\sigma) = \sum_{i=0}^{n} c_{C\sigma\sigma, i} \left(\frac{\gamma_{C\sigma\sigma} x_\sigma^2}{1 + \gamma_{C\sigma\sigma} x_\sigma^2} \right)^i$$

with $\gamma_{C\alpha\beta} = 0.0031$ and $\gamma_{C\sigma\sigma} = 0.06$. The quantity

$$D_\sigma = 1 - \frac{x_\sigma^2}{4(z_\sigma + C_F)}$$

vanishes for any one-electron system. One important carryover from the M05 and M05-2X [68] functionals is the use of the relationships:

$$E_{C\alpha\beta}^{\text{UEG}}(\rho_\alpha, \rho_\beta) = E_C^{\text{LSDA}}(\rho_\alpha, \rho_\beta) - E_C^{\text{LSDA}}(\rho_\alpha, 0) - E_C^{\text{LSDA}}(0, \rho_\beta) \text{ and}$$

$$E_{C\sigma\sigma}^{\text{UEG}}(\rho_\alpha) = E_C^{\text{LSDA}}(\rho_\alpha, 0)$$

where UEG stands for uniform electron gas. The correlation energy is then

$$E_C = E_C^{\alpha\beta} + E_C^{\alpha\alpha} + E_C^{\beta\beta}$$

The values for the α_X, $\alpha_{\sigma\sigma}$, and $\alpha_{\sigma\sigma'}$ are taken from earlier work by Voorhis and Scuseria [69] while the values of $d_0, d_1, d_2, d_3, d_4, d_5$ correspond to a, b, c, d, e, f, respectively, in the same paper.

Zhao and Truhlar test their method with a broad spectrum of data including atomization energies, ionization potentials, electron and proton affinities, barrier heights, and noncovalent interactions. In view of the great importance of DFT methods in transition metal systems, they include data from metal–metal and metal–ligand bond energies. Databases including key chemical properties (bond length and vibrational frequencies) are also used in this broad evaluation. We quote a small sample of their data (Table 11.3).

The most rigorous test of the M06-L is not the comparison to the nonlocal M05 method but rather to the even more costly (N^5 scaling) fifth-rung (virtual–orbital dependent) method of Grimme [70]. This method uses Kohn–Sham

TABLE 11.3

Mean Absolute Deviation (MAD) for Various Functionals

Method	Thermochemistry	Barriers	Noncovalent	Composite
Database	MGAE109/05	HTBH38/04	HB6/04, CT7/04, DI6/04, PPS5/05	
Pure DFT				
BP86	3.96	8.86	1.51	6.16
BLYP	2.02	8.52	1.63	6.29
mPWPW	2.33	8.27	1.30	5.90
PBE	3.09	8.64	1.14	5.53
TPSS	1.52	8.54	1.22	5.86
Hybrid				
B3LYP	1.47	4.50	1.13	5.22
TPSSh	1.53	6.62	1.05	5.52
M06-L	1.39	3.88	0.55	3.34
M05	1.12	2.19	0.52	3.44

Energy values in kcal/mol.

Source: Zhao, Y. and Truhlar, D.G., *J. Chem. Theor. Comput.* 1:415, 2005.

orbitals in an MP2 calculation, and thus can deal with some problems which are very challenging for DFT as now formed. Although the B2-PLYP method of Grimme seems more accurate for the AE6 database of six atomization energies and the BH6 set of six barrier heights, Truhlar points out that for the key standard of the isomerization of tetramethylbutane to *n*-octane the M06-L result of +0.6 kcal/mol is very close to the experimental value of +1.9 kcal/mol. The B2-PLYP value is −3.5 kcal/mol; three wave function calculations gave values ranging from −11.5 to +4.6 kcal/mol. The results for this important standard case speak well for the accuracy of the M06-L functional. Truhlar estimates the M06-L is about 12 times faster than B2-PLYP (Table 11.4).

Zhao and Truhlar [67] make a convincing case that M06-L is the best overall local functional now available for thermochemistry and chemical kinetics for sizeable organic and organometallic systems. It provides reliable characterizations of bond lengths, noncovalent interactions, and vibrational frequencies. Overall, M06-L is superior to B3LYP for a wide variety of properties and should be widely used. To this end, the work is to be incorporated into several commercially available software suites.*

Since addition of HF exact exchange has always improved accuracy dramatically, justifying its computational cost, Zhao and Truhlar did investigate the effect of added nonlocal exchange [67,68]. They found that adding the Hartree–Fock exchange did not improve the description of transition metal complexes but was important in characterization of weak interactions.

*According to Prof. Truhlar's web page M06-L, M06-2X, and M06-HF are available in a development version of GAUSSIAN [71] and in QCHEM [72]. Incorporation in GAMESS software is anticipated.

TABLE 11.4

Mean Unsigned Error: Comparison
of Performance in Two Data Sets

Method	AE6	BH6
B2-PLYP	3.4	1.9
M06-L	3.8	4.3

Note: AE = atomization   energy;   BH =
barrier height.

Adding various amounts (X percent) of Hartree–Fock exchange to the M06-
L functional can enhance accuracy for particular kinds of chemical applications: X is chosen for specific types of interactions in formulations M06-HF
and M06-2X.

$$E_{XC}^{hyb} = \left(\frac{X}{100}\right)E_X^{HF} + \left(1 - \frac{X}{100}\right)E_X^{DFT} + E_C^{DFT}$$

Position-Dependent Exchange

Since 2003, a new strategy for functional design has emerged. This type of
functional allows the amount of Hartree–Fock type exact exchange to vary
with distance from the atomic nucleus within each atomic cell. Scuseria and
his associates [73–76] use what they called a "screened coulomb potential."
A range parameter ω is used to divide space into short-range zones for
which electrons are in close proximity, where a local exchange functional is
suitable, and longer range zones for which Hartree–Fock exchange is preferable. The regions are defined by a parameter ω and joind smoothly by the
error function $\text{erf}(\omega r_{12})$

$$\frac{1}{r_{12}} = \underbrace{\frac{\text{erfc}(\omega r_{12})}{r_{12}}}_{SR} + \underbrace{\frac{\text{erf}(\omega r_{12})}{r_{12}}}_{LR} \quad \text{where} \quad \text{erf}(x) = \frac{2}{\sqrt{\pi}}\int_0^x e^{-t^2}dt \quad \text{and}$$

$$\text{erfc}(x) = \frac{2}{\sqrt{\pi}}\int_x^\infty e^{-t^2}dt$$

The exact exchange is modified in quite a simple way:

$$V_X = \sum_{ab}\left\langle a(1)b(2)\left|\frac{\text{erf}(\omega r_{12})}{r_{12}}\right|b(1)a(2)\right\rangle$$

The PBE exchange is developed from the hole already described:

$$J^{\omega PBE}(\rho, s, y) = J^{PBE}(s, y) \, \mathrm{erfc}(\omega y / k_F)$$

The enhancement factor is

$$F_X^{\omega PBE}(\rho, s) = -\frac{8}{9} \int\limits_0^\infty dy \, J^{\omega PBE}(\rho, s, y)$$

Then,

$$\varepsilon_X^{\omega PBE} = \varepsilon_X^{LDA}(\rho) F_X^{\omega PBE}(\rho, s)$$

$$E_X^{\omega PBE}(SR: \omega) = \int dr \, \rho(r) \varepsilon_X^{\omega PBE}$$

defines the short-range exchange. The long-range complement is

$$E_X^{\omega PBE}(LR: \omega) = \int dr \, \rho \left[\varepsilon_X^{PBE} - \varepsilon_X^{\omega PBE} \right]$$

The error function and its complement add to unity. If $\omega = 0$, $\mathrm{erf}(\omega r)$ vanishes and there is no long-range exchange; all exchange is represented by the local form. If ω grows to infinity, $\mathrm{erf}(\omega r)$ approaches unity; then there is no short-range exchange and all exchange is represented by the exact (or Hartree–Fock) exchange expression. Thus, ω controls the relative amounts of local and exact exchange. The Gaussian form in the error function and its complement is a practical advantage. Heyd et al. [73] employed the PBE0 functional [77] with the parameter $a = 1/4$ as suggested by MP4.

$$E_{XC}^{PBE0} = a E_X^{HF} + (1-a) E_X^{PBE} + E_C^{PBE}$$

They merged the long- and short-range exchange forms by splitting both the Hartree–Fock and local (PBE) terms into long and short range part:

$$E_{Xc}^{\omega PBEh} = a E_X^{HF}(SR:\omega) + (1-a) E_X^{PBE}(SR:\omega) + E_X^{PBE}(LR:\omega) + E_c^{PBE}$$

$$E_X^{PBE0} = a E_X^{HF}(SR:\omega) + \underbrace{a E_X^{HF}(LR:\omega)}_{HF, LR} + (1-a) E_X^{PBE}(SR:\omega) + E_X^{PBE}(LR:\omega)$$

$$- \underbrace{a E_X^{PBE}(LR:\omega)}_{PBE, LR}$$

According to Heyd et al., the long-range Hartree–Fock and PBE long-range contributions are small and tend to cancel one another; they are neglected to produce the working expression

$$E_{Xc}^{\omega PBEh} = aE_X^{HF}(SR{:}\omega) + (1-a)E_X^{PBE}(SR{:}\omega) + E_X^{PBE}(LR{:}\omega) + E_c^{PBE}$$

The effect is to confine the exact exchange to the short range. A simpler form is

$$E_{Xc}^{SR\text{-}HF} = aE_X^{HF}(SR{:}\omega) + (1-a)E_X^{PBE} + E_c^{PBE}$$

Figure 11.4 shows the MAD for the G2-1 database achieved with these two functional assemblies. The minimum in the MAD occurs near $\omega = 0.2\,\text{bohr}^{-1}$. A two-dimensional search in (a, ω) space showed a very broad minimum; thus, for the rest of their studies they used $a = 1/4$ and $\omega = 0.15\,\text{bohr}^{-1}$.

For atoms, the ωPBEh functional tends to undershoot exact total energies. Consider the value for $E(H, \omega PBEh) = -0.511$ hartree, for which the exact value is precisely -0.5 hartree. This might be unnerving to someone who relies on the security of a variational lower bound, though in fairness it is a feature in powerful and reliable perturbation and coupled cluster expansions. The strength of this functional is that it yields optimized bond lengths which are more accurate than those obtained by B3LYP, PBE, and PBE0 functionals. In addition, it converges more readily than other DFT methods for solids, probably owing to its smooth crossover to long-range exchange.

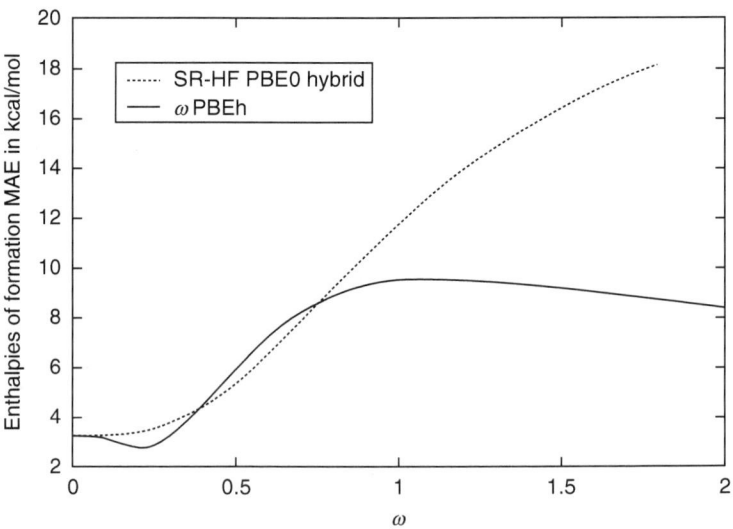

FIGURE 11.4

The mean absolute deviation from heats of atomization in the G2 data set depends on the local screening parameter ω. The ωPBE modification shows best performance with the parameter about 0.2, whereas a simplified variant (see text) shows no improvement for nonzero ω. (From Heyd, J., Scuseria, G.E., and Enzerhof, M., *J. Chem. Phys.*, 118, 8207, 2003. With permission.)

The MAD of atomization energies for the G2-1 set of 55 molecules, 2.93 kcal/mol, comes close to the B3LYP standard of 2.46 kcal/mol, and improves on both PBE (8.19 kcal/mol) and PBE0 (3.01 kcal/mol). This is not yet "chemical accuracy," i.e., MAD about 0.1 kcal/mol. No DFT functional has achieved that level of accuracy.

A related investigation of a Coulomb attenuated method [CAM-B3LYP] has been presented by Yanai et al. [78], who used the range partition to allow use of distinct hybrids of B3LYP form at short and long interelectronic distances. This achieved near-B3LYP quality for thermochemistry and, undoubtedly owing to the special treatment of the outer reaches of the density distribution, also produced much improved representation of charge-transfer UV excitations. We will return to this idea when we describe Baerends' SAOP functional.

The Scuseria group refined their ω formulation [73,75] but we will not pursue this further development. Instead, we turn to the notion that the extent of exact exchange required might differ in various regions of the molecule. Jaramillo et al. [79] took a long step toward characterizing this position, i.e., this position-dependent variation in exact exchange in their definition of "local hybrid functionals." They defined an exchange–correlation functional of the form

$$E_{Xc} = \int d^3\mathbf{r}\rho(\mathbf{r})\left[f(\mathbf{r})\varepsilon_X^{DFT} + (1 - f(\mathbf{r}))\varepsilon_X^{HF} + \varepsilon_c^{DFT}\right]$$

in which the function f is $\mathbf{r}$-dependent. They used the computationally efficient "resolution of the identity" to evaluate the HF exchange, following Della Salla and Görling [80]. For the DFT exchange–correlation they used two alternatives: Becke88 exchange with LYP correlation and PBE exchange with the PKZB correlation which includes a self-interaction correction to the PBE correlation. In the latter case, the formulation belongs on the fourth rung of Perdew's ladder.

The local mixing function depends on the ratio of the Weizsäcker kinetic variable τ_W to the scaled kinetic variable τ.

$$f(r) = 1 - \frac{\tau_W}{\tau}$$

This is Becke's "self-correlation detection function." In regions where the wave function behaves exponentially

$$\psi \sim \exp(-\alpha r)$$

(this is called the one-electron region) the two kinetic variables are equal. In that case, the local mixing vanishes and the HF exchange dominates. In the homogeneous region, the Weizsäcker variable vanishes and pure density functional exchange is used.

In the first implementation, the local hybrid energy was applied as a single correction to LSDA or HF results. The authors were particularly interested in treatment of cationic dimers, a considerable challenge to

DFT. In these systems HF underestimates binding and both pure and hybrid density functional theories (BLYP and B3LYP) overestimate binding. The local hybrid method strikes a balance and gives binding energies and bond lengths comparable to MP4. The heats of atomization in the small G2 set were, however, not satisfactory. As a first expression of a new idea, this marked a promising beginning.

The Iterative Localized Local Hybrid Method, LLH

Arbuznikov, Kaupp, and Bahmann [81] (AKB) pursued this idea in detail, and incorporated the local hybrid expression in each Kohn–Sham iteration, a considerable computational burden. They wished to compare the effectiveness of the add-on correction of Jaramillo et al. [79] with the fully self-consistent local hybrid scheme.

The first test of the AKB local hybrid scheme (which they call the iterative localized local hybrid method, LLH) was conducted with Becke88 exchange and Perdew–Wang (PW91) correlation. (This is not quite consistent with the tests by Jaramillo et al. [79].) Atomization energies were not much different; both single-point correction and fully iterative LLH underbound molecules in the test set.

It is possible to make two kinds of comparisons of the local hybrid methods with the successful global hybrid methods such as B3LYP which assign a fixed fraction of exact exchange. First, is the position-dependent local admixture nearly equal to a fixed value of admixture throughout a molecule? And second, is the average of the local admixture roughly equal to the global value? In response to the first question, AKB plotted the ratio of the Weizsäcker kinetic variable to the exact kinetic variable (and the square of that ratio) along the interatomic axis for several molecules. Results for CO are shown below (Figure 11.5).

As a reminder, the mixing function is $1 - f(\mathbf{r})$ where

$$f(\mathbf{r}) = \frac{\tau_W(\mathbf{r})}{\tau(\mathbf{r})} \quad \text{where} \quad \tau(\mathbf{r}) = (1/2) \sum_i^{\text{occ}} \left| \nabla \varphi_i(\mathbf{r}) \right|^2 \quad \text{and} \quad \tau_W(\mathbf{r}) = \frac{1}{8} \frac{|\nabla \rho(\mathbf{r})|^2}{\rho(\mathbf{r})}$$

Note first that the ratio τ / τ_W varies from 0 to 1. The ratio is close to unity in "the one-electron region" near the nuclei and at the periphery of the molecule. This is where the density is roughly described as exponential. The ratio is close to zero in the "valence region" where we expect to see bonds and lone pairs. Here the charge distribution is much more uniform.

At least in this implementation, the mean admixture of exact exchange is not always close to the values suggested by the global admixture employed in well-established hybrid methods [82]. The global percent admixture is approximately 25% for energetics, with some exceptions for functionals calibrated for specific molecular properties. AKB found

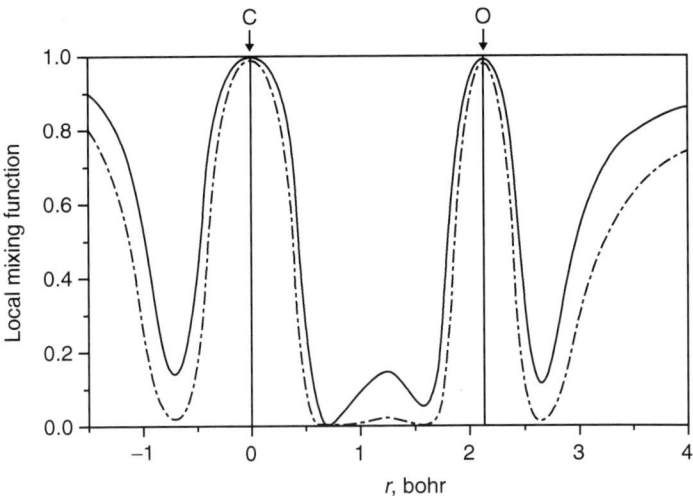

FIGURE 11.5

Arbuznikov et al. show that the extent of mixing of exact exchange varies with position in CO. The mixing function $1 - x$ is defined by the accuracy of the Weizsäcker kinetic variable τ_W, compared with the kinetic energy τ computed with Kohn–Sham orbitals; it vanishes when the two agree. The solid line is the result of setting x to the ratio τ_W/τ; the broken line results from setting x to the square of this ratio. (From Arbuznikov, A.V., Kaupp, M., and Bahmann, H., *J. Chem. Phys.*, 124, 204102, 2006. With permission.)

1. Large mean exact-exchange admixtures—near 100%—for systems of light elements, especially H.

2. Large mean admixtures of exact exchange for molecules containing first-row atoms other than H; values are scattered over the range 60% to 90%.

3. Lesser mean admixtures for molecules containing second-row atoms, for which the LLH calculations show better agreement with experimental atomization energies than for molecules with first-row atoms.

4. Large admixtures of exact exchange and especially poor performance for systems with multiple bonds.

The authors considered that refinement of the local mixing functions is likely to remedy all these problems. Arbuznikov and Kaupp [83] pursued this lead, mixing only local and exact exchange and including in the mixing function the variable s, defined by

$$s(\mathbf{r}) = \frac{|\nabla\rho|}{2k_F\rho} \equiv \frac{|\nabla\rho|}{2(3\pi^2)^{1/3}\rho^{4/3}}$$

Mixing functions of s-type included

$$g(s) = \left(\frac{s}{\lambda + s}\right)^2, \; (1 - e^{-\lambda s})^2 \; \text{and} \; [\text{erf}(\lambda s)]^2$$

Their best choices of mixing function and density functionals compete with B3LYP in representation of energetics in the full G2 database of 148 molecules. Not only are the energetics good, the choice of these forms of $g(s)$ makes their method simpler than the complicated method originally tested [81].

Gridless!

Over the past 30 years investigators have developed better and better grids for the numerical calculation of $E_{XC}[\rho]$. (We provide a discussion of grid design in the technical appendix.) However, even the largest grids with thousands of points produce slight discontinuities in potential energy surfaces calculated for conformational analysis or for reaction paths. Worse, a small grid can display departures from rotational invariance—i.e., the energy computed for molecule oriented along an axis might differ from the energy computed for the same molecule with no change in internal coordinates, but oriented at an angle to that axis. As long as those errors are less than microhartrees in size there is no adverse effect on computed results. However, achieving that level of numerical accuracy requires a vast number of grid points. Avoiding the unsatisfactory results of approximate numerical methods and also avoiding the computational burden of very accurate numerical analysis are obviously desirable. Hence, there has been a considerable interest in gridless methods. These methods all rely on an auxiliary basis. Almlöf's pioneering work is discussed in the appendix. Here we provide an introduction to Gordon's gridless methods, incorporated in GAMESS [84].

Resolution of the Identity

Glaesemann and Gordon [85–87] used the techniques of Zheng and Almlöf [88,89] to obtain a matrix representation of the density matrix:

$$M[\rho]_{i,j} = \int \chi_i \rho(r) \chi_j \, d\mathbf{r} = \sum_{kl} D_{kl} \int \chi_i \chi_k \chi_l \chi_j \, d\mathbf{r} = \sum_{kl} D_{kl}(iklj)$$

These four-index overlap integrals are easy to evaluate for Gaussian functions, but one may wish to adopt the concept of the resolution of the identity (RI) to reduce the problem to evaluation of (more) three-index integrals:

$$(iklj) = \sum_m (ik\theta_m)(\theta_m lj)$$

This permits a matrix representation of ρ; in density functional theory we need a matrix representation of fractional powers of the density, such as $\rho^{1/3}$. As explained in the appendix, such a representation $\mathbf{M}(\rho^{1/3})$ can be obtained by a sequence of matrix manpulations beginning with $\mathbf{M}(\rho)$.

A key principle that Gordon has used over and over is that the matrix representation $\mathbf{M_0}$ of the product $(f \cdot g)$ is the product of the matrix representations $\mathbf{M_1}$ of f and $\mathbf{M_2}$ of g.

$$\mathbf{M_0}[f \cdot g]_{i,j} = \int \chi_i (f \cdot g) \chi_j \, d\mathbf{r} \approx \sum_m \int \chi_i f \theta_m \, d\mathbf{r} \cdot \int \theta_m g \chi_j \, d\mathbf{r} = \sum_m \mathbf{M_1}[f]_{i,m} \mathbf{M_2}[g]_{m,j}$$

In practice, it seems that a large auxiliary basis is needed. The requirement on the auxiliary basis, to represent the identity, is different from the requirements on the "molecular" basis, to represent the electron distribution. Thus, we can imagine the two basis sets to be of different size and composition.

The matrix representation of the scaled kinetic variable requires separate treatment. Here the auxiliary basis is the orthonormal set $\{\theta_m\}$.

$$\int \chi_\mu \left(\rho^{-4/3} \frac{d\rho}{dx} \right) \chi_\nu \, d\mathbf{r} = -3 \sum_m^{\text{orth}} \left(\int \chi_\mu \frac{d}{dx} \theta_m \, d\mathbf{r} \cdot \int \theta_m \rho^{-1/3} \chi_\nu \, d\mathbf{r} \right.$$
$$\left. + \int \chi_\nu \frac{d}{dx} \theta_m \, d\mathbf{r} \cdot \int \theta_m \rho^{-1/3} \chi_\mu \, d\mathbf{r} \right)$$

Besides the integrals of powers of the density, only dipole velocity integrals are needed.

Glaesemann and Gordon [85–87] extend their analysis to the construction of the Kohn–Sham analogy to the Fock operator which is essential to the completion of the DFT energy and also to the gradients of the total energy. The gridless methods may find their most important use in this latter application since energy derivatives in grid-based computations should, in principle, include a term referring to changes in the grid as the geometry changes. This is not commonly implemented.

The GAMESS [84] program offers several functionals for calculations within the gridless scheme:

1. Local exchange K-alpha for $\alpha = (2/3)$ and 0.7
2. Local correlation as VWN5 or PW
3. Three forms of the Wigner type of correlation
4. DePristo–Kress [90] functionals, Becke88, and two modifications (CAMA and CAMB)
5. Gradient-corrected LYP

The gridless method offers smooth energy surfaces without the grainy pattern often evident in grid-based calculations. There is also no issue of rotational invariance. Werpetinski and Cook [91,92] demonstrate that gridless techniques produce smoother potential energy surfaces than can be obtained using the more common grid-point integration.

Total energies depend on the auxiliary basis set used. In the early imple-mentations, the set of auxiliary functions was identical with the atomic orbital basis, but it proved that a more complete set of auxiliary functions was required for an effective resolution of the identity. Glaesemann and Gordon [87] found that very large auxiliary basis sets are necessary for convergence to the answers obtained from the grid-mesh integration method. Even with the large auxiliary basis sets now in use in GAMESS, energies obtained by its gridless methods are higher than the grid-based energies obtained with the same molecular basis set. Predicted bond lengths and angles generally agree closely with results of analogous grid-based calculation; especially for larger auxiliary basis sets. So far, systematic investigations with the databases developed as training sets and criteria for performance by DFT formulations and the challenge of weak inter-actions have not been conducted with gridless methods. However, it seems reasonable to expect behavior characteristic of the functional employed whether a grid-based or gridless integration is performed.

A Gridless DFT Method Using Least-Squares Slater Exchange

Dunlap's study of the implications of the original Slater K-alpha method [20,93–98] illustrates the power of the density fitting techniques. Following the notation of Zope and Dunlap [93,94], the Kohn–Sham energy is

$$E^{KS}[\rho] = \sum_i^N \langle \phi_i | f_1 | \phi_i \rangle + E_{ee} + E_{xc}[\rho_\alpha, \rho_\beta]$$

Here the one-electron terms can be evaluated analytically

$$f_1 = -\frac{\nabla_1^2}{2} - \sum_A^M \frac{Z_A}{|\mathbf{r}_1 - \mathbf{R}_A|}$$

The two-electron coulomb terms E_{ee}

$$E_{ee} = \frac{1}{2} \iint \frac{\rho(\mathbf{r})\rho(\mathbf{r}')}{|\mathbf{r} - \mathbf{r}'|} d\mathbf{r}\, d\mathbf{r}'$$

are reexpressed with the use of an auxiliary basis, which permits analytic evaluation of this part of the energy. Zope and Dunlap [93,94] extended the Slater exchange

$$E_x = -\alpha \frac{9}{8} \left(\frac{3}{\pi}\right)^{1/3} \int \rho^{4/3}\, d^3\vec{r}$$

to unpaired spins:

$$E_x[\rho_\alpha, \rho_\beta] = -\frac{9}{8}\alpha \left(\frac{6}{\pi}\right)^{1/3} \int d^3\mathbf{r} \left[\rho_\alpha^{4/3} + \rho_\beta^{4/3}\right]$$

The unusual part of Dunlap's method is his use of linear combinations of Gaussian functions to fit fractional powers of the density:

$$\rho^{1/3}(\vec{r}) \cong \bar{\rho}^{1/3} = \sum_i e_i E_i \quad \text{and} \quad \rho^{2/3}(\vec{r}) \cong \bar{\rho}^{2/3} = \sum_i f_i F_i$$

The basis functions E and F are adapted from the members of the original Gaussian basis sets. The fractional powers of the density are expressed in basis functions with exponents scaled by (2/3) for $\rho^{1/3}$ and by (4/3) for its square $\rho^{2/3}$. Dunlap chooses these eccentric powers for good reason. Dunlap defines calculations as "robust" if the fitting procedures contain no first-order error; direct calculation of $\rho^{4/3}$ would not necessarily fulfill that condition. One can express $\rho^{4/3}$ in terms of $\rho^{1/3}$ and $\rho^{2/3}$; if

$$x = (\rho^{1/3} + \Delta x) \quad \text{and} \quad y = (\rho^{2/3} + \Delta y)$$

and

$$E_X[\rho, x, y] = -\kappa \left\langle \frac{4}{3}\rho x - \frac{2}{3}xxy + \frac{1}{3}yy \right\rangle$$

direct substitution produces

$$\frac{4}{3}\rho x - \frac{2}{3}x^2 y + \frac{1}{3}y^2 = \rho^{4/3} - \frac{2}{3}y(\Delta x)^2 - \frac{4}{3}x(\Delta x)(\Delta y) + \frac{1}{3}(\Delta y)^2 + \frac{4}{3}(\Delta x)^2(\Delta y)$$

This is a "robust" approximation in the sense that departures from its estimate of $\rho^{4/3}$ are at least quadratic in the error. At each step in the iterative refinement of the density, the constraint

$$4\rho x - 2x^2 y + y^2 = 0$$

requires a solution for two quantities x and y; this is obtained by a Newton–Raphson step,* explained by Dahlquist and Bjorck [99]. The advantage of the robust approximation is that the Dunlap formulation [93,94,97] requires a smaller basis than that needed for the Glaesemann–Gordon–Zheng–Almlöf [85–88] method.

Slater–Roothaan SCF

Dunlap and his group have extended the X-α method [20,98,100] which constrains all elements to use a single value of α to now permit each element

---

*In a private communication Dunlap reports that the program merely applies the Newton–Raphson process for one iteration per SCF iteration and both processes converge together.

to have a distinct optimum value. They express their "Slater–Roothaan" energy by

$$E^{SR} = \sum_i \langle \phi_i | f_1 | \phi_i \rangle + 2\langle \rho \| \bar{\rho} \rangle - \langle \bar{\rho} \| \bar{\rho} \rangle$$

$$- \sum_{\sigma=\alpha,\beta} C_x \left[\frac{4}{3}(g_\sigma \bar{g}_\sigma^{1/3}) - \frac{2}{3}(\bar{g}_\sigma^{1/3}\bar{g}_\sigma^{1/3}\bar{g}_\sigma^{2/3}) + (\bar{g}_\sigma^{2/3}\bar{g}_\sigma^{2/3}) \right]$$

The coulomb repulsion E_{ee} is variationally fit by the methods of Dunlap et al. [100]. Here ρ is the density, g is a version of the density that includes factors allowing for distinct α values for each atom, and the robust approximations to powers of the density are denoted by the superior bars. Zope and Dunlap find element-specific values of α range from 0.55 to 0.85; the common values around 0.7 are a reasonable average. A glance at the values shows that the atom-specific α values are greater for second-row atoms (Na–Cl) than for first-row atoms (Li–F), and the widely adopted global value of 2/3 is reasonable for atoms important in organic systems, BCNO but interestingly not H (Figure 11.6).

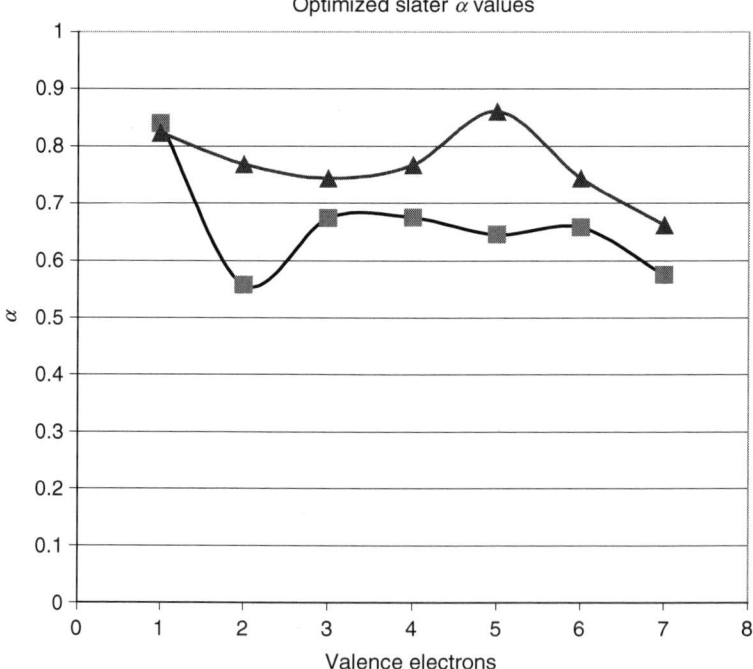

FIGURE 11.6
Optimized values of the Slater exchange functional's parameter α for neutral first-row atoms (lower curve, solid squares) and second-row atoms (upper curve, solid triangles) with varying number of valence electrons (horizontal axis). The generally used value 0.7 is nearly optimum for most atoms.

Zope and Dunlap show that fits to the G2 data set of atomization and ionization energies obtained by optimizing the α value for each element are comparable with estimates by pure density functional methods such as PBE on the second rung of Perdew's ladder. The method produces bond distances which are slightly too short, consistent with its systematic overbinding. The performance is remarkable considering that there appears to be no explicit treatment of correlation in Dunlap's method.

We have seen above that the best results of the LLH method of Arbuznikov and Kaupp [83] employ Slater exchange and LYP correlation. This suggests that Dunlap's method may produce excellent results if a suitable correlation functional can be included. Alternatively, its innovative techniques for representing the density may improve the efficiency of the LLH method.

Two DFT-Focussed Programs

Both GAUSSIAN-03 [71] and GAMESS [84] have added options for DFT calculations to their array of choices of model chemistry. The fact that Kohn–Sham calculations employ much of the Hartree–Fock technology made it feasible to incorporate DFT alternatives. We turn now to discussion of two programs which were designed entirely to deal with the special demands and opportunities of density functional theory.

The Amsterdam Density Functional Suite

The Amsterdam density functional (ADF) program [101,102], developed at the Vrije Universiteit in Amsterdam and and Calgary University and more recently at the University in Groningen, is exceptional in that it uses Slater-type orbitals in the molecular and the fitting basis sets. This is possible only because the expansion techniques described above and in the appendix allow avoidance of the impossible four-index four-center repulsion integrals. The program is also unusual in that it reports energy increments relative to a defined reference of spherical atoms rather than separate electrons and bare nuclei. The developers confine the suite to density functional methods; hybrid methods are a relatively recent addition. A particular strength of this software is its treatment of very heavy atoms, including incorporation of relativistic corrections. Since the summary of its capabilities [102] appeared in 2001, the active development group has incorporated versatile description of optical properties including not only the time-dependent DFT methods for estimating excitation energies but also methods for optical rotatory dispersion, natural and magnetic circular dichroism, and polarizabilities. The group led by Ziegler, Baerends, and many others has led the application of all these methods to transition metal complexes and condensed matter. One of the features of

ADF which makes it especially effective in the description of optical properties is its incorporation of a functional based on the statistical average of orbital potentials. We will make use of this functional in our discussion of excited states and electronic spectra. Here we sketch the assumptions behind SAOP.

SAOP: A Fourth-Rung Exchange–Correlation Model for Excitations

Gritsenko et al. [103] addressed the question of the long-range behavior of the Kohn–Sham exchange–correlation potential V_{SC}.

$$\lim_{r \to \infty} V_{XC} = -r^{-1} + V_{XC}(\infty)$$

The asymptotic coulomb behavior may well be critical for adequate treatment of excitation energies, transition intensities, polarizabilities, and rotational strengths. There is more to be said about the behavior of V_{SC} however. Besides being identifiable as the functional derivative of the exchange–correlation energy

$$V_{SC} = \frac{\delta E_{SC}}{\delta \rho}$$

the potential is related to the system virial integral.

$$E_{XC} + T_C = I_V = \int d\mathbf{r}[3\rho + \mathbf{r} \cdot \nabla \rho]V_{SC}$$

The term T_C is the kinetic contribution to the correlation energy. The ionization energy is also

$$I_p = V_{XC}(\infty) - \varepsilon_N$$

Here ε_N is the energy of the highest-energy occupied Kohn–Sham orbital. (Note that this is not exactly Koopmans' theorem, unless $V_{XC}\infty$ can be set to zero.) This can be done without impact on energy differences. In widely available functionals, the asymptotic behavior is not coulombic, with the consequence that $|\varepsilon_N|$ is always too small. Figure 11.7 shows the behavior of exact V_{XC} for Ne atom and several approximations.

The GGA functional is composed of the Becke88 exchange and the Perdew–Wang (PW91) correlation functionals. The LB94 functional, which incorporates correct asymptotic behavior of V_{XC}, is from van Leeuwen and Baerends [104]. Intended to improve calculations of properties dependent on the outer reaches of V_{XC}, it fits the asymptote better than the B88PW91 GGA, but does not display the variations in slope characteristic of the exact V_{XC} and modeled fairly well by the GGA. The variant by Gritsenko, van Leeuwen, and Baerends (GLB) [104] is a merger of the LB asymptotic form

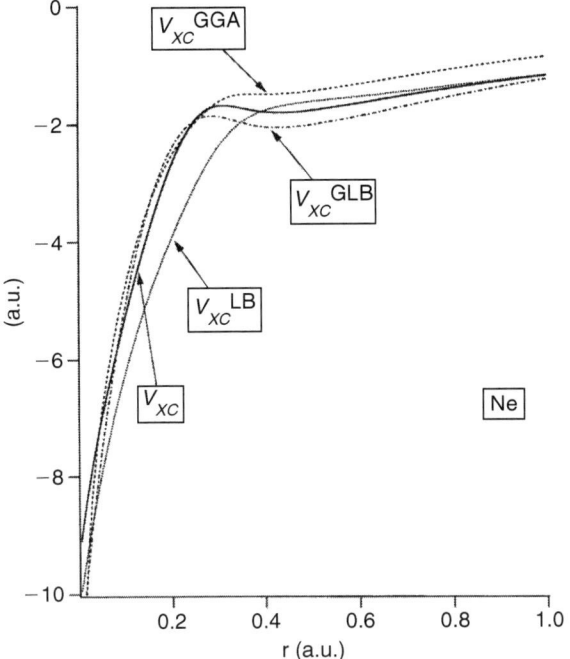

FIGURE 11.7
Gritsenko et al. describe the exchange–correlation potential for Ne atom. The horizontal axis is distance from the nucleus in bohrs; the vertical axis is energy in hartrees. The proper asymptotic form is $-1/r$ as displayed by the exact potential (the heavier solid line). Models LB and GLB match well at short range and long range, respectively. (From Gritsenko, O.V., Schipper, P.R.T., and Baerends, E.J., *Chem. Phys. Lett.*, 302, 199, 1999. With permission.)

with the Becke exchange at shorter range. Its slow approach to the proper asymptote leads to underestimates of polarizabilities.

SAOP, the statistical average of orbital potentials functional, employs different models for the outermost occupied orbital and for the inner orbitals. The statistical averaging is a way to interpolate between the two. The asymptotic behavior is determined by the exchange hole, so it must be that

$$V_{XC}^{(\text{asymptote})} = \rho^{1/3} f(r)$$

The factor f must depend on the gradient x and for large x

$$f(x) \sim -\frac{1}{3} \frac{x}{\ln(x)}$$

For densities characteristic of atoms, f should be compatible with the gradient expansion:

$$f(x) \sim -\beta x^2$$

This explains Becke's choice of form

$$f^{B88}(x) = -\frac{\beta x^2}{1 + 3\beta \sinh^{-1}(x)}$$

The function $\sinh^{-1}(x)$ can be written as a logarithm

$$\sinh^{-1}(x) = \ln\left[x + (x^2 + 1)^{1/2}\right]$$

Baerends suggests the alternative, the exponential integral with argument $z = 1/x$:

$$E_1(z) = \int_z^\infty \frac{1}{t} \exp(-t)\, dt$$

E_1 vanishes for small x, and for large r and x, it approaches $\ln(x)$ (Figure 11.8).

The behavior at long range is much improved for the V_{XC} based on exponential–integral interpolation. This potential is suitable for the highest occupied orbital, which defines the ionization energy.

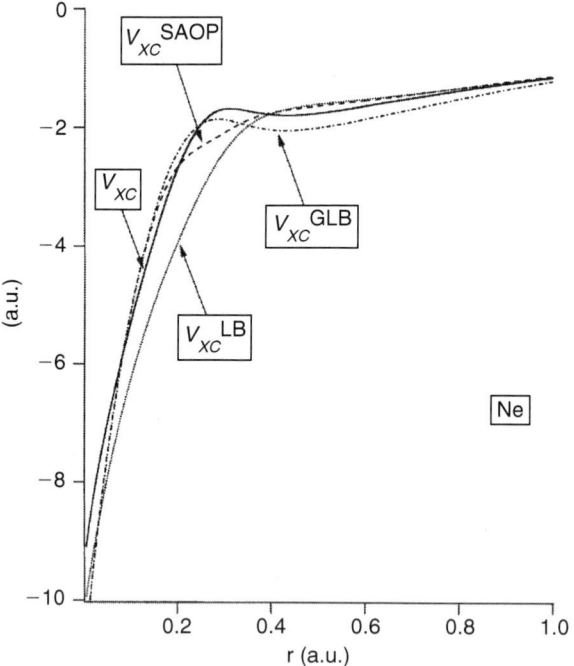

FIGURE 11.8

The proper asymptotic behavior is captured by the statistical average of outer potentials or SAOP. Axes are identical to those of Figure 11.7. (From Gritsenko, O.V., Schipper, P.R.T., and Baerends, E.J., *Chem. Phys. Lett.*, 302, 199, 1999. With permission.)

The SAOP model is constructed from the two cases by weighting the form suitable to the highest occupied Kohn–Sham MO by the orbital density, and the form suitable to the lower-energy MOs by the remainder of the density.

$$V_{XC}^{SAOP} = V_{XC}^{GLB} \sum_{i=1}^{N-1} \frac{|\psi_i|^2}{\rho} + V_{XC}^{Ei} \frac{|\psi_N|^2}{\rho}$$

Since there is explicit reference in the averaging to the occupied MOs, this puts SAOP at the fourth rung of Perdew's ladder. The averaging assures that the functional with the best asymptotic form will be used when the density is dominated by the contribution from the highest occupied KS MO.

SAOP was tested first by its ability to estimate the energy of the highest occupied MO and the value of the virial integral. It was subsequently adapted to computations of polarizabilities and excitation energies, and more recently optical rotation tensors [105], where it displayed considerably better accuracy than more familiar functionals.

deMon2k

The program deMon2k [21] is one of the most sophisticated DFT test beds. It contains some of the newest functionals occupying the highest rungs of the Perdew ladder. The main features of deMon2k [21] are documented by Köster [106–113] in unusually clear fashion. Since the program is available for download, we leave it to the ambitious reader to explore some of these new developments. The deMon2k program can be found at http://www.demon-software.com/public_html/index.html.

We will work our way through a VWN treatment of formaldehyde showing some details of the deMon2k output. The structure of the program would not be changed if we were to choose a more sophisticated functional.

Credit those who did the work

```
*********************************************************************
*********************************************************************
***                                                               ***
***                     PROGRAM deMon2k                           ***
***                                                               ***
***                 (VERSION 2.00, Jul. 2005)                     ***
***                                                               ***
***   COPYRIGHT (C) BY THE INTERNATIONAL DEMON DEVELOPERS COMMUNITY ***
***                                                               ***
***   AUTHORS: Andreas M. Köster, Patrizia Calaminici, Mark E. Casida, ***
***            Roberto Flores-Moreno, Gerald Geudtner, Annick Goursot, ***
***            Thomas Heine, Andrei Ipatov, Florian Janetzko, Serguei ***
***            Patchkovskii, J. Ulises Reveles, Alberto Vela and   ***
***            Dennis R. Salahub                                   ***
***                                                               ***
*********************************************************************
*********************************************************************
```

Specify the molecule, density functional, basis, and task

```
TITLE CH2O
#
VXCTYPE VWN       Vosko–Wilk–Nusair functionals
BASIS (DZVP)      Double-zeta basis with polarization in the valence shell
AUXIS (A2)        Auxiliary set specified
#
SYMMETRY ON
#
OPTIMIZATION CARTESIAN   Optimize in Cartesian coordinates, not internal coordinates
PRINT MOS
GUESS TB
#
DIPOLE
POPULATION MULLIKEN
VISUALIZATION MOLEKEL FULL
#
GEOMETRY

C      0.000000      0.000000    0.0
O      0.000000      1.208000    0.0
H      0.948993     -0.587255    0.0
H     -0.948993     -0.587255    0.0
#
END
```

The input is complete; now the program responds.

```
*** MOLECULE DATA ***

MOLECULAR CHARGE: 0
MOLECULAR MULTIPLICITY: 1
NUMBER OF ATOMS: 4
NUMBER OF ELECTRONS: 16

*** SCF OPTIONS ***

SCF THEORY: RESTRICTED KOHN-SHAM (RKS)
DIAGONALIZER: DSYEV
START GUESS: HUECKEL DENSITY
SCF TOLERANCE: 0.100E-05
POTENTIAL DENSITY: CHARGE DENSITY
NUMBER OF SCF ITERATIONS: 100
DIIS ACCELERATION: ON

XC-POTENTIAL: LSD - VWN
```

If the input is silent, choose the default quadrature

```
GRID TECHNIQUE: ADAPTIVE
GRID METHOD: CONVENTIONAL
GRID GENERATING FUNCTION: SCF
GRID TOLERANCE: 0.100E-04
ATOMIC WEIGHT FUNCTION: BECKE SCREENED
RADIAL QUADRATURE TYPE: GAUSS-CHEBYSHEV
SCF METHODE: CONVENTIONAL
FREEZE AUXILIARY FUNCTIONS: OFF
DECOMPOSITION: SVD
SVD THRESHOLD: 0.100E-05
SCF SCREENING: 0.625E-15
SYMMETRY USE: ON
ROTATION OF THE ANGULAR GRIDS: ON
SCF PRINTING UNTIL CYCLE 0
SPHERICAL ORBITALS WILL BE USED

*** PROPERTY OPTIONS ***

CALCULATE ELECTROSTATIC MOMENTS
CALCULATE MULLIKEN POPULATION

*** PRINT OPTIONS ***

PRINT MOLECULAR ORBITALS

*** GEOMETRY ***

INPUT ORIENTATION IN ANGSTROM   This is the optimized structure from previous iterations.
```

NO.	ATOM	X	Y	Z	Z-ATOM	MASS	OPTXYZ
1	C	0.000003	−0.001690	−0.000002	6	12.011	
2	O	−0.000002	1.212476	0.000001	8	15.999	
3	H	0.958460	−0.593257	0.000003	1	1.008	
4	H	−0.958463	−0.593251	0.000003	1	1.008	

Note loss of symmetry

```
INPUT                    STEP TIME:   0.050    TOTAL TIME:    0.060

*** THE STANDARD ORIENTATION IS USED ***

POINT GROUP: C2v

GROUP ORDER: 4
```

Reorient the molecule so that the symmetry axis is "z"

```
STANDARD ORIENTATION IN ANGSTROM
```

NO.	ATOM	X	Y	Z	Z-ATOM	MASS
1	C	−0.000001	0.000001	−0.607248	6	12.011
2	O	0.000000	0.000000	0.606918	8	15.999
3	H	0.000005	0.958455	−1.198819	1	1.008
4	H	0.000005	−0.958468	−1.198806	1	1.008

```
DEGREES OF FREEDOM: 3

SYMMETRY ANALYSIS        STEP TIME:   0.000    TOTAL TIME:    0.060
```

In symmetry C_{2v}, one may alter the HCH angle, the CH bond distance, and the CO bond distance—three degrees of freedom

```
NUMBER OF TB ORBITALS: 12
START GUESS                        STEP TIME:   0.480   TOTAL TIME:   0.540

GENERAL BASIS SET: (DZVP)

NUMBER OF SHELLS: 18
NUMBER OF ORBITALS: 40
NUMBER OF PRIMITIVE GAUSSIAN EXPONENTS: 42

GENERAL AUXIS SET: (A2)

NUMBER OF AUXILIARY FUNCTION SETS: 22
NUMBER OF AUXILIARY FUNCTION SHELLS: 34
NUMBER OF AUXILIARY FUNCTION: 76
```

The auxiliary basis is larger than the basis in which MOs are expressed

```
SCF INTEGRALS                      STEP TIME:   0.010   TOTAL TIME:   0.550

TOTAL NUMBER OF INTEGRATED ERIS: 62320

ERIS ARE HOLD IN CORE

ERI CALCULATION                    STEP TIME:   0.040   TOTAL TIME:   0.590

NULLSPACE                          STEP TIME:   0.000   TOTAL TIME:   0.600

TIGHT-BINDING MO COEFFICIENTS OF CYCLE 1
```

```
*****************************************************************************
```

Note: We suppress over four pages of the initial 38 × 38 first-guess orbitals to save space. The final orbitals of Cycle 11 are printed in full below.

```
*****************************************************************************
```

```
ADAPTIVE GRID GENERATION          STEP TIME:   0.050  TOTAL TIME:   0.670

GRID RETRIEVE                     STEP TIME:   0.020  TOTAL TIME:   0.690

CYCLE:   1    EUPPER =   -130.805204065205   CDMIX =     0.300000000000
CYCLE:   1    ELOWER =   -213.992729939573   SHIFT =     0.000000000000
CYCLE:   1       EXC =    -13.423595922253   ERCDC =    11.072846430143
CYCLE:   1    ETOTAL =   -144.228799987458   ERROR =    83.187525874368
```

Several iterations suppressed

```
CYCLE:  11    EUPPER =   -130.892060019272   CDMIX =     0.300000000000
CYCLE:  11    ELOWER =   -130.892060020510   SHIFT =     0.000000000000
CYCLE:  11       EXC =    -13.709658913272   ERCDC =     0.000027735060
CYCLE:  11    ETOTAL =   -144.601718932544   ERROR =     0.000000001238

*** SCF CONVERGED ***

MO COEFFICIENTS OF CYCLE 11
```

Two 1s cores of O and C: a_1 COσ, a_1, b_2 CH2

				1	2	3	4	5
				−18.63798	−9.89668	−0.97235	−0.57039	−0.44437
				2.00000	2.00000	2.00000	2.00000	2.00000
1	1	C	1s	0.00005	0.99735	−0.12532	−0.17597	0.00000
2	1	C	2s	−0.00057	0.01092	0.23803	0.38265	0.00000
3	1	C	3s	0.00006	−0.00366	0.09647	0.25938	0.00000
4	1	C	2py	0.00000	0.00000	0.00000	0.00000	−0.51413
5	1	C	2pz	−0.00083	0.00068	0.21343	−0.24897	0.00000
6	1	C	2px	0.00000	0.00000	0.00000	0.00000	0.00000
7	1	C	3py	0.00000	0.00000	0.00000	0.00000	−0.09105
8	1	C	3pz	0.00023	0.00042	−0.01075	−0.06089	0.00000
9	1	C	3px	0.00000	0.00000	0.00000	0.00000	0.00000
10	1	C	3d−2	0.00000	0.00000	0.00000	0.00000	0.00000
11	1	C	3d−1	0.00000	0.00000	0.00000	0.00000	0.00416
12	1	C	3d+0	−0.00058	0.00027	0.03305	−0.01487	0.00000
13	1	C	3d+1	0.00000	0.00000	0.00000	0.00000	0.00000
14	1	C	3d+2	−0.00003	0.00019	0.00563	−0.00943	0.00000
15	2	O	1s	−0.99738	0.00025	0.21097	−0.09471	0.00000
16	2	O	2s	−0.01053	0.00018	−0.43968	0.20923	0.00000
17	2	O	3s	−0.00127	0.00024	0.31635	−0.21352	0.00000
18	2	O	2py	0.00000	0.00000	0.00000	0.00000	−0.31704
19	2	O	2pz	−0.00163	−0.00022	−0.19518	−0.10903	−0.00001
20	2	O	2px	0.00000	0.00000	0.00000	0.00000	0.00000
21	2	O	3py	0.00000	0.00000	0.00000	0.00000	−0.09957
22	2	O	3pz	0.00029	0.00007	−0.00082	−0.04798	0.00000
23	2	O	3px	0.00000	0.00000	0.00000	0.00000	0.00000
24	2	O	3d−2	0.00000	0.00000	0.00000	0.00000	0.00000
25	2	O	3d−1	0.00000	0.00000	0.00000	0.00000	0.02339
26	2	O	3d+0	0.00044	0.00065	0.01797	0.00749	0.00000
27	2	O	3d+1	0.00000	0.00000	0.00000	0.00000	0.00000
28	2	O	3d+2	0.00005	0.00001	0.00432	−0.00231	0.00000
29	3	H	1s	−0.00006	−0.00001	0.03246	0.18550	−0.19792
30	3	H	2s	0.00013	0.00109	−0.00196	0.05975	−0.11160
31	3	H	2py	−0.00005	−0.00041	−0.00679	−0.01698	0.00706
32	3	H	2pz	−0.00003	0.00025	0.00659	0.00514	−0.00825
33	3	H	2px	0.00000	0.00000	0.00000	0.00000	0.00000
34	4	H	1s	−0.00006	−0.00001	0.03246	0.18550	0.19792
35	4	H	2s	0.00013	0.00109	−0.00196	0.05975	0.11160
36	4	H	2py	0.00005	0.00041	0.00679	0.01698	0.00706
37	4	H	2pz	−0.00003	0.00025	0.00659	0.00514	0.00825
38	4	H	2px	0.00000	0.00000	0.00000	0.00000	0.00000

#6 a_1, O-σlp; #7 b_1 π; #8 b_2 Op-π lp; #9 b_1 π*—note this vacant orbital has $\varepsilon < 0$.

				6	7	8	9	10
				−0.40251	−0.37436	−0.22890	−0.10372	0.07559
				2.00000	2.00000	2.00000	0.00000	0.00000
1	1	C	1s	0.02896	0.00000	0.00000	0.00000	0.13526
2	1	C	2s	−0.08552	0.00000	0.00000	0.00000	−0.33455
3	1	C	3s	0.05834	0.00000	0.00000	0.00001	−2.11331
4	1	C	2py	−0.00001	0.00000	0.16420	0.00000	0.00001
5	1	C	2pz	−0.39480	0.00000	0.00000	0.00000	0.30265
6	1	C	2px	0.00000	0.43032	0.00000	−0.61013	0.00000
7	1	C	3py	0.00000	0.00000	−0.05567	0.00000	0.00001
8	1	C	3pz	0.01215	0.00000	0.00000	0.00000	0.38361
9	1	C	3px	0.00000	0.12217	0.00000	−0.49907	−0.00001
10	1	C	3d−2	0.00000	0.00000	0.00000	0.00000	0.00000
11	1	C	3d−1	0.00000	0.00000	−0.06713	0.00000	0.00000
12	1	C	3d+0	−0.01014	0.00000	0.00000	0.00000	−0.00784
13	1	C	3d+1	0.00000	0.03173	0.00000	0.03770	0.00000
14	1	C	3d+2	−0.00969	0.00000	0.00000	0.00000	−0.02274
15	2	O	1s	0.09226	0.00000	0.00000	0.00000	0.01761
16	2	O	2s	−0.20210	0.00000	0.00000	0.00000	−0.01478
17	2	O	3s	0.31264	0.00000	0.00000	0.00000	0.29001
18	2	O	2py	−0.00001	0.00000	−0.66728	0.00000	0.00000
19	2	O	2pz	0.58759	−0.00001	0.00000	0.00000	−0.11416
20	2	O	2px	0.00000	0.58434	0.00000	0.51634	0.00000
21	2	O	3py	0.00000	0.00000	−0.32839	0.00000	0.00000
22	2	O	3pz	0.19776	0.00000	0.00000	0.00000	−0.21816
23	2	O	3px	0.00000	0.22382	0.00000	0.40143	0.00000
24	2	O	3d−2	0.00000	0.00000	0.00000	0.00000	0.00000
25	2	O	3d−1	0.00000	0.00000	0.01260	0.00000	0.00000
26	2	O	3d+0	−0.03266	0.00000	0.00000	0.00000	0.00620
27	2	O	3d+1	0.00000	−0.03144	0.00000	0.00613	0.00000
28	2	O	3d+2	0.00215	0.00000	0.00000	0.00000	−0.00789
29	3	H	1s	0.07764	0.00000	0.17610	0.00000	0.11462
30	3	H	2s	0.07536	0.00000	0.28949	0.00000	1.40566
31	3	H	2py	−0.00495	0.00000	0.00000	0.00000	0.01390
32	3	H	2pz	−0.00542	0.00000	0.00117	0.00000	0.00136
33	3	H	2px	0.00000	0.00970	0.00000	−0.02756	0.00000
34	4	H	1s	0.07765	0.00000	−0.17610	0.00000	0.11462
35	4	H	2s	0.07536	0.00000	−0.28949	0.00000	1.40567
36	4	H	2py	0.00495	0.00000	0.00000	0.00000	−0.01390
37	4	H	2pz	−0.00542	0.00000	−0.00117	0.00000	0.00136
38	4	H	2px	0.00000	0.00970	0.00000	−0.02756	0.00000

				11	12	13	14	15
				0.11962	0.17827	0.24042	0.24076	0.40184
				0.00000	0.00000	0.00000	0.00000	0.00000
1	1	C	1s	0.02458	0.00000	0.00000	0.00000	−0.00728
2	1	C	2s	−0.04227	0.00000	0.00000	0.00000	−0.52744
3	1	C	3s	−0.16392	−0.00001	−0.00001	0.00002	2.98376
4	1	C	2py	−0.00002	−0.88259	0.00000	−0.04576	0.00000
5	1	C	2pz	−0.11487	0.00000	0.00000	0.00000	0.74011
6	1	C	2px	0.00000	0.00000	0.86159	0.00000	0.00000
7	1	C	3py	−0.00003	−0.08284	0.00000	3.17527	−0.00002
8	1	C	3pz	−1.61863	0.00004	0.00000	−0.00005	−0.81408
9	1	C	3px	0.00000	0.00000	−1.17127	0.00000	0.00000
10	1	C	3d−2	0.00000	0.00000	0.00000	0.00000	0.00000
11	1	C	3d−1	0.00000	−0.03608	0.00000	0.00466	0.00000
12	1	C	3d+0	0.05080	0.00000	0.00000	0.00000	−0.04427
13	1	C	3d+1	0.00000	0.00000	0.00939	0.00000	0.00000
14	1	C	3d+2	0.00702	0.00000	0.00000	0.00000	0.08095
15	2	O	1s	0.09260	0.00000	0.00000	0.00000	−0.07816
16	2	O	2s	−0.14985	0.00000	0.00000	0.00000	0.21842
17	2	O	3s	1.12305	−0.00002	0.00000	0.00002	−0.67560
18	2	O	2py	0.00001	0.18397	0.00000	−0.31759	0.00000
19	2	O	2pz	−0.12978	0.00000	0.00000	0.00000	0.04239
20	2	O	2px	0.00000	0.00000	0.00169	0.00000	0.00000
21	2	O	3py	0.00001	0.10410	0.00000	−0.60692	0.00000
22	2	O	3pz	−0.21153	0.00000	0.00000	0.00000	1.03550
23	2	O	3px	0.00000	0.00000	0.13691	0.00000	0.00000
24	2	O	3d−2	0.00000	0.00000	0.00000	0.00000	0.00000
25	2	O	3d−1	0.00000	0.03313	0.00000	0.00855	0.00000
26	2	O	3d+0	0.00951	0.00000	0.00000	0.00000	0.01097
27	2	O	3d+1	0.00000	0.00000	−0.04243	0.00000	0.00000
28	2	O	3d+2	0.00501	0.00000	0.00000	0.00000	0.01558
29	3	H	1s	−0.00622	0.00368	0.00000	−0.11164	−0.31042
30	3	H	2s	−0.54468	0.87430	0.00001	−2.13613	−0.72780
31	3	H	2py	−0.01296	0.03911	0.00000	0.03146	0.03714
32	3	H	2pz	−0.02997	−0.05009	0.00000	0.00670	−0.03153
33	3	H	2px	0.00000	0.00000	0.01209	0.00000	0.00000
34	4	H	1s	−0.00622	−0.00368	0.00000	0.11164	−0.31042
35	4	H	2s	−0.54473	−0.87426	0.00001	2.13609	−0.72781
36	4	H	2py	0.01296	0.03912	0.00000	0.03146	−0.03714
37	4	H	2pz	−0.02997	0.05009	0.00000	−0.00670	−0.03153
38	4	H	2px	0.00000	0.00000	0.01209	0.00000	0.00000

ELECTRONIC HARTREE ENERGY = −130.892110355

Kinetic, nuclear–electron attraction, and "classical" coulomb repulsion (hartrees)

```
EXCHANGE-CORRELATION ENERGY   =     -13.709628773
```

These terms are stabilizing!

```
NUCLEAR-REPULSION ENERGY      =      30.975730929
TOTAL ENERGY                  =    -113.626008199

SCF ITERATION          STEP TIME:   0.650  TOTAL TIME:   1.320
```

```
* * * * * * * * * * * * * * * * * * * * * * * * * * * * * * * * * * * * * * * * * * * * * * * * * * * * * * * * *
***              CALCULATION OF MOLECULAR PROPERTIES              ***
* * * * * * * * * * * * * * * * * * * * * * * * * * * * * * * * * * * * * * * * * * * * * * * * * * * * * * * * *
```

```
MOLECULE ORIENTATION FOR PROPERTY CALCULATION IN ANGSTROM
```

NO.	ATOM	X	Y	Z	Z-ATOM	MASS
1	C	-0.000001	0.000001	-0.607248	6	12.011
2	O	0.000000	0.000000	0.606918	8	15.999
3	H	0.000005	0.958455	-1.198819	1	1.008
4	H	0.000005	-0.958468	-1.198806	1	1.008

```
*** ELECTROSTATIC MOMENTS IN ESU ***
```

AXIS	NUCLEAR PART	ELECTRONIC PART	TOTAL MULTIPOLE	TOTAL TRACELESS
DIPOLE				
x	0.2200605E-04	-.2318662E-04	-.1180570E-05	-.1180570E-05
y	-.2917040E-04	0.5394611E-04	0.2477571E-04	0.2477571E-04
z	-.5695504E+01	0.3330256E+01	-.2365248E+01	-.2365248E+01

Note small departures from symmetry

QUADRUPOLE				
xx	0.2455785E-09	-.1183505E+02	-.1183505E+02	-.1994699E+00
xy	-.2101493E-09	0.2036542E-04	0.2036521E-04	0.3054782E-04
xz	-.2636240E-04	0.2104826E-04	-.5314138E-05	-.7971207E-05
yy	0.8824913E+01	-.2032360E+02	-.1149869E+02	0.3050738E+00
yz	0.4518661E-05	-.5627521E-04	-.5175655E-04	-.7763482E-04
zz	0.3858701E+02	-.5035948E+02	-.1177247E+02	-.1056039E+00
OCTUPOLE				
xxx	0.8991372E-15	-.3478898E-04	-.3478898E-04	-.4428912E-04
xxy	-.4945662E-15	0.1767723E-04	0.1767723E-04	-.3910930E-04
xxz	-.2662503E-09	0.1702642E+01	0.1702642E+01	0.1712379E+01
xyy	0.4069818E-04	-.4179308E-04	-.1094898E-05	0.1149053E-04
xyz	-.3809211E-10	-.1480796E-04	-.1480800E-04	-.3702001E-04
xzz	0.5544290E-04	-.4801458E-04	0.7428323E-05	0.3279859E-04
yyy	-.1619787E-03	0.2284300E-03	0.6645137E-04	-.8377867E-04
yyz	-.1057941E+02	0.1070523E+02	0.1258146E+00	-.2229689E+01
yzz	0.7358912E-04	0.8887014E-05	0.8247613E-04	0.1228880E-03
zzz	-.1441357E+02	0.1767357E+02	0.3259994E+01	0.5173097E+00

```
                    *** MULLIKEN POPULATION ANALYSIS ***

                        ATOMIC NET CHARGES

                    C        -0.0454
                    O        -0.2074
                    H         0.1264
                    H         0.1264


                         VALENCE MATRIX

                 1          2          3          4
     1   C    3.97192    2.17842    0.89675    0.89675
     2   O    2.17842    2.24632    0.03395    0.03395
     3   H    0.89675    0.03395    0.93377    0.00307
     4   H    0.89675    0.03395    0.00307    0.93377
```

Mayer bond index: C to O double bond, C to H single bonds; tetravalent C, divalent O, monovalent Hs

```
ELECTROSTATIC ANALYSIS        STEP TIME:   0.010  TOTAL TIME:   1.330
PROPERTY CALCULATION          STEP TIME:   0.010  TOTAL TIME:   1.330

TIMING STATISTICS (SECONDS)

TIMER ENTRY                 NUMBER OF CALLS   CUMUL. CPU TIME   CONTRIBUT.

INITIALIZATION                    1                0.010           0.8 %
INPUT                             1                0.050           3.8 %
DIAGONALIZATION                  62                0.010           0.8 %
SYMMETRY ANALYSIS                 2                0.000           0.0 %
START GUESS                       1                0.480          36.1 %
MATRIX MULTIPLICATION           198                0.010           0.8 %
SCF INTEGRALS                     1                0.010           0.8 %
ORTHOGONALIZATION MATRIX          1                0.000           0.0 %
AUXILIARY FUNCTION FIT            1                0.000           0.0 %
CORE HAMILTONIAN                  1                0.010           0.8 %
ERI CALCULATION                   1                0.040           3.0 %
SCF ITERATION                     1                0.650          48.9 %
NULLSPACE                         1                0.000           0.0 %
GRID INITIALIZATION               2                0.010           0.8 %
ADAPTIVE GRID GENERATION          2                0.170          12.8 %
GRID RETRIEVE                     2                0.030           2.3 %
XC POTENTIAL                     14                0.390          29.3 %
PROPERTY CALCULATION              1                0.010           0.8 %
ELECTROSTATIC ANALYSIS            1                0.010           0.8 %
```

Judge the timing—notice that for this single-point calculation the most demanding operation is the treatment of the VWN exchange–correlation functional.

Visualize the results

From Hartree–Fock theory we can ascribe meaning to the orbital energies—Koopmans' theorem associates the eigenvalues of the Fock operator with the system's ionization energies (occupied orbitals) or electron affinities (virtual

FIGURE 11.9
Isosurface representation of the A_2 pseudo-pi lone pair for formaldehyde: a Kohn–Sham orbital in a VWN approximation. Courtesy of Ulises Reveles, Virginia Commonwealth University.

orbitals). The highest occupied orbital's energy ε_{HO} defines the easiest ionization, and the lowest vacant orbital's energy ε_{LU} defines the maximum electron affinity. Density functional theory does not associate this meaning with the eigenvalues of the Kohn–Sham effective operator. There are roughly analogous interpretation of η_{HO} and η_{LU} relating their average to the change in energy of the system with respect to adding or extracting electrons. The negative value of the Kohn–Sham η_{LU} does not necessarily imply that an electron would be bound to neutral formaldehyde (Figures 11.9 through 11.13).

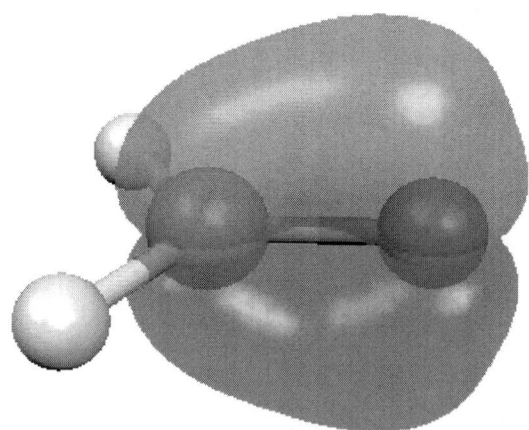

FIGURE 11.10
Isosurface representation of the B_1 pi bonding MO for formaldehyde: a Kohn–Sham orbital in a VWN approximation. Courtesy of Ulises Reveles, Virginia Commonwealth University.

FIGURE 11.11
Isosurface representation of the A_1 sigma lone pair MO for formaldehyde: a Kohn–Sham orbital in a VWN approximation. Courtesy of Ulises Reveles, Virginia Commonwealth University.

The low-lying virtual orbitals of the Kohn–Sham effective Hamiltonian do present practical complications; convergence of the self-consistent process often fails. One of the standard simple ways to hasten convergence is "level shifting." Changing the energy of a virtual orbital cannot alter the energy (this is defined by occupied orbitals) but it can reduce the mixing of the shifted level. This can prevent "root flipping" where HOMO and LUMO change order on successive iterations.

In principle, all means of anlysis and interpretation of DFT results should deal with the density rather than the Kohn–Sham orbitals. These are after all

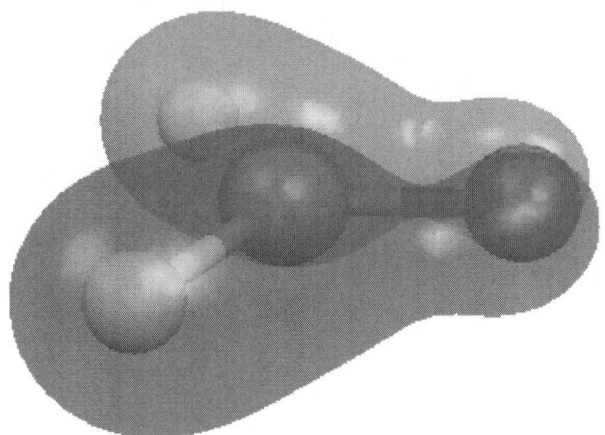

FIGURE 11.12
Isosurface representation of the B_2 sigma bonding MO for formaldehyde: a Kohn–Sham orbital in a VWN approximation. Courtesy of Ulises Reveles, Virginia Commonwealth University.

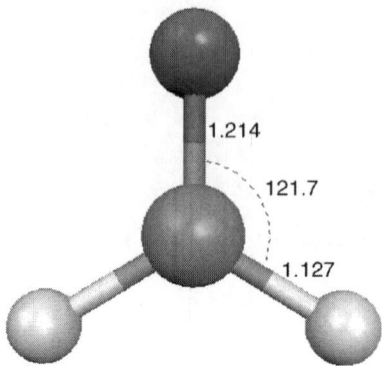

FIGURE 11.13
Representation of the equilibrium structure for formaldehyde in a VWN approximation. Courtesy of Ulises Reveles, Virginia Commonwealth University.

only a kind of scaffolding by which one constructs the density, which is the key and meaningful object in the theory. Perhaps a part of the appeal of DFT in the Kohn–Sham formulation is that the hard-won means of understanding developed when wave function theory was capable of only qualitative description of chemical systems can be applied in the otherwise new and strange setting of DFT. Although the entire program was designed to generate the energy from density, the results are interpretable with familiar orbital concepts.

Summary, Conclusions, and Overview

We had little idea when density functional theory made its first appearance at the Sanibel conferences that Johnson and Wahlgren [114] and Slater [20] had anything valuable to offer chemists. The theory had an unpromising beginning; the result of the thought of the greatest physicists failed to confirm that molecules could even exist. Even though the Sambe–Felton paper [98] anticipated many of the features of modern DFT calculations, including the orbital starting point, only a few had the vision [100] to see the potential of DFT methods at that time.

Fifty years makes a difference, and the fledgling theory has developed into a sophisticated and capable guide to all sorts of chemical phenomena. Now it is the method of choice for description of the structures and energetics of organometallic systems as we mentioned in our introduction. It is now the common language for chemical reactivity and catalysis, in solution, on surfaces, and in the solid state. Its cubic scaling, comparable with SCF methods—and its incorporation of post-HF corrections to the energy otherwise requiring methods which scale as N^5 or worse—make feasible its use for large systems. No electronic structure modeling package can overlook its power and popularity.

For this reason it is particularly important to know more than the key words which invoke DFT options in a modeling suite. In this chapter, we sketched the developments in theory of the past two decades; we hope we have made clear that DFT is developing rapidly. Remarkably, some software packages are content to provide the 1994 B3LYP technique. To be sure B3LYP is an entirely suitable starting point. To illustrate this point, here are some grossly oversimplified generalizations for B3LYP/6–31G(d); with just a few exceptions, this is the standard.

1. Bond distances have a MAD of <0.01 Å for molecules with first-row atoms; TZ bases may be needed for difficult systems. Performance is better than MP2 except for weak interactions, energy barriers, and hydrogen bonds.

2. Bond distances for transition metal complexes have a MAD of <0.02 Å.

3. Mean absolute errors in vibrational frequencies for organic systems and metal complexes are a few tens of wave numbers, systematically too large. This is about half the error of MP2 estimates, and is attributed to the systematic lengthening and thus weakening of bonds. Scott and Radom recommend scale factors of 0.8953 for HF in 6–31G(d), 0.9434 for MP2, and 0.9614 for B3LYP. Pure DFT formulations, BLYP and BP86, need no scaling at all.

4. For the G2 set, MAD in atomization energies is about half the values for MP2; large basis sets (diffuse functions in particular) improve the B3LYP values by a factor of 2, to about 3 kcal/mol.

5. Ionization energies have a MAD of about 0.2 eV.

6. Electron affinities should, in principle, be mistrusted since currently available functionals must underbind anions. However, in practice for all but the smallest systems this seems not to be a practical problem; affinities are also predictable within 0.2 eV, so long as diffuse functions are included in the basis.

These rules of thumb are adapted from a variety of studies; three excellent sources are Hehre's [115] compendium of results, Foresman and Frisch's guide to wise use of GAUSSIAN software [116], and the invaluable resource by Koch and Holthausen [117]. This latter work outlines the theory of DFT but is focused on assessing the performance of turn-of-the-century DFT methods. Topics treated include:

1. Molecular structures and vibrational frequencies

2. Relative energies and thermochemistry

3. Electric and magnetic properties

4. Weakly bound systems (H-bonding)
5. Potential energy surfaces

These sources are not yet out of date, despite the advances in DFT since about 2000. The wide use of 1990s functionals in applications to chemical problems, especially those involving transition metal systems, assures that these summaries will continue to be broadly useful. As a complement reaching to about 2005, we recommend reference [65] by Zhao and Truhlar.

B3LYP, still the method of choice, has well-recognized shortcomings. It is unreliable for weak interactions dominated by dispersion forces, for charge-transfer processes, and for description of excited states. Like almost all DFT-based methods, it underestimates reaction barriers and overemphasizes delocalization. Despite its broad success and now ease of use and interpretation, alternatives to B3LYP now available should be explored in any modeling study.

While we lack the much sought-after universal functional, the most promising implementations now available are Truhlar's M06-L [67] functional M06–2X or his M06-nX functionals which incorporate some exact exchange. The L (local) functional package is fast; special cases such as delocalized pi systems may require the slower alternatives with nonlocal exact exchange.

For weak interactions such as van der Waals potentials or H-bonding it would appear that one might use Truhlar's fourth-rung MPWB1K [66] functional although the widely available MP2 method is an appealing alternative to any DFT choice. Zhang and Salahub's recalibrated TPSS [63] functional may also be a good choice for weak interactions.

We can expect rapid development of new functionals, and further ascent of Perdew's ladder. For example, Grimme's [70] dispersion corrections to PBE occupy the fifth rung, incorporating dependence on virtual as well as occupied orbitals. We have a notion that the last step will be the hardest to develop further and the method may require a heavy computational burden.

We want to thank the experts in DFT who were so helpful and patient with our questions. First, Dr. Brett Dunlap was very helpful in the overall history of DFT technology going all the way back to the Sambe–Felton paper [101]. Prof. Mark Gordon graciously responded to our inquiries on methods in GAMESS and gridless DFT. Prof. Don Truhlar has provided key insights in a few words regarding modern methods. Prof. Scuseria offered valuable assessment of several functionals. Our understanding of the new LLH position-variable-exact-exchange method is entirely due to helpful guidance by Dr. Alexei Arbuznikov with brief comments by Prof. Martin Kaupp. Dr. Ulises Reveles kindly supplied the formaldehyde output from deMon2k. Finally, we can thank the developers of the software package MathType™. It is impossible for attention to wander as you work slowly through the construction of key equations with its help. Our comprehension was much enhanced by this change of pace.

Appendix: Technical Aspects of DFT Calculations

Grid-Mesh Integration in DFT Programs

The very complicated expressions for exchange–correlation density functionals require speedy and accurate numerical evaluation of integrals. In general, this is accomplished by evaluation of the integrands at specific points and weighting those values. In general, some integral I_A needs to be evaluated; consider

$$I = \int F(\mathbf{r}) \, d^3\mathbf{r}$$

Becke [A1] divided the space of the integration into zones

$$I = \sum_n F_n(r)$$

and assigned each nucleus a weight function w_n

$$\sum_n w_n = 1$$

He defined Voronoi polyhedra by constructing planes perpendicular to and bisecting interatomic vectors. Each of these polyhedra contains the set of points closer to the central atom than any other. Becke defines elliptical coordinates from each interatomic distance R from nucleus i to j with the nuclear positions as foci. Then, the distances from a point to nucleus i and nucleus j appear in the definitions of the new coordinate system:

$$\mu_{ij} = \frac{r_i - r_j}{R}; \quad \lambda_{ij} = \frac{r_i + r_j}{R}$$

Becke defines the step function

$$s(\mu_{ij}) = 1 \text{ if } -1 \leq \mu_{ij} \leq 0; \quad 0 \text{ if } 0 \leq \mu_{ij} \leq 1$$

and specifies each polyhedron by the "cell function"

$$P_i(\mathbf{r}) = \prod_{j \neq i} s(\mu_{ij})$$

This has the value 1 if $\mathbf{r}$ lies within the cell, and 0 otherwise. The discontinuous steps of the s functions are to be replaced by a smoothed step to permit easier integration. Beginning with the simplest polynomial which is suitably smooth

$$p(\mu) = \frac{3}{2}\mu - \frac{1}{2}\mu^3$$

one obtains a steeper and steeper transition by recursion

$$f_1 = p(\mu)$$
$$f_2 = p[p(\mu)]$$

etc.

$$f_k$$

For each f_k there is a smoothed step function s_k:

$$s_k(\mu) = \frac{1}{2}[1 - f_k(\mu)]$$

Once the cell function has been calculated, the weight function (value for a given grid point) can be calculated as

$$P_A(r) = \prod_{B \neq A} s(\mu_{AB})$$

so that finally we obtain the value of the weight for the particular grid point as

$$w(r) = \frac{P_A(r)}{\sum_B P_B(r)}$$

Weights for the grid points decrease to half value at the boundary of an atomic "cell" and rapidly decrease to zero outside the boundary.

Now consider the integral I_A centered on nucleus A:

$$I_A = \int_0^\infty r^2\, dr \int_0^{2\pi} d\phi \int_0^\pi \sin\theta \cdot f_A(r, \phi, \theta)\, d\theta$$

Integrating this form is a familiar problem.

$$I_A \cong \sum_{pR=1}^{NR} W_R^A(p_R) \sum_{p\Omega=1}^{N\Omega} W_\Omega^A(p_\Omega) \cdot f_A(r_p, \phi_p, \theta_p)$$

The values $W_R^A(p_R)$ and $W_\Omega^A(p_\Omega)$ are the weighting factors at the points p_R and p_Ω. The subscripts refer to radial and angular values, respectively. One standard set called SG-1 was documented by Gill et al. [A2]. They employed Lebedev weights and points for the angular quadrature on the surface of a sphere; the octahedral symmetry of the array of points seemed consistent with the functional forms of standard basis functions. The angular integration scheme of points and weights of order l is capable of exact integration of the components of angle-dependent functions representable by spherical harmonics of order l. Gill et al. adopted an Euler–Maclaurin scheme suggested by Handy and Boys [A3] for radial quadrature. They sought to guarantee that any error owing to the integration process would be less than 300 microhartree (ca. 0.2 kcal/mol) for medium-sized molecules. A very large grid of 96 Euler–Maclaurin points for the radial quadrature, 32 Legendre points for the azimuthal angle, and 64 evenly spaced points for the axial angle proved to be accurate to 0.1 microhartree. The authors investigated grids of NR Euler–Maclaurin points and $N\Omega$ Lebedev points (specified as $[NR, N\Omega]$). They introduced a device of assigning distinct Lebedev sets to particular spheres—called "pruning"—based on the observation that the charge distribution becomes more nearly spherical in the neighborhood of a nuclear center. Errors from pruning are judged to be greatest near the nucleus, which would have minimal impact on chemical properties. They found that the large [50,194] p grid called "SG-1" behaved well in a trying case, the isomerization of pentane. The larger [75,302] p grid termed "FineGrid" was also introduced into GAUSSIAN-94.

Krack and Köster [A4] have developed "adaptive" grids which represent improvements in accuracy and flexibility to SG-1 for a given number of points, or alternatively can achieve a prescribed accuracy with a smaller number of points than the available fixed grids.

Integration can be carried out for very complex functions using these grids; no investigator need hesitate to try some new functional for fear of difficulty in analytic integration. However, a large number of points are required for chemical accuracy; it is not unusual for each atom in a DFT calculation to involve several thousand grid-point evaluations. Grids however generated may have a symmetry lower than the molecular point group, may show departures from rotational invariance, and may produce irregularities in computed potential surfaces. This has prompted careful studies of various grids and also motivated research into gridless methods. This requires effective representation of the density and its functionals.

Representation of the Density by Auxiliary Basis Functions

The repulsion integral bottleneck plagues SCF theory, requiring effort that scales formally as the fourth power of the number of basis functions. DFT does not escape this formal requirement. Representation of the density in an auxiliary basis set has immediate appeal since it reduces the collection of four-index repulsion integrals to a (larger) set of three-index integrals. The key idea is the expansion of the density—ordinarily expressed as a sum of products of basis functions—in a series of single functions in an auxiliary basis. Köster [A5] gives a concise review of the implementation. We can trace this idea at least as far back as Sambe and Felton's work on DFT's exchange–correlation functionals [A6], and Dunlap et al.'s fitting of the coulomb potential with three-index integrals [A7]. This device is particularly common in programs which give DFT special emphasis, such as deMon2k, DGAUSS, and ADF. Vahtras et al. [A8] wrote

$$\chi_p\chi_q \equiv |pq) \approx |\theta_{pq}) = \sum_u C_u^{pq}\alpha_u$$

One can determine the weighting coefficients by minimizing the norm of the error R_{pq}

$$R_{pq} = \theta_{pq} - \chi_p\chi_q$$

This produces

$$C_u^{pq} = \sum_t (pqt)S_{tu}^{-1}; \quad \mathbf{C^{pq}} = \mathbf{S}^{-1}\mathbf{a^{pq}}$$

Alternatively, one may minimize the self-repulsion $(R_{pq}\|R_{pq})$; then

$$C_u^{pq} = \sum_t (pq\|t)V_{tu}^{-1}; \quad \mathbf{C^{pq}} = \mathbf{V}^{-1}\mathbf{b^{pq}}$$

Here the integrals are two- and three-index repulsion integrals, much easier to evaluate than the four-index integrals. In the first case, the general four-index integral becomes

$$(pq\|rs) \approx \mathbf{a^{pq}}^{\dagger}\mathbf{S}^{-1}\mathbf{V}\mathbf{S}^{-1}\mathbf{a^{rs}}$$

An alternative proves much preferable:

$$(pq\|rs) \approx \mathbf{b^{pq}}^{\dagger}\mathbf{V}^{-1}\mathbf{b^{rs}}$$

The nature of the auxiliary basis is a matter of considerable study. Broadly, it must be considerably more flexible than the basis used to represent the SCF HF or Kohn–Sham MOs.

DFT can be formulated with auxiliary basis representations of the density, as Zheng and Almlöf [A9] show. Let the matrix representation of the density in a basis be written

$$M_{pq}[\rho] = \left\langle \chi_p|\rho|\chi_q \right\rangle = \sum_{rs} D_{rs}(pqrs)$$

If the density is expanded in an orthonormal auxiliary basis

$$\rho(r) \approx \sum_{m=1}^{K} C_m \theta_m$$

and the error in the representation is minimized

$$C_m = \sum_{rs} (rs\|m)D_{rs}; \text{ then } M_{pq} = \sum_{m} (pq\|m)C_m$$

The key to further progress is to reexpress $\mathbf{M}$ in an orthonormal basis and then diagonalize it.

$$\bar{\mathbf{M}}[\rho] = \mathbf{V}^\dagger \mathbf{M} \mathbf{V}; \quad \bar{\mathbf{M}} \mathbf{U} = \Lambda \mathbf{U}$$

The diagonal matrix Λ can be manipulated as a scalar. The matrix representation in the orthonormal basis of any function $f(\rho)$ can be obtained immediately:

$$\bar{\mathbf{M}}[f(\rho)] = \mathbf{U}[f(\lambda)]\mathbf{U}^\dagger$$

Then, the matrix representation in the original basis is

$$\mathbf{M}[f(\rho)] = (\mathbf{V}^\dagger)^{-1}\bar{\mathbf{M}}[f(\rho)](\mathbf{V})^{-1}$$

Given the matrix representation we can write

$$\rho = \sum_{pq} \chi_p M_{pq}(\rho)\chi_q; \quad \rho^{4/3} = \sum \chi_p M_{pq}\left(\rho^{4/3}\right)\chi_q$$

$$\mathbf{M}\left[\rho^{4/3}\right] = (\mathbf{V}^\dagger)^{-1}\mathbf{U}\Lambda^{4/3}\mathbf{U}^\dagger(\mathbf{V})^{-1}$$

Zheng and Almlöf use this technique to evaluate the integral of the $X\alpha$ exchange:

$$E_X = -3\alpha \int \frac{\rho^{4/3}}{4\pi} d\mathbf{r}$$

For more advanced functionals, reference is made to the square of x, the dimensionless gradient variable

$$x = \rho^{-4/3}|\nabla\rho|$$

Representation of this quantity is discussed by the authors in a later work [A10].

DFT functionals refer to fractional powers of the density. It is not at all clear that a basis set optimized for energy will appropriately span the functions ρ and $\rho^{1/3}$. Usually, additional auxiliary basis functions are employed for powers of the density. The auxiliary functions may be rescaled members of the LCAO basis. That is, since

$$\rho = |\psi^* \psi|$$

it might be effective to use basis functions with exponent scaled by the factor $2/3$.

$$\rho^{1/3} \simeq \sum x_k G_k(2\alpha_k/3{:}\mathbf{r})$$

It is also possible to use separate sets of optimized orbitals for ρ and $\rho^{1/3}$ as shown by the Ahlrichs group [A11]. In some programs (ADF) a representation of the density is obtained by a least-squares error fit to the auxiliary basis.

In the case of the deMon2k program [A12], Hermite polynomials are used as auxiliary functions instead of the basis Gaussians while in the original Sambe–Felton algorithm only s-type Gaussian orbitals were used in auxiliary basis sets. The original reasoning seems to have been that inner shells would be spherical and that most molecules form bonds in such a way as to lead to nearly filled valence shells which would be spherical as well. In deMon2k, Hermite polynomials are used with clear recursion relationships while in Ahlrich's work special Gaussian auxiliary orbitals are used. It is not surprising that the Ahlrichs group found that adding higher angular momentum functions as p, d, and f orbitals to the auxiliary basis improved the results because in the process described above a single set of auxiliary functions is being fitted to charge density products of two orbitals in the original basis. A multipole expansion of those charge densities surely requires angular shapes more complex than spheres.

Appendix References

A1. Becke, A.D., *J. Chem. Phys.* 88:2547, 1988.
A2. Gill, P.M.W., Johnson, B.G., and Pople, J.A., *Chem. Phys. Lett.* 209:506, 1993.
A3. Handy, N.C. and Boys, S.F., *Theor. Chem. Acta* 31:195, 1973.
A4. Krack, M. and Köster, A.M., *J. Chem. Phys.* 108:3226, 1998.
A5. Köster, A.M., *J. Chem. Phys.* 104:4114, 1996.
A6. Sambe, H. and Felton, R.H., *J. Chem. Phys.* 61:3862, 1975; ibid 62:1122.
A7. Dunlap, B.I., Connolly, J., and Sabin, J.R., *J. Chem. Phys.* 71:3396, 1979.
A8. Vahtras, O., Almlöf, J., and Feyereisen, M.W., *Chem. Phys. Lett.* 213:514, 1993.
A9. Zheng, Y.C. and Almlöf, J., *Chem. Phys. Lett.* 214:397, 1993.
A10. Zheng, Y.C. and Almlöf, J.E.J., *Mol. Struct. (Theochem)* 38:277, 1996.

A11. Eichkorn, K., Treutler, O., Öhm, H., et al., *Chem. Phys. Lett.* 242:652, 1995.
A12. Köster, A.M., Calaminici, P., Casida, M.E., et al., deMon2k V. 1.8; The deMon developers, 2005.

References

1. Hohenberg, P. and Kohn, W., *Phys. Rev. B* 136:864, 1964.
2. Kohn, W. and Sham, L.J., *Phys. Rev. A* 140:1133, 1965.
3. Levy, M., *Proc. Natl. Acad. Sci. USA* 76:6062, 1979; see also Yang, W., Ayers, P.W., Wu, Q., *Phys. Rev. Lett.* 92:146404-1, 2004.
4. Gill, P.M.W., DFT, HF and the self-consistent field, *The Encyclopedia of Computational Chemistry*, Wiley, New York, 1998.
5. Koch, W. and Holthausen, M.C., *A Chemist's Guide to Density Functional Theory*, 2nd Ed., Wiley, New York, 2001.
6. Foresman, J.B. and Frisch, Æ, *Exploring Chemistry with Electronic Structure Methods*, 2nd Ed., Gaussian Inc., Pittsburg, PA, 1996.
7. Perdew, J.P. and Schmidt, K., in *Density Functional Theory and its Applications to Materials*, Eds. Van Doren, V.E., Van Alseoy, K., Geerlings, P., American Institute of Physics, 2001.
8. Jensen, F., *Introduction to Computational Chemistry*, 2nd Ed., Wiley, New York, 2007.
9. Thomas, L.H., *Proc. Camb. Phil. Soc.* 23:542, 1927.
10. Fermi, E., *Rend Accad Lincei* 6:602, 1927.
11. Dirac, P.A.M., *Proc. Camb. Phil. Soc.* 26:376, 1930.
12. Hartree, D.R., *Proc. Camb. Phil. Soc.* 24:89, 1928.
13. Fock, V., *Z. Physik* 61:126, 1930.
14. von Weizsäcker, E., *Z. Physik* 96:431, 1935.
15. Wigner, E., *Trans. Faraday Soc.* 34:678, 1938.
16. Pauling, L., *J. Chem. Phys.* 1:280, 1933.
17. Mulliken, R.S., *J. Chem. Phys.* 1:492, 1933.
18. Roothaan, C.C.J., *Rev. Mod. Phys.* 23:69, 1951.
19. Hall, G.G., *Proc. R. Soc. (London)* A205:541, 1951.
20. Slater, J.C., *Phys. Rev.* 81:385, 1951.
21. Köster, A.M., Calaminici, P., Gomez, Z, et al., in *Reviews of Modern Quantum Chemistry: A Celebration of the Contributions of R.G. Parr*, K.D. Ser, Ed., World Scientific, Singapore, 2002.
22. Wang, Y. and Perdew, J.P., *Phys. Rev. B* 44:13298, 1991.
23. Ceperley, D.M., *Phys. Rev. B* 18:3126, 1978.
24. Ceperley, D.M. and Alder, B.J., *Phys. Rev. Lett.* 45:566, 1980.
25. Vosko, S.H., Wilk, L., and Nusair, M., *Can. J. Phys.* 58:1200, 1980.
26. Ernzerhof, M. and Perdew, M., *J. Chem. Phys.* 109:3313, 1998.
27. Perdew, J.P. and Wang, Y., *Phys. Rev. B* 45:13244, 1992.
28. Becke, A.D. and Dickson, R.M., *J. Chem. Phys.* 92:3610, 1990.
29. Becke, A.D., *J. Chem. Phys.* 85:7184, 1986.
30. Becke, A.D., *J. Chem. Phys.* 88:2547, 1987.
31. Becke, A.D., *J. Chem. Phys.* 88:1053, 1988.
32. Becke, A.D., *Phys. Rev. A* 38:3098, 1988.

33. Becke, A.D., *J. Chem. Phys.* 96:2155, 1992.
34. Becke, A.D., *J. Chem. Phys.* 97:9173, 1992.
35. Pople, J.A., Head-Gordon, M., Fox, D.J., et al., *J. Chem. Phys.* 90:5622, 1989.
36. Curtiss, L.A., Jones, C., Trucks, G.W., et al., *J. Chem. Phys.* 93:2537, 1990.
37. Curtiss, L.A., Raghavachari, K., Trucks, G.W., et al., *J. Chem. Phys.* 94:7221, 1991.
38. Colle, R. and Salvetti, D., *Theor. Chim. Acta* 37:329, 1975.
39. Colle, R. and Salvetti, D., *Theor. Chim. Acta* 53:55, 1979.
40. Colle, R. and Salvetti, D., *J. Chem. Phys.* 79:1404, 1983.
41. Lee, C., Yang, W., and Parr, R.G., *Phys. Rev. B* 37:785, 1988.
42. Parr, R.G. and Yang, W., *Density-Functional Theory of Atoms and Molecules,* Oxford University Press, Oxford, 1989.
43. Miehlich, B., Savin, A., Stoll, H., et al., *Chem. Phys. Lett.* 157:200, 1989.
44. Becke, A.D., *J. Chem. Phys.* 98:1372, 5648, 1993.
45. Perdew, J.P. and Wang, Y., *Phys. Rev. B* 45:13244, 1992.
46. Becke, A.D., *J. Chem. Phys.* 104:1040, 1996.
47. Curtiss, L.A., Raghavachari, K., Redfern, P.C., et al., *J. Chem. Phys.* 93:1063, 1997.
48. Curtiss, L.A., Redfern, P.C., Raghavachari, K., et al., *J. Chem. Phys.* 109:42, 1998.
49. Curtiss, L.A., Raghavachari, K., Redfern, P.C., et al., *J. Chem. Phys.* 109:7764, 1998.
50. Becke, A.D., *J. Chem. Phys.* 107:8554, 1997.
51. Perdew, J.P., Burke, K., and Ernzerhof, M., *Phys. Rev. Lett.* 77:3865, 1996; ibid. 78:1396, 1997.
52. Lieb, E.H. and Oxford, S., *Int. J. Quantum Chem.* 19:427, 1981.
53. Becke, A.D., *J. Chem. Phys.* 84:4524, 1986.
54. Zhang, Y. and Yang, W., *Phys. Rev. Lett.* 80:890, 1998.
55. Hammer, B., Hansen, L.B., and Norskov, J.K., *Phys. Rev. B* 59:7413, 1999.
56. Ernzerhof, M. and Scuseria, G.E., *J. Chem. Phys.* 110:5029, 1999.
57. Gill, P.M.W., *Aust. J. Chem.* 54:661, 2001.
58. Perdew, J.P., Kurth, S., Zupan, A., and Blaza, P., *Phys. Rev. Lett.* 82:2544, 1999.
59. Brack, M., Jennings, B.K., and Chu, Y.H., *Phys. Lett.* 65B:1, 1976.
60. Svendsen, P.S. and von Barth, U., *Phys. Rev. B* 54:17402, 1996.
61. Tao, J., Perdew, J.P., Staroverov, V.N., and Scuseria, G.E., *Phys. Rev. Lett.* 91:146401, 2003.
62. Perdew, J.P., Tao, J., Staroverov, V.N., and Scuseria, G.E., *J. Chem. Phys.* 120:6898, 2004.
63. Zhang, Y. and Salahub, D.R., *Chem. Phys. Lett.* 436:394, 2007.
64. Zhao, Y., Pu, J., Lynch, B.J., et al., *Phys. Chem. Chem. Phys.* 6:673, 2004.
65. Zhao, Y. and Truhlar, D.G., *J. Chem. Theor. Comput.* 1:415, 2005.
66. Zhao, Y. and Truhlar, D.G., *J. Phys. Chem. A* 108:6908, 2004.
67. Zhao, Y. and Truhlar, D.G., *J. Chem. Phys.* 125:194101, 2006.
68. Zhao, Y., Schultz, N.E., and Truhlar, D.G., *J. Chem. Theor. Comput.* 2:364, 2006.
69. Voorhis, T.V. and Scuseria, G.E., *J. Chem. Phys.* 109:400, 1998.
70. Grimme, S., *J. Chem. Phys.* 124:034108, 2006.
71. Frisch, M.J., Trucks, G.W., Schlegel, H.B., et al., Gaussian 03, Revision C.02, Gaussian Inc., Wallingford, CT, 2004.
72. Shao, Y., Fusti-Molnar, L., Jung, Y., et al., *Phys. Chem. Chem. Phys.* 8:3172, QCHEM 3.1, 2006.
73. Heyd, J., Scuseria, G.E., and Ernzerhof, M., *J. Chem. Phys.* 118:8207, 2003.
74. Heyd, J. and Scuseria, G.E., *J. Chem. Phys.* 120:7274, 2004.
75. Heyd, J. and Scuseria, G.E., *J. Chem. Phys.* 121:1187, 2004.
76. Vydrov, O.A. and Scuseria, G.E., *J. Chem. Phys.* 125:234109, 2006.

77. Perdew, J.P., Ernzerhof, M., and Burke, K., *J. Chem. Phys.* 105:9982, 1996.
78. Yanai, T., Tew, D.P., and Handy, N.C., *Chem. Phys. Lett.* 393:51, 2004.
79. Jaramillo, J., Scuseria, G.E., and Ernzerhof, M., *J. Chem. Phys.* 118:1068, 2003.
80. Della Salla, F. and Görling, A., *J. Chem. Phys.* 115:5718, 2001.
81. Arbuznikov, A.V., Kaupp, M., and Bahmann, H., *J. Chem. Phys.* 124:204102, 2006.
82. Bahmann, H., Rodenberg, A., Arbuznikov, A.V., et al., *J. Chem. Phys.* 126:11103, 2007.
83. Arbuznikov, A. and Kaupp, M., *Chem. Phys. Lett.* 440:160, 2007.
84. Schmidt, M.W., Baldridge, K.K., Boatz, J.A., et al., *J. Comput. Chem.* 14:1347, 1993 (*General Atomic and Molecular Electronic Structure System* (*GAMESS*)).
85. Glaesemann, K.R. and Gordon, M.S., *J. Chem. Phys.* 108:9959, 1998.
86. Glaesemann, K.R. and Gordon, M.S., *J. Chem. Phys.* 110:6580, 1999.
87. Glaesemann, K.R. and Gordon, M.S., *J. Chem. Phys.* 112:10738, 2000.
88. Zheng, Y.C. and Almlöf, J., *Chem. Phys. Lett.* 214:397, 1993.
89. Vahtras, O., Almlöf, J., and Feyereisen, M.W., *Chem. Phys. Lett.* 213:514, 1993.
90. DePristo, A.E. and Kress, J.D., *J. Chem. Phys.* 86:1425, 1987.
91. Werpetinski, K.S. and Cook, M., *Phys. Rev. A* 52:R3397, 1995.
92. Werpetinski, K.S. and Cook, M., *J. Chem. Phys.* 106:7124, 1997.
93. Zope, R.R. and Dunlap, B.I., *J. Chem. Phys.* 124:44107, 2006.
94. Dunlap, B.I., Karna, S.P., and Zope, R.R., *J. Chem. Phys.* 125:214104, 2006.
95. Dunlap, B.I., *J. Mol. Struct.* (*Theochem*) 501–502:221, 2000.
96. Dunlap, B.I., *J. Mol. Struct.* (*Theochem*) 529:37, 2000.
97. Dunlap, B.I., *Int. J. Quantum Chem.* 69:317, 1997.
98. Sambe, H. and Felton, R.H., *J. Chem. Phys.* 61:3862, 1975; ibid. 62:11221975.
99. Dahlquist, G. and Bjorck, A., *Numerical Methods*, translated by Ned Anderson, Prentice-Hall, Dover, Mineola, New York, 1974, p. 184 and p. 250.
100. Dunlap, B.I., Connolly, J.W.D., and Sabin, J.R., *J. Chem. Phys.* 71:3396, 1979.
101. te Velde, G., Bickelhaupt, F.M., Baerends, E.J., et al., *J. Comp. Chem.* 22:931, 2001.
102. ADF software and documentation from http://scm.com
103. Gritsenko, O.V., Schipper, P.R.T., and Baerends, E.J., *Chem. Phys. Lett.* 302:199, 1999.
104. van Leeuwen, R. and Baerends, E.J., *Phys. Rev. A* 49:2421, 1994.
105. Autschbach, J., Patchkovskii, S., and Ziegler, T.J., *Chem. Phys.* 117:581, 2002.
106. Köster, A.M., *J. Chem. Phys.* 104:4114, 1995.
107. Krack, M. and Köster, A.M., *J. Chem. Phys.* 108:3226, 1998.
108. Jug, K., Zimmermann, B., Calaminici, P., et al., *J. Chem. Phys.* 116:4497, 2002.
109. Köster, A.M., *J. Chem. Phys.* 118:9943, 2003.
110. Köster, A.M., Reveles, J.U., and del Campo, J.M., *J. Chem. Phys.* 121:3417, 2004.
111. Köster, A.M., Flores-Moreno, R., and Reveles, J.U., *J. Chem. Phys.* 121:681, 2004.
112. Reveles, J.U. and Köster, A.M., *J. Comp. Chem.* 25:1109, 2004.
113. Reveles, J.U., Heine, T., and Köster, A.M., *J. Phys. Chem.* 109:7068, 2005.
114. Johnson, K.H. and Wahlgren, U., *Int. J. Quantum Chem. Symp.* 6:243, 1972.
115. Hehre, W.J., *A Guide to Molecular Mechanics and Quantum Chemical Calculations*, Wavefunction Inc., Irvine, CA, 1997.
116. Foresman, J.F. and Frisch, Æ., *Exploring Chemistry with Electronic Structure Methods: A Guide to Using Gaussian*, Gaussian Inc., Pittsburg, PA, 1996.
117. Koch, W. and Holthausen, M.C., *A Chemist's Guide to Density Functional Theory*, 2nd Ed., Wiley-VCH, New York, NY, 2001.

12

Calculation of Nuclear Magnetic Resonance Shielding/Shifts

Introduction

Perhaps the most powerful and versatile means of investigation of molecular structure is nuclear magnetic resonance. The central phenomenon is the response of a magnetic moment $\mathbf{m}_A$ carried by a nucleus A to an externally imposed magnetic field $\mathbf{B}$.

$$E = \mathbf{B} \cdot \mathbf{m}_A$$

In molecules, each nucleus is surrounded by an electronic charge distribution. While an electric field E imposes a colinear force on any charge q

$$\mathbf{F}_E = q\mathbf{E}$$

the magnetic field exerts a perpendicular force on a current J of charge q moving at velocity v:

$$\mathbf{F}_B = \mathbf{r} \times \mathbf{J}$$

This force accelerates the charge so to trace a circular path. The circulating charge defines a magnetic field which contributes to the field at nucleus A.

$$\mathbf{B}_A = \mathbf{B}(1 - \boldsymbol{\sigma}_A)$$

The quantity σ is called the shielding constant. Though it is defined as a tensor in all theories of NMR, most experimental reports refer to the isotropic average

$$\sigma = (1/3)(\sigma_{xx} + \sigma_{yy} + \sigma_{zz})$$

The shielding is dependent on details of the charge distribution and is generally reported as the extent in parts per million of departure from a standard value for a given nucleus in a well-defined electronic environment.

Ramsey Theory

The quantum theory of magnetic resonance is venerable. Ramsey described a perturbative method for calculation of chemical shifts in 1950 [1,2]. The first comprehensive guide to NMR theory and practice by Pople et al. [3] appeared in 1959, during the early stages of the long and fruitful development of the field. Interpretation of NMR spectra at this time owed little to electronic structure methods—the first all-electron treatment of formaldehyde appeared in 1960 [4]. Rather, the work by Pople et al. [3] describes the coupling of angular momentum, with the help of "effective Hamiltonians" which absorb all electronic features of the system into parameters. This permits informative modeling of NMR spectra with minimal computational effort, as exemplified in the pioneering computer programing text by Wiberg [5]. The parametrized spin algebra is so simple that the Laocoon program [6] needs only to diagonalize matrices no larger than 7×7.

Reliable computational methods for the fundamental quantities of NMR, such as full shielding and coupling tensors, were slow to emerge. We can appreciate the difficulties of the task by a glance at the definitions of the shielding tensor.* In second-order perturbation theory the response of a magnetic moment $\boldsymbol{\mu}$ of a nucleus A to a magnetic field $\mathbf{B}$ is an interaction energy:

$$E^{(2)}(B,\mu_A) = \sum_{\alpha\beta} B_\alpha \sigma_{\alpha\beta} \mu_{A\beta}$$

Here the indices α, β refer to Cartesian components of the field and magnetic moment vectors, and label the components of the shielding tensor.

$$\sigma_{\alpha\beta} = \left[\frac{\partial^2 E^{(2)}}{\partial B_\alpha \partial \mu_{A\beta}} \right]_{\mu_A = \mathbf{B} = 0}$$

The perturbation analysis identifies paramagnetic and diamagnetic components of the shielding tensor.

$$\sigma_{\alpha\beta} = \sigma_{\alpha\beta}^{DIA} + \sigma_{\alpha\beta}^{PARA}$$

The diamagnetic term labeled DIA is a first-order property defined entirely by the ground state.

$$\sigma_{\alpha\beta}^{DIA} = C \left\langle 0 \left| \sum_k r_k^{-3} (r_k^2 \delta_{\alpha\beta} - r_{k\alpha} r_{k\beta}) \right| 0 \right\rangle$$

---

*For a clear development of magnetic phenomena in molecules which leads from the classical phenomena of magnetism to the quantum mechanical expressions for the shielding and coupling tensors, we recommend the second edition of Atkins' *Molecular Quantum Mechanics* (1983), Oxford University Press.

In these sums the index k identifies a particular atom; r_k is the distance of atom k from an origin of coordinates. The constant C is $\mu_0 e^2 / 8\pi m_e$ where μ_0—the permittivity of free space—is associated with the SI electromagnetic system of units, e is the charge of an electron, and m_e is the mass of the electron. It is assumed that the origin of the vector potential is situated at the nuclear center—if not, additional complications must be addressed.

The paramagnetic part of the shielding tensor labeled PARA is a second-order property, and in Ramsey's formulation requires a sum over states.

$$\sigma_{\alpha\beta}^{PARA} = -D\sum_{n\neq 0}(E_n - E_0)^{-1}\left[\left\langle 0\left|\sum_k r_k^{-3}L_{k\alpha}\right|n\right\rangle\langle n|L_{k\beta}|0\rangle + \left\langle 0\left|\sum_k r_k^{-3}L_{k\beta}\right|n\right\rangle\langle n|L_{k\alpha}|0\rangle\right]$$

Here E_n and E_0 are the energies of ground and excited states of the unperturbed Hamiltonian; r_k is the distance from the origin of coordinates to the kth nucleus; and $L_{k\alpha}$ is the α-component of the angular momentum relative to the k-nucleus. The constant D is c/m_e.

Challenges to Use of Ramsey's Form

According to a useful 1977 review by Ebraheem and Webb [7] of the state of affairs in the mid-1970s, two major problems beset any attempt to employ these equations. These include the difficulty of a sum over states and the subtleties of choice of gauge in the computational description of electromagnetic phenomena.

Excitations

The evaluation of the energies and wave functions of all the excited states is implied in Ramsey's formula. This was well beyond reach for those attempting applications before the 1970s, and for that matter still is nearly impossible. A popular measure was to introduce an averaged excitation energy as a parameter, and simplify second-order expressions by the closure rule; thus,

$$\sum_{n\neq 0}\frac{\langle 0|A|n\rangle\langle n|B|0\rangle}{E_0 - E_n} \approx \frac{\sum_{n\neq 0}\langle 0|A|n\rangle\langle n|B|0\rangle}{\Delta E} = \frac{\langle 0|AB|0\rangle - \langle 0|A|0\rangle\langle 0|B|0\rangle}{\Delta E}$$

This proved useful in some carefully selected cases but was generally unsatisfactory. In a landmark report, Ditchfield [8] formulated a theory employing the coupled perturbed HF method to obtain the system response to the magnetic field perturbation. We will work through his paper in some detail.

Gauge Origin

The second problem was the location of the origin of the vector field, that is to say gauge dependence. To address this problem London [9] defined gauge-including basis functions, a more proper description than the often-encountered "gauge-independent atomic orbitals" [10,11].

$$\chi_n = \exp\left(-\frac{i}{c}\mathbf{A}_n \cdot \mathbf{r}\right)\varphi_n$$

With this choice we find all integrals depend on the differences $\mathbf{A}_n - \mathbf{A}_m$ and not the gauge origin. Finite-perturbation-theory INDO calculations (in which the effect of the field was estimated by finite differences between energies computed in the presence and absence of the magnetic field, thus avoiding the need for a sum over states) with standard parameters and London orbitals did not produce good agreement with observed ^{13}C shifts. Reparametrization improved performance [12].

Semiempirical Adaptions of Ramsey's Formalism

The rapid development of experimental ^{13}C NMR and the simplicity in these spectra arising from the weak coupling between ^{13}C nuclei made NMR of this nucleus an appealing target for computational description. Ebraheem and Webb [7] began with Ramsey's framework [1,2] and simplified the computational task by adopting the approximations of semiempirical CNDO/S [13] and INDO [14] SCF methods to predict trends in the NMR shifts of ^{13}C, ^{15}N, ^{17}O, and ^{19}F isotopes. The expressions for the shielding tensor component $\sigma_{\alpha\beta}$ with single-determinant wave functions become

$$\sigma_{\alpha\beta}^A = \sum_j \left\langle j \left| (\mathbf{r} \cdot \mathbf{r}_A \delta_{\alpha\beta} - r^\alpha r_A^\beta)/r_A^3 \right| j \right\rangle$$

$$- \sum_b \sum_j \frac{\langle j|L_\alpha|b\rangle\langle b|L_{\beta A}/r_A^3|j\rangle + \langle j|L_{\alpha A}/r_A^3|b\rangle\langle b|L_{\beta A}|j\rangle}{\varepsilon_j - \varepsilon_b}$$

Here $\mathbf{L}$ is again the angular momentum operator, $\mathbf{r}_A$ is the vector from the origin of coordinates to nucleus A, and $\mathbf{r}$ is the vector to the electron. The set of orbitals occupied in the SCF determinant carry labels $i, j, \ldots$ and the virtual orbitals are labeled by $a, b, \ldots$. It is worth noting that the angular momentum operator can effect only one-electron excitations, so the excited states in question are within the CIS space.

Ditchfield's Formalism

Ditchfield [8] avoided the need for a sum over states by applying the coupled perturbed Hartree–Fock method. The central assumption of this method is that the response of a system to a perturbation retains the single-determinant form but mixes the originally unperturbed virtual and occupied MOs. We will follow his development closely.

As shown elsewhere in this text the Hamiltonian for electrons in an electromagnetic field should include the augmented momentum

$$\mathbf{p}_{\text{Field}} = \mathbf{p} + \frac{1}{c}\mathbf{A}$$

This expression includes the vector potential $\mathbf{A}$ from which the magnetic field derives

$$\mathbf{A}'(\mathbf{r}_j) = (1/2)\mathbf{H} \times \mathbf{r}_j + \sum_B (\boldsymbol{\mu}_B \times \mathbf{r}_{jB})/r_{jB}^3 = \mathbf{A}(\mathbf{r}_j) + \sum_B (\boldsymbol{\mu}_B \times \mathbf{r}_{jB})/r_{jB}^3$$

Here $\mathbf{H}$ is the magnetic field, $\boldsymbol{\mu}_B$ is the magnetic moment of nucleus B, $\mathbf{r}_j$ is the distance to the jth electron from the center of coordinates, and $\mathbf{r}_{jB}$ is the distance between the jth electron and nucleus B. The Hamiltonian is in atomic units and in Ditchfield's notation [8] becomes

$$H = (1/2)\sum_j \left\{ \left[\mathbf{p}_j + (1/c)\mathbf{A}'(\mathbf{r}_j)\right]^2 - 2\sum_B (Z_B/r_{jB}) \right\}$$
$$+ \sum_{j<l}\sum_l (1/r_{jl}) + \sum_{B<D}\sum_D (Z_B Z_D/R_{BD})$$

$\mathbf{A}$ is the vector potential and c is the speed of light in a vacuum. Z_B is the nuclear charge. H is the Hamiltonian while H is the magnetic field strength. It is common to choose the Coulomb gauge where $\nabla \mathbf{A} = 0$ and use $\mathbf{p} = -i\nabla$ for the momentum in atomic units. The Hamiltonian is dependent on both the magnetic field and the magnetic moments of all the nuclei. Ditchfield expands the Hamiltonian through quadratic terms in these quantities:

$$H = (H, \mu_B) = H^{(0)} + \sum_\alpha H_\alpha H_\alpha^{(1,0)} + \sum_\alpha \mu_{B\alpha} H_{B\alpha}^{(0,1)}$$
$$+ (1/2)\sum_\alpha \sum_\beta H_\alpha H_{\alpha\beta}^{(2,0)} H_\beta + \sum_\alpha \sum_\beta H_\alpha H_{B\alpha\beta}^{(1,1)} \mu_{B\beta}$$

The superscripts specify derivatives of the Hamiltonian with respect to magnetic field and moment. For example,

$$H^{(1,1)}_{B\alpha} = \frac{\partial^2 H_{B\alpha}}{\partial H \partial \mu}$$

To recover the quantities of interest it is necessary to evaluate the associated derivative of the total energy for the molecule in the presence of the magnetic field. For example, the shielding tensor is defined by the collection of second derivatives of the energy in the magnetic field. The ji element of this tensor for nucleus B is

$$\sigma^B_{ji} = \frac{\partial^2 E(H, \mu_B)}{\partial H_i \partial \mu_{Bj}}$$

Ditchfield defines the terms in the Hamiltonian individually. First

$$H^{(0)} = \sum_j \left[-(1/2)\nabla^2_j - \sum_B (Z_B/r_{jB}) \right] + \sum_{j<l} \sum_l r_{jl}^{-1} + \sum_{B<D} \sum_D (Z_B Z_D/R_{BD})$$

This is just the molecular Hamiltonian in the absence of a field. First-order responses follow

$$H^{(1,0)}_\alpha = -(1/2c) \sum_j L_{j\alpha}$$

The external magnetic field interacts with the jth electron's angular momentum.

$$H^{(0,1)}_{B\alpha} = -(1/c) \sum_j L_{jB\alpha}/r^3_{jB}$$

The angular momentum of motion of an electron j with respect to the nuclear center B interacts with the magnetic moment of that nucleus. The quadratic response to the magnetic field is

$$H^{(2,0)}_{\alpha\beta} = \left(\frac{1}{4c^2} \right) \sum_j (r^2_j \delta_{\alpha\beta} - r_{j\alpha} r_{j\beta})$$

There is no quadratic response to the magnetic moment of nucleus B, but the mixed term exists in the form:

$$H^{(1,1)}_{B\alpha\beta} = \left(\frac{1}{2c^2} \right) \sum_j [(\mathbf{r}_j \cdot \mathbf{r}_{jB})\delta_{\alpha\beta} - r_{j\alpha} r_{jB\beta}]/r^3_{jB}$$

In these expressions, $\mathbf{r}_j$ is the distance of an electron from the origin of coordinates and $\mathbf{r}_{jB}$ is the distance of the electron from nucleus B; the angular momentum components are

$$L_{j\alpha} = (\mathbf{r}_j \times \boldsymbol{\nabla}_j)_\alpha; \quad L_{jB\alpha} = (\mathbf{r}_{jB} \times \boldsymbol{\nabla}_j)_\alpha$$

and as usual the subscripts α and β refer to (x,y,z) components. Fortunately, all the matrix elements can be computed in a Gaussian basis set.

The Schroedinger equation in the presence of the magnetic field is

$$H(H, \mu_B)\Psi(H, \mu_B) = E(H, \mu_B)\Psi(H, \mu_B)$$

Both the wave function and the energy need to include at least first-order terms of both the magnetic field and the magnetic moment of each nucleus.

$$\Psi(H, \mu_B) = \Psi^{(0)} + \sum_\alpha (\partial\Psi(H, \mu_B)/\partial H_\alpha)_0 H_\alpha + \sum_\alpha (\partial\Psi(H, \mu_B)/\partial\mu_{B\alpha})_0 \mu_{B\alpha} + \cdots$$

$$= \Psi^{(0)} + \sum_\alpha \Psi_\alpha^{(1,0)} H_\alpha + \sum_\alpha \Psi_{B\alpha}^{(0,1)} \mu_{B\alpha} + \cdots \text{ and}$$

$$E(H, \mu_B) = E^{(0)} + \sum_\alpha E_\alpha^{(1,0)} H_\alpha + \sum_\alpha E_{B\alpha}^{(0,1)} \mu_{B\alpha} + (1/2)\sum_\alpha \sum_\beta H_\alpha E_{\alpha\beta}^{(2,0)} H_\beta$$

$$+ \sum_\alpha \sum_\beta H_\alpha E_{B\alpha\beta}^{(1,1)} \mu_{B\beta} + (1/2)\sum_\alpha \sum_\beta \mu_{B\alpha} E_{B\alpha\beta}^{(0,2)} \mu_{B\beta} + \cdots$$

Take note of the notation by which derivatives are designated. For example,

$$\sum_\alpha \sum_\beta H_\alpha E_{B\alpha\beta}^{(1,1)} \mu_{B\beta} = \sum_\alpha \sum_\beta H_\alpha \frac{\partial^2 E}{\partial H_\alpha \partial\mu_{B\beta}} \mu_{B\beta}$$

As is generally the case in perturbation expansions, first-order corrections to the wave function define second-order corrections in the energy. In the energy expansion, the term linear in H will reduce to a sum of products of the field with the nuclear magnetic moments for most organic molecules, which have a singlet ground state with zero angular momentum. We recognize the second derivative with respect to field $(2, 0)$ as a polarization defining the diamagnetic shielding, and see that the $(1, 1)$ mixed second derivative is the shielding derived from secondary magnetic fields exerted by the induced motion of electrons.

Now introduce the GIAO basis; use of these complex functions requires that the molecular orbital coefficients and density matrix elements may also be complex numbers.

$$\psi_j(\mathbf{H}, \boldsymbol{\mu}_B) = \sum_\nu c_{\nu j}(\mathbf{H}, \boldsymbol{\mu}_B)\chi_\nu(\mathbf{H}) \text{ with } \chi_\nu(\mathbf{H}) = \exp\left(-(i/c)\mathbf{A}_\nu \bullet \mathbf{r}\right)\phi_\nu$$

The usual derivation of the Roothaan-SCF equations can be followed, and leads to

$$\sum_\lambda \left[F_{\nu\lambda}(H, \mu_B) - \varepsilon_j(H, \mu_B) S_{\nu\lambda} \right] c_{\lambda j}(H, \mu_B) = 0$$

The modified Fock matrix contains one- and two-electron terms:

$$F_{\nu\lambda}(H, \mu_B) = h_{\nu\lambda}(H, \mu_B) + g_{\nu\lambda}(H, \mu_B)$$

Integrals in the one-electron term refer to the modified momentum operator including the vector potential:

$$h_{\nu\lambda}(\mathbf{H},\{\boldsymbol{\mu}_B\}) = \langle \chi_\nu | (1/2)(-i\boldsymbol{\nabla} + (1/c)\mathbf{A}'(\mathbf{r}))^2 - \sum_B (Z_B/r_B) | \chi_\lambda \rangle$$

The sum includes all nuclei B with magnetic moments. The two-electron terms are also modified by the effects of the field, through the basis functions and the perturbed MO coefficients:

$$G_{\nu\lambda}(H, \mu_B) = \sum_{\rho\sigma} P_{\rho\sigma}(H, \mu_B) \left\{ (\chi_\nu^* \chi_\lambda | \chi_\rho^* \chi_\sigma) - (1/2)(\chi_\nu^* \chi_\sigma | \chi_\rho^* \chi_\lambda) \right\}$$

Following Ditchfield we are using the "charge distribution" notation for two-electron integrals:

$$(\chi_\kappa^* \chi_\lambda | \chi_\rho^* \chi_\sigma) = \int d\tau_1 d\tau_2 \chi_\kappa^*(1) \chi_\lambda(1) (1/r_{12}) \chi_\rho^*(2) \chi_\sigma(2)$$

The density matrix is

$$P_{\rho\sigma} = \sum_i c_{i\rho}^* c_{i\sigma}$$

The sum extends over all occupied spin-orbitals. The two-electron part is futher abbreviated in Ditchfield's development:

$$G_{\nu\lambda}(\mathbf{H}, \mu_B) = \sum_{\rho\sigma} P_{\rho\sigma}(\mathbf{H}, \mu_B) G_{\nu\lambda\rho\sigma}(\mathbf{H})$$

Coupled Perturbed Hartree–Fock Method

One key part of Ditchfield's success was his use of the coupled perturbed Hartree–Fock (CPHF) method to describe the perturbation of the RHF orbitals by the magnetic field. The CPHF method is reviewed by Jensen [15] and can be used for other properties as well as magnetic perturbations. Its central assumption is that the single-determinant form of the wave function can be retained in the presence of the perturbation, though the orbitals are transformed. The perturbed Fock equation is, briefly

$$\mathbf{F}(\lambda)\mathbf{C}(\lambda) = \varepsilon(\lambda)\mathbf{S}(\lambda)\mathbf{C}(\lambda)$$

The parameter λ defines the order of the perturbation. Then, we can expand all the quantities in powers of λ:

$$\mathbf{F} = \mathbf{F}^{(0)} + \lambda\mathbf{F}^{(1)} + \lambda^2\mathbf{F}^{(2)} + \cdots$$

$$\mathbf{C} = \mathbf{C}^{(0)} + \lambda\mathbf{C}^{(1)} + \lambda^2\mathbf{C}^{(2)} + \cdots$$

$$\varepsilon = \varepsilon^{(0)} + \lambda\varepsilon^{(1)} + \lambda^2\varepsilon^{(2)} + \cdots$$

$$\mathbf{S} = \mathbf{S}^{(0)} + \lambda\mathbf{S}^{(1)} + \lambda^2\mathbf{S}^{(2)} + \cdots$$

Collecting terms that have the same power of λ we are first reminded of the unperturbed solution

$$\mathbf{F}^{(0)}\mathbf{C}^{(0)} = \varepsilon^{(0)}\mathbf{S}^{(0)}\mathbf{C}^{(0)}$$

And then we encounter the first-order equation

$$(\mathbf{F}^{(0)}\mathbf{C}^{(0)} + \mathbf{S}^{(0)}\varepsilon^{(0)})\mathbf{C}^{(1)} = (\mathbf{F}^{(1)} + \mathbf{S}^{(0)}\varepsilon^{(1)} + \mathbf{S}^{(1)}\varepsilon^{(0)})\mathbf{C}^{(0)}$$

In this equation $\mathbf{F}^{(1)} = \mathbf{h}^{(1)} + \mathbf{P}^{(0)}\mathbf{G}^{(1)} + \mathbf{P}^{(1)}\mathbf{G}^{(0)}$. The density matrix $\mathbf{P}^{(0)} = \mathbf{C}^{\dagger(0)}\mathbf{C}^{(0)}$ and its first derivative is $\mathbf{P}^{(1)} = \mathbf{C}^{\dagger(1)}\mathbf{C}^{(0)} + \mathbf{C}^{\dagger(0)}\mathbf{C}^{(1)}$. The necessary orthonormalization of the corrected wave functions where the zeroth-order condition can be expressed in matrix form as

$$\mathbf{C}^{\dagger}(\lambda)\mathbf{S}(\lambda)\mathbf{C}(\lambda) = \mathbf{I}$$

The expansion in λ leads to a reminder of the unperturbed orbitals' orthonormality

$$\mathbf{C}^{\dagger(0)}\mathbf{S}^{(0)}\mathbf{C}^{(0)} = \mathbf{I}$$

and defines the first-order requirement for continued orthonormality of the perturbed orbitals:

$$\mathbf{C}^{\dagger(1)}\mathbf{S}^{(0)}\mathbf{C}^{(0)} + \mathbf{C}^{\dagger(0)}\mathbf{S}^{(1)}\mathbf{C}^{(0)} + \mathbf{C}^{\dagger(0)}\mathbf{S}^{(0)}\mathbf{C}^{(1)} = 0$$

If we recognize that the response of the orbitals to a perturbation will be to mix virtual orbitals into the occupied set so to correct the occupied MOs, we can write the process as the effect of a unitary transform

$$\chi(H) = \mathbf{U}\chi(0)$$
$$\mathbf{U} \approx \mathbf{I} + \mathbf{u}(H)$$

If the matrix $\mathbf{U}$ is to be unitary

$$\mathbf{U}^{\dagger}\mathbf{U} = \mathbf{1} = \mathbf{U}^{Tr}\mathbf{U}$$

$$(\mathbf{I} + \mathbf{u})^{Tr}(\mathbf{I} + \mathbf{u}) \approx \mathbf{I}$$

$$\mathbf{u}^{Tr} + \mathbf{u} = 0$$

We require the derivatives of the overlap matrix elements with respect to the perturbation:

$$S_{ij}^{(1)} = \frac{\partial}{\partial \lambda} \langle \chi_i | \chi_j \rangle$$

To define the first correction to the Fock matrix we need derivatives of familiar integrals with respect to the perturbation λ:

$$h_{ij}^{(1)} = \frac{\partial}{\partial \lambda} \langle \chi_i | h | \chi_j \rangle$$

$$G_{ijkl}^{(1)} = \langle \chi_i \chi_j | g | \chi_k \chi_l \rangle^{(1)} = \frac{\partial}{\partial \lambda} \langle \chi_i \chi_j | g | \chi_k \chi_l \rangle$$

Noting that the first-order terms are really derivatives with respect to the perturbation, a matrix equation can be set up involving the values of the unperturbed quantities in a matrix $\mathbf{A}^{(0)}$ and the resulting perturbed quantities in $\mathbf{B}^{(1)}$ as

$$\mathbf{A}^{(0)}\mathbf{U}^{(1)} = \mathbf{B}^{(1)}$$

This sort of iterative approach will usually converge to make the off-diagonal elements of the perturbed Fock matrix zero; then $\mathbf{B}^{(1)}$ defines a diagonal (perturbed) Fock matrix.

First Results of Calculations in Ditchfield's Formulation

Table 12.1 displays some of Ditchfield's results. The "MB" basis is a minimal basis of Slater-type orbitals represented by a series of five Gaussian functions (STO-5G); the series was defined by a best (least-squares) match to

TABLE 12.1a

Ditchfield's Mean Absolute Deviations of Chemical Shifts,
in Parts Per Million Relative to CH_4

Isotope: Basis	STO-5G	LEMAO-5G	4-31G
^{13}C	22.6	22.4	4.1
^{1}H	1.31	0.83	0.43

TABLE 12.1b

Absolute Values of the Isotropic Shift Relative to Methane (^{13}C)
and Water (^{17}O)

Species: Model	HF/MB	HF/SV	Exp.	HF/6-311G(d,p)[a]
Formaldehyde ^{13}C	−147.9	−199.6	−197	−188.1
Formaldehyde ^{17}O	−757.6	−858.8	−590	−656.6

[a] Implementation of Ditchfield method in GAMESS.

the Slater single simple exponential form. The LEMAO expansion was chosen to give the variationally optimum energy, while the SV basis is a split-valence 4-31G basis. The large basis 6-311G(d,p) was employed in Freitag's GAMESS implementation [16,17]. Reported values are departures from ^{13}C in methane and ^{17}O in water in ppm. Gaussian's HF/6-311G(d,p) value for C in methane is 197.5 ppm and for O in water is 255.1 ppm. The experimental value of the ^{17}O shift for formaldehyde was not well established at the time these calculations were reported, but is estimated now near 320 ppm.

As our mention of Freitag's recent installation of the Ditchfield formulation in GAMESS suggests, the method is still of central interest. In the next section we illustrate Freitag's implementation.

Ditchfield's Formulation in GAMESS

Mark Freitag [16] implemented Ditchfield's formalism [8] in the GAMESS software suite. His independent check of the pioneering work of Beveridge provides invaluable verification of the formulation. Commentary on the output appears in the following section.

* *

(Thanks to Ilhan Yavuz and the molecular physicists at Marmara University for this GAMESS result.)

```
---------------------- GAMESS execution script----------------------
This job is running on host altun
under operating system SunOS at Fri Jun 15 12:16:15 EEST 2007
```

```
**********************************************************
*           GAMESS VERSION = 22 FEB 2006 (R5)           *
*                FROM IOWA STATE UNIVERSITY             *
*   M.W.SCHMIDT, K.K.BALDRIDGE, J.A.BOATZ, S.T.ELBERT,  *
*     M.S.GORDON, J.H.JENSEN, S.KOSEKI, N.MATSUNAGA,    *
*           K.A.NGUYEN, S.J.SU, T.L.WINDUS,             *
*         TOGETHER WITH M.DUPUIS, J.A.MONTGOMERY        *
*            J.COMPUT.CHEM. 14, 1347-1363(1993)         *
***********  SUN MICROSYSTEMS INC. VERSION  ***********
```

SINCE 1993, STUDENTS AND POSTDOCS WORKING AT IOWA STATE UNIVERSITY
AND ALSO IN THEIR VARIOUS JOBS AFTER LEAVING ISU HAVE MADE IMPORTANT
CONTRIBUTIONS TO THE CODE:
IVANA ADAMOVIC, CHRISTINE AIKENS, YURI ALEXEEV, POOJA ARORA, ROB BELL,
PRADIPTA BANDYOPADHYAY, BRETT BODE, GALINA CHABAN, WEI CHEN,
CHEOL HO CHOI, PAUL DAY, TIM DUDLEY, DMITRI FEDOROV, GRAHAM FLETCHER,
MARK FREITAG, KURT GLAESEMANN, GRANT MERRILL, TAKESHI NAGATA,
HEATHER NETZLOFF, BOSILJKA NJEGIC, RYAN OLSON, MIKE PAK, JIM SHOEMAKER,
LYUDMILA SLIPCHENKO, JIE SONG, TETSUYA TAKETSUGU, SIMON WEBB.

ADDITIONAL CODE HAS BEEN PROVIDED BY COLLABORATORS IN OTHER GROUPS:
IOWA STATE UNIVERSITY: JOE IVANIC, KLAUS RUEDENBERG
UNIVERSITY OF TOKYO: KIMIHIKO HIRAO, HARUYUKI NAKANO, TAKAHITO
  NAKAJIMA, TAKAO TSUNEDA, MUNEAKI KAMIYA, SUSUMU YANAGISAWA,
  KIYOSHI YAGI
UNIVERSITY OF SOUTHERN DENMARK: FRANK JENSEN
UNIVERSITY OF IOWA: VISVALDAS KAIRYS, HUI LI
NATIONAL INST. OF STANDARDS AND TECHNOLOGY: WALT STEVENS, DAVID GARMER
UNIVERSITY OF PISA: BENEDETTA MENNUCCI, JACOPO TOMASI
UNIVERSITY OF MEMPHIS: HENRY KURTZ, PRAKASHAN KORAMBATH
UNIVERSITY OF ALBERTA: MARIUSZ KLOBUKOWSKI
UNIVERSITY OF NEW ENGLAND: MARK SPACKMAN
MIE UNIVERSITY: HIROAKI UMEDA
MICHIGAN STATE UNIVERSITY:
  KAROL KOWALSKI, MARTA WLOCH, PIOTR PIECUCH
UNIVERSITY OF SILESIA: MONIKA MUSIAL, STANISLAW KUCHARSKI
FACULTES UNIVERSITAIRES NOTRE-DAME DE LA PAIX:
  OLIVIER QUINET, BENOIT CHAMPAGNE
UNIVERSITY OF CALIFORNIA - SANTA BARBARA: BERNARD KIRTMAN
INSTITUTE FOR MOLECULAR SCIENCE: KAZUYA ISHIMURA AND SHIGERU NAGASE
UNIVERSITY OF NOTRE DAME: DAN CHIPMAN
KYUSHU UNIVERSITY:
  FENG LONG GU, JACEK KORCHOWIEC, MARCIN MAKOWSKI, AND YURIKO AOKI
PENNSYLVANIA STATE UNIVERSITY:
  TZVETELIN IORDANOV, CHET SWALINA, SHARON HAMMES-SCHIFFER

Runtype

The first part of the output is the usual GAMESS result for a fixed geometry
RHF calculation with a medium to large 6-311G(d,p) basis set. The key word
near the top of the data echo is "RUNTYP = NMR."

```
                    ECHO OF THE FIRST FEW INPUT CARDS -
INPUT CARD> $CONTRL SCFTYP = RHF RUNTYP = NMR COORD = ZMT $END
INPUT CARD> $SYSTEM TIMLIM = 400 MWORDS = 50 MEMDDI = 50 PARALL = .F. $END
INPUT CARD> $BASIS GBASIS = N311 NGAUSS = 6 NDFUNC = 1 NPFUNC = 1 DIFFSP = .T.
            $END
INPUT CARD> $DATA
INPUT CARD>
INPUT CARD>Cn 1
INPUT CARD>
INPUT CARD> c
INPUT CARD> h           1 hc2
INPUT CARD> o           1 oc3              2 och3
INPUT CARD> h           1 hc4              2 hch4        3 dih4
INPUT CARD>
INPUT CARD>hc2                   1.095011
INPUT CARD>oc3                   1.178326
INPUT CARD>och3             122.095
INPUT CARD>hc4                   1.094992
INPUT CARD>hch4             115.805
INPUT CARD>dih4             180.000
INPUT CARD> $end


RUN TITLE
---------


THE POINT GROUP OF THE MOLECULE IS CN
THE ORDER OF THE PRINCIPAL AXIS IS             1


YOUR FULLY SUBSTITUTED Z-MATRIX IS

C
H      1    1.0950110
O      1    1.1783260    2    122.0950
H      1    1.0949920    2    115.8050   3    180.0000    0


THE MOMENTS OF INERTIA ARE (AMU-ANGSTROM**2)

IXX = 1.734      IYY = 12.481       IZZ = 14.216
```

ATOM	ATOMIC	COORDINATES (BOHR)		
	CHARGE	X	Y	Z
C	6.0	−1.1129358377	0.0000055551	0.0000000000
H	1.0	−2.2123924613	−1.7530137097	0.0000000000
O	8.0	1.1137774265	0.0000029314	0.0000000000
H	1.0	−2.2125222262	1.7529010431	0.0000000000

Basis

Note that GAMESS uses six Cartesian GTO orbitals for the d-shell and as a result there will be one linear combination of the six d-like orbitals which forms an additional s-type orbital for each atom having d-orbitals.

ATOMIC BASIS SET

THE CONTRACTED PRIMITIVE FUNCTIONS HAVE BEEN UNNORMALIZED
THE CONTRACTED BASIS FUNCTIONS ARE NOW NORMALIZED TO UNITY

SHELL	TYPE	PRIMITIVE	EXPONENT	CONTRACTION	COEFFICIENT(S)
C					
1	S	1	4563.2400000	0.001966650249	
1	S	2	682.0240000	0.015230601932	
1	S	3	154.9730000	0.076126909656	
1	S	4	44.4553000	0.260801033080	
1	S	5	13.0290000	0.616462078191	
1	S	6	1.8277300	0.221006028032	
2	L	7	20.9642000	0.114660080729	0.040248692673
2	L	8	4.8033100	0.919999647749	0.237593956746
2	L	9	1.4593300	−0.003030682134	0.815853851473
3	L	10	0.4834560	1.000000000000	1.000000000000
4	L	11	0.1455850	1.000000000000	1.000000000000
5	L	12	0.0438000	1.000000000000	1.000000000000
6	D	13	0.6260000	1.000000000000	
H					
7	S	14	33.8650000	0.025493814541	
7	S	15	5.0947900	0.190373108582	
7	S	16	1.1587900	0.852161486043	
8	S	17	0.3258400	1.000000000000	
9	S	18	0.1027410	1.000000000000	
10	P	19	0.7500000	1.000000000000	
O					
11	S	20	8588.5000000	0.001895150083	
11	S	21	1297.2300000	0.014385900631	
11	S	22	299.2960000	0.070732003103	
11	S	23	87.3771000	0.240001010530	
11	S	24	25.6789000	0.594797026097	
11	S	25	3.7400400	0.280802012320	
12	L	26	42.1175000	0.113889012440	0.036511397380
12	L	27	9.6283700	0.920811100576	0.237152982984
12	L	28	2.8533200	−0.003274470358	0.819701941186
13	L	29	0.9056610	1.000000000000	1.000000000000
14	L	30	0.2556110	1.000000000000	1.000000000000
15	L	31	0.0845000	1.000000000000	1.000000000000
16	D	32	1.2920000	1.000000000000	
H					
17	S	33	33.8650000	0.025493814541	
17	S	34	5.0947900	0.190373108582	
17	S	35	1.1587900	0.852161486043	

18	S	36	0.3258400	1.000000000000
19	S	37	0.1027410	1.000000000000
20	P	38	0.7500000	1.000000000000

```
TOTAL NUMBER OF BASIS SET SHELLS            = 20
NUMBER OF CARTESIAN GAUSSIAN BASIS FUNCTIONS = 58
NUMBER OF ELECTRONS                         = 16
CHARGE OF MOLECULE                          =  0
SPIN MULTIPLICITY                           =  1
NUMBER OF OCCUPIED ORBITALS (ALPHA)         =  8
NUMBER OF OCCUPIED ORBITALS (BETA)          =  8
TOTAL NUMBER OF ATOMS                       =  4
THE NUCLEAR REPULSION ENERGY IS    31.8963171723
```

```
         ----------------------    -----------------------------
         RHF GIAO CHEMICAL SHIFTS   PROGRAM WRITTEN BY M.A.FREITAG
         ----------------------    -----------------------------
```

```
                    ------------------
                    NMR INPUT PARAMETERS
                    ------------------

         INMEM = F      ANGINT = T
          PDIA = F       PPARA = F       PEVEC = F
        POEINT = F      PTEINT = F      PRMAT = F       PITER = F
```

```
                    * * * * * * * * * * * * * * * * * * *
                    1 ELECTRON INTEGRALS
                    * * * * * * * * * * * * * * * * * * *

. . . . . . END OF ONE-ELECTRON INTEGRALS . . . . . .
STEP CPU TIME          = 0.00 TOTAL CPU TIME = 0.0 (0.0 MIN)
TOTAL WALL CLOCK TIME = 0.0 SECONDS, CPU UTILIZATION IS 100.00%
```

```
                    -------------
                    GUESS OPTIONS
                    -------------

         GUESS  = HUCKEL    NORB   = 0          NORDER = 0
         MIX    = F         PRTMO  = F          PUNMO  = F
         TOLZ   = 1.0E-08   TOLE   = 1.0E-05
         SYMDEN = F         PURIFY = F
```

```
INITIAL GUESS ORBITALS GENERATED BY HUCKEL ROUTINE.
HUCKEL GUESS REQUIRES 28096 WORDS.
```

```
SYMMETRIES FOR INITIAL GUESS ORBITALS FOLLOW. BOTH SET(S).
8 ORBITALS ARE OCCUPIED (2 CORE ORBITALS).
    3 = A      4 = A     5 = A      6 = A      7 = A      8 = A      9 = A
   10 = A     11 = A    12 = A     13 = A     14 = A     15 = A     16 = A
   17 = A     18 = A
```

```
. . . . . . END OF INITIAL ORBITAL SELECTION . . . . . .
STEP CPU TIME        = 0.01   TOTAL CPU TIME = 0.0 (0.0 MIN)
TOTAL WALL CLOCK TIME = 0.0 SECONDS, CPU UTILIZATION IS 100.00%


                    ---------------------
                    AO INTEGRAL TECHNOLOGY
                    ---------------------


   S,P,L SHELL ROTATED AXIS INTEGRALS, REPROGRAMMED BY
      KAZUYA ISHIMURA (IMS) AND JOSE SIERRA (SYNSTAR).
   S,P,D,L SHELL ROTATED AXIS INTEGRALS PROGRAMMED BY
      KAZUYA ISHIMURA (INSTITUTE FOR MOLECULAR SCIENCE).
   S,P,D,F,G SHELL TO TOTAL QUARTET ANGULAR MOMENTUM SUM 5,
      ERIC PROGRAM BY GRAHAM FLETCHER (ELORET AND NASA ADVANCED
      SUPERCOMPUTING DIVISION, AMES RESEARCH CENTER).
   S,P,D,F,G,L SHELL GENERAL RYS QUADRATURE PROGRAMMED BY
      MICHEL DUPUIS (PACIFIC NORTHWEST NATIONAL LABORATORY).


                    ------------------
                    RHF SCF CALCULATION
                    ------------------


   NUCLEAR ENERGY = 31.8963171723
   MAXIT = 30 NPUNCH = 2
   EXTRAP = T DAMP = F SHIFT = F RSTRCT = F DIIS = F DEM = F SOSCF = T
   DENSITY MATRIX CONV = 1.00E-05
   SOSCF WILL OPTIMIZE 400 ORBITAL ROTATIONS, SOGTOL = 0.250
   MEMORY REQUIRED FOR RHF STEP = 55960 WORDS.


ITER EX DEM   TOTAL ENERGY       E CHANGE      DENSITY CHANGE   ORB. GRAD
  1   0  0  -113.5000653633  -113.5000653633   0.251154270    0.000000000

---------------------- START SECOND ORDER SCF ----------------------
 11  10  0  -113.9029590480    -0.0000000001   0.000001240    0.000000520


                    ----------------
                    DENSITY CONVERGED
                    ----------------

      TIME TO FORM FOCK OPERATORS = 0.5 SECONDS (0.0 SEC/ITER)
      TIME TO SOLVE SCF EQUATIONS = 0.0 SECONDS (0.0 SEC/ITER)

FINAL RHF ENERGY IS -113.9029590480 AFTER 11 ITERATIONS


                    -----------
                    EIGENVECTORS
                    -----------
```

Already Familiar!

```
. . . . . . END OF RHF CALCULATION . . . . . .
STEP CPU TIME        = 0.52   TOTAL CPU TIME = 0.7 (0.0 MIN)
TOTAL WALL CLOCK TIME = 0.0 SECONDS, CPU UTILIZATION IS 100.00%


          ------------------------------------------------------------
          PROPERTY VALUES FOR THE RHF   SELF-CONSISTENT FIELD WAVEFUNCTION
          ------------------------------------------------------------


                         ----------------
                         ENERGY COMPONENTS
                         ----------------

          WAVEFUNCTION NORMALIZATION =    1.0000000000
                 ONE ELECTRON ENERGY = -218.4605107846
                 TWO ELECTRON ENERGY =   72.6612345642
             NUCLEAR REPULSION ENERGY =   31.8963171723
                                      ----------------
                        TOTAL ENERGY = -113.9029590480

   ELECTRON-ELECTRON POTENTIAL ENERGY =   72.6612345642
    NUCLEUS-ELECTRON POTENTIAL ENERGY = -332.3390216075
    NUCLEUS-NUCLEUS POTENTIAL ENERGY =   31.8963171723
                                      ----------------
             TOTAL POTENTIAL ENERGY = -227.7814698710
               TOTAL KINETIC ENERGY =  113.8785108229
                 VIRIAL RATIO (V/T) =    2.0002146869

          ---------------------------------------
          MULLIKEN AND LOWDIN POPULATION ANALYSES
          ---------------------------------------
```

Charge Analysis

In the split and polarized basis, the Mulliken population analysis is more difficult to untangle; the total atom populations for each atom are more meaningful than the individual basis function populations. The dipole moment of 2.8 debye is high, as is common for RHF dipole moments.

```
            TOTAL MULLIKEN AND LOWDIN ATOMIC POPULATIONS
        ATOM   MULL.POP.    CHARGE     LOW.POP.    CHARGE
     1    C    5.891157    0.108843   5.933045   0.066955
     2    H    0.905386    0.094614   0.914775   0.085225
     3    O    8.298079   -0.298079   8.237408  -0.237408
     4    H    0.905378    0.094622   0.914772   0.085228


                    --------------------
                    ELECTROSTATIC MOMENTS
                    --------------------
```

```
POINT     1              X            Y        Z (BOHR)       CHARGE
                     0.000000    -0.000000   0.000000      -0.00 (A.U.)

           DX          DY           DZ       /D/ (DEBYE)
        -2.806992   0.000033    0.000000     2.806992
```

```
. . . . . . END OF PROPERTY EVALUATION . . . . . .
STEP CPU TIME        =0.01  TOTAL CPU TIME =0.7 (0.0 MIN)
TOTAL WALL CLOCK TIME=0.0 SECONDS, CPU UTILIZATION IS 100.00%
NMR CHEMICAL SHIFT CALCULATION REQUIRES 5902893 WORDS OF MEMORY.
```

```
CALCULATING GIAO ONE-ELECTRON INTEGRALS USING MCMURCHIE-DAVIDSON
   METHOD . . .
STEP CPU TIME       =0.27  TOTAL CPU TIME =1.0 (0.0 MIN)
TOTAL WALL CLOCK TIME=1.0 SECONDS, CPU UTILIZATION IS 100.74%
```

Timing

The RHF calculation takes a much smaller time than the NMR shielding calculation printed near the end of the output. Considering that the CPHF equations required new integrals and had to be solved more than once to fill out the various tensor components, it is not surprising that this calculation takes longer than the RHF step.

```
CALCULATING GIAO TWO-ELECTRON INTEGRALS USING MCMURCHIE-DAVIDSON METHOD
CHOOSING OUT OF MEMORY OPTION USING TWO INTEGRAL PASSES . . .
   WORKING ON (^I J| K L) INTEGRALS . . .
   WORKING ON ( I J|^K L) INTEGRALS . . .
STEP CPU TIME         =151.51  TOTAL CPU TIME=152.5 (2.5 MIN)
TOTAL WALL CLOCK TIME=152.0 SECONDS, CPU UTILIZATION IS 100.34%
```

NMR Report

The output acknowledges the work of Dr. Mark Freitag [16] who adapted the Ditchfield GIAO routines. The full 3×3 shielding tensors are reported, as well as the isotropic average. In liquids (and more rarely, gases) one can measure only the average owing to raid rotational motion. Experimental spectra are characterized by chemical shifts, which we can estimate as the difference between the absolute isotropic shielding for a specific nucleus and the corresponding value for the value computed for ^{13}C, ^{17}O, or ^{1}H in standard systems such as methane for C or H, water for O. For experimental convenience the standard tetramethylesilane (TMS) is often employed. Not illustrated in this report, but a key part of the GAMESS implementation, is the capability to estimate solvent effects.

```
                    GIAO CHEMICAL SHIELDING TENSOR (PPM):

                        X             Y             Z        ISOTROPIC
                                                             SHIELDING
                                                             (ANISOTROPY)

    1  C   X       -1.9400        0.0021        0.0000
           Y        0.0029      -93.8984        0.0000
           Z        0.0000        0.0000      123.9437
                                                             9.3684
    EIGENVALS:      -1.9400      -93.8984      123.9437
                                                             (171.8629)

    2  H   X       23.6306        1.5445        0.0000
           Y       -1.5993       22.9476        0.0000
           Z        0.0000        0.0000       23.2702
                                                             23.2828
    EIGENVALS:      23.2891       23.2891       23.2702
                                                             (0.0095)

    3  O   X    -1119.9796        0.0297        0.0000
           Y        0.0058     -511.5728        0.0000
           Z        0.0000        0.0000      427.1020
                                                            -401.4835
    EIGENVALS:   -1119.9796     -511.5728      427.1020
                                                            (1242.8782)

    4  H   X       23.6297       -1.5434        0.0000
           Y        1.6001       22.9479        0.0000
           Z        0.0000        0.0000       23.2709
                                                             23.2828
    EIGENVALS:      23.2888       23.2888       23.2709
                                                             (0.0089)

. . . . . DONE WITH NMR SHIELDINGS. . . . .
STEP CPU TIME        = 9.76  TOTAL CPU TIME = 162.3 ( 2.7 MIN)
TOTAL WALL CLOCK TIME = 162.0 SECONDS, CPU UTILIZATION IS 100.17%
```

Ditchfield's Formulation in GAUSSIAN-03

The Ditchfield formulation is also the foundation for the implementation of NMR shielding tensor calculation in GAUSSIAN-03 [18]. We will explore the capabilities of this implementation in later sections. Here we merely point out that advances in computer capacity and coding have permitted much larger calculations than could be attempted in the 1970s. Enhancing the basis seems to have a substantial effect. Within the HF framework even small changes in geometry and basis seem to have a serious impact on values of the isotropic average (i.e., observable) shielding (Table 12.2).

Use of Localized Orbitals in NMR-Shielding Calculations

The methods used by Ditchfield were successful in predicting gross features of the chemical shifts even in modest basis sets. His conception survives in

TABLE 12.2

Absolute Shielding Constant (isotropic average) for C and O

Model Chemistry: Nucleus	HF/6-311G(d,p) GAMESS	HF/6-311G(d,p) GAUSSIAN-03	HF/6-311++G(2df,2p) GAUSSIAN-03
Formaldehyde C	9.37	11.14	7.11
Formaldehyde O	−401.48	−421.35	−388.12

the organization of most implementations, including those in GAMESS [17], GAUSSIAN [18], and SPARTAN [19]. Before the Ditchfield method gained such wide acceptance, however, it seemed that its computational demands would inhibit its use in molecules large enough to be of chemical interest. Here we describe two methods intended to overcome the size obstacle by taking advantage of the simplifications of orbital localization.

Individual Gauge for Local Orbitals

Chemists, of course, have recognized that large molecules are often well represented as clusters of identifiable fragments, bonds, lone pairs, and cores. Localized orbitals resemble these chemical constructs. Schindler and Kutzelnigg [20,21] developed a method in which orbitals localized by the Boys criterion [22], i.e., transformed so to maximize separation of orbital centroids, played a key role. They defined a gauge origin specific to each local orbital in their protocol, which they termed individual gauge localized orbitals (IGLO). The immediate advantage of the IGLO formulation over use of any common gauge origin is that two major difficulties of the Ramsey theory [1,2] are largely overcome. These are (first) the near-cancelation of the oppositely signed diamagnetic and paramagnetic contributions to the shielding, which makes accurate estimates of the result difficult, and (then) the relative inaccuracy of the estimate of the paramagnetic part compared to the diamagnetic component of the shielding. The authors arrive at the expression for the spatially averaged chemical shift σ_B for nucleus B at position $\mathbf{R}_B$:

$$\sigma = 2\sum_k \frac{e}{mc^2}\left\langle k\left| \frac{[\mathbf{B} \times (\mathbf{r} - \mathbf{R}_k)] \cdot [\boldsymbol{\mu} \times (\mathbf{r} - \mathbf{R}_B)]}{|\mathbf{r} - \mathbf{R}_B|^3}\right| k\right\rangle$$
$$- \frac{2e}{mci}\left\langle k\left| \frac{\boldsymbol{\mu} \times (\mathbf{r} - \mathbf{R}_B) \cdot \mathbf{p}}{|\mathbf{r} - \mathbf{R}_B|^3}(1 - P_0)\right| k'\right\rangle$$
$$- \frac{e}{mci}\sum_{j,k}\left\langle k\left| \frac{\boldsymbol{\mu} \times (\mathbf{r} - \mathbf{R}_B) \cdot \mathbf{p}}{|\mathbf{r} - \mathbf{R}_B|^3}\right| j\right\rangle \langle j|\mathbf{A}_j - \mathbf{A}_k|k\rangle$$

Here $\mathbf{R}_k$ refers to the centroid of the kth (occupied) localized MO, and reference to any gauge origin is removed. That is, the quantities $\mathbf{A}_j - \mathbf{A}_k$ are defined entirely by differences in gauge origins.

TABLE 12.3

Basis Set Dependence of Isotropic Average of the Shielding Tensor

Species: Basis	DZ	I	II	III	IV	Exp.
Methane (C)	222.2	195.9	194.8	196.4	193.7	197.35
Methane (H)	44.2	27.5	24.7	24.6	19.9	18.7

TABLE 12.4

IGLO Estimates of Shieding Constants for Formaldehyde

Species		Exp. (error)
Formaldehyde (C)	−10.4	−1 (±10)
Formaldehyde (O)	−456.1	−375 (±150)

Note: TZP basis in Huzinaga form 1-172 (4, 7 × 1; 2, 5 × 1; 1, 1) for C and O; 52 (2, 3 × 1; 1, 1).

Using contracted Huzinaga sets (roughly, DZ and a range of TZP sets), Schindler and Kutzelnigg evaluated magnetic properties for a range of small molecules. It appears that any TZP basis gives reasonable values for heavy atoms, but protons are more demanding of the basis (Tables 12.3 and 12.4).

One of the appealing results of the Schindler–Kutzelnigg IGLO method is that the diamagnetic terms for each bond appeared to be transferable between similar environments. That is, carbon atoms in CC single bond in ethane would have diamagnetic shielding similar to that of carbon atoms in a CC single bond in a much larger molecule. Although desirable, transferability is not general, as a chemist would immediately expect. For example, the pi antibonding level in formaldehyde made the isotropic average of the shielding tensor for C and O in that molecule strikingly different, and highly anisotropic compared with C and O in reference species methane and water. The main variation between chemical environments was found to be in the paramagnetic term [20,21].

RPAC Program (LORG)

One serious complication in the methods developed by Ditchfield [8] and by Schindler and Kutzelnigg [20,21] relates to the use of the complex gauge factor used to form the GIAO basis or the IGLO molecular orbitals. In 1985, Hansen and Bouman [23,24] published a method called RPAC capable of computing ^{13}C chemical shifts which did not require such intensive complex arithmetic. In place of the CPHF procedure Hansen and Bouman employed the random phase approximation within linear response theory as developed by Dunning and McKoy [25]. This method represents responses to the electromagnetic perturbation in a way that incorporates some double excitations, so it can be considered a step beyond CIS. They applied this approximation to the paramagnetic terms of the Ramsey equation [1,2], using localized real orbitals.

The Hansen–Bouman method evaluates the Ramsey shielding terms.

$$\sigma_{ij}^{tot} = \sigma_{ij}^{d} + \sigma_{ij}^{p}$$

They expressed the diamagnetic components as

$$\sigma_{ij}^{d} = (1/2c^2)\Big\langle 0 \Big| \sum_{B} [\boldsymbol{\varepsilon}_j \times (\mathbf{r}_B - \mathbf{R})] \bullet [\boldsymbol{\varepsilon}_i \times \mathbf{r}_B] r_B^{-3} \Big| 0 \Big\rangle$$

Here the centroid of the local orbital $\mathbf{R}$ is the origin of the gauge and $\mathbf{r}_B$ is the position of nucleus B. The unit vectors $\boldsymbol{\varepsilon}_i$ and $\boldsymbol{\varepsilon}_j$ are directed along the nuclear magnetic moment and the applied magnetic field, respectively. In due course an average overall orientation of $\boldsymbol{\varepsilon}_i$ and $\boldsymbol{\varepsilon}_j$ must be accomplished. The paramagnetic contribution to the shielding tensor is

$$\sigma_{ij}^{p} = -(1/c^2)\sum_{q}\Big[\boldsymbol{\varepsilon}_i \bullet \Big\langle 0 \Big| \sum_{s} \mathbf{L}_s r_s^{-3} \Big| q \Big\rangle \Big\langle q \Big| \boldsymbol{\varepsilon}_j \bullet \sum_{s} \mathbf{L}'_s \Big| 0 \Big\rangle\Big](E_q - E_0)^{-1}$$

This expression employs atomic units ($c = 137$). $\mathbf{L}_s = \mathbf{r}_s \times \mathbf{p}_s$; the unprimed angular momentum refers to an electron labeled s, relative to the nucleus for which the shift is being calculated. $\mathbf{L}'_s = \mathbf{r}_s - \mathbf{R} \times \mathbf{p}_s$ is the angular momentum with respect to a gauge origin at $\mathbf{R}$. The energies E_q are from the RPA calculation of electronic states, which includes an approximate appreciation of correlation in ground and excited states. The term "0" refers to the ground state. Again this is the formula for the (i,j) part of the tensor.

The wave function is assumed to take the form of a restricted Slater determinant; thus, the matrix elements of any of the one-electron operators (such as those in the definition of the shielding) connecting the ground determinant with a singlet excitation removing an electron from the ith occupied orbitals and placing it in the pth virtual orbital of the HF manifold are reduced in a familiar way to

$$\langle 0|\hat{f}|\psi_i^p\rangle = \sqrt{2}\langle i|\hat{f}|p\rangle$$

The Hansen–Bouman method localizes the occupied molecular orbitals by the Boys method [22], which relies on the fact that any unitary transform of the orbitals in the HF determinant leaves the wave function unaltered.

Each of these orbitals ψ_p has a well-defined centroid position, $\mathbf{R}_p$. The assignment of the gauge origins to the centroids of Boys-localized MOs is the source of the name of the "Local ORigin Gauge" (LORG) method. Reference to the new gauge origin is accomplished through the expansion

$$\mathbf{L} = (\mathbf{r} - \mathbf{R}_p) \times \mathbf{p} + (\mathbf{R}_p - \mathbf{R}) \times \mathbf{p}$$

Here the first term is the angular momentum relative to $\mathbf{R}_p$ and the second is a simple correction. We do not trace the steps by which Bouman and Hansen rearrange the RPA equations to isolate the references to the shift in gauge origin; they find

$$\sigma_{ij}^d = (1/c^2) \sum_p \left\langle p \left| [\varepsilon_j \times (\mathbf{r} - \mathbf{R}_p)] \bullet [\varepsilon_i \times \mathbf{r}] r^{-3} \right| p \right\rangle$$
$$+ (1/c^2) \sum_p \left\langle p \left| [\varepsilon_j \times (\mathbf{R}_p - \mathbf{R})] \bullet [\varepsilon_i \times \mathbf{r}] r^{-3} \right| p \right\rangle$$

The sum extends over the set of occupied localized orbitals. Writing

$$\sigma_{ij}^d \equiv \sigma_{ij}^d(1) + \sigma_{ij}^d(2)$$

we see that only the second term refers to the shift. Manipulation of the paramagnetic term (which incorporates the RPA state coefficients) produces a similar separation into two terms. The first is the RPA expression for the paramagnetic term of the Ramsey formulation and the second is the part that depends on the centroid positions of the localized orbitals.

$$\sigma_{ij}^p = \sigma_{ij}^p(1) + \sigma_{ij}^p(2)$$

Remarkably, gauge-dependent terms of the diamagnetic and paramagnetic terms can be combined in a way which leads to cancelation of dependence on any arbitrary gauge origin, requiring only distances between the centroids of the Boys orbitals.

$$\sigma_{ij}^d(2) + \sigma_{ij}^p(2) = (1/c^2) \sum_p \sum_q [\varepsilon_i \bullet \langle p | \mathbf{L}/r^3 | q \rangle] \times [\varepsilon_j \bullet \{(\mathbf{R}_\alpha - \mathbf{R}_\beta) \times \langle q | \mathbf{r} | p \rangle\}]$$

Likewise the authors' final expression for isotropic shielding depends on the positions of the centroids of the localized orbitals, but makes no reference to any unique gauge origin.

$$\sigma^{\text{Isotropic}} = (2/3c^2) \sum_p \langle p | \mathbf{r} \bullet (\mathbf{r} - \mathbf{R}_p) | p \rangle - (i/3c^2) \sum_{p \neq q} (\mathbf{R}_p - \mathbf{R}_q) \bullet \left\langle p \left| \frac{\mathbf{L}}{r^3} \right| q \right\rangle \times \left\langle q \left| \frac{\mathbf{L}}{r^3} \right| p \right\rangle$$

$$- (2/3c^2) \sum_p \sum_{mn} \left\langle p \left| \frac{\mathbf{L}}{r^3} \right| m \right\rangle [\mathbf{A} - \mathbf{B}]^{-1}_{pm,pn} \left\langle n \left| \mathbf{L}^{(p)} \right| p \right\rangle$$

$$- (2/3c^2) \sum_{p \neq q} \sum_{mn} \left\langle p \left| \frac{\mathbf{L}}{r^3} \right| m \right\rangle [\mathbf{A} - \mathbf{B}]^{-1}_{pm,qn} \langle n | \mathbf{L}^{(p)} | q \rangle$$

Here p and q are occupied (localized) MOs and m and n are virtual orbitals. The four terms are, respectively, one- and two-center contributions to the diamagnetic shielding and one- and two-center contributions to the paramagnetic shielding. The paramagnetic terms require quantities constructed during the RPA calculation:

$$(\mathbf{A} - \mathbf{B})_{pm,qn} = F_{mn}\delta_{mn} - F_{pq}\delta_{pq} + (mq|pn) - (nm|pq)$$

Here F_{pq} are elements of a Fock matrix and the two-electron integrals are expressed in the charge-density (Mulliken) order. The inversion of the matrix $\mathbf{A} - \mathbf{B}$ can be accomplished by diagonalizing it, inverting the diagonal elements, and back-transforming the result.

Bouman and Hansen applied their method to a variety of small molecules ranging in size to benzene and pyridine, using basis sets called A and B, of DZ and TZP quality, respectively. The paramagnetic term is more sensitive to basis quality. Use of TZ valence shell splitting seems to be as important as any degree of polarization. One major advantage of the expression in local orbitals and the choice of gauge origin at local orbital centroids is that long-range shieldings (terms with $p \neq q$) are well represented. This is significant for the paramagnetic terms (these have important contributions from interactions between different bonds), while the diamagnetic terms are insensitive to choice of gauge origin. Their ^{13}C shifts match observed values within MAD ca. 5 ppm in the TZP basis. This level of accuracy is experimentally useful as an aid in assignments. We quote their results for the ^{13}C nucleus in formaldehyde (Table 12.5).

TABLE 12.5

LORG Values for Absolute Shielding Constants (ppm)

Model Chemistry: Nucleus and Reference	HF/DZ LORG	HF/TZP LORG
C (formaldehyde, relative to CH_4)	204	192
C (formaldehyde) absolute	−11	4
C (methane) absolute	193	196

Subtleties of Orbital Localization

Hansen and Bouman [23,24] analyzed the impact of the orbital localization on the quality of shielding estimates. Ramsey [1,2] noted that the diamagnetic and paramagnetic terms in the shielding made differing demands on the quality of a basis set. Errors in the paramagnetic term are more serious than errors in the diamagnetic term owing to the long range $(1/R)$ dependence of the contributions to the paramagnetic term. This is exaggerated for small basis sets. If localized orbitals are used, the contributions to paramagnetic coupling seem to decline more drastically with distance; according to Hansen and Bouman, as $1/R^2$.

The RPAC program [24] was designed to be used following a GAMESS [16] calculation and is still available as of this writing from Prof. Hansen at Copenhagen University in Denmark via aage.hansen@post.cybercity.dk. Copies of earlier versions, QCPE No. 459 and No. 556, may also be available from the Quantum Chemistry Program Exchange at the chemistry department of Indiana University by special request.

The Hansen–Bouman LORG method does not use complex GIAO orbitals and still leads to quite good results. Its natural interface to GAMESS means that it could be adapted to other programs. Because renewed use of RPAC may reopen the discussion of the relation between IGLO and LORG formulations we pass along a clarification from Prof. Hansen [26] of the relation between LORG and IGLO given in a summary paper [27].

The shielding tensor may be written

$$
c^2\sigma_{ij} = c^2\sigma_{ij}(\text{CHF, } \mathbf{R} = 0) - \sum_p \left\langle p \left| \mathbf{R}_p \bullet \mathbf{r}\delta_{ij} - [\mathbf{u}_i \bullet \mathbf{R}_p][\mathbf{u}_j \bullet \mathbf{r}]r^{-3} \right| p \right\rangle
$$

$$
+ i\sum_{pq} \left[\mathbf{u}_i \bullet \left\langle p \left| \frac{\mathbf{L}}{r^3} \right| q \right\rangle \right] [\mathbf{u}_j \bullet (\mathbf{R}_p - \mathbf{R}_q) \times \langle q|\mathbf{r}|p\rangle]
$$

$$
- 2i\sum_{p,m} \left[\mathbf{u}_i \bullet \left\langle p \left| \frac{\mathbf{L}}{r^3} \right| m \right\rangle \right] [\mathbf{u}_j \bullet \mathbf{R}_p \times \langle m|\mathbf{Z}|p\rangle]
$$

The leading term in this expression refers to the value of the shielding obtained by coupled Hartree–Fock description of the perturbation with a gauge origin at $\mathbf{R} = 0$; the centroids of local MOs are called $\mathbf{R}_p$, and the electronic coordinate is $\mathbf{r}$. The occupied (local) orbitals are labeled p, q, and virtual orbitals are m, n. $\mathbf{L}$ is the angular momentum operator.

The difference between IGLO and LORG resides in the definition of $\mathbf{Z}$, a transition moment. In LORG

$$
\sum_{p,m} (\mathbf{A} - \mathbf{B})_{qn,pm}\langle p|\mathbf{Z}|m\rangle = i\langle q|\mathbf{p}|n\rangle
$$

where $\mathbf{p}$ is the momentum operator (i.e., the dipole velocity) and $\mathbf{A}$ and $\mathbf{B}$ are the RPA matrices defined above. In IGLO

$$
\langle m|\mathbf{Z}|p\rangle = \langle m|\mathbf{r}|p\rangle
$$

In a complete basis set the two definitions for **Z** agree. Also in that limit

$$\sum_{m=\text{virtual}} |m\rangle\langle m| = 1 - \sum_{p=\text{occ}} |p\rangle\langle p|$$

Imposing this condition leads to cancellation of all terms but the first with **R** = 0 in the definition of the shielding tensor; i.e., in the complete basis the gauge origin is not an issue. Therefore, we can look on the terms beyond that first one as approximate corrections for basis set inadequacy. Hansen and Bouman distinguish "FULL LORG" (in which all orbitals are local and gauge origin for each is at that orbital's centroid) from "LORG" (in which the gauge origin is at **R** = 0 for local orbitals bonding to the magnetic nucleus in question). In small basis sets LORG's performance is slightly superior to FULL LORG, apparently a consequence of slightly different estimates of close-range contributions to the shielding. In both LORG and IGLO the long-range basis set errors noted by Ramsey are damped as R^{-2}, according to Hansen.

Correlation Corrections and NMR Shielding: DFT

All the examples we have quoted so far have been based on the Hartree–Fock model, with the exception of Bouman and Hansens' LORG method which employed the RPA approximation for excitation energies. We might expect that correlation-corrected theory might produce improved descriptions of shielding. Correlation correction can be obtained by perturbation expansions such as MP2 [28] and the coupled cluster hierarchy [29,30], or within the density functional theory [31–33]. Before describing such theories in detail, we give a brief summary of the results for formaldehyde. The first section of Table 12.6 displays shielding constants from a calculation in the Hartree–Fock formalism which by definition makes no correlation correction, a calculation in PBE's "pure" density functional scheme, and a calculation in the popular hybrid method B3LYP which incorporates a degree of exact (HF) exchange with Becke's density functional exchange and the Lee–Yang–Parr correlation functional. The basis is 6-311++G(2df,2p) in each case; geometry optimization is done in the model chemistry.

The density functional estimates of shielding constants are consistently shifted to the negative of the HF values. In contrast the perturbation-based methods have little impact on the values of ^{13}C and ^{17}O shielding constants. The CCSDT results from Bartlett's ACES-II [34] are probably the best available, but surprisingly its results are only slightly different from the HF values. The experimental value for ^{13}C in formaldehyde is -4.4 ± 3 ppm. The ^{17}O value is not so well defined. These calculations take no account of vibrational and temperature effects. The sensitivity of shielding constants to the quality of the geometry suggests that vibrational effects may be significant.

TABLE 12.6

DFT Values of Absolute Shielding Constants for C and O

Model: Nucleus	HF/6-311++G(2df,2p) GAUSSIAN-03	B3LYP6-311++G(2df,2p) GAUSSIAN-03	PBE1PBE6-311++G(2df,2p) GAUSSIAN-03
C	7.11	−15.85	−11.26
O	−388.12	−425.64	−420.32

Model: Nucleus	HF/6-311++G(2df,2p) GAUSSIAN-03	MP26-311++G(2df,2p) GAUSSIAN-03	CCSDT Near-Complete Basis ACES-II
C	7.11	9.37	4.03
O	−388.12	−406.79	−383.47

With this first impression, we turn to a widely tested implementation of the formalism devised by Ditchfield.

GAUSSIAN-03 NMR Shifts: DFT-Based Values of Shielding

Cheeseman et al. [33] described a variant of Ditchfield's formulation, but extended to accommodate a choice of density functional methods. They present a compact formalism as developed by Johnson and Frisch [35] in which the HF calculation of the shielding tensor appears as a limiting case. The energy is written generally as

$$E = (\mathbf{hP}) + (1/2)(\mathbf{PG}_{2e}) + \varepsilon_{XC} + V$$

The notation is compressed, so that

$$(\mathbf{hP}) = \sum_{\mu\nu} h_{\mu\nu} P_{\mu\nu}$$

Here $\mathbf{h}$ is the core Hamiltonian, $\mathbf{P}$ is the density matrix, ε_{XC} is the exchange–correlation functional, and V is the nuclear repulsion. $\mathbf{G}_{2e}$ is a specially defined two-electron matrix

$$G_{2e}(P)_{\mu\nu} = \sum_{\lambda} \sum_{\sigma} P_{\lambda\sigma}(\mu\nu \parallel \lambda\sigma)$$

The round-bracket notation is explicitly

$$(\mu\nu \parallel \lambda\sigma) = \int \chi_\mu^*(1)\chi_\lambda^*(2) \frac{1}{r_{12}} [\chi_\nu(1)\chi_\sigma(2) - C_{HFX}\chi_\sigma(1)\chi_\nu(2)] d\tau_1 d\tau_2$$

The constant C allows variable admixture of the "exact exchange," and varies from zero if the exchange is defined entirely by ε_{XC} to unity for full HF exchange. Intermediate values define hybrid methods such as B3LYP. (This is discussed more thoroughly in our chapter on density functional theory.) The effective Fock matrix is defined by

$$F_{\mu\nu} = \frac{\partial E}{\partial P_{\mu\nu}} = h_{\mu\nu} + [G_{2e}(\mathbf{P})]_{\mu\nu} + [G_{XC}]_{\mu\nu}$$

The general exchange–correlation density is defined indirectly. The exchange–correlation energy is defined by

$$\varepsilon_{XC} = \int f(\rho, \gamma) \, d\mathbf{r}$$

with reference to densities for α and β spins

$$\rho_\alpha = \sum_\mu \sum_\nu P^\alpha_{\mu\nu} \chi_\mu \chi_\nu \quad \text{and} \quad \rho_\beta = \sum_\mu \sum_\nu P^\beta_{\mu\nu} \chi_\mu \chi_\nu$$

and associated gradients

$$\gamma_{\alpha\alpha} = \nabla \rho_\alpha \cdot \nabla \rho_\alpha \quad \text{and} \quad \gamma_{\alpha\beta} = \nabla \rho_\alpha \cdot \nabla \rho_\beta$$

Finally

$$(G^\alpha_{XC})_{\mu\nu} = \int \left[\frac{\partial f}{\partial \rho_\alpha} \chi_\mu^* \chi_\nu + \left(2 \frac{\partial f}{\partial \gamma_{\alpha\alpha}} \nabla \rho_\alpha + \frac{\partial f}{\partial \gamma_{\alpha\beta}} \nabla \rho_\beta \right) \cdot \nabla(\chi_\mu^* \chi_\nu) \right] d\mathbf{r}$$

There is a corresponding expression for $(G^\beta_{XC})_{\mu\nu}$. If the function $f = 0$ and $C = 1$ the Hartree–Fock expressions are recovered. The advantage of this formulation is that much of the coding is common to both HF and a broad range of DFT methods.

CSGT Method (Continuous Set of Gauge Transformations)

Keith and Bader [36] work within the frame of density functional theory, in which numerical methods for integration of the complicated functionals are of central importance. Their shielding formula refers to the current density induced by an external magnetic field. The shielding at a nucleus N is given by

$$\sigma^N_{ji} = \frac{\partial^2 E}{\partial B_i \partial m_{Nj}} = -\frac{1}{Bc} \int d\mathbf{r}_N \left[\mathbf{r}_N \times \mathbf{J}_i^{(1)}(\mathbf{r}_N)/r_N^3 \right]_j$$

Here B is the magnitude of the field, c is the speed of light, and $\mathbf{r}_N$ is the distance of the electron from nucleus N. The idea of the current $\mathbf{J}$ was adopted in the early discussion of aromatic molecules for which one assumed that circulation of pi electrons defined "ring currents" [37]. The Keith–Bader method might be considered to treat all electrons' contribution to the currents around nuclei.

Using a single-determinant function composed of Kohn–Sham molecular orbitals

$$\varphi_i = \sum_\mu c_{\mu i} \chi_\mu$$

Keith and Bader [36,38] introduce perturbations $V(\mathbf{L})$ and $V(\mathbf{p})$ arising from angular and linear momentum

$$V(L_i) = -\frac{i}{2c}(\mathbf{r} \times \nabla)_i$$

$$V(p_i) = -\frac{i}{2c}\nabla_i$$

According to CPHF one defines increments in the mixing coefficients $\mathbf{C^L}$ and $\mathbf{C^P}$ induced by the perturbations, which alter the MOs:

$$\psi_i^L = \sum_\mu C_{\mu i}^L \chi_\mu$$

$$\psi_i^p = \sum_\mu C_{\mu i}^p \chi_\mu$$

The first-order current is

$$\mathbf{J}^{(1)}(\mathbf{r}) = (1/c) \sum_{i-1}^{n/2} B\left[\psi_i^* \hat{\mathbf{p}} \psi_i^{L_x} + \psi_i^{L_x^*} \hat{\mathbf{p}} \psi_i - d_y(\mathbf{r})\left(\psi_i^{p_z} + \psi_i^{p_z^*} \hat{\mathbf{p}} \psi_i\right)\right.$$
$$\left. + d_z(\mathbf{r})\left(\psi_i^* \mathbf{p} \psi_i^{p_y} + \psi_i^{p_y^*} \mathbf{p} \psi_i\right)\right] - B(\mathbf{r} - \mathbf{d})\rho(\mathbf{r})$$

The momentum operator is

$$\hat{\mathbf{p}} = -i\nabla$$

The function $\mathbf{d}(\mathbf{r})$ defines a separate gauge for each point $\mathbf{r}$ in the DFT grid. If $\mathbf{d}$ is a constant, a single origin is defined for the gauge. Keith and Bader use the exponential form

$$\mathbf{d}(\mathbf{r}) = \mathbf{r} - \sum_\Omega (\mathbf{r} - \mathbf{R}_\Omega) \exp\left(-\alpha_\Omega[\mathbf{r} - \mathbf{R}_\Omega]\right)$$

TABLE 12.7

Isotropic Shielding for C, H, and O for Formaldehyde

Model	Single Ref.	IGAIM	CSGT	GIAO
HF				
C	6.35	1.33	1.32	7.11
O	−414.52	−413.01	−413.02	−388.12
H	21.74	22.58	22.58	22.95
B3LYP				
C	−9.30	−14.04	−14.05	−11.26
O	−422.44	−421.14	−421.15	−454.17
H	20.74	21.56	20.91	21.46
PBE1PBE				
C	−5.76	−10.51	−10.52	−15.85
O	−423.08	−421.77	−421.78	−425.64
H	20.63	21.42	21.42	20.53

Note: Basis 6-311++G(2df,2p).

The Keith–Bader CSGT method is an elaboration of the assumption of the "individual gauges for atoms in molecules"—IGAIM—which uses a distinct nuclear-centered gauge origin for each atom [38]. As we can see from our GAUSSIAN-03 calculations on formaldehyde, IGAIM and CSGT give virtually indistinguishable results (Table 12.7).

Practical Use of Calculated Chemical Shifts

Two questions must present themselves now: Can computational methods for shielding tensors (and more immediately to the purpose, chemical shifts) be applied to real problems? If so, is there a preferred method?

Cheeseman et al. [33] compared GIAO and CGST descriptions of shielding, testing HF and MP2 model chemistries as well as several DFT functionals including LSDA, BPW91, BLYP, B3PW91, and B3LYP. The purpose was to identify the least demanding calculational method(s) capable of providing guidance to the problem of assignments of the observed NMR C-13 signal. It appears that quite large basis sets—6-311+G(2d,p) for example—are required for numerical accuracy in absolute values of the isotropic shielding. Larger basis sets—quadruple zeta with double polarization or better—seem to afford little advantage.

Experimental values are invariably chemical shifts relative to a standard (tetramethylsilane for ^{13}C). At a common MP2/TZ2P optimized geometry, with a QZ2P basis for the single-point MNR calculation, Cheeseman et al. found that pure DFT and hybrid DFT models gave closely comparable isotropic averaged shielding values and chemical shifts for ^{13}C. HF and MP2 showed a degree of inconsistency with experimental values for shielding in

TABLE 12.8

Effect of Functional Choice on Chemical Shifts

Molecule	Nucleus	HF	LSDA	BPW91	BLYP	B3PW91	B3LYP	MP2	Exp.
C_2H_6	C	11.7	17.0	16.1	17.9	14.8	16.0	13.5	14.2
H_2CO	C	205	235	214	215	215	215	195	
	O	788	841	778	784	791	796	687	
CH_4	C	196	194	190	188	192	190	202	195
H_2O	O	327	332	325	325	326	326	345	344

Note: Absolute shielding in ppm.

reference compound's shielding constant, but as the results in Table 12.8 show, shifts in general can be represented well by less extensive basis sets.

The selection recorded in Table 12.9, drawn from the report by cheesman et al. [33], refers only to HF-GIAO and the authors' recommended B3LYP-GIAO method; it also shows the basis set sensitivity of this model. It seems fair to say that for this very small sample there is no obvious advantage to the use of a correlation-corrected model chemistry, or even a considerably extended basis set. The results are only mildly responsive to both these choices.

Cheeseman et al. reported isotropic shielding values for 23 organic molecules containing ^{13}C, ^{15}N, ^{17}O, and ^{11}B. This more general data set also showed no clear preference either for method or basis set. The key point would seem to be that all calculations permit secure assignments of experimental peaks. Even calculations in very small basis sets may be useful for ^{13}C if not for ^{15}N (Table 12.10).

The recommended level of treatment for large organic molecules is to optimize the geometry at the B3LYP/6-31G* level and run the Hartree–Fock GIAO calculation with the 6-31G* basis.

The discussion so far has focused almost entirely on heavy nuclei with relatively weak coupling. For protons, both chemical shielding and coupling tensors are of significance. Rablen et al. [39] pursued a very thorough investigation of NMR computational procedures for proton shielding, testing

TABLE 12.9

Absolute Isotropic Shielding Constants for Reference Molecules (ppm)

Molecule	Nucleus	6-31G*		6-311+G(2d,p)		Exp.
		HF	B3LYP	HF	B3LYP	
TMS	^{13}C	195.1	183.7	188.5	177.2	188.1
NH_3	^{15}N	260.8	255.0	260.3	258.4	264.5
H_2O	^{17}O	323.1	316.6	321.8	320.0	344.0

Note: Geometries optimized by B3LYP/6-31G(d).

TABLE 12.10

HF Root-Mean-Square Errors for ^{13}C and ^{15}N Shifts (ppm)

Isotope	STO-3G	3-21G(d)	6-31G(d)
^{13}C	18.4	12.6	11.1
^{15}N	36.1	33.1	

Note: Results from GIAO calculations using geometries optimized with B3LYP/6-31G*.

several density functionals in both the GIAO and IGAIM formulations. Besides compiling absolute shielding constants for 80 small- to medium-sized organic species, they imposed a linear scaling adjustment on the computed values to improve the representation of experimental values for species in solution. The combination of calculations in large basis sets and scaling was very successful in the most practical sense. They recommended GIAO/B3LYP/6-311++G(d,p)//B3LYP/6-31+G(d) procedure as the best compromise between accuracy and expense. This choice yielded proton chemical shifts having a root-mean-square error of 0.15 ppm. (Note the relatively narrow span of proton shielding and chemical shift values.) One important outcome of this work is that GIAO-based calculations for proton shifts seem to have value even in modest basis sets. This is consistent with the observations of Cheeseman et al. for heavy atom shielding and shifts.

Chemical Shifts of Taxol

The reader may still be skeptical that such elaborate calculations can be of value in the assignment of ^{13}C (or other) NMR spectra for large systems of chemical (especially synthetic) interest. Cheeseman et al. [33] reported GIAO values of shielding for a system of undoubted interest, taxol.

B3LYP-based calculations with the modest 6-31G(d) basis seem to behave well for both ^{13}C and ^{15}N—RMS errors of approximately 10 ppm are found for this large system (Table 12.11). Enhancing the basis actually degrades the B3LYP performance for both ^{13}C and ^{15}N—doubtless another expression of the fortunate cancelation of errors that makes this method such a useful workhorse in so many chemical settings (Figure 12.1).

To guide ^{13}C assignments in spectra of even larger molecules, the authors recommend using HF/6-31G* basis as a usefully accurate method with

TABLE 12.11

RMS Errors in Shielding Constants for Taxol

Isotope	6-31G*		6-311±G(2d,p)	
	HF	B3LYP	HF	B3LYP
^{13}C	14.3	9.3	9.6	11.0
^{15}N	17.2	9.2	40.6	23.2

Observed and predicted ^{13}C chemical shifts for taxol

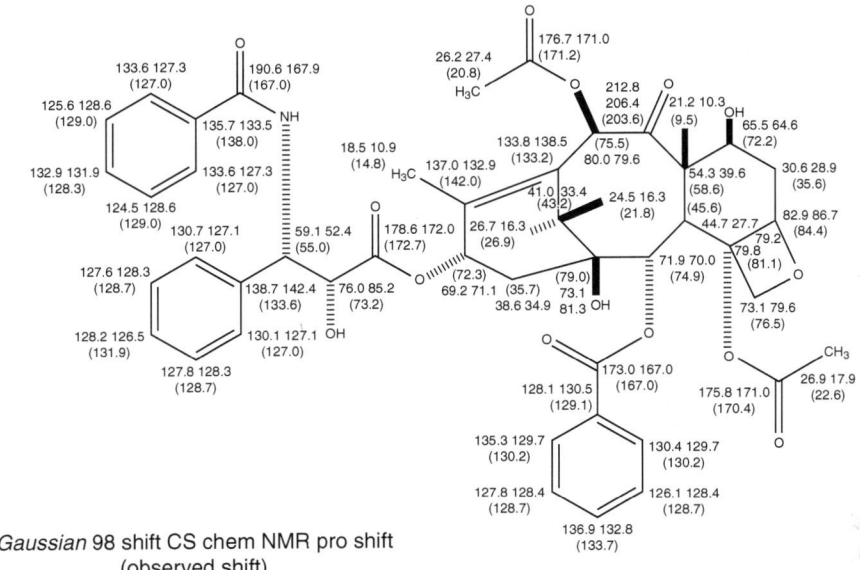

Gaussian 98 shift CS chem NMR pro shift
(observed shift)

Chemical shifts are with respect to TMS

FIGURE 12.1

Characterization of taxol chemical shifts (ppm relative to standard tetramethysilane). The results of computational methods incorporated in GAUSSIAN, the empirical laws encoded in Cambridge Scientific's ChemOffice, and the experimental values are displayed. For conventionally bound systems, roughly equivalent results can be expected. For strained or otherwise unusual cases, the empirical rules may fail. (From Cheeseman, J.R., Trucks, G.W., Keith, T.A., et al., *J. Chem. Phys.*, 104, 5497, 1996. With permission.)

> At the end of this chapter, following the references, we quote the analysis of performance of GAUSSIAN and the empirical method incorporated in CS chemNMR Pro, from Cheeseman, J.R., Frisch, Æ., http://gaussian.com/nmrcomp.htm

modest computational burden. If one is mainly interested in calculating the most accurate NMR tensors and coupling constants at any cost, one would employ the CCSDT method in the ACES II [34] program.* This very demanding methods give superb results for small molecules.

Advances in DFT-Based NMR

Keal and Tozer

Keal and Tozer [40] were interested in developing a simple gradient-corrected density functional expressly to fit selected NMR shielding. They

---

*http://www.qtp.ufl.edu/Aces2/or http://www.asc.hpc.mil/software/info/aces.

chose three molecules as their test set, CO, N_2, and PN, representing NMR resonances for ^{13}C, ^{15}N, ^{17}O, and ^{31}P. In their exceptionally clear notation, the shielding tensor for nucleus A is

$$\sigma^A_{\alpha\beta} = \sum_j \left\langle j \left| (\mathbf{r} \cdot \mathbf{r}_A \delta^{\alpha\beta} - r^\alpha r^\beta_A)/r^3_A \right| j \right\rangle$$

$$- \sum_b \sum_j \frac{\langle b|L^\alpha|j\rangle\langle j|L^\beta_A/r^3_A|b\rangle + \langle b|L^\alpha_A/r^3_A|j\rangle\langle j|L^\beta|b\rangle}{\varepsilon_b - \varepsilon_j}$$

Here α and β are the Cartesian components x, y, and z; $\mathbf{r}_A$ is the distance of an electron from the specific nucleus. L^α and L^β are components of the angular momentum operator; j and b are labels for occupied and virtual orbitals, respectively. The energy values of the occupied and virtual orbitals are ε_j and ε_b. This is an orbital representation of the Ramsey formula [1,2]. Keal and Tozer [40] used the standard Kohn–Sham formulation (see our DFT chapter) to obtain DFT orbitals:

$$\left[-\tfrac{1}{2}\nabla^2 + V_{NE}(r) + V_J(r) + V_{xc}(r) - \varepsilon_\rho \right] \varphi_\rho(r) = 0$$

They were interested in developing a simple gradient formula to fit selected NMR shifts. Observing that the familiar LDA, BLYP, and HCTH functionals gave relatively poor NMR shifts, they designed their own, beginning with the form

$$E_{XC}[\rho_\alpha, \rho_\beta] = E^{LDA}_{XC}[\rho_\alpha, \rho_\beta] + \gamma \sum_\sigma \int d\mathbf{r}\, \frac{|\nabla\rho_\sigma(r)|^2}{\rho^{4/3}_\sigma(r) + \delta}$$

This will be recognized as the standard reference local density approximation (LDA) for exchange and correlation, with a correction referring to the density gradient. The Vosko–Wilks–Nusair (VWN) form for the LDA term was used. Two disposable parameters γ and δ make the gradient correction flexible. With values for γ and δ chosen by nonlinear optimization to be -0.006 and 0.1, respectively, the authors obtained excellent values of atomization energies, ionization potentials, and especially the sought-after NMR parameters.

Keal and Tozer [40] computed the NMR shielding tensors of 21 small compounds including the original set of CO, N_2, and PN as well as formaldehyde. Their best results (shown in Table 12.12) were obtained with a modified functional incorporating adjustable portions of the VWN exchange–correlation functional.

$$E_{XC}[\rho_\alpha, \rho_\beta] = aE^{LDA}_X[\rho_\alpha, \rho_\beta] + bE^{LDA}_C[\rho_\alpha, \rho_\beta] + \gamma \sum_\sigma \int \frac{|\nabla\rho_\sigma(\mathbf{r})|^2}{\rho^{4/3}_\sigma(\mathbf{r}) + \delta}\, d\mathbf{r}$$

TABLE 12.12

Performance of a DFT Model Specifically Designed for NMR

a. Molecular Properties for 21 Small Molecules

Property	MAD Values for Each Functional Choice		
	LDA	HCTH	KT2
Atomization energies	35.3	3.2	6.4
Ionization energies	3.1	1.6	3.9
Total energies (cations)	241.5	4.2	461.8
AH bond lengths	0.022	0.013	0.010
AB bond lengths	0.017	0.037	0.020
Shielding: isotropic averages	52.2	32.4	13.2
Shielding: anisotropicity	54.3	28.4	8.8

Note: Energies in kilocalories per mole, bond lengths in angstroms, shielding in ppm.

b. Shielding Constants for Formaldehyde

Molecule	Atom	LDA	HCTH	KT2	CCSD(T)	Exp.
H_2CO	C	−40.0	−17.7	−3.0	4.7[a]	−4.4 ± 3[a]
	O	−494	−407	−380	−383[a]	−375 ± 100[a]

[a] Wu, G., Lumsden, M.D., Ossenkamp, G.C., Eichele, K., and Wasylishen, R.E., *J. Phys. Chem.*, 99:15806, 1995.

Here *a* weights the exchange part and *b* weights the correlation part of the VWN-LDA functional. If $a = b = 1$ the expression reduces to the authors' first choice, the KT1 functional. The optimized values are 1.07173 and 0.576727, respectively for paramertes *a* and *b*.*

The estimates by Keal and Tozer of isotropic shielding constants and their anisotropicities are among the best so far reported for DFT calculations, thanks to the flexibility accompanying the four disposable parameters. Optimization for NMR had adverse effects on total energies, though geometric parameters are well represented by Keal and Tozer's functional. It is characteristic of semiempirical schemes that requiring excellent representation of one kind of molecular property degrades the quality of representation of other properties. We can look on the KT2 choice of functional as useful primarily for its special purpose. The "universal functional" is still elusive.

Arbuznikov et al.

Almost every effective DFT scheme employs HF "exact exchange" to complement the local expressions of exchange and correlation typical of "pure" DFT.

---

*We notice that the experimental ^{17}O values have quite large estimated errors. Fitting the functional to minimize MAD for shielding constants should give greatest weight to the most accurate data.

$$E_{XC}^{hybr} = a_0 E_X^{exact} + E_{XC}^{DFT}$$

The Hartree–Fock form for exchange is employed, with Kohn–Sham orbitals taking the place of HF-SCF orbitals.

$$E_X^{exact} = -\frac{1}{2} \sum_{\sigma=\alpha,\beta} \sum_{i,j}^{occ} \int \frac{\phi_{i_\sigma}^*(\mathbf{r})\phi_{j_\sigma}^*(\mathbf{r}')\phi_{j_\sigma}(\mathbf{r})\phi_{i_\sigma}(\mathbf{r}')}{|\mathbf{r}-\mathbf{r}'|} d\mathbf{r}\, d\mathbf{r}'$$

Evaluation of this nonlocal "exact exchange" is much more demanding than evaluation of any local form of exchange, so it is tempting to apply a single correction after orbitals are defined by the Kohn–Sham process and by the pure DFT effective Hamiltonian. In our DFT chapter, we discuss work by Jaramillo et al. [41] and its development by Arbuznikov et al. [42]. These investigators proposed that the amount of exact exchange be dependent on position in space.

$$E_{XC}^{loc-hyb} = \sum_{\sigma=\alpha,\beta} \int \left\{ g_\sigma(\mathbf{r})\varepsilon_{X,\sigma}^{exact}(\mathbf{r}) + [1 - g_\sigma(\mathbf{r})]\varepsilon_{X,\sigma}^{DFT} \right\} d\mathbf{r} + E_C^{DFT}$$

They carried through the chain rule derivatives required to optimize the orbital exchange simultaneously with the electron density. Finding that optimizing the amount of exact exchange at every iteration was computationally burdensome, they turned to use of a single correction at the end of the Kohn–Sham iteration. Their functionals produced thermochemical atomization energies comparable with the best established empirically optimized DFT functionals. However, it appears that to calculate NMR shielding it is necessary to include evaluation of the exact exchange at every iteration.

The function $g(\mathbf{r})$ took two different forms. The first refers to the ratio of the Weizsäcker kinetic energy τ_W to the kinetic energy computed with the Kohn–Sham orbitals, and is scaled by a factor $\lambda_1 < 1$, as shown below:

$$\frac{\tau_{W,\sigma}(\mathbf{r})}{\tau_\sigma(\mathbf{r})} \equiv t_\sigma(\mathbf{r}) \quad \text{where } \tau_\sigma(\mathbf{r}) = \frac{1}{2}\sum_{i_\sigma}^{occ}|\nabla\phi_{i_\sigma}|^2 \text{ and } \tau_{W,\sigma}(\mathbf{r}) = \frac{1}{8}\frac{|\nabla\rho_\sigma(\mathbf{r})|^2}{\rho_\sigma(\mathbf{r})}$$

$$g_{\sigma,1}(\mathbf{r}) = \lambda_1 t_\sigma(\mathbf{r}); \quad \lambda_1 < 1$$

The value for λ_1 which was found best to match atomization energies (0.48) was not quite optimum for fitting the isotropic chemical shifts of the 22 light molecules in their test set (a list essentially the same as used by Keal and Tozer [40]).

A second form for $g(\mathbf{r})$ was also considered: defining $s(\mathbf{r})$ as the magnitude of a dimensionless density gradient scaled to the Fermi wave number $k_F(\mathbf{r})$, the authors introduce a simpler form, with parameter λ_2:

$$s_\sigma(\mathbf{r}) = \frac{|\nabla\rho_\sigma(\mathbf{r})|}{2k_F(\mathbf{r})\rho_\sigma(\mathbf{r})} = \frac{|\nabla\rho_\sigma(\mathbf{r})|}{2(3\pi^2)^{1/3}\rho_\sigma^{4/3}(\mathbf{r})}; \quad g_{\sigma,2}(\mathbf{r}) = \left(\frac{s_\sigma(\mathbf{r})}{\lambda_2 + s_\sigma(\mathbf{r})}\right)^2$$

TABLE 12.13

Absolute Shielding Constants and Errors for Nuclei in ppm. The Lh(L)-S Functional Is by Arbuznikov et al. [42]

Molecule and Nucleus		Lh(L)-S-VWN; $g(r) = \lambda \, t(r)$				Lh(L)-S-LYP; $g(r) = [s(r)/(\lambda + s\,(r))]^2$				Expt.
		$\lambda = 0.48$	$\lambda = 0.60$	$\lambda = 0.80$	$\lambda = 0.90$	$\lambda = 0.73$	$\lambda = 0.60$	$\lambda = 0.50$	$\lambda = 0.40$	
C_2H_2	C	110.6	112.9	116.6	118.4	111.8	112.9	113.8	114.8	117.2
H_2CO	C	−22.7	−18.4	−11.4	−8.0	−21.2	−19.4	−18.0	−16.5	−4.4 ± 3
H_2O	O	334.9	335.0	335.0	335.0	338.4	338.7	339.2	339.7	324 ± 1.5
CH_4	C	190.2	189.8	189.2	188.9	194.7	194.8	194.9	195.0	195.1
MAD (ppm)		22.5	18.5	14.7	14.9	17.0	15.2	14.5	15.0	

Source: Arbuznikov, A.V., Kaupp, M., and Bahmann, H., *J. Chem. Phys.* 124:204102, 2006.

This parameter took the value 0.73 in the authors' study of atomization energies. But once again it was not possible to find a single parameter which produced excellent NMR shielding and energetic properties simultaneously.

Arbuznikov et al. [42] introduced their functional Lh(L) into the ReSpect [43] NMR shielding program, which employed the Kutzelnigg–Schindler IGLO [20,21] method. The results appearing in Table 12.13 were obtained with the large IGLO-IV basis [20], developed and tested for IGLO-NMR calculations. The errors refer to a test set of 22 compounds. The quality of the fit for slected values of the parameters λ_1 and λ_2 is expressed by the mean absolute deviation (MAD) and by the index of fit to a linear regression.

Each of the two methods can reduce the MAD for the isotropic shielding to 15 ppm, which means the technique is perfectly adequate to guide assignments in ^{13}C spectra (Table 12.14). Whether the method can be applied to large systems effectively has not been established.

We emphasize that the story of the quest for a versatile functional is not finished (as of 2007). The notion of spatially variable amounts of exact exchange is sufficiently important in DFT research that one should apply a version of the Lh(L) approach to a GIAO basis and other ways to treat the gauge issue. We look forward to a broader testing of the method.

TABLE 12.14

Summary of Best Results from Localized Local Hybrid NMR Calculations

Type	λ	DFT Model	MAD (ppm)	Slope	Intercept (ppm)	R
$t(\vec{r})$	0.90	Lh(L)-S-VWN	14.9	1.000	−1.0	0.9988
$s(\vec{r})$	0.40	Lh(L)-S-LYP	15.0	1.007	−1.8	0.9985

TABLE 12.15

Total Coupling in Hz for Nuclei in Formaldehyde

Atoms	HF	B3LYP	PBE1PBE
CO	−58	31	30
CH	219	163	154
OH	18	−26	−24
HH	26	41	38

The Question of Coupling and Fine Structure

In proton NMR spectra especially, the coupling constants are as important a part of the experimental data as the chemical shifts. Below we present Gaussian's report for formaldehyde, in HF, B3LYP, and PBE1PBE. The basis set is 6-311++G(2df,2p) and is composed of 88 members (Table 12.15).

The total coupling shows considerable response to correlation corrections by density functional theory, even changing sign of the CO and (long range) OH coupling constants. The hybrid B3LYP and pure PBE density functional method produce very consistent results.

Coupling is made up of several contributions, the most important of which is the Fermi "contact" term. This contact is defined by the density at the locations of particular nuclei. It is plain that correlation is essential to the proper treatment of this part of the coupling. HF gets signs wrong for CO and OH coupling, as is the case for total coupling. DFT is however perfectly equal to this task, giving values closely compatible with the coupled cluster treatment from ACES-II, which used a near-complete basis with 170 members. This calculation was executed for us by Ajith Perera of Prof. Bartletts' research group; we are grateful for his gracious gesture. The foundation for these CCSDT calculations is due to Gauss and Stanton [44,45] who have recently extended the formalism to full triples, i.e., CCSDT [46] (Table 12.16).

Other terms seem far less significant, if not negligible. The pattern already established persists for spin–dipole interactions and paramagnetic

TABLE 12.16

Fermi Contact Contribution to Coupling in Formaldehyde (Hz)

Atoms	HF	B3LYP	PBE1PBE	ACES-II
CO	−50	11.1	11.0	10.0
CH	220.2	162.3	154.0	167.5
OH	16.0	−5.7	−5.4	−5.6
HH	25.0	40.8	38.2	36.9

TABLE 12.17

Spin–Dipole Contribution to Coupling
in Formaldehyde (Hz)

Atoms	HF	B3LYP	PBE1PBE	ACES-II
CO	−2.92	−2.24	−2.51	−1.74
CH	−0.32	0.46	0.45	0.31
OH	−1.30	−0.80	−0.92	−0.72
HH	0.84	0.40	0.43	0.35

spin–orbit (PSO) interactions; DFT provides far better estimates of this part of coupling between nuclei. Note that the PSO coupling from C to O is not small, ca. 20 Hz. The HF method is reliable for this particular aspect of coupling (Tables 12.17 and 12.18).

The ACES-II values are the best we can imagine computing at this stage of the development of theory and algorithms. The coupled cluster methods deserve closer attention, as the latest development of correlated methods for evaluation of shielding tensors apart from density functional theory.

Perturbative Correlation-Corrected Methods for Shielding

We draw this perspective from a survey by Gauss and Stanton [47] of the terrain of non-DFT correlation-corrected methods for NMR shielding. An earlier and more wide-ranging review by Helgaker et al. [48] is highly recommended. Gauss and Stanton divide their overview into three categories: (1) methods based on perturbation theory including the Moller–Plesset series [27]; (2) methods based on the coupled cluster expansion [44–46]; and (3) methods employing multiconfigurational SCF wave functions [49]. Only the latter is variational. Most use GIAO basis functions, with the continuous set of gauge transforms being the major alternative. The relatively favorable

TABLE 12.18

Paramagnetic Spin–Orbit Contribution to Coupling
in Formaldehyde (Hz)

Atoms	HF	B3LYP	PBE1PBE	ACES-II
CO	21.97	21.66	21.53	20.56
CH	−0.94	−0.76	−0.83	−0.74
OH	3.31	3.39	3.44	2.97
HH	3.00	2.97	3.01	3.35

TABLE 12.19

The Difficult Case of Ozone Shielding: Absolute
Isotropic Shielding Values in ppm

Method	O (Terminal)	O (Central)
IGLO-HF	−2814	−2929
GIAO-HF	−2862	−2768
MC-IGLO	−1153	−658
GIAO-MCSCF	−1126	−703
GIAO-MP2	+1248	+2875
GIAO-CCSD	−1403	−968
GIAO-CCSD(T)	−1183	−724
GIAO-CCSDT	−1261	−775
Exp.	−1289 ± 170	−625 ± 240

scaling of GIAO-MP2 (N^5, where N is a measure of size such as the number of basis function) along with its generally useful appreciation of correlation recommends it for molecules of medium size. CCSD(T), scaling as N^7, is more restricted in application, though in principle it must be superior to GIAO-MP2. The full CCSDT formulation is available but its N^8 scaling restricts its use still further. In certain cases in which the ground state must be described by more than one determinant, adaptations of multiconfigurational SCF can be especially valuable.

Certain molecules are especially hard cases; ozone is difficult in several ways (the asymmetric O–O stretch is especially elusive, for example) and also serves as a severe test of methods for estimation of NMR shielding. Its pathology stems from a high degree of nondynamical correlation. In Table 12.19, quoted from [47], we see the MCSCF methods (whether or not local orbitals are used) do very well, while MP2 fails dramatically. Values from the best calculations now possible, CCSDT and CCSD(T), fall within the considerable error bounds of experimental values. DFT does not treat the context-specific nondynamical correlation well. B3LYP/6-311++G (2df,2p) produces values of −2430 and −2415 ppm for terminal and central oxygen atoms, respectively. These are not much different from HF values.

ACES-II Coupled Cluster Calculation of NMR Properties

We have already quoted the values of the shielding constant estimated by CCSDT in a large basis of 174 functions. ACES-II also presents the anisotropy. The HF shielding in the same basis shows the impact of the correlation effect—considerable for the heavy atoms. ACES-II analyzes the diamagnetic

TABLE 12.20

CCSDT and HF Shieldings and Anisotropies
for Formaldehyde

	CCSDT	
Atomic Center	Isotropic Shielding, ppm	Anisotropy, ppm
C	4.030	155.422
O	−383.466	1208.648
H	21.852	0.613
H	21.852	0.613
	Hartree–Fock	
Atomic Center	Isotropic Shielding, ppm	Anisotropy, ppm
C	−8.671	184.554
O	−447.923	1310.049
H	22.290	1.505
H	22.290	1.505

Note: Near complete basis set of 174 members.

(the first-order or ground state portion) and paramagnetic (the second-order or response-derived portion) terms as Ramsey prescribes. The HF values and the correlation-corrected values of both parts of the shielding are presented in Table 12.20.

TABLE 12.21

Diamagnetic Part of Shielding Tensors for Formaldehyde (HF)

			HF diamagnetic part		
			Bx	By	Bz
C	#1	x	226.213120	0.000000	0.000000
C	#1	y	0.000000	249.504326	0.000000
C	#1	z	0.000000	0.000000	258.802939
O	#2	x	387.076810	0.000000	0.000000
O	#2	y	0.000000	383.600677	0.000000
O	#2	z	0.000000	0.000000	416.213338
H	#31	x	22.116955	0.000000	0.000000
H	#31	y	0.000000	34.025045	6.778343
H	#31	z	0.000000	6.711212	29.764049
H	#32	x	22.116955	0.000000	0.000000
H	#32	y	0.000000	34.025045	−6.778343
H	#32	z	0.000000	−6.711212	29.764049

TABLE 12.22

Diamagnetic Part of Shielding Tensors for Formaldehyde with CCSDT Correction

			Total Diamagnetic part of shielding tensor		
			Bx	By	Bz
C	#1	x	228.529566	0.000000	0.000000
C	#1	y	0.000000	251.051727	0.000000
C	#1	z	0.000000	0.000000	261.529837
O	#2	x	390.354341	0.000000	0.000000
O	#2	y	0.000000	386.332971	0.000000
O	#2	z	0.000000	0.000000	416.856038
H	#31	x	21.422904	0.000000	0.000000
H	#31	y	0.000000	33.775862	7.232921
H	#31	z	0.000000	6.939620	29.398905
H	#32	x	21.422904	0.000000	0.000000
H	#32	y	0.000000	33.775862	−7.232921
H	#32	z	0.000000	−6.939620	29.398905

Tables 12.21 and 12.22 shows that the correlation effects on the diamagnetic part of the shielding are modest.

Correlation corrections to the paramagnetic part of the shielding tensor, determined by response of the system to the perturbation, are highly significant, as shown in Tables 12.23 and 12.24.

TABLE 12.23

Paramagnetic Part of Shielding Tensors for Formaldehyde (HF)

			HF-SCF contribution to paramagnetic part of shielding tensor		
			Bx	By	Bz
C	#1	x	−111.847838	0.000000	0.000000
C	#1	y	0.000000	−373.152937	0.000000
C	#1	z	0.000000	0.000000	−275.531971
O	#2	x	38.366023	0.000000	0.000000
O	#2	y	0.000000	−965.999552	0.000000
O	#2	z	0.000000	0.000000	−1603.026154
H	#31	x	0.168820	0.000000	0.000000
H	#31	y	0.000000	−12.655896	−5.464984
H	#31	z	0.000000	−8.800691	−6.548467
H	#32	x	0.168820	0.000000	0.000000
H	#32	y	0.000000	−12.655896	5.464984
H	#32	z	0.000000	8.800691	−6.548467

TABLE 12.24

Paramagnetic Part of Shielding Tensors for Formaldehyde with CCSDT Correction

HF-SCF contribution to paramagnetic part of shielding tensor

			Bx	By	Bz
C	#1	x	−120.884971	0.000000	0.000000
C	#1	y	0.000000	−334.764712	0.000000
C	#1	z	0.000000	0.000000	−275.531971
O	#2	x	21.945044	0.000000	0.000000
O	#2	y	0.000000	−857.004367	0.000000
O	#2	z	0.000000	0.000000	−1603.026154
H	#31	x	−0.092571	0.000000	0.000000
H	#31	y	0.000000	−11.676893	−5.464984
H	#31	z	0.000000	−8.331639	−6.548467
H	#32	x	−0.092571	0.000000	0.000000
H	#32	y	0.000000	−11.676893	5.464984
H	#32	z	0.000000	8.331639	−6.548467

The total shielding tensor in CCSDT, reported in Table 12.25, is the sum of the diamagnetic and paramagnetic components. The isotropic shielding constants for each atom (as quoted in Table 12.20 above) are each the average of diagonal elements of that atom's 3×3 tensor.

TABLE 12.25

Total Shielding Tensor at the CCSDT Level

			Bx	By	Bz
C	#1	x	107.644695	0.000000	0.000000
C	#1	y	0.000000	−83.712985	0.000000
C	#1	z	0.000000	0.000000	−14.002134
O	#2	x	422.299385	0.000000	0.000000
O	#2	y	0.000000	−470.671396	0.000000
O	#2	z	0.000000	0.000000	−1186.170116
H	#31	x	21.330333	0.000000	0.000000
H	#31	y	0.000000	22.098969	1.767937
H	#31	z	0.000000	−1.392019	22.850438
H	#32	x	21.330333	0.000000	0.000000
H	#32	y	0.000000	22.098969	−1.767937
H	#32	z	0.000000	1.392019	22.850438

Conclusion

The task of computing usefully accurate NMR shielding and coupling tensors for molecules of chemical interest is challenging. Theory and algorithms developed especially in the past 20 years have gone far to make the task—if not routine—within reach. As is often the case in electronic structure modeling, "success" consists in choosing a method capable of responding meaningfully to the question posed. Most investigators will want a computation to provide an independent and trustworthy view of a system, whether in experimental or conceptual settings. At present it seems that DFT methods—as so often—provide generally reliable estimates of subtle properties for systems which cannot be treated by other methods; the key NMR example is the taxol system. Even HF methods in a modest basis can give useful guidance for very large molecules. At the very minimum one can advise that such calculations be tried, but that calibrations on well-understood systems accompany the study of new species.

References

1. Ramsey, N.F., *Phys. Rev.* 77:567, 1950; ibid. 78:699.
2. Ramsey, N.F., *Phys. Rev.* 86:243, 1952.
3. Pople, J.A., Schneider, W.G., and Bernstein, H.J., *High Resolution Nuclear Magnetic Resonance*, McGraw Hill, New York, 1959.
4. Foster, S. and Boys, S.F., *Rev. Mod. Phys.* 32:303, 1960.
5. Wiberg, K.B., *Computer Programming for Chemists*, Frontiers in Chemistry, Benjamin, New York, 1965.
6. Bothner-By, A.A. and Castellano, S., QCPE 11,111 (LAOCN3, Program 111), 1967.
7. Ebraheem, K.A.K. and Webb, G.A., *Prog. NMR Spectros.* 11:149, 1977.
8. Ditchfield, R., *Mol. Phys.* 27:789, 1974.
9. London, F., *J. Phys. Radium* 8:397, 1937.
10. Pople, J.A., *J. Chem. Phys.* 37:53, 1962.
11. Pople, J.A., *Discuss. Faraday Soc.* 34:7, 1962.
12. Ellis, P.D., Maciel, G.E., and McIver, J.W., *J. Am. Chem. Soc.* 94:4069, 1972; Maciel, G.E., Dallas, J.L., Elliott, R.L., et al., *J. Am. Chem. Soc.* 95:5857, 1973.
13. Del Bene, J. and Jaffe, H.H., *J. Chem. Phys.* 48:1807, 1968.
14. Pople, J.A. and Beveridge, D.L., *Approximate Molecular Orbital Theory*, McGraw-Hill, New York, 1970.
15. Jensen, F., *Introduction to Computational Chemistry*, Wiley, New York, Chapter 10, Section 5, 1999.
16. Freitag, M.A., Hillman, B., Agrawal, A., et al., *J. Chem. Phys.* 120:1197, 2004.
17. Schmidt, M.W., Baldridge, K.K., Boatz, J.A., et al., *J. Comput. Chem.* 14:1347, *General Atomic and Molecular Electronic Structure System (GAMESS)*, 1993.
18. Gaussian 03, Revision C.02, Frisch, M.J., Trucks, G.W., Schlegel, H.B., et al., Gaussian Inc., Wallingford, CT, 2004.

19. Hehre, W.J., *A Guide to Molecular Mechanics and Quantum Chemical Calculations w/CD*, Wavefunction Inc., Irvine, CA, 2007.
20. Schindler, M. and Kutzelnigg, W., *J. Chem. Phys.* 76:1919, 1982.
21. Schindler, M. and Kutzelnigg, W., *J. Am. Chem. Soc.* 105:1360, 1983.
22. Boys, S.F., *Rev. Mod. Phys.* 32:296, 1960.
23. Hansen, A.E. and Bouman, T.D., *J. Chem. Phys.* 82:5035, 1985.
24. Bouman, T.D., Hansen, A.E., Bak, K.L., et al., *RPAC version 11.0*, H.C. Orsted Institute, University of Copenhagen, Denmark, <aage.hansen@post.cybercity. dk>, 1996.
25. Dunning, T.H. and McKoy, V., *J. Chem. Phys.* 47:1735, 1967.
26. Hansen, A., Private communication.
27. Hansen, A.E. and Bouman, T.D., *Nuclear Magnetic Shieldings and Molecular Structure*, Ed., Tossel, J.A., Kluwer, Dordrecht, 1993.
28. Gauss, J., *J. Chem. Phys.* 99:3629, 1993.
29. Gauss, J. and Stanton, J.F., *J. Chem. Phys.* 102:251, 1995.
30. Gauss, J. and Stanton, J.F., *J. Chem. Phys.* 104:2574, 1996.
31. Malin, V.G., Makina, O.L., and Casida, M.R.J., *Am. Chem. Soc.* 116:5898, 1994.
32. Schreckenbach, G. and Ziegler, T., *J. Phys. Chem.* 99:606, 1995; Schreckenbach, G. and Ziegler, T., *Theor. Chem. Acc.* 99:71, 1998.
33. Cheeseman, J.R., Trucks, G.W., Keith, T.A., et al., *J. Chem. Phys.* 104:5497, 1996.
34. Stanton, J.F., Gauss, J., Watts, J.D., et al., *Int. J. Quantum Chem. Symp.* 26:879, 1992.
35. Johnson, B.G. and Frisch, M.J., *J. Chem. Phys.* 100:7429, 1994.
36. Keith, T.A. and Bader, R.F.W., *Chem. Phys. Lett.* 210:223, 1993.
37. Salem, L., *The Molecular Orbital Theory of Conjugated Systems*, Benjamin, New York, Chapter 4, 1966.
38. Keith, T.A. and Bader, R.F.W., *Chem. Phys. Lett.* 194:1, 1992.
39. Rablen, R.R., Shoshannah, A., Pearlman, S.A., et al., *J. Phys. Chem. A*, 103:7357, 1999.
40. Keal, T.W. and Tozer, D.J., *J. Chem. Phys.* 119:3015, 2003.
41. Jaramillo, J., Scuseria, G.E., and Ernzerhof, M., *J. Chem. Phys.* 118:1068, 2003.
42. Arbuznikov, A.V., Kaupp, M., and Bahmann, H., *J. Chem. Phys.* 124:204102, 2006.
43. Malkin, V.G., Malkina, O.L., Reviakine, R., et al., ReSpect software v. 1.1, 2002.
44. Gauss, J. and Stanton, R.E., *J. Chem. Phys.* 103:3561, 1996.
45. Gauss, J. and Stanton, R.E., *J. Chem. Phys.* 104:2574, 1996.
46. Gauss, J., *J. Chem. Phys.* 116:4773, 2002.
47. Gauss, J. and Stanton, R.E., *Calculation of NMR and EPR Parameters: Theory and Applications*, Eds., Kaupp, M., Bühl, M., Malkin, V.G., Wiley, New York, 2004.
48. Helgaker, T., Jaszuński, M., and Ruud, K., *Chem. Rev.* 99:293, 1999.
49. Jaszuński, M., Helgaker, T., Ruud, K., et al., *Chem. Phys. Lett.* 220:154, 1994.

Predicting Magnetic Properties with ChemDraw and Gaussian

For taxol CS ChemNMR Pro's mean absolute error with respect to the observed ^{13}C chemical shifts is 3.8 ppm, with a standard deviation of 4.6, and the largest error is 19.0 ppm. *Gaussian* also does well for this molecule: its mean absolute error is 4.2, with a standard deviation of 3.8 and a maximum error of 23.6.

CS ChemNMR Pro uses a heuristically driven procedure in order to estimate chemical shifts; it starts with a base value determined from the molecular mechanics atom type of the atom in question, and then applies corrections for each of the groups to which it is bonded in order to compute its final value. This process is illustrated in this excerpt from the textual output from a calculation on taxol:

```
              Protocol of the 13C NMR Estimation

Node   Estim.   Base    Incr.   Comment (ppm rel. to TMS)
C      138.5    123.3                  1-ethylene
                        -8.9           1 -C-C-C-C
                        -7.4           1 -C
                        17.3           1 -C-C-C-C
                        14.2           1 -C-O
```

The main advantage of this approach to computing chemical shifts is its speed: chemical shifts can be computed almost instantaneously even for very large molecules. However, the method has an important weakness which must be kept in mind. Since it relies on a fixed set of parameters corresponding to atom types and subgroups, the method will be reliable only for molecules for which parameters are available and for which the assumptions about molecular structure and bonding which are built-in to the parameters are valid.

In simple terms, this NMR estimation method is appropriate only for ordinary organic molecules. It produces reasonable results for such systems, but becomes quite unreliable for systems with any unusual features: unusual bonding, strained systems, systems for which electron correlation is important for accurate modeling of the molecular structure or properties, and so on. In these cases, a more accurate computational method is required.

13

The Representation of Electronically Excited States

Introduction

Excited states are of great significance in all of the chemistry initiated by the absorption of light (photochemistry) or by the light-emitting reactions of high-energy systems (chemiluminescence). However, the rapid and reliable description of excited states has been a considerable challenge to computational chemistry. Variational calculations inevitably favor the ground state, and the representation of excited states by excitations from such a favored ground configuration must systematically overestimate excitation energies. In semiempirical approximation, one feels justified in scaling parameters so to match some reference system's longest-wavelength transition. In ab initio calculations, reasonable treatment of excitations requires so flexible a basis set that it is capable of simultaneous description of the ground state and (at least several low-lying) excited states as well. Despite such difficulties and complications, however, informative calculations are feasible. We will explore in some detail the simplest model chemistry suitable to characterization of low-lying electronically excited states, called configuration interaction (singles) or CIS. We will provide a brief introduction and some guidelines for use of more subtle and demanding methods such as complete active space MCSCF and linear-response techniques.

CI-Singles

The simplest treatment of excited states employs the space of single excitations from a reference ground state restricted SCF determinant.

$$\Psi_{\text{CIS}} = \sum A_{Ki}^a \psi_i^a; \; \psi_i^a = (\| \, \varphi_1^\alpha \varphi_1^\beta \cdots \varphi_i^\alpha \varphi_a^\beta \cdots \| \pm \| \, \varphi_1^\alpha \varphi_1^\beta \cdots \varphi_a^\alpha \varphi_i^\beta \cdots \|)/\sqrt{2}$$

Here K is one of the excited states; the single excitation removes an electron from orbital i in the HF sea and places that electron in previously vacant orbital a. The upper sign in the $\pm$ choice refers to a triplet excited state, the lower to a singlet excited state. The weighting coefficients are found by a linear variation calculation, which requires the solution of a secular determinant

$$|< \psi_i^a |H| \psi_j^b > -E\delta_{ij,ab}| = 0$$

for some of the lower roots and associated eigenvectors. This CI-singles (CIS) method has the convenience that all these singly excited states are orthogonal to the reference configuration. However, it cannot be expected to represent excited states not reasonably described by one-electron excitations. For this reason it is considered uncorrelated, and is thus the counterpart in excited states to the SCF theory of ground states. It can be considered an adaptation of the Tamm–Dancoff method from nuclear physics, and is also related to the random-phase-approximation methods developed from the physics of the electron gas.

CIS was described systematically by Foresman et al. [1]; we will follow their exposition of the formalism and illustrate the method with a detailed discussion of formaldehyde. We begin with orbitals satisfying the SCF equations

$$\sum (F_{\mu\nu} - \varepsilon_p S_{\mu\nu})C_{p\nu} = 0$$

in which

$$F_{\mu\nu} = H_{\mu\nu} + \sum_i^{2N} \sum_{\mu\nu}^{\text{Basis}} C_{i\mu}C_{i\nu}(\langle \mu\nu|\rho\sigma\rangle - \langle \mu\nu|\sigma\rho\rangle)$$

Here the index i refers to spin-orbitals while μ, ν refer to members of the basis set. The core and repulsion integrals have already been thoroughly discussed. The total energy for spin-orbitals is

$$E_{\text{HF}} = \sum_{\mu\nu} P_{\mu\nu}H_{\mu\nu} + \frac{1}{2}\sum_{\mu\nu\rho\sigma} P_{\mu\nu}P_{\rho\sigma}(\langle \mu\nu|\rho\sigma\rangle - \langle \mu\nu|\sigma\rho\rangle) + V_{\text{NN}}$$

with the density matrix

$$P_{\mu\nu} = \sum_i^{\text{occupied}} C_{i\mu}C_{i\nu}$$

and V_{NN} defining the internuclear repulsion. According to Szabo and Ostlund [2], the matrix elements for singlet-coupled excitations are

$$\langle \psi_i^m |H| \psi_j^n \rangle = \delta_{ij}\delta_{mn}[E_{\text{HF}} + (\varepsilon_m - \varepsilon_i)] - \langle im|jn\rangle + 2\langle im|nj\rangle$$

in which the two-electron integrals refer to molecular spin-orbitals

$$\langle pq \parallel rs \rangle = \sum C_{p\mu} C_{q\nu} C_{r\rho} C_{s\sigma} \langle \mu\nu \parallel \rho\sigma \rangle$$

One may follow Pople [3] more closely by considering that the basis is composed of single determinants rather than spin-coupled excitations, in which case

$$\Psi_K^{\mathrm{CIS}} = \sum B_{im}^K D_i^m$$

and the coefficients are obtained by solving the secular equation

$$\| \langle D_i^m | H | D_j^n \rangle - E \langle D_i^m | D_j^n \rangle \| = 0 \quad \text{and} \quad \sum_{nj} (\langle D_i^m | H | D_j^n \rangle - E^K \langle D_i^m | D_j^n \rangle) B_{nj}^K = 0$$

with

$$\langle D_i^m | H | D_j^n \rangle = \delta_{ij} \delta_{mn} [E_{\mathrm{HF}} + (\varepsilon_m - \varepsilon_i)] - \langle jm \parallel in \rangle + \langle jm \parallel ni \rangle$$

Practical Use of the CIS Equations

To gain insight into the details of a CIS calculation, we appeal to a very simple example. Consider the case of diatomic hydrogen with internuclear distance D and the 1G basis introduced in the SCF discussion. The orbitals are dictated by symmetry.

$$\sigma_b = N_b (G_A + G_B)$$
$$\sigma_a = N_a (G_A - G_B)$$

where

$$G_M = \left(\frac{\pi}{2\alpha} \right)^{3/2} \exp\left(-\alpha [R - R_M]^2 \right)$$

and the normalization constants are

$$N_b^2 = \frac{1}{2[1 + W]} \quad \text{and} \quad N_a^2 = \frac{1}{2[1 - W]}$$

where $W = \exp(-\alpha D^2 / 2)$. There are two excited determinants with spin-angular momentum component $m_S = 0$, and the 2×2 matrix is

$$\left[\begin{array}{cc} \langle b^\alpha a^\beta | H | b^\alpha a^\beta \rangle - E & \langle b^\alpha a^\beta | H | a^\alpha b^\beta \rangle \\ \langle b^\alpha a^\beta | H | a^\alpha b^\beta \rangle & \langle a^\alpha b^\beta | H | a^\alpha b^\beta \rangle - E \end{array} \right] = 0$$

Each matrix element can be reduced by the explicit formulas already presented. Here the diagonal elements are chosen to be the energy increment beyond the ground state SCF energy.

$$\left[\begin{array}{cc} (\varepsilon_a - \varepsilon_b) - \langle b^\alpha a^\beta \| b^\alpha a^\beta \rangle & -\langle a^\alpha b^\beta \| b^\alpha a^\beta \rangle + \langle a^\alpha b^\beta \| a^\beta b^\alpha \rangle \\ \quad + \langle b^\alpha a^\beta \| a^\beta b^\alpha \rangle - E & \\ -\langle b^\alpha a^\beta \| a^\alpha b^\beta \rangle + \langle \sigma_b^\alpha \sigma_a^\beta \| b^\beta a^\alpha \rangle & (\varepsilon_a - \varepsilon_b) - \langle a^\alpha b^\beta \| a^\alpha b^\beta \rangle \\ & \quad + \langle a^\alpha b^\beta \| b^\beta a^\alpha \rangle - E \end{array} \right] = 0$$

The repulsion integrals can be simplified by performing the spin integration.

$$\left[\begin{array}{cc} (\varepsilon_a - \varepsilon_b) - \langle b^\alpha a^\beta \| b^\alpha a^\beta \rangle - E & -\langle a^\alpha b^\beta \| b^\alpha a^\beta \rangle \\ \quad - \langle b^\alpha a^\beta \| a^\alpha b^\beta \rangle & (\varepsilon_a - \varepsilon_b) - \langle a^\alpha b^\beta \| a^\alpha b^\beta \rangle - E \end{array} \right] = 0$$

The off-diagonal elements are the exchange integrals similar to those we encountered in the Hartree–Fock energy expression, but referring now to exchange between an electron excited to the antibonding MO and the electron remaining in the bonding MO. The diagonal elements are orbital energy differences corrected by coulomb repulsion between electrons in those orbitals. A simplification in notation shows the structure of the result.

$$\left\| \begin{array}{cc} \Delta - J - E & -K \\ -K & \Delta - J - E \end{array} \right\| = (\Delta - J - E)^2 - K^2 = 0$$

or $E_\pm = \Delta - J \pm K$. The difference between the states, $2K$, is the singlet–triplet gap for the simplest representation of those states. The eigenvectors for the singlet and triplet are

$$\psi_S = (|a^\alpha b^\beta| - |a^\beta b^\alpha|)/\sqrt{2}$$
$$\psi_T = (|a^\alpha b^\beta| + |a^\beta b^\alpha|)/\sqrt{2}$$

The other components of the triplet are those with $S_z = 1$ and -1.

$$|a^\alpha b^\alpha|, \; |a^\beta b^\beta|$$

Singlet and Triplet State Energies for the 1G Model

Expansion of the coulomb integral generates

$$\frac{2\langle AA \parallel AA \rangle - 4\langle AA \parallel BB \rangle + 2\langle AB \parallel AB \rangle}{4(1 - W^2)^2}$$

Expansion of the exchange integral produces the expression

$$\frac{2\langle AA \parallel AA \rangle - 2\langle AB \parallel AB \rangle}{4(1 - W^2)}$$

The one-center integrals are

$$\langle AA \parallel AA \rangle = \frac{1}{4}\left(\frac{\pi^5}{\alpha^3}\right)^{1/2}\left(\frac{2\alpha}{\pi}\right)^3 = 2\left(\frac{\alpha}{\pi}\right)^{1/2}$$

and the two-center coulomb integrals are

$$\langle AB \parallel AB \rangle = 2\left(\frac{\alpha}{\pi}\right)^{1/2}F_0(\alpha R_{AB}^2)$$

The remaining integral is

$$\langle AB \parallel BA \rangle = 2\left(\frac{\alpha}{\pi}\right)^{1/2}W^2$$

The coulomb integral becomes

$$\left(\frac{\alpha}{\pi}\right)^{1/2}\frac{[1 + F_0(\alpha R_{AB}^2) - 2W^2]}{2(1 - W^2)}$$

and the exchange integral becomes

$$\left(\frac{\alpha}{\pi}\right)^{1/2}\frac{[1 - F_0(\alpha R_{AB}^2)]}{2(1 - W^2)}$$

The diagonal elements depend on the difference in Fock energies for bonding and antibonding levels

$$\varepsilon_p = \sum C_{p\mu}F_{\mu\nu}C_{p\nu} = \sum_{\mu\nu} C_{p\mu}h_{\mu\nu}C_{p\nu} + \sum_{\mu\nu\rho\sigma,q} C_{p\mu}C_{q\sigma}\langle \mu\rho \parallel \nu\sigma \rangle C_{q\rho}C_{p\nu}$$

Here p can be either the bonding orbital or the antibonding orbital, but q can only be the bonding orbital, the only orbital which defines the ground charge distribution.

We can borrow some values from the SCF1s exercise in a previous chapter and our Maple treatment of ground state diatomic hydrogen to estimate the

TABLE 13.1

Energies (Hartrees) for the 1G Representation of H_2 (1G basis)

Total energy (including nuclear repulsion)	-0.9761695
Orbital energy (bonding MO φ_b)	-0.5123222
Orbital enegy (antibonding MO φ_a)	$+0.5099470$
Coulomb integral J_{ab}	$+0.47335$
Exchange integral K_{ab}	$+0.24030$
Singlet excitation energy	$+1.73592$
Triplet excitation energy	$+1.25532$
Singlet–triplet gap	$+0.48060$ (ca. 13 eV)

excitation energy for H_2; we choose the same rather arbitrary scale factor of 0.4, and an HH distance of 1.4 bohr (Table 13.1).

As we expect, the excited triplet is substantially more stable than the excited singlet (0.48 hartree or ca. 13 eV), but both excited states of diatomic hydrogen are very unstable relative to separate H atoms. Improving the basis permits a more balanced treatment of ground and excited states, depressing both singlet and triplet. In HF/6-311G(d,p) the singlet–triplet gap drops to about 4 eV.

PCLOBE Illustration of CIS

PCLOBE uses the formalism of Pople [3] to set up the configuration interaction matrix. The row and column labels refer only to singly excited configurations. Once again we use formaldehyde as our example, and trace the execution of a CIS calculation. The SCF calculation is the source of the formaldehyde MO energies and coefficients, as described above, and quoted below. We have already characterized the high-lying occupied MOs in our discussion of the SCF procedure. In particular, #7 is the pi bonding MO of symmetry b_1 and #8 is predominantly the pseudo-pi O lone pair of symmetry b_2. The lowest lying vacant MO is #9, the CO pi antibonding level of symmetry b_1. Levels #10 and #11 are CH sigma antibonding MOs of symmetry a_1 and b_2, and the highest lying level #12 is largely a CO sigma antibonding orbital of symmetry a_1. The optical spectrum is dominated by excitations to the CO pi antibonding level from the occupied pi bonding level and the pseudo-pi lone pair.

```
* * * * * * * * * * * * * * * * * * * * * * * * * * * * * * * * * * * * * * * * * * * * * * * * * * * *
PCLOBE   PCLOBE   PCLOBE   PCLOBE   PCLOBE   PCLOBE   PCLOBE   PCLOBE
* * * * * * * * * * * * * * * * * * * * * * * * * * * * * * * * * * * * * * * * * * * * * * * * * * * *
          Gaussian-Lobe Program for Organic Molecules
          Adapted to Personal Computers by Don Shillady
                 Virginia Commonwealth University
                        Richmond Virginia
                    One-Electron Energy Levels
                 E(  1)  =  -20.756914192746 a1
                 E(  2)  =  -11.473757332662 a1
                 E(  3)  =   -1.344589661060 a1
                 E(  4)  =   -0.832659560808 a1
                 E(  5)  =   -0.673507585901 b2
                 E(  6)  =   -0.601856695616 a1
                 E(  7)  =   -0.497583334056 b1
                 E(  8)  =   -0.420825519200 b2
                 E(  9)  =    0.177111638238 b1
                 E(10)  =    0.544086554928 a1
                 E(11)  =    0.627636330169 b2
                 E(12)  =    0.699311195798 a1
```

```
      Alpha-spin Orbitals for  8 Filled Orbitals by Column

#   At-Orb    1       2       3       4       5       6      7       8

1   C   1s  -0.001   0.994  -0.107   0.187   0.000   0.022 0.000   0.000
2   C   2s   0.008   0.027   0.227  -0.679   0.000  -0.087 0.000   0.000
3   C   2px  0.006   0.000   0.127   0.208   0.000  -0.497 0.000   0.000
4   C   2py  0.000   0.000   0.000   0.000  -0.563   0.000 0.000  -0.247
5   C   2pz  0.000   0.000   0.000   0.000   0.000   0.000 0.623   0.000
6   O   1s  -0.995   0.000  -0.228  -0.107   0.000  -0.099 0.000   0.000
7   O   2s  -0.024  -0.005   0.801   0.474   0.000   0.525 0.000   0.000
8   O   2px  0.005   0.000  -0.200   0.180   0.000   0.662 0.000   0.000
9   O   2py  0.000   0.000   0.000   0.000  -0.475   0.000 0.000   0.856
10  O   2pz  0.000   0.000   0.000   0.000   0.000   0.000 0.661   0.000
11  H   1s  -0.001  -0.006   0.025  -0.208  -0.241   0.143 0.000  -0.333
12  H   1s  -0.001  -0.006   0.025  -0.208   0.241   0.143 0.000   0.333
```

```
           #   At-Orb      9       10      11      12
           1   C   1s    0.000    0.212   0.000  -0.075
           2   C   2s    0.000   -1.451   0.000   0.549
           3   C   2px   0.000    0.304   0.000   1.263
           4   C   2py   0.000    0.000   1.170   0.000
           5   C   2pz   0.812    0.000   0.000   0.000
           6   O   1s    0.000   -0.051   0.000   0.127
           7   O   2s    0.000    0.331   0.000  -0.918
           8   O   2px   0.000   -0.321   0.000   0.901
           9   O   2py   0.000    0.000  -0.319   0.000
          10   O   2pz  -0.781    0.000   0.000   0.000
          11   H   1s    0.000    0.964  -0.881   0.244
          12   H   1s    0.000    0.964   0.881   0.244
```

The basis of excitations is listed below; there are 24 of these, corresponding to all possible singlet-coupled transfers of an electron from the valence set of six occupied MOs (#3 to #8) into the virtual manifold of four MOs (#9 to #12). In the output given, a typical excitation could be denoted as $8 \rightarrow 9$, for instance, to indicate a single excitation of an electron from orbital 8, the b_2-symmetric pseudo-pi lone pair on oxygen, to orbital 9, the b_1-symmetric pi* virtual orbital. The symmetry of the excitation would be $b_2 \times b_1 = a_2$.

```
* * * * * * * * * * * * * * * * * * * * * * * * * * * * * * * * * * * * * * * * * * * * * * * * * * * * * *
              Single-Excitation Configuration Interaction
         J. A. Pople, Proc. Phys. Soc., A, Vol. LXVIII, p81, 1955
* * * * * * * * * * * * * * * * * * * * * * * * * * * * * * * * * * * * * * * * * * * * * * * * * * * * * *
           CORE = 2          HOMO = 8      LUMO = 9  NBS = 12

           Excitation No.    1 = orbital (8 ->  9) b2 × b1 = a2
           Excitation No.    2 = orbital (8 -> 10) b2 × a1 = b2
           Excitation No.    3 = orbital (8 -> 11) b2 × b2 = a1
           Excitation No.    4 = orbital (8 -> 12) b2 × a1 = b2
           Excitation No.    5 = orbital (7 ->  9) b1 × b1 = a1
           Excitation No.    6 = orbital (7 -> 10) b1 × a1 = b1
           Excitation No.    7 = orbital (7 -> 11) b1 × b2 = a2
           Excitation No.    8 = orbital (7 -> 12) b1 × a1 = b1
           Excitation No.    9 = orbital (6 ->  9) a1 × b1 = b1
           Excitation No.   10 = orbital (6 -> 10) a1 × a1 = a1
           Excitation No.   11 = orbital (6 -> 11) a1 × b2 = b2
           Excitation No.   12 = orbital (6 -> 12) a1 × a1 = a1
           Excitation No.   13 = orbital (5 ->  9) b2 × b1 = a2
           Excitation No.   14 = orbital (5 -> 10) b2 × a1 = b2
           Excitation No.   15 = orbital (5-> 11) b2 × b2 = a1
           Excitation No.   16 = orbital (5 -> 12) b2 × a1 = b2
           Excitation No.   17 = orbital (4 ->  9) a1 × b1 = b1
           Excitation No.   18 = orbital (4 -> 10) a1 × a1 = a1
           Excitation No.   19 = orbital (4 -> 11) a1 × b2 = b2
           Excitation No.   20 = orbital (4 -> 12) a1 × a1 = a1
           Excitation No.   21 = orbital (3 ->  9) a1 × b1 = b1
           Excitation No.   22 = orbital (3 -> 10) a1 × a1 = a1
           Excitation No.   23 = orbital (3 -> 11) a1 × b2 = b2
           Excitation No.   24 = orbital (3 -> 12) a1 × a1 = a1
```

No excitations from the cores of C or O are included in this basis of single excitations.

A portion of the Hamiltonian matrix in the basis of single valence excitations is shown below. Please note that each diagonal element is the difference between the expectation values of the Hamiltonian and the ground state SCF energy. Thus, all values are positive. We see that this matrix is very sparse; since the Hamiltonian is totally symmetric in the group C_{2v}, only excitations of like symmetry can be mixed. If PCLOBE had listed the excitations within sets of like symmetry, the Hamiltonian matrix would appear as blocks for each symmetry species.

SINGLET Configuration Interaction Matrix

	1	2	3	4	5	6	7
1	0.1430	0.0000	0.0000	0.0000	0.0000	0.0000	-0.0017
2	0.0000	0.6871	0.0000	0.0658	0.0000	0.0000	0.0000
3	0.0000	0.0000	0.7804	0.0000	0.0827	0.0000	0.0000
4	0.0000	0.0658	0.0000	0.6672	0.0000	0.0000	0.0000
5	0.0000	0.0000	0.0827	0.0000	0.5119	0.0000	0.0000
6	0.0000	0.0000	0.0000	0.0000	0.0000	0.7212	0.0000
7	-0.0017	0.0000	0.0000	0.0000	0.0000	0.0000	0.7420
8	0.0000	0.0000	0.0000	0.0000	0.0000	0.0414	0.0000
9	0.0000	0.0000	0.0000	0.0000	0.0000	-0.0031	0.0000
10	0.0000	0.0000	0.0312	0.0000	0.0480	0.0000	0.0000
11	0.0000	0.0094	0.0000	0.0182	0.0000	0.0000	0.0000
12	0.0000	0.0000	-0.0473	0.0000	-0.2008	0.0000	0.0000
13	0.0695	0.0000	0.0000	0.0000	0.0000	0.0000	-0.0085
14	0.0000	-0.0033	0.0000	-0.0411	0.0000	0.0000	0.0000
15	0.0000	0.0000	0.0045	0.0000	-0.0876	0.0000	0.0000
16	0.0000	-0.0367	0.0000	0.0640	0.0000	0.0000	0.0000
17	0.0000	0.0000	0.0000	0.0000	0.0000	0.0305	0.0000
18	0.0000	0.0000	-0.0063	0.0000	0.0447	0.0000	0.0000
19	0.0000	0.0075	0.0000	-0.0019	0.0000	0.0000	0.0000
20	0.0000	0.0000	-0.0024	0.0000	0.0144	0.0000	0.0000
21	0.0000	0.0000	0.0000	0.0000	0.0000	0.0020	0.0000
22	0.0000	0.0000	-0.0070	0.0000	-0.0440	0.0000	0.0000
23	0.0000	0.0024	0.0000	-0.0119	0.0000	0.0000	0.0000
24	0.0000	0.0000	0.0305	0.0000	0.1135	0.0000	0.0000

The Hamiltonian matrix is strongly dominated by its diagonal. Thus, it is reasonable to refer to each eigenvector by its dominant excitation. For example, eigenvector #1 is almost entirely the $8 \rightarrow 9$ a_2-symmetry excitation generally called an $n \rightarrow \pi^*$ transition. Eigenvector #3 is primarily the $7 \rightarrow 9$ a_1-symmetry excitation, a $\pi \rightarrow \pi^*$ transition. The second eigenvector is the best representation of a Rydberg state possible in this limited basis; the experimental spectrum displays a number of Rydberg states between the $n \rightarrow \pi^*$ and $\pi \rightarrow \pi^*$ excitation energies.

SINGLET-C.I. Eigenvectors by Column

	1	2	3	4	5	6	7
1	0.9742	0.0000	0.0000	-0.2258	0.0000	0.0000	0.0000
2	0.0000	0.0000	0.0000	0.0000	0.5956	0.0000	0.0000
3	0.0000	0.0000	-0.1504	0.0000	0.0000	0.0000	0.0000
4	0.0000	0.0000	0.0000	0.0000	-0.7635	0.0000	0.0000
5	0.0000	0.0000	0.9233	0.0000	0.0000	0.0000	0.0000
6	0.0000	-0.0020	0.0000	0.0000	0.0000	-0.6116	-0.1596
7	-0.0005	0.0000	0.0000	-0.0290	0.0000	0.0000	0.0000
8	0.0000	-0.0056	0.0000	0.0000	0.0000	0.5351	0.6380

9	0.0000	-0.9934	0.0000	0.0000	0.0000	-0.0435	0.0497
10	0.0000	0.0000	-0.1078	0.0000	0.0000	0.0000	0.0000
11	0.0000	0.0000	0.0000	0.0000	0.0226	0.0000	0.0000
12	0.0000	0.0000	0.3089	0.0000	0.0000	0.0000	0.0000
13	-0.2259	0.0000	0.0000	-0.9737	0.0000	0.0000	0.0000
14	0.0000	0.0000	0.0000	0.0000	-0.1242	0.0000	0.0000
15	0.0000	0.0000	0.0940	0.0000	0.0000	0.0000	0.0000
16	0.0000	0.0000	0.0000	0.0000	0.2149	0.0000	0.0000
17	0.0000	-0.0720	0.0000	0.0000	0.0000	0.5812	-0.7515
18	0.0000	0.0000	-0.0508	0.0000	0.0000	0.0000	0.0000
19	0.0000	0.0000	0.0000	0.0000	-0.0081	0.0000	0.0000
20	0.0000	0.0000	-0.0004	0.0000	0.0000	0.0000	0.0000
21	0.0000	0.0890	0.0000	0.0000	0.0000	0.0046	-0.0171
22	0.0000	0.0000	0.0316	0.0000	0.0000	0.0000	0.0000
23	0.0000	0.0000	0.0000	0.0000	-0.0127	0.0000	0.0000
24	0.0000	0.0000	-0.0738	0.0000	0.0000	0.0000	0.0000

The eigenvalues of the CIS secular equation are approximate values for excitation energies. The energies are reported in hartrees relative to the ground state. The energies can be reexpressed in more widely used units, i.e., electron volts or the corresponding wavelength in nanometers.

SINGLET State Excitation Energy Values (experimental)

E(1) = 0.12689758 au   3.45306 ev 27850.80 1/cm 359.1 nm   (4.1 ev nπ*)

E(2) = 0.31082499 au   8.45798 ev 68218.20 1/cm 146.6 nm   (9.0 ev σπ*)

E(3) = 0.40367770 au   10.98463 ev 88597.01 1/cm 112.9 nm (10.7 ev ππ*)

E(4) = 0.44245885 au   12.03992 ev 97108.49 1/cm 103.0 nm

There are 4 transitions less than 12.4 ev.

The oscillator strength, a measure of the intensity of the absorption, is proportional to the square of the transition moment defined by the integral

$$\boldsymbol{\mu}_{0K} = \langle \Psi_0 | \mathbf{r} | \Psi_K \rangle$$

This integral is subject to symmetry-based selection rules. The components of $\mathbf{r}$ (that is, the coordinates x, y, z) transform as a_1, b_1, and b_2 in C_{2v}, and thus can mix excitations of symmetry a_1, b_1, or b_2 into the ground state. Excitations of symmetry a_2 cannot mix into the ground state under the influence of a perturbation which transforms as $\mathbf{r}$, so transitions to such states have zero transition moment—they are "forbidden." If the ground state is distorted, by vibrational motion, this prohibition can be relaxed. We take advantage of this loophole to describe the MCD of formaldehyde (Chapter 14).

These energies are quite inaccurate, and agreement with observed values should be considered fortunate, if not suspicious. This arises in part from

```
State Energy=   3.4531 ev    0.126898 au    359.06 nm
Symmetry=a2    Osc. Strength(f) =   0.00000
M= 0.00000    Mx= 0.00000    My= 0.00000    Mz=  0.00000 in Debyes
Transition Direction Angles    (0.0, 0.0, 0.0)    in degrees.

State Energy=   8.4580 ev    0.310825 au    146.59 nm
Symmetry=a1    Osc. Strength(f) = 0.00955
M= 0.54564    Mx= 0.00000    My= 0.00000    Mz=  0.54564 in Debyes
Transition Direction Angles    (90.0, 90.0, 0.0) in degrees.

State Energy=   10.9846 ev    0.403678 au    112.87 nm
Symmetry=a1    Osc. Strength(f) = 0.27197
M= 2.55521  Mx= -2.55521  My= 0.00000  Mz= 0.00000  in Debyes
Transition Direction Angles    (180.0, 90.0, 90.0) in degrees.

State Energy=   12.0399 ev    0.442459 au    102.98 nm
Symmetry=a2    Osc. Strength(f) = 0.00000
M= 0.00000  Mx= 0.00000  My= 0.00000  Mz= 0.00000 in Debyes
Transition Direction Angles    (0.0, 0.0, 0.0) in degrees.
```

the inflexibility of the basis which in this case is being asked to represent both the ground state and excited states, and in part from the intrinsic limitations of the single-excitation representation. Here the small STO-4GL basis is probably the greater problem. It cannot possibly represent the Rydberg-type absorptions (n → 3s,p) which are part of the formaldehyde spectrum (see following). Foresman et al. [1] explore the substantial effects of basis set extension.

As Table 13.2 suggests, CIS excitations in very small basis sets are systematically red for triplets and blue for singlets compared with experimental values. This can be understood as a consequence of the fact that the CIS representation of the excited state is less adequate than the RHF

TABLE 13.2

Basis Set Dependence of Computed Absorption in CIS: Transition Energies in Electron Volts (eV)

Transition	STO3G	6-31G(d)	6-31+G(d)	Exp. [4]
$^3A_2\, n \to \pi^*$	3.142	3.807	3.789	3.5
$^1A_2\, n \to \pi^*$	4.249	4.614	4.566	4.1
$^3A_1\, \pi \to \pi^*$	4.010	4.752	4.805	6.0
$^1A_1\, \pi \to \pi^*$	12.274	10.148	9.732	10.7

Source: From Foresman, J.B., Head-Gordon, M., Pople, J.A., et al. *J. Phys. Chem.* 96:135, 1992.

Note: Rydberg transitions observed between 7 and 9 eV are not well represented in the valence-based basis sets used here. (Estimated energies are blue by more than 1 eV.)

representation of the ground state. The orbitals forms used for the ground and excited states are defined by a variational calculation of the ground state energy.

Further, in the excited state there is no analogy to Brillouin's theorem. That is, whereas the RHF ground state has been chosen so that mixing of singly excited states is zero, the CIS state can admix further single excitations, i.e., double excitations from the SCF reference state. These excitations, were they included, would have an impact both on the SCF ground state and the CIS excited states, as we will show in our treatments of methods which incorporate such doubles (see CIS(D) and TDHF discussion below).

Structural Relaxation in the Excited State

The vertical excitation to a low-lying excited state will not produce an equilibrium structure; the excited molecule will seek a new form. According to Foresman et al. [1], a component of the gradient of the energy in the excited state can be written as the partial derivative of the CIS state energy:

$$E_{CIS} = E_{HF} + \sum_{ia} B_{ia}^2 (\varepsilon_a - \varepsilon_i) - \sum B_{ia} B_{jb} (\langle ja|ib \rangle - \langle ja|bi \rangle)$$

Here the B_{ia} are coefficients of the ith eigenfunction of the secular equation, and the two-electron integrals arise from the mixing integral for two single excitations $i \to a$ and $j \to b$.

The derivative with respect to any imposed change y, which could be an external field or a displacement of a nucleus, can be written

$$\frac{\partial E_{CIS}}{\partial y} = \frac{\partial E_{HF}}{\partial y} + \sum_{ia} B_{ia}^2 \left(\frac{\partial \varepsilon_a}{\partial y} - \frac{\partial \varepsilon_i}{\partial y} \right) - \sum B_{ia} B_{jb} R_{ijab}$$

The derivative of the Hartree–Fock energy has already been discussed. The CIS coefficients, **B**, have been chosen variationally, so their derivatives vanish, but the LCAO-MO coefficients, **c**, are no longer variational in the excited states. The derivatives of the two-electron terms in the CIS energy are collected in

$$R_{ijab} = [\langle (\partial \phi_j / \partial y) \phi_a | \phi_i \phi_b \rangle - \langle (\partial \phi_j / \partial y) \phi_a | \phi_b \phi_i \rangle]$$

$$+ \text{ similar forms involving } \frac{\partial \phi_a}{\partial y}, \frac{\partial \phi_b}{\partial y}, \text{ and } \frac{\partial \phi_i}{\partial y}$$

In order to evaluate the derivative of the Fock eigenvalues, we need derivatives of the overlap and Fock matrices:

$$F_{pp} = \varepsilon_p S_{pp} \text{ or } \sum_{\mu\nu} c_{p\mu} F_{\mu\nu} c_{p\nu} = \varepsilon_p \sum_{\mu\nu} c_{p\mu} S_{\mu\nu} c_{p\nu}$$

Before imposing the perturbation, S_{pq} (the overlap matrix in the basis of MOs) was the unit matrix. However, the perturbation may have an impact by its influence on the MO coefficients.

$$\frac{\partial \varepsilon_p}{\partial y} = \frac{\partial F_{pp}}{\partial y} - \varepsilon_p \frac{\partial S_{pp}}{\partial y}$$

To complete the energy derivative also requires the derivatives of the two-electron integrals and the MO coefficients. Following Pople et al. [5] we write

$$\mathbf{c} = \mathbf{U}\mathbf{c}_0; \quad \mathbf{U} = \mathbf{I} + \mathbf{u}; \quad c_{\mu p} = c_{\mu p}^0 + \sum_q u_{pq} c_{\mu q}^0; \quad \frac{\partial c_{\mu p}}{\partial y} = \sum_q c_{\mu q} \frac{\partial u_{qp}}{\partial y}$$

The authors now require solutions of coupled perturbed HF equations which define the $\mathbf{U}$ matrix.

Considerable manipulation—and insight—is required to put the final gradient expressions into a form efficient for computation. Foresman et al. [1] followed the pioneering work of Handy and Schaefer [6] and others [7] who developed general gradient techniques for more complex correlation-corrected schemes. Instead of following their argument further we will work through an example for the geometry optimization of our simple polyatomic system.

Case Study for Formaldehyde

PCLOBE does not optimize the geometry of excited states, so we are using GAUSSIAN-03. We begin with the ground state geometry optimized in RHF/6-31G(d,p): we determine the ground state structure and vibrational frequencies, and compute vertical excitation energies. Then we reoptimize the structure of formaldehyde in the lowest energy excited state, and reevaluate both the vibrational frequencies and excitation energies in this case pertaining to fluorescent emissions (Table 13.3).

Formaldehyde relaxes in the B_2 excited state by pyramidalizing, and in accord with the weakened CO bonding attending population of the CO π^* MO, extending the CO bond length. The CO stretching frequencies also decline sharply, by about 20%. The CH stretches are actually blue-shifted, but the wag and rock of the CH_2 fragment, and the HCH angle bend, are all made easier. As we have seen, the CIS computed excitation energies are very blue compared with experimental values. However the fluorescence spectrum is shifted substantially (by more than 1 eV) to the red of the absorption spectrum, owing to the geometry change.

TABLE 13.3

Changes in Molecular Parameters of Formaldehyde in the A_2 Excited State

State	Ground State	CIS Excited State	Exp.	
Structural parameters			Ground: Excited	
R(CO)	1.1843	1.2576	1.203	1.323
R(CH)	1.0933	1.0852	1.101	1.103
A(HCO)	122.1	117.6		
A(HCH)	115.7	117.9	116.3	118
D(HHCO)	180.0	150.2	0	144
Frequencies				
CH$_2$ wag	$B_1 = 1335$	$A' = 523$	1167	NA
(pyramidalization)				
CH$_2$ rock	$B_2 = 1377$	$A'' = 982$	1249	899
HCH bend	$A_1 = 1668$	$A' = 1425$	1500	1293
CO stretch	$A_1 = 2026$	$A' = 1662$	1746	1183
CH sym. stretch	$A_1 = 3121$	$A' = 3187$	2782	2846
CH antisym. stretch	$B_2 = 3195$	$A'' = 3280$	2850	2968
Excitations	Absorption	Emission	Absorption	Emission
A$_2$ (triplet)	3.94	$A''(T) = 3.10$	3.50	
A$_2$ (singlet)	4.76	$A'(T) = 3.64$	4.07	
A$_1$ (triplet)	5.24	$A''(S) = 3.99$	5.86	
B$_1$ (triplet)	8.88	$A'(T) = 7.41$		
B$_2$ (triplet)	9.85			
B$_1$ (singlet)	10.24	$A'(S) = 8.57$		
A$_1$ (singlet)	10.58	$A'(S) = 9.40$		

Note: Experimental values of the vibrational frequencies of ground state and excited state of formaldehyde are quoted in Clouthier and Ramsay [8]. Absorption energies are quoted from Ref. [4].

Correlation Corrections

One deficiency of the CIS model chemistry is that it includes no correlation correction to either ground or excited state. A variety of methods have been developed which go beyond this starting point. Perhaps the simplest correction, CIS(D) [9] adds two-electron corrections to both states. The key advance in CIS(D) is that the computed energies are size-extensive (the energy for N isolated systems is assured to be the energy of a single system $\times N$. Following Head-Gordon et al., the CIS wave function is

$$\Phi_{CIS} = U_1 \Phi_0 = \sum_{ia} b_i^a \Phi_i^a$$

U_1 induces all single excitations, $i \to b$. The MP2 ground state energy correction is generated by an operator T_2

$$E^{MP2} = \langle \Phi_0 | V | T_2 \Phi_0 \rangle = \frac{1}{4} \sum_{ijmn} a_{ij}^{mn} (ij \| ab); \quad a_{ij}^{mn} = \frac{(mn \| ij)}{\Delta_{ij}^{mn}}$$

The denominator Δ is composed of orbital energies $\varepsilon_m + \varepsilon_n - \varepsilon_i - \varepsilon_j$. The possibility of double excitations from the CIS space means that triple excitations from the ground state enter the wave function. If only those triples that can be represented as a double excitation accompanied by a separate single excitation are retained, the result

$$E^{CIS(D)} = \langle \Phi^{CIS} | V | U_2 \Phi_0 \rangle + \langle \Phi^{CIS} | V | T_2 U_1 \Phi_0 \rangle$$

is shown to be size consistent and to scale as the fifth power of the number of MOs. The corrected excitation energy is the difference between the CIS(D) energy and the ground MP2 energy.

More detailed CI can also yield a correlation-corrected description of excited states, and the coupled cluster and density functional methods can estimate excitations within the response-theoretic framework. We provide a first orientation to response theory in the following section.

Time-Dependent Hartree–Fock Treatments of Excitations

The family of time-dependent methods arise from the variational form of the time-dependent Schroedinger equation

$$\delta \int (\Phi^* \hat{H} \Phi - \Phi^* (i \partial \Phi / \partial t)) dr dt = 0$$

The central assumption is that the single-determinant form is adequate to the description of the response of the system to a time-varying perturbation. The MOs for the determinant Φ are expressed in a basis, $\varphi = \mathbf{u}\mathbf{C}(t)$ with the time dependence confined in the coefficients. For an orthogonal basis, we recover finally

$$\mathbf{h}^F \mathbf{C} = i \partial \mathbf{C} / \partial t$$

This is the time-dependent HF (TDHF) equation. The further processing of the equation toward a perturbative solution (i.e., the response to an oscillating effect with a specific frequency ω) is outlined by Cook [10]. Inspection of his equations can confirm that the major obstacle to use of the TDHF method is the transformation of certain two-electron integrals from an atomic to a molecular basis. In principle, a separate matrix inversion is required for each value of the frequency of the perturbation; given all these, one may construct the response to any perturbation expressible in Fourier components.

For certain frequencies characteristic of the molecular system, one finds a disproportionate response, that is to say resonance. Cook develops the matrix eigenvalue equation which identifies these special values of the frequency.

$$\begin{pmatrix} \mathbf{A} & \mathbf{B} \\ \mathbf{B} & \mathbf{A} \end{pmatrix} \begin{pmatrix} \mathbf{X} \\ \mathbf{Y} \end{pmatrix} = \omega \begin{pmatrix} 1 & 0 \\ 0 & -1 \end{pmatrix} \begin{pmatrix} \mathbf{X} \\ \mathbf{Y} \end{pmatrix}$$

or, after manipulation,

$$(\mathbf{A} + \mathbf{B})(\mathbf{X} + \mathbf{Y}) = \omega^2 (\mathbf{A} - \mathbf{B})^{-1}(\mathbf{X} + \mathbf{Y})$$

This is an eigenvalue problem of familiar form. The matrices $\mathbf{A}$ and $\mathbf{B}$ are defined

$$A_{ia,jb} = A_{jb,ia} = \delta_{ij}\varepsilon_{ii} - \delta_{ab}\varepsilon_{aa} + < ia \,||\, jb > - < ij \,||\, ab >$$
$$B_{ia,jb} = B_{jb,ia} = + \langle ia \,||\, jb \rangle - \langle ib \,||\, ja \rangle$$

(i, j are occupied MOs, and a, b are virtual orbitals). $\mathbf{A}$ and $\mathbf{B}$ require only quantities from the HF calculation besides the transformed two-electron integrals. Some, but not all, of these are done in an MP2 calculation. $\mathbf{X}(\omega)$ and $\mathbf{Y}(\omega)$ are defined by quantities generated in the SCF calculation, and refer to the perturbation through $\mathbf{\Delta}$.

$$X_{jb} = -\frac{\Delta_{bj}}{\varepsilon_b - \varepsilon_j - \omega}; \quad Y_{jb} = -\frac{\Delta_{bj}}{\varepsilon_j - \varepsilon_b + \omega}$$

The matrix $\mathbf{\Delta}$ is defined by properties of the SCF solution:

$$\Delta_{bj} = c_b^{\dagger}(\mathbf{f} + \mathbf{G})c_j$$

Here the $\mathbf{c}$ vectors are eigenvectors of the Fock matrix, $\mathbf{G}$ is the two-electron part of the Fock operator, and $\mathbf{f}$ is the spatial factor of the perturbation, all in the MO representation.

Our purpose here is to see what sort of performance can be expected from CIS, CIS(D), and TDHF methods. All these are implemented in GAUSSIAN [11]. One naturally expects the crudest and least consistent results from the CIS method, but such results may nonetheless be of practical value.

Formaldehyde Again

We see that these calculations describe only the low-lying (valence) excitations. The formaldehyde spectra have considerable Rydberg character,

TABLE 13.4

Vertical Excitations for Formaldehyde in CIS, TDHF, and RPA

	Formaldehyde Ground State				
State Label	Exp. (eV) [4]	TDHF (eV)	CIS (eV)	f_{CIS}	RPA[a] (eV)
$^3A_2$	3.50	3.649	3.938	0	2.13
$^1A_2$	4.07	4.598	4.765	0	3.47
$^3A_1$	5.86	2.679	5.236	0	3.88
$^3B_1$	6.83	8.503	8.881	0	6.53
$^3B_2$		9.601	9.850	0	NR
$^1B_1$	7.11	10.005	10.238	0.0012	8.56
$^1A_1$	8.14	10.054	10.579	0.2064	11.22

[a] Dunning and McKoy [12]. Basis set is 6-31+G(d) except for the early RPA report, which cites an otherwise undefined "minimal basis set." NR means not reported.

which is very hard for these models (especially in the modest 6-31+G(d) basis) to represent. TDHF is just a little better than CIS for the two leading transitions. The pioneering RPA calculation is not fairly compared here, owing to the small basis available then (Table 13.4).

A Case Study—Sulfur Dioxide

Hinchliffe et al. [13] applied CIS and CIS(D) to sulfur dioxide. Our HF/6-31+G(d) calculations on ground state sulfur dioxide put the bond angle at 118.6° and bond length at 1.415 Å. MP2 extends the bond lengths and reduces the stiffness of vibrations. The bond length seems to be difficult to capture.

Excitation energies respond to choices of basis and method. Table 13.5 shows excitation energies at the HF/6-31+G(d) geometry.

The vibrational frequencies are shifted upon vertical excitation to the $^1B_1$ state to 1368, 1395, and 388 cm^{-1}. The geometry changes in the lowest excited state—$^1B_1$ according to CIS/6-31+G(d)—extending the SO distance to 1.4874 and opening the OSO angle slightly to 121.48° (Table 13.6).

TABLE 13.5

Structures of Sulfur Dioxide in Several Model Chemistries

Model Chemistry	Angle (°)	Bond (Å)	Vibrations (cm^{-1})
HF/6-31+G(d)	118.60	1.4150	1569, 1359, 592
MP2/6-31+G(d)	119.30	1.4823	1271, 1060, 479
MP2/6-311++G(3d)	119.91	1.4550	1307, 1089, 488
CCSD/6-311++G(3d)	119.38	1.4340	1398, 1193, 523

TABLE 13.6

Vertical Exciations of Sulfur Dioxide in Several Models. (Oscillator Strengths)

Symmetry	CIS—Vertical		CIS(D)—Vertical	TDHF—Vertical	
$^1B_1$	4.832	(0.0182)	4.604	4.487	(0.0114)
$^1A_2$	5.564	(-0-)	4.889	5.301	(-0-)
$^1B_2$	7.221	(0.1298)	6.791	6.821	(0.1105)
$^1B_1$	9.940	(0.0038)	9.287	9.818	(0.0032)
$^1A_2$	10.036	(-0-)	9.029	9.829	(-0-)
$^1A_1$	10.064	(0.0284)	8.411	9.994	(0.0284)

Note: Basis is 6-31+G(d); structure optimized in HF/6-31+G(d) Excitation energies in eV.

The experimental characterizations of the long-wavelength transitions [14] are that the formally forbidden $^1A_2$ state lies at 340 nm (3.62 eV) and the $^1B_1$ state lies at 315 nm (3.91 eV) [14]. The CIS/6-31+G(d) calculation fails both quantitatively, blue by almost 2 eV, and qualitatively; it places the $^1B_1$ state below the $^1A_2$ state. The CIS(D) correction improves matters somewhat, bringing the two states within 0.3 eV of one another, but the excitations are still too blue by about 1 eV. TDHF is only slightly better than CIS uncorrected.

Since the CIS(D) is said to be the counterpart in excited states to MP2 in ground states (corrected to be size-extensive), we computed the structure of sulfur dioxide in the model chemistry MP2/6-311++G(3d). In that large basis, the vertical excitations become as shown in Table 13.7. This extension does not improve the description substantially. Hinchliffe et al. [13] report the CIS and CIS(D) values in Table 13.8.

It may seem disappointing that even quite ambitious calculations within the CIS and TDHF framework are far from reliable in any quantitative sense. About the only virtue that can be claimed for these calculations is that they are consistent in the ordering of low-lying excited states, though apparently the ordering is inconsistent with experimental spectra.

This disappointing outcome should not discourage the user, since the behavior of CIS and CIS(D) is generally satisfactory in qualitative aspects

TABLE 13.7

Vertical Excitations of Sulfur Dioxide in Several Models in eV

Symmetry	CIS—Vertical		CIS(D)—Vertical	TDHF—Vertical	
$^1B_1$	4.580	(0.0126)	4.369	3.958	(0.0114)
$^1A_2$	5.244	(-0-)	4.632	4.707	(-0-)
$^1B_2$	6.738	(0.1045)	6.367	6.146	(0.1105)
$^1B_1$	9.208	(0.0060)	8.708	9.038	(0.0032)
$^1A_1$	9.581	(0.0347)	8.228	9.533	(0.0149)
$^1A_2$	10.036	(-0-)	9.029	9.152	(-0-)

Note: Extended basis, 6-311++G(3d); structure optimized in MP2/6-311++G(3d).

TABLE 13.8

Vertical Excitations of Sulfur Dioxide

Symmetry	CIS—Vertical Energy (oscillator strength)		CIS(D)—Vertical Energy
$^1B_1$	5.214	(0.0202)	4.933
$^1A_2$	6.003	(-0-)	5.331
$^1B_2$	7.592	(0.1005)	7.111
$^1A_1$	10.004	(0.0440)	8.534
$^1B_1$	10.283	(0.0073)	9.629
$^1A_2$	10.519	(-0-)	9.396

Source: From Hinchliffe, A., Machado, H.J.S., Mkadmh, A., et al., *J. Mol. Struct. (Theochem)*, 717, 231, 2005.

Note: Excitation energies in electron volts. CIS(D) does not correct the oscillator strengths.

in more forgiving environments. Application of the CIS and TDHF to pyrrole, for example, yields a semiquantitative representation of the excitation spectrum, generally and consistently overestimating their energies by less than 0.5 eV. Foresman and Frish [15] have a very helpful description of how one may use CIS representations of excited states to interpret optical spectra, and also show how CAS calculations—"not for the faint of heart"— can help in the description of photochemical reaction pathways.

Adaptation of the Time-Dependent Formalism to DFT

TD-DFT provides the most accurate treatment of low-lying excitations now available for medium-sized molecules. The formalism borrows heavily from the TDHF response theory already mentioned. One might consult Bauernschmitt and Ahlrichs [16], who began with the time-dependent version of the Kohn–Sham equations developed by Runge and Gross [17] to estimate excitations within the DFT framework. This was transformed to the code adapted to GAUSSIAN by Stratmann et al. [18]. Gisbergen et al. [19] developed a unified formalism which is the foundation for the ADF excitation package. We will follow the presentation of Hirata and Head-Gordon [20].

Beginning with the Kohn–Sham equation

$$\sum (K_{pq}P_{qr} - P_{pq}K_{qr}) = i\frac{\partial P_{pr}}{\partial t}$$

it is required that the density matrix P obeys the idempotency condition

$$\sum_q P_{pq}P_{qr} = P_{pr}$$

K is the Kohn–Sham energy operator. An orthonormal basis ϕ_p of Kohn–Sham orbitals is generated by the time-independent DFT problem.

$$K^0_{pq} = \delta_{pq}\varepsilon_p; \quad P^0_{ij} = \delta_{ij}; \quad P^0_{ia} = P^0_{ai} = 0$$

The perturbation is

$$g_{pq} = \frac{1}{2}[f_{pq}e^{-i\omega t} + f^*_{qp}e^{i\omega t}]$$

The response of the density matrix is

$$P^1_{pq} = \frac{1}{2}[d_{pq}e^{-i\omega t} + d^*_{qp}e^{i\omega t}]$$

The Kohn–Sham energy operator changes partly because the perturbation adds a term and partly because the density matrix responds to the perturbation.

$$K_{pq}(g) = K^0_{pq} + g_{pq} + \sum_{rs} \frac{\partial K_{pq}}{\partial P_{rs}} P^1_{rs}$$

The partial derivative of K is

$$\frac{\partial K_{pq}}{\partial P_{rs}} = (pq \| rs) + (pq|V_{XC}|rs)$$

Hirata and Head-Gordon recover the eigenvalue equation similar to the TDHF equation:

$$\begin{pmatrix} \mathbf{A} & \mathbf{B} \\ \mathbf{B} & \mathbf{A} \end{pmatrix} \begin{pmatrix} \mathbf{X} \\ \mathbf{Y} \end{pmatrix} = \omega \begin{pmatrix} 1 & 0 \\ 0 & -1 \end{pmatrix} \begin{pmatrix} \mathbf{X} \\ \mathbf{Y} \end{pmatrix}$$

Here the matrices $\mathbf{A}$ and $\mathbf{B}$ are defined:

$$A_{ai,bj} = \delta_{ij}\delta_{ab}(\varepsilon_a - \varepsilon_i) + (ai \| jb) + (ai|V_{XC}|jb)$$

Hirata and Head-Gordon introduce a Tamm–Dancoff simplification; this reduces the problem to

$$\mathbf{AX} = \omega\mathbf{X}$$

The problem is now very similar in computational demand to the CIS calculation, apart from the evaluation of the density functional.

TABLE 13.9

Formaldehyde Optical Excitations in a Variety of Density Functionals

CH_2O	State	CIS	SVWN	LB	B3LYP	RPA	SVWN	LB	B3LYP	Exp. [4]
Singlets	V^1A_2	4.48	3.65	3.52	3.87	4.30	3.63	3.49	3.85	4.07
	R^1B_2	8.63	5.90	7.45	6.44	8.62	5.89	7.43	6.43	7.11
	R^1B_2	9.36	6.68	8.63	7.20	9.35	6.68	8.62	7.20	7.97
	V^1A_1	9.45	6.54	8.76	7.16	9.07	6.54	8.73	7.15	–
	R^1A_1	9.66	7.22	10.09	7.94	9.58	7.22	9.58	7.94	8.14
Triplets	V^3A_2	3.67	3.03	2.88	3.19	3.35	3.00	2.86	3.12	3.50
	V^3A_1	4.65	6.16	6.01	5.62	1.24	6.02	5.86	5.20	5.86
	R^3B_2	8.28	5.84	7.15	6.33	8.20	5.84	7.14	6.32	6.83

Overall performance for nitrogen, carbon monoxide, water, ethylene, and formaldehyde

| Deviation | V | 0.72 | 0.23 | 0.36 | 0.28 | 1.51 | 0.24 | 0.36 | 0.42 | |
| (eV) | R | 0.86 | 1.11 | 0.77 | 0.79 | 0.84 | 1.11 | 0.73 | 0.80 | |

Source: From Hirata, S. and Head-Gordon, M., *Chem. Phys. Lett.*, 314, 291, 1999.

Note: V = valence; R = Rydberg; excitation energies in electron volts.

Hirata and Head-Gordon employed a variety of functionals and a triple-zeta basis with extensive polarization and augmentation with diffuse functions. This should embrace several valence excitations and also transitions with Rydberg character. Here is a selection from their extensive table (Table 13.9). It shows there is little to be gained with the full-fledged TD-DFT calculation.

We will give most attention to the practical application and performance of TD-DFT as implemented in GAUSSIAN and ADF. This is the approach taken by Koch and Holthausen [21], but we will consider somewhat larger molecules than they describe.

Applications

Sulfur Dioxide in TD-DFT

First, we look back at sulfur dioxide; B3LYP/6-31G(d) produces the ground state geometry and vibrational frequencies quoted in Table 13.10.

The structure is still sensitive to the completeness of the basis. The TD-B3LYP vertical excitations in these basis sets are shown in Table 13.11.

TABLE 13.10

Sulfur Dioxide Structure and Vibrations in B3LYP with Two Basis Sets

Model Chemistry	Angle (°)	Bond (Å)	Vibrations (cm^{-1})
B3LYP/6-31+G(d)	119.12	1.4636	1338, 1141, 502
B3LYP/6-311++G(3d)	119.41	1.4403	1359, 1161, 509

TABLE 13.11

Excitation Energies (eV) and Oscillator Strengths for Sulfur Dioxide in B3LYP with Two Basis Sets

Symmetry	B3LYP/6-31+G(d)		B3LYP/6-311++G(3d)	
$^1B_1$	3.84	(0.0053)	4.42	(0.0051)
$^1A_2$	4.19	(-0-)	5.54	(-0-)
$^1B_2$	6.19	(0.0543)	6.37	(0.0566)
$^1B_1$	7.99	(0.0014)	8.25	(0.0032)
$^1A_2$	8.15	(-0-)	8.42	(-0-)
$^1A_1$	8.78	(0.0323)	8.25	(0.0355)

Paradoxically the smaller basis set performs better than the large basis set. This is not an uncommon outcome in DFT, but is not well understood. Another fortunate quirk of the method is that performance is sometimes better in larger systems.

Ahlrichs et al. not only developed a formalism for TD-DFT but also surveyed the performance of a number of pure and hybrid density functionals, compared with CIS and TDHF results. Their statistical analysis of results for nitrogen, ethylene, formaldehyde, and pyridine showed that the B3LYP functional produced the best results for low-energy excitations (MAD on the order of 0.3 eV for excitation energies up to half of the ionization energy) but degrades for higher energy excitations. It is however always better than CIS or TDHF.

Pyridine in TD-B3LYP

Bauernschmitt and Ahlrichs [16] used an extended basis owing to Sadlej [22], optimized to describe static polarizabilities. A sample of their results appears in Table 13.12. B3LYP is a considerable improvement over the

TABLE 13.12

Excitation Energies for Species Treated in B3LYP with Two Basis Sets

Species	Symmetry	B3LYP	CIS	RPA	Experimental Value
N_2	$^1\Delta$	9.73	9.06	8.78	10.27
Formaldehyde	$^1B_1$	8.93	9.83	9.59	9.0
Pyridine	$^1A_1$	6.19	6.45	6.12	6.38
	$^1A_2$	5.07	6.79	6.79	5.43
	$^1B_22$	5.47	6.10	5.85	4.99
	$^1B_1$	4.76	6.16	5.99	4.59
	$^3A_2$	4.93	6.66	6.64	5.40
	$^3B_2$	4.47	4.63	4.30	4.84
	$^3A_1$	3.91	3.53	4.79	4.1
	$^3B_1$	4.05	5.12	4.82	4.1
MAD		0.29	0.90	0.88	

Source: From Bauernschmitt, R. and Ahlrichs, R., *Chem. Phys. Lett.*, 256, 454, 1996.

TABLE 13.13

Singlet Excitation Energies for Pyrrole in Four Model Chemistries

State	CIS	SAC-CI [24]	TD-B3LYP	CCSD [25]	Exp.
$^1A_2$	5.37	5.11	4.71	5.12	5.22
	5.99	5.82	5.34	5.83	
	6.52	6.38	5.52	6.40	
$^1B_1$	5.92 (0.0346)	5.80	5.32 (0.0749)	5.82	5.7
	6.54 (0.0030)	6.39	5.54 (0.0017)	6.43	6.4
	6.66 (0.0003)	6.05 [?]	5.59	5.97 [?]	
$^1B_2$	6.04 (0.0603)	5.88	5.52 (0.0243)	5.96	5.86
	6.64 (0.0416)	6.48	5.83	6.61	6.3 ± 0.2
	6.91 (0.0900)	6.78	5.86	6.86	6.78
$^1A_1$	6.73 (0.0000)	6.41	5.82	6.53	7?
	7.33 (0.0009)	6.64	6.29	6.73	
	7.42 (0.0026)	6.86	6.37	6.89	

Source: From Zhan, C.G. and Dixon, D.A., *J. Mol. Spectrosc.*, 216, 81, 2000.

random phase approximation (TDHF) and CIS, reducing the mean absolute deviation (MAD) from 0.9 to 0.3 eV.

Pyrrole in B3LYP-DFT

Zhan and Dixon [23] studied pyrrole, using the cc-pVTZ basis augmented with diffuse functions, with TD-B3LYP as well as CIS and SAC-CI (symmetry-adapted cluster [24]) techniques (see Table 13.13). Their hopeful conjecture is that CIS and TD-DFT can be used to estimate upper and lower limits, respectively, for excitation energies. The underestimate seems to be a systematic error of density functional methods—ionization energy estimates using the analog to Koopmans' theorem also produce low estimates. This has been attributed to the incorrect behavior of the exchange functionals at large distance; they decay faster than the proper coulombic limiting form.

Their results for triplets are shown below (Table 13.14). We estimated the spectrum with TD-B3LYP/6-31+G(d) for pyrrole, *N*-methylpyrrole, and 2,5-dimethylpyrrole (Table 13.15).

TABLE 13.14

Triplet Excitation Energies for Pyrrole in Three Model Chemistries

State	CIS (eV)	SAC-CI (eV)	TD-B3LYP (eV)	Exp. (eV) [23]
$^3B_2$	3.76	4.58	4.11	4.2
$^3A_2$	5.31	5.08	4.70	5.10
$^3A_1$	5.15	5.60	5.24	
$^3B_1$	5.85	5.82	5.31	

Source: From Zhan, C.G. and Dixon, D.A., *J. Mol. Spectrosc.*, 216, 81, 2000.

TABLE 13.15

Excitation Energies for Pyrrole Derivatives in TD-B3LYP/6-31+G(d) in eV

State	[a]Pyrrole 6-31+G(d)	Pyrrole 6-31+G(d)	2,5-Dimethylpyrrole 6-31+G(d)	State	N-methylpyrrole 6-31+G(d)
$^3B_2$	4.11	4.093	3.859	$^3A''$	4.019
$^3A_2$	4.87	4.920	4.431	$^3A'$	4.976
$^3A_1$	5.23	5.182	5.066	$^3A''$	5.027
$^3B_1$	5.64	5.626	5.000	$^3A'$	5.526
$^1A_2$	4.89 (0000)	4.943 (0000)	4.462 (0000)	$^1A''$	5.058 (0.0000)
$^1B_1$	5.72 (0.0121)	5.716 (0.0045)	5.599 (0.0007)	$^1A''$	5.674 (0.0060)
$^1B_2$	6.04 (0.1841)	6.054 (0.1864)	5.697 (0.1086)	$^1A''$	5.586 (0.1160)
$^1A_1$	6.43 (0.0001)			$^1A'$	5.7058 (0.0000)

Note: [a]Pyrrole results in left column are by Zahn and Dixon [23].

The shifts upon methyl substitution may be trustworthy, though the absolute values are not impressively accurate. We might take note of the fact that the order of the second and third singlet excited states is switched for *N*-methylpyrrole and 2,5-dimethylpyrrole.

A hopeful suggestion comes out of Dixon and Khan's work, that the TD-DFT method is not very sensitive to basis quality. Even the modest 6-31+G(d) basis allows informative calculations, even though it contains no Rydberg orbitals and many of the excited states have considerable Rydberg character. It may be worth recalling that the standard for useful work in this context is precision within a few tenths to about half an electron volt.

Substituted Aniline

Zhan and Dixon's study of substituted aniline introduces the exchange–correlation functional PBE1PBE, which is said to be an improvement on more familiar functionals (Table 13.16).

From this small example it seems the improvement for a large system is marginal.

Recent work correcting the exchange functional to display proper (coulombic) long-range behavior promises to effect further improvement in the description of properties involving the outer reaches of the electronic distribution including polarizability and optical excitations. Particularly appealing is the SAOP model, discussed in more detail in Chapter 11. We

TABLE 13.16

Excitation Energies for *N,N*-dimethyl-2,6-dimethyl-4-cyano-aniline

State	S_1(CT)	S_3	T_1 (CT)	T_2
Exp.	4.27	5.00	3.48	4.31
TD-B3LYP	3.91	4.98	3.08	3.94
TD-PBE1PBE	4.09	5.16	3.02	4.01

TABLE 13.17

Excitation Energies for Formaldehyde with PBE/DZP and SAOP/ATZP

			Singlet–Singlet			
No.	SAOP/ATZP	f	PBE/DZP	F	Symmetry	Exp. [4]
1:	4.1814	0.000	3.94490	0.000	A_2	4.07
2:	7.2057	0.560E-01	7.09386	0.6883E-01	B_1	7.11
5:	8.3012	0.220E-01			B_1	7.97
4:	8.4299	0.589E-01	9.46038	0.1103E-02	A_1	8.14
3:	8.9541	0.529E-04	8.99471	0.4577E-04	B_2	
			Singlet–Triplet			
1:	3.5855	0.000	3.29443	0.000	A_2	3.50
2:	6.1916	0.000	6.15479	0.000	A_1	5.86
3:	6.9639	0.000	6.70781	0.000	B_1	6.83

will return to formaldehyde to get a first idea of the value of these methods. Table 13.17 contains data on excitations in formaldehyde.

The relatively simple (!) PBE functional performs pretty well for the lowest (A_2 valence) transitions and even gives a sensible view of the leading B_1 Rydberg transition at 7.1 eV. The SAOP density functional, which has correct asymptotic behavior at large distances from the nuclear center, seems to blue-shift the low-lying transitions. It seems to give a more coherent view of the two B_1 singlet Rydberg states and the mixed-character $\pi \rightarrow \pi^*$ transition. We recommend use of this method with an augmented basis with triple-zeta valence splitting and polarization.

CCSD-EOM Treatment of Excited States

We are grateful to Prof. Bartlett's group member Ajith Perera who provided us the results of an ACES-II calculation with the CCSD-EOM representation of excited states. We quote parts of the output from a calculation with 170 symmetry adapted basis functions. The variant called "similarity transformed equations of motion for coupled cluster theory" or STEOM-CC theory, introduced by Nooijen and Bartlett [26], employs a similarity transform of the Hamiltonian which strongly reduces the coupling between singly excited determinants and more highly excited configurations. Consequently, excitation energies can be obtained to a good approximation by diagonalizing the transformed Hamiltonian in the space of single excitations only. The accuracy of STEOM is considered to be comparable to equation-of-motion coupled cluster theory and MCSCF-based methods.

The best density functional theory gives excitations at 7.3, 8.3, and 8.4 eV, to be compared with experimental 7.11, 7.97, and 8.14 eV. The EOM method quoted here looks to enjoy much better accuracy than even SAOP-based DFT,

ST-EOM-CC Singlets

Multiplet	root	irrep	energy diff (eV)	osc.	strength %	singles	total energy
Singlet	1	[2]	7.01940886	0.0000E+00		92.90	−113.99111190
Singlet	2	[2]	7.97389777	0.0000E+00		93.29	−113.95603509
Singlet	1	[1]	8.03751989	0.0000E+00		92.98	−113.95369703
Singlet	2	[1]	10.42553087	0.0000E+00		92.91	−113.86593927

EOM-CCSD Singlets

Multiplet	root	irrep	energy diff (eV)	osc.	strength %	active %	singles	total energy
Singlet	1	[2]	6.95089681	0.2264E-01		99.88	92.42	−113.99362967
Singlet	2	[2]	7.90427359	0.4320E-01		99.91	93.04	−113.95859373
Singlet	1	[1]	7.96673616	0.6744E-01		99.86	92.43	−113.95629828
Singlet	2	[1]	9.50903302	0.1847E+00		98.38	91.09	−113.89961993

estimating excitation energies at ca. 6.95, 7.90, and 8.14 eV, within 0.2 eV of observed values. STEOM seems to be a significant further improvement.

Conclusions

The computational modeling of excited states is much more demanding than the study of ground states, for which structure and vibrational spectra can be accurately estimated. The quickest appreciation of the changes in structure and properties attending excitation is the CIS method. This very inexpensive method can generally provide semiquantitative guidance suitable for interpretation of optical spectra and even some aspects of photochemistry. Correlation corrections can be achieved by more extensive configuration interaction, either in variational form or by perturbation expansions including especially the coupled cluster "equations of motion" variant. Alternatively, density functional theory in its linear-response, "time-dependent" formulation promises high accuracy estimates of the energies of excited states at modest cost. So far, calculations are most successful in systems with familiar chromophores, and are most reliable for low-energy transitions dominated by single excitations. Higher energy excitations, like the states for systems lacking chromophores, are hard to model consistently owing to their partial Rydberg character. Familiar basis sets, optimized for variational calculations of ground state energies, are not ideally suited for description of the outer reaches of the molecular charge distributions. The same can be said for most of the exchange–correlation functionals so far developed for DFT description of ground state properties.

References

1. Foresman, J.B., Head-Gordon, M., Pople, J.A., et al., *J. Phys. Chem.* 96:135, 1992.
2. Szabo, A. and Ostlund, N.S., *Modern Quantum Chemistry: Introduction to Advanced Electronic Structure Theory*, Macmillan, NY, 1982.
3. Pople, J.A., *Proc. Phys. Soc. A*, LXVIII(68):81, 1955.
4. Duncan, J.L. and Mallinson, P.D., *Chem. Phys. Lett.* 23:597, 1973; Nakanaga, T., Kondo, Sh., and Saeki, Sh., *J. Chem. Phys.* 76:3860, 1982.
5. Pople, J.A., Krishnan, R., Schlegel, H.B., et al., *Int. J. Quantum Chem. Symp.* 13:225, 1979.
6. Handy, N.C. and Schaefer, III H.F., *J. Chem. Phys.* 81:5031, 1984.
7. Yamaguchi, Y., Osamura, Y., Goddard, J.D., et al., *A New Dimension to Quantum Chemistry: Analytic Derivative Methods in Ab Initio Molecular Electronic Structure Theory*, Oxford University Press, Oxford, UK, 1994.
8. Clouthier, D.J. and Ramsay, D.A., *Ann. Rev. Phys. Chem.* 34:31, 1983; see: http://www.cem.msu.edu/~reusch/VirtualText/Spectrpy/InfraRed/infrared.htm
9. Head-Gordon, M., Maurice, D., and Oumi, M., *Chem. Phys. Lett.* 246:114, 1995.
10. Cook, D.B., *Handbook of Computational Quantum Chemistry*, Oxford University Press, Oxford, UK, 1998.
11. Frisch, M.J., Trucks, G.W., Schlegel, H.B., et al., Gaussian 03, Revision C.02, Gaussian Inc., Wallingford, CT, 2004.
12. Dunning, T. and McKoy, V.J., *Chem. Phys.* 48:5263, 1968.
13. Hinchliffe, A., Machado, H.J.S., Mkadmh, A., et al., *J. Mol. Struct. (Theochem)* 717:231, 2005.
14. Okabe, H., *J. Am. Chem. Soc.* 93:7095, 1971; Cosofret, D.R., Dylewski, S.M., and Houston, P.L., *J. Phys. Chem.* 104:10240, 2002.
15. Foresman, J.B. and Frisch, Æ., *Exploring Chemistry with Electronic Structure Methods*, 2nd Ed., Gaussian Inc., Pittsburgh, PA, 1996.
16. Bauernschmitt, R. and Ahlrichs, R., *Chem. Phys. Lett.* 256:454, 1996.
17. Runge, E. and Gross, E.K.U., *Phys. Rev. Lett.* 52:997, 1984.
18. Stratmann, E., Scuseria, G.E., and Frisch, M.J., *J. Chem. Phys.* 109:8218, 1998.
19. Gisbergen, S.J.A., Snijders, J.G., and Baerends, E.J., *J. Chem. Phys.* 103:9347, 1995.
20. Hirata, S. and Head-Gordon, M., *Chem. Phys. Lett.* 314:291, 1999.
21. Koch, W. and Holthausen, M.C., *A Chemist's Guide to Density Functional Theory*, 2nd Ed., Wiley-VCH, New York, NY, 2001.
22. Sadlej, A., *Theor. Chim. Acta* 79:123, 1991.
23. Zhan, C.G. and Dixon, D.A., *J. Mol. Spectrosc.* 216:81, 2000.
24. Wan, J., Meller, J., Hada, M., et al., *J. Chem. Phys.* 113:7853, 2000.
25. Christiansen, O., Gauss, J., Stanton, J.F., et al., *J. Chem. Phys.* 111:525, 1999.
26. Nooijen, M. and Bartlett, R.J.J., *Chem. Phys.* 107:6812, 1997.

14

Circular Dichroism and Optical Rotatory Dispersion

The Phenomenon of Optical Rotation

Optical rotation refers to the effect on a plane polarized beam of light as it passes through a chiral medium. A plane polarized beam of light can be thought of as the superposition of two circularly polarized beams, of equal amplitude and opposite senses of polarization. The most commonly observed phenomenon is a reorientation of the plane of polarization, "circular birefringence." This may be considered a consequence of a different index of refraction for the left- and right-polarized components, n_L and n_R. Moscowitz [1] writes

$$\alpha = \frac{\pi}{\lambda}(n_L - n_R)$$

where α is the optical rotation due to circular birefringence in radians per centimeter and λ is the wavelength of light used in the measurement. The optical rotation is dependent on wavelength and temperature; the specific rotation is defined at a given temperature T (usually 25°C) and wavelength λ (usually the Na D-line at 589 nm) as

$$[\alpha]_\lambda^T = \frac{\alpha}{c} \cdot \frac{1800}{\pi}$$

Here c is the concentration in grams per milliliter of solution and $1800/\pi$ is the conversion giving specific rotation in degrees per decimeter. The sample cells are usually cylindrical tubes 1 or 2 decimeters in length, long enough to allow observably large rotation. The rotation also depends on the concentration of the chiral material. This factor appears in the definition of the molecular rotation

$$[\Phi]_\lambda^T = \frac{[\alpha]_\lambda^T \cdot M}{100}$$

where M is the molecular weight of the optically active compound.

Circular dichroism (CD) is associated with a difference in absorption of left and right circularly polarized light. Accordingly, CD is observed near an absorption band. Define the absorption coefficient k by

$$I = I_0 \exp(-kl) = I_0 \exp\left(-\left[\frac{4\pi\kappa}{\lambda}\right]l\right)$$

I_0 and I are respectively the intensity of light entering and emerging from a chiral medium, following a path of length l. If the absorption coefficient differs for right- and left-polarized components of light, the "ellipticity" θ is given by

$$\tan\theta = \tanh\left[\frac{(k_L - k_R)l}{4}\right]$$

A useful linear approximation produces an expression for the ellipticity per centimeter

$$\theta = \frac{k_R - k_L}{4} = \left(\frac{\pi}{\lambda}\right)(\kappa_L - \kappa_R)$$

The molecular ellipticity is

$$[\theta] = \left(\frac{1800}{\pi}\right)\left(\frac{M}{C'}\right)\left(\frac{k_L - k_R}{4}\right)$$

The quantum mechanical description of optical activity was provided first by Rosenfeld [2] and Condon [3]. The accessible reinterpretation by Moscowitz [1] is highly recommended. Atkins and Friedman's [4] brief discussion is also an excellent starting point. Here we point out some landmarks in the development.

For a homogeneous isotropic optically inactive medium, the electric displacement vector and the magnetic induction vector are defined by

$$\mathbf{D} = \varepsilon\mathbf{E}$$
$$\mathbf{B} = \mu\mathbf{H}$$

These equations make reference to the dielectric constant ε and electric field intensity $\mathbf{E}$ and also the magnetic permeability μ and magnetic field intensity $\mathbf{H}$. In a chiral medium these equations become

$$\mathbf{D} = \varepsilon\mathbf{E} - g\left(\frac{\partial\mathbf{H}}{\partial t}\right)$$
$$\mathbf{B} = \mu\mathbf{H} + g\left(\frac{\partial\mathbf{E}}{\partial t}\right)$$

That is, the electric displacement vector **D** depends in part on the time dependence of the magnetic field and the magnetic induction **B** depends in part on the time dependence of the electric field.

Caldwell and Eyring [5] rederive Rosenfeld's results by working through the perturbation theoretic description of the expectation value of an observable Q in an oscillating vector potential, and apply the general results to the electric and magnetic dipole. Upon averaging overall orientations with respect to the laboratory frame, they obtain

$$\langle\bar{\boldsymbol{\mu}}\rangle_{nn} = \alpha_n \mathbf{E} + \frac{\beta}{c}\frac{\partial\mathbf{B}}{\partial t}$$

$$\langle\bar{\mathbf{m}}\rangle_{nn} = \kappa_n \mathbf{B} + \frac{\beta}{c}\frac{\partial\mathbf{E}}{\partial t}$$

Rosenfeld's quantum treatment takes into account the fact that an incident light wave can induce an electric transition in a molecule simultaneously with a magnetic transition. His second-order perturbation treatment produces a contribution to the optical rotation from each excitation

$$\beta_{0\to j} = \left(\frac{c}{3\pi h}\right)\frac{\mathrm{Im}\{\langle 0|\boldsymbol{\mu}|j\rangle\bullet\langle j|\mathbf{m}|0\rangle\}}{\nu_j^2 - \nu^2}$$

where $\boldsymbol{\mu}$ and $\mathbf{m}$ are the electric and magnetic dipole operators, respectively. The total rotation depends on the number density (N_1) of the chiral species and the sum of all such transition terms [1].

$$\phi = (16\pi^3 N_1)(\nu^2/c^2)\sum_j \beta_{0\to j} = \sum_j \phi_j$$

Note that $\beta_{0\to j}$ is in fact a 3×3 tensor with components $\beta_{\alpha\beta}\,(0\to j)$ with the indices defined by the three choices of Cartesian component for the electric and magnetic dipole operators. For later use we introduce another tensor: the rotational strength of a transition, $\mathbf{R}_j$

$$\mathbf{R}_j = \mathrm{Im}\{\langle 0|\boldsymbol{\mu}|j\rangle\bullet\langle j|\mathbf{m}|0\rangle\}$$

Caldwell and Eyring derive the sum rule

$$\sum_m \mathbf{R}_m = 0$$

This relation also applies to the experimentally accessible orientational average of the rotational strength (the average of the diagonal elements of the tensor). It is easy to recognize that

$$\beta \to 0 \text{ as } \nu \to 0 \text{ or } \infty$$

This means that the optical rotation also approaches zero for frequencies that are low and high compared with system resonant absorptions.

Kauzmann [6] gives the expression for the molar optical activity at a given wavelength λ and temperature T

$$[M]_\lambda^T = \frac{48N_0}{\hbar c} \left(\frac{n^2 + 2}{3}\right) \sum_k \frac{\lambda_{k0}^2 \operatorname{Im}(m_{0k} \cdot \mu_{k0})}{(\lambda^2 - \lambda_{k0}^2)}$$

where

N_0 is Avogadro's number

n is the refractive index of the medium

λ_{k0} is the resonant wavelength of the kth transition

Optical dispersion is one aspect of the response of a charge distribution to light. The complex polarizability

$$\beta = \theta + i\phi$$

contains information on both dispersion ϕ (rotation, or distinct refractive indices for left and right circularly polarized light) and absorption θ (dichroism, or distinct absorption coefficients for left and right circularly polarized light). The Kramers–Kronig [1] relations link the real and imaginary parts

$$[\phi(\nu)] = \left(\frac{2\nu^2}{\pi}\right) \int_0^\infty \frac{[\theta(n)]}{n(n^2 - \nu^2)} dn \text{ and } [\theta(\nu)] = \left(\frac{-2\nu^3}{\pi}\right) \int_0^\infty \frac{[\phi(n)]}{n^2(n^2 - \nu^2)} dn$$

If the absorption spectrum is a series of very sharply peaked forms (delta functions), then

$$\theta(\nu) = \sum_k \theta_k \delta(\nu - \nu_k)$$

and the integration is much simplified:

$$\phi(\nu) = 2\pi\nu_0^2 \sum_k \frac{\theta_n}{\nu_k(\nu_k^2 - \nu^2)}$$

Interpreting θ_n as the area under the absorption (circular dichroism) curve, Caldwell and Eyring [5] show that the (scalar) rotational strength R_n is proportional to θ_n and can be estimated from the experimental circular dichroism spectrum.

Moscowitz [1] specialized the Kramers–Kronig relation to CD line shapes of Gaussian form, which afford a more realistic representation of CD spectra. The transformation could also be effected entirely numerically, using a digitized trace of the spectrum. In principle, if one measures the CD spectrum, the Kramers–Kronig transform can be used to produce the ORD spectrum. The limitation of this procedure is that, in general, one cannot obtain the full spectrum of the high-energy transitions to realize the upper limit on the integration. However, experience with the MOSCOW [7] program implementation shows that using the CD bands to 200 nm will give a very good approximation to the ORD curve in the visible; where vacuum–UV CD spectra are available, the added bands from that spectral region will improve the transformed representation of the ORD curve.

PCLOBE Modeling of CD Spectra

The most direct way to model optical rotation would seem to be to compute excited states of the molecule in question, evaluate the several rotational strengths, and carry out the sum of terms refering to the transitions. The CD spectrum would be a welcome by-product. We can accomplish this within PCLOBE.

As we have already seen, PCLOBE represents excited states by mixing states obtained by single excitations from the SCF ground state determinant. This configuration interaction with singles (CIS) method is simple, size-extensive, and well defined as a model chemistry. All single excitations are orthogonal to the SCF ground state determinant, so a linear variation calculation in the CIS space is appropriate. The drawback is that CIS typically overestimates excitation energies. We will see what we can learn from the simplest exercise.

1. Construct the singlet spin single excitations

$$\Psi_i^a = \left(\left| \chi_i^\alpha \chi_a^\beta \right| - \left| \chi_i^\beta \chi_a^\alpha \right| \right) / \sqrt{2}$$

Within PCLOBE we limit the number of excitations to 75.

2. Construct the matrix elements needed to define elements of the secular equation

$$\left\langle \Psi_i^a | H | \Psi_j^b \right\rangle$$

Here i, j orbitals are occupied and a, b are vacant in the SCF ground determinant. Szabo and Ostlund's [8] expression for the Hamiltonian matrix elements is

$$\left\langle \Psi_i^a | H | \Psi_j^b \right\rangle = (F_{aa} - F_{ii})\, \delta_{ab}\delta_{ij} - (ab|ij) + 2(ai|jb)$$

One must add the ground state energy to the diagonal elements. A similar expression is available for triplet states, but singlet–triplet transitions are spin-forbidden from singlet ground states and are not considered here.

3. Extract energy eigenvalues and associated eigenvectors from the secular equation. Davidson's method [9] is preferred since one generally needs only very few roots. For our small system of 75 or fewer excitations, we will not require this computational advantage. The Lth excited state from the CIS solution is represented as a linear combination of excited state configurations.

$$\Psi_L = \sum_{ia} c_{Li} \psi_i^a$$

4. Evaluate the oscillator and rotational strengths. Here we need elements of the magnetic and electric dipole moments to define the rotational strength.

$$R_{0k} = \mathrm{Im}(\langle 0 | \boldsymbol{\mu} | k \rangle \bullet \langle k | \mathbf{m} | 0 \rangle)$$

Here 0 refers to the SCF ground state, k is a state involving single excitations $i \to a$ and the integrals have the form

$$\left\langle \psi_{\mathrm{SCF}} | O | \psi_i^a \right\rangle = \sqrt{2} \langle i | O | a \rangle$$

where i and a are MO labels.

The lowest few excitation energies ΔE_j and their rotational strengths R_j are sufficient to define the accessible CD spectra, while ORD values require many such energies and R values. Because the individual contributions to the rotation may differ in sign, even semiquantitative quality may be difficult to achieve. The following section contains some applications of the simple model built into PCLOBE and provides a basis for judging the performance of small-scale calculations.

Examples of CD/ORD Computations with PCLOBE

> Occam's Razor:
> *Numquam Ponenda est pluralitas sine necessitate*
> (Plurality ought never be posited without necessity)
> —Franciscan Monk William of Ockham (ca. 1285–1349)

With inspiration from William of Ockham, let us try to use the minimum quantum mechanical description of CD, ORD, and MCD as a learning exercise. We can begin with a well-understood system, D(+)-camphor.

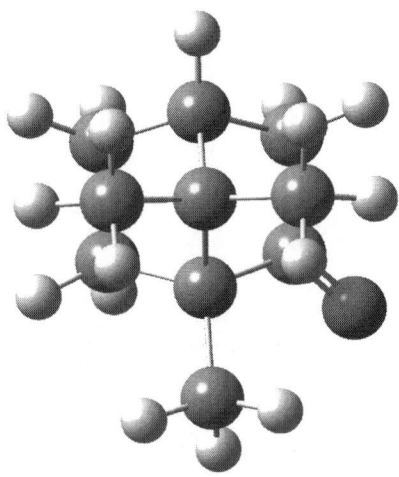

FIGURE 14.1
Structure of camphor. This is the mirror image of the structure shown in the Merck Index (12th Ed.).

In this case, the structure and the optical spectrum are well known, and both the CD and ORD are dominated by an $n \rightarrow \pi^*$ transition near 300 nm (Figure 14.1).

The structure of D(+)-camphor was first modeled using the parametrized force field of the CHEMSITE modeling program [10]; the CIS transition energies and rotational strengths were computed with PCLOBE. The PCLOBE calculation took a little over 2 h to produce the following graph (Figure 14.2).*

This isolated positive feature at the lowest energy suggests that the ORD observed in the visible region—well to the red of the transition—would be dominated by the positive feature which we place at about 230 nm. This is just what our calculation produces (Figure 14.3).

This is not always the case; when several CD bands of opposite signs and varying intensities are close together, the net ORD—which is the result of a sum of contributions from several transitions—may not have the same sign as the CD [11].

Crabbe [12] gives both CD and ORD curves for D(+)-camphor. The positive CD we compute is in agreement with Crabbe's experimental report. The CIS representation of excited states often places them well to the blue, but in this case CIS represents the $n \rightarrow \pi^*$ transition well. Enhancing the basis to 6-31GL(d,p) hardly changes the CIS prediction: transitions at 240 and 125 nm are computed. But TD-DFT with the MPW functional

---

* The CD spectrum is simulated by a sum of Gaussian forms $R_j \exp(-a[\lambda - \lambda_j]^2)$ with half-width of 15 nm.

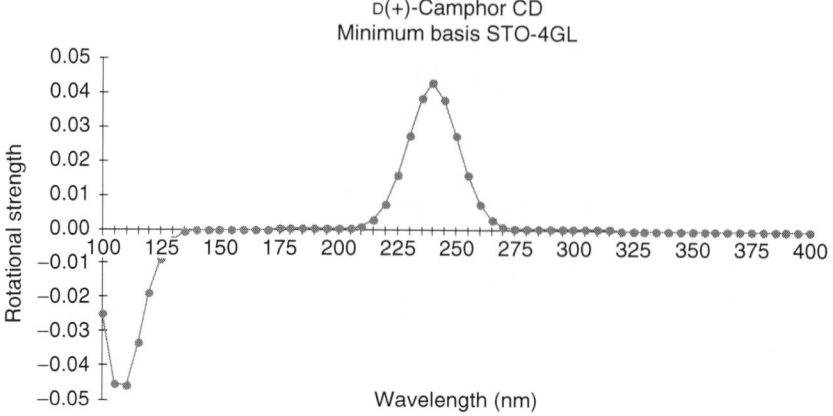

FIGURE 14.2

The simulated CD spectrum of D(+)-camphor using PCLOBE with a minimal STO-4GL basis set and 75 × 75 CIS. The vertical axis is the rotational strength (scaled); the horizontal axis is wavelength in nanometers. The curve is obtained by summing Gaussian forms with 15 nm half-width, multiplied by the rotational strengths computed with dipole velocity integrals. No wavelength shift is used here. Compare to the experimental spectrum in Figure 14.4.

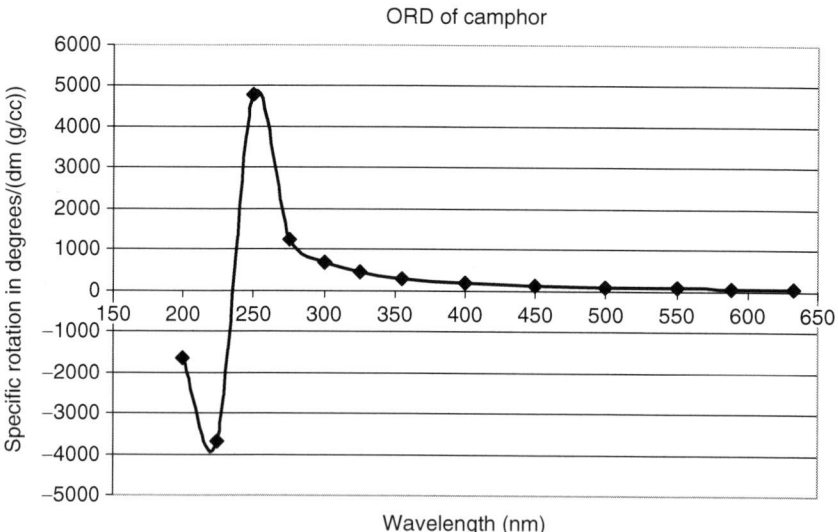

FIGURE 14.3

The computed optical rotation of D(+)-camphor using dipole length rotational strengths from a 75 × 75 CIS treatment. The vertical axis is specific rotation; the horizontal axis is wavelength in nanometers. The smooth curve is a spline-fit to calculated points. Compare to the experimental spectrum in Figure 14.5.

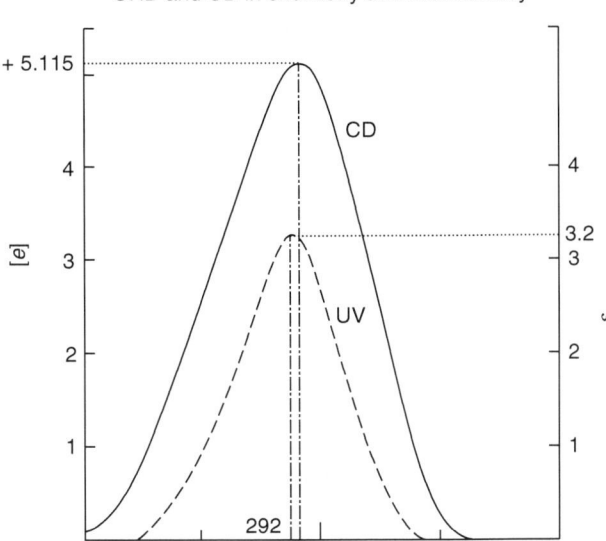

ORD and CD in chemistry and biochemistry

FIGURE 14.4
The experimental CD spectrum of D(+)-camphor. The vertical axis is the rotational strength (left) or extinction coefficient (right). The horizontal axis is wavelength in nanometers. (From Crabbe, P., *Optical Rotatory Dispersion and Circular Dichroism in Organic Chemistry*, Holden-Day, San Francisco, 1965. With permission.)

shifts the transitions to 283 and 206 nm. The predicted rotational strength is 0.7 cgs units. Crabbe's CD curve has a pronounced positive feature centered at 292 nm (Figure 14.4).

By detailed numerical analysis Crabbe shows that an isolated positive CD band is accompanied by a positive ORD curve which passes through extrema at about 310 and 275 nm and through zero at 294 nm (Figure 14.5).

This example shows that the handedness of a specific structure is expressed in both the CD spectra and the ORD curve. It is encouraging—maybe a bit surprising—that such a modest computational effort can capture such a subtle physical phenomenon.

We can anticipate the results of more sophisticated calculation using the capability built into the ADF program. The graph below was obtained by use of time-dependent density functional theory, the SAOP model functional which imposes correct asymptotic behavior on the exchange–correlation functional, and a basis set which includes diffuse functions in a polarized triple-zeta basis (Figure 14.6).

The horizontal axis is now excitation energy: the leading CD feature is at 4.36 eV (285 nm) while the experimental feature is at 4.23 eV (293 nm). The spectrum is simulated by Gaussians weighted by the computed rotational strengths, each with a breadth parameter of $50 \ eV^{-2}$. We see a very

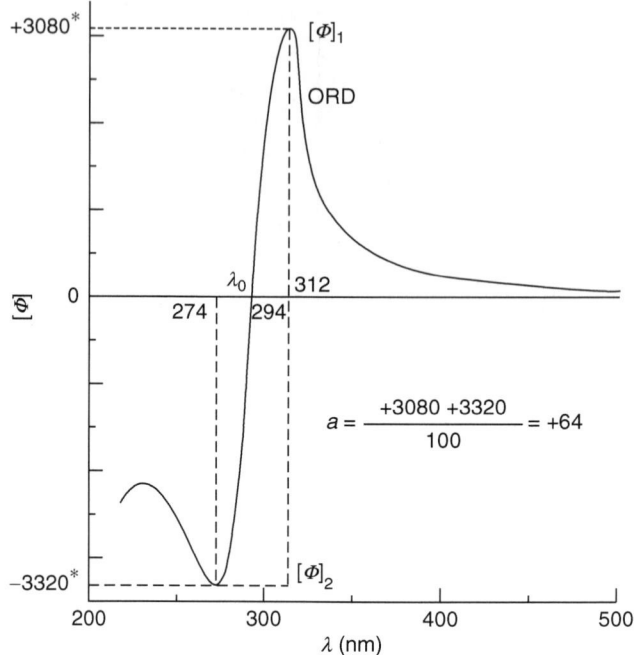

FIGURE 14.5
The experimental ORD spectrum of D(+)-camphor. The vertical axis is specific rotation; the horizontal axis is wavelength in nanometers. (From Crabbe, P., *Optical Rotatory Dispersion and Circular Dichroism in Organic Chemistry*, Holden-Day, San Francisco, 1965. With permission.)

close agreement for the low-energy reaches of the computed spectra, but more serious disagreement at high energy.

We have just quoted features of spectra both in wavelength and energy (or frequency) units. Since the sign of a feature in CD and related spectra is the critical quantity, one should keep in mind that

$$dv = d\left(\frac{c}{\lambda}\right) = -\frac{c}{\lambda^2}d\lambda$$

Overlooking that change in sign can complicate comparisons.

α-Pinene

A common test for calibration of optical activity instrumentation is (1R,5R)-α-pinene. Therefore, it presents itself as a system that any model with pretensions to utility should be able to represent.

Once again the structure of this rigid system was optimized using the force field of CHEMSITE. This is very fast and accurate for molecules with conventional bonding. We have found that whereas the $n \rightarrow \pi^*$ transitions are well characterized in CIS, there is advantage in scaling $\pi \rightarrow \pi^*$

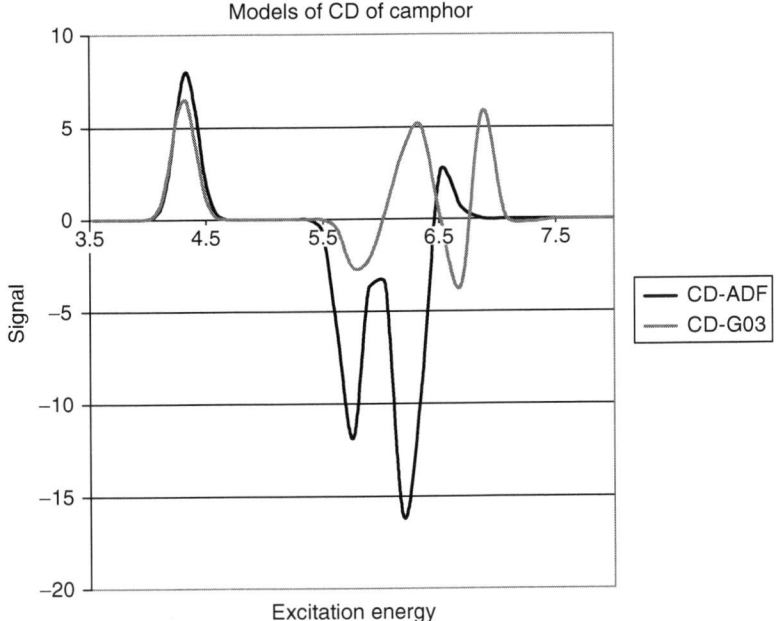

FIGURE 14.6
CD spectrum of D(+)-camphor according to ADF and Gaussian models. The vertical axis is the circular dichroism in cgs units $\times 10^{40}$; the horizontal axis is excitation energy in electron volts. ADF results are from the SAOP/ATZP model while the Gaussian results are from B3LYP/6-311+G(d,p) calculations. The curve is composed of Gaussian forms centered at the absorption frequencies, weighted by the rotational strength, and with a width of 50 eV^{-2}.

transition energies to match the first $A_{1g} \rightarrow B_{2u}$ absorption in benzene which lies at 262 nm. All computed transition energies are multiplied by 0.662 but the CIS mixing coefficients are unchanged (Figure 14.7).

The simulated CD spectrum shows that the long wavelength edge of the CD spectrum is positive (Figure 14.8).

The experimental spectrum from Mason and Schnepp [13] is shown in Figure 14.9 for what they called $\alpha(-)$pinene. The vertical scale at left defines the absorption (the solid line) while the vertical scale at right defines the circular dichroism (the dashed line). The horizontal wavelength scale is reversed from the scale of the calculated spectrum.

The small positive features near 225 nm may be due to Rydberg transitions which cannot be captured in the STO-4G (valence only) basis. The negative band near 205 nm on the experimental spectrum corresponds to the positive band at 170 nm in the simulated spectrum. The sign disagreement of the calculated spectrum for (1R,5R)-α-pinene strongly suggests that Mason and Schnepp measured the spectrum of the (1S, 5S) isomer of α-pinene. This confusion may arise from a change in nomenclature from (+) and (−) to R and S since Mason and Schnepp reported their

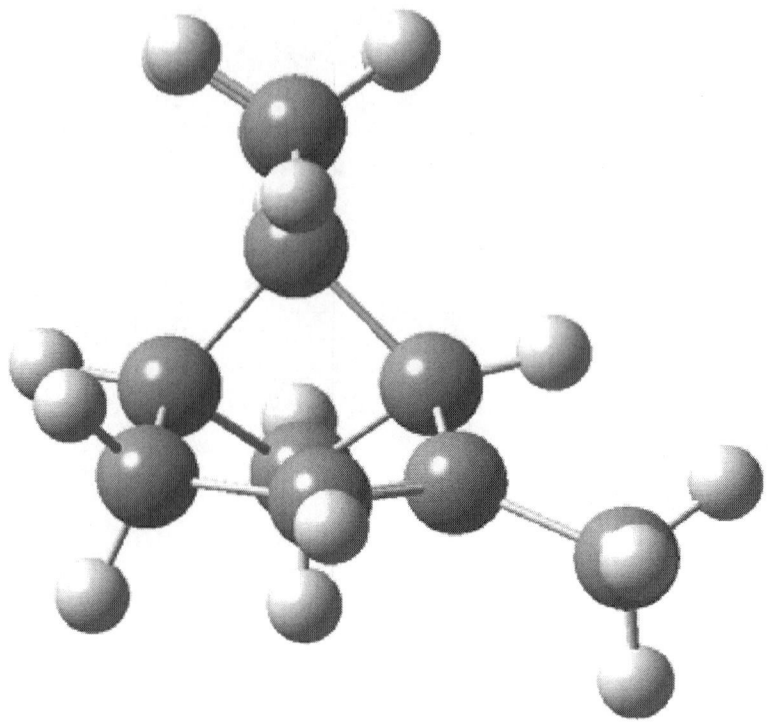

FIGURE 14.7
Structure of α-pinene used in all computations.

measurements. We should bring other evidence to bear, in order to resolve the issue.

This case is a good example of how calculated simulation of CD can be used together with experimental spectra to assign the absolute configuration of an optically active compound. The Merck Index [14] describes both (1R, 5R) and (1S, 5S) forms of pinene. The first of these displays positive optical rotation in the range 589 to 633 nm, while the second shows the opposite sign. Mason and Schnepp reported that the compound they studied exhibited a specific rotation of $[\alpha]_D$ −42.8° at 589 nm. This agrees with the recent cavity ring-down polarimetry (CRDP) of Wilson et al. [15] which gave specific rotations of −187° at 355 nm and −46° at 633 nm for the (1S, 5S)-α-pinene. On the basis of the specific rotation measured by Mason and Schnepp and the recent CRDP measurements by Wilson et al. [15] it would seem that the spectrum measured by Mason and Schnepp and labeled as α(−)pinene is for (1S, 5S)-α-pinene. The important point here is that the simulated CD spectrum tells us the same thing.

We can see if our conjecture can be supported if the leading edge derives from a transition with substantial Rydberg character as suggested by Robin et al. [16]. A diffuse sp set of basis functions was added to the

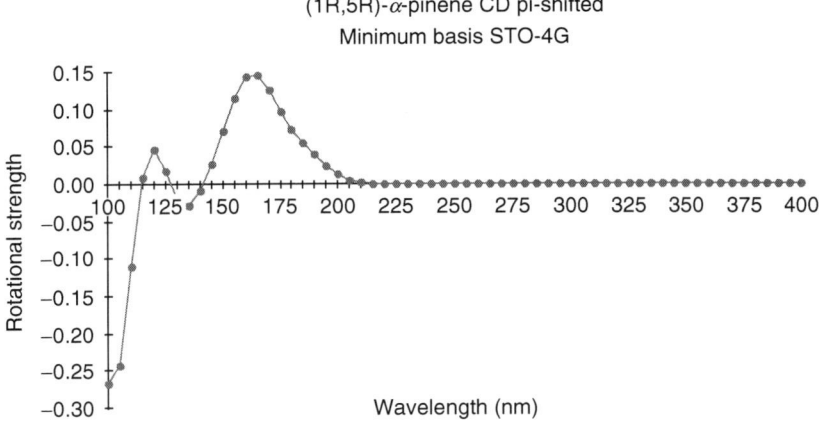

FIGURE 14.8
The simulated CD spectrum of (1R,5R)-α-pinene obtained using an STO-4G minimum basis with a 75×75 CIS treatment. The vertical axis is the rotational strength (scaled); the horizontal axis is wavelength in nanometers. The wavelengths have been pi-shifted by an amount to fit the first $\pi \rightarrow \pi^*$ transition of benzene in the same basis set. Two small positive bands on the red edge observed experimentally do not occur with this treatment.

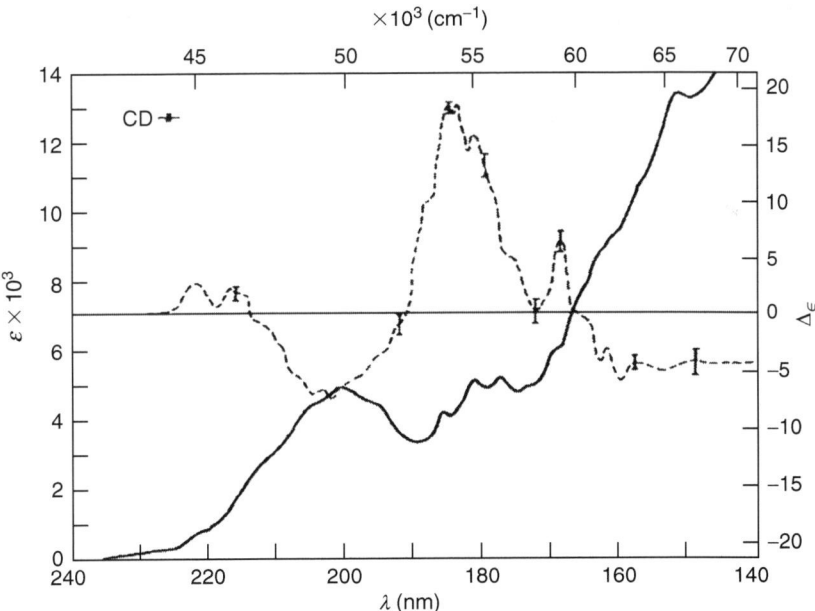

FIGURE 14.9
The experimental CD and absorption spectra of α-pinene. The vertical axis is the extinction coefficient ($\times 1000$) for absorption (left), and the difference in extinction coefficient for left and right circularly polarized light for CD (right). The horizontal axis is wavelength in nanometers (bottom) or frequency in wave numbers/1000 (top). (From Mason, M.G. and Schnepp, O., *J. Chem. Phys.*, 59, 1092, 1973. With permission.)

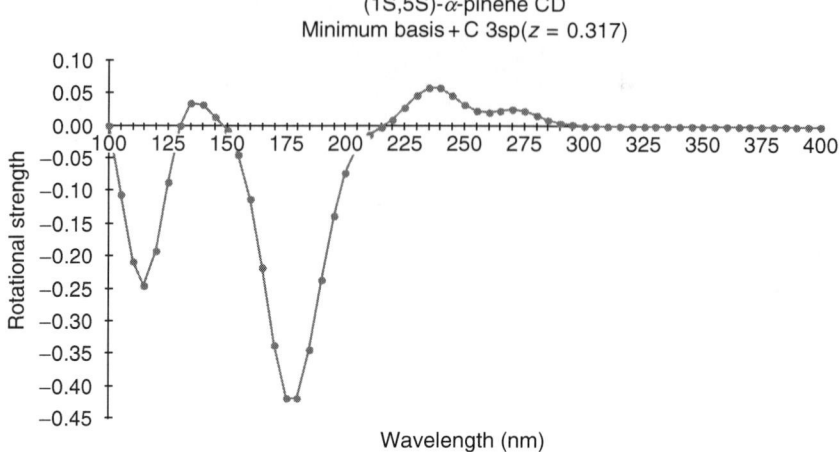

FIGURE 14.10

The simulated CD spectrum of (1S, 5S)-α-pinene. The vertical axis is rotational strength (scaled); the horizontal axis is wavelength in nanometers. In this case, 3sp orbitals are added to the STO-4G basis only on the carbon atoms with double bonds. The 3sp Rydberg orbitals produce the two small positive features on the red edge of the experimental spectrum.

minimum basis for the two C atoms involved in the double bond. Here we see that the augmentation of the basis does produce the two small positive bands in the spectrum (Figure 14.10).

This minimally demanding calculation gives a qualitatively helpful representation of the experimental data. To anticipate a later discussion, we show how a more sophisticated model might treat the pinene problem. In Tables 14.1 and 14.2 we present the features of a CD spectrum predicted by a TD-B3LYP/6-311+G(d,p)//B3LYP/6-31G(d,p) model, and by a SAOP/ATZP//PBE/DZ model. Both calculations capture the positive feature of the CD spectrum at the red edge near 220–230 nm, followed by the stronger negative feature at higher energy (shorter wavelength). Neither shows the

TABLE 14.1

TD-B3LYP Computation of UV–Visible Excitations and Rotational Strengths for α-Pinene

Transition	Exp.	Energy, eV (wavelength, nm)	Rotational Strengths (v, l)	
1	5.66	5.48 (226)	7.60	7.11
2	6.10	5.89 (210)	−22.47	−22.88
4		5.98 (207)	21.34	19.82
5		6.12 (203)	−8.09	−6.34

Note: v, l refer to velocity and length dipole formulations respectively.

TABLE 14.2

SAOP/ATZP Computation of UV–Visible Excitations and
Rotational Strengths for α-Pinene

Transition	Exp.	Energy, eV (wavelength, nm)	Rotational Strengths (v, I)	
1	5.66	5.50 (226)	7.67	7.71
2	6.1	5.77 (210)	−18.26	−18.32
3		6.01 (207)	−5.82	−5.84
4		6.09 (203)	7.19	7.15
5		6.17 (195)	13.91	13.97
6		6.26	−1.82	−1.75

Note: v, I refer to velocity and length dipole formulations respectively.

splitting in the low-energy positive band. The results from a SAOP/ATZP//PBE/DZ model are shown in Table 14.2.

Treatment of large molecules in these demanding models may be inconvenient, so it is of interest that useful information—the sign and sequence of features in the CD spectrum—can be captured in a low-level representation (Figure 14.11).

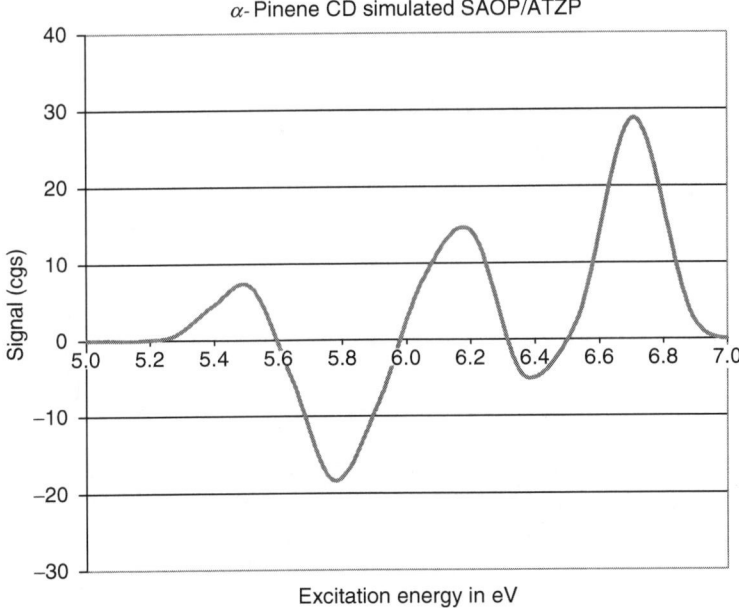

FIGURE 14.11
CD spectrum of (1R, 5R)-α-pinene according to density functional theory obtained with the Amsterdam Density Functional software suite. The vertical axis is the circular dichroism in cgs units $\times 10^{40}$; the horizontal axis is excitation energy in electron volts. ADF results are from the SAOP/ATZP model.

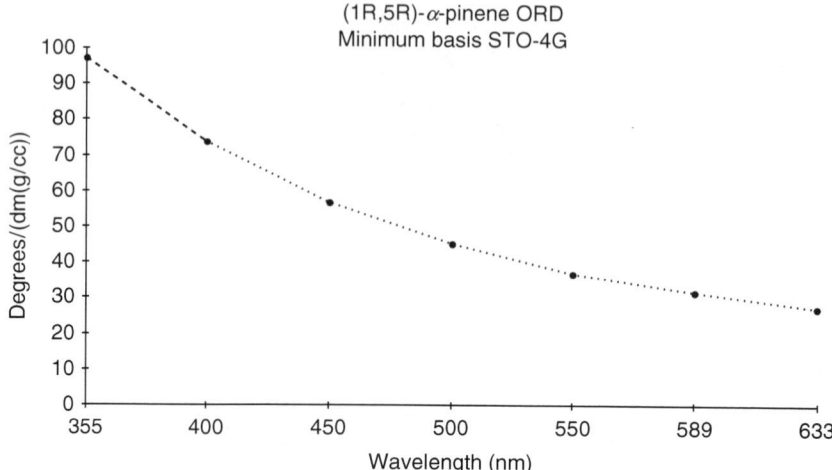

FIGURE 14.12

The calculated ORD spectrum of (1R,5R)-α-pinene. The vertical axis is specific rotation; the horizontal axis is wavelength in nanometers. The calculation uses the dipole length formula and the STO-4G basis in a 75×75 CIS treatment.

The conclusion that the leading edge of the CD spectrum shows a positive feature is borne out by the fact that recently measured gas-phase ORD values for (1R,5R)-α-pinene are also positive in the range of 355–633 nm. This is captured in our PCLOBE results, which rely on a small basis, a limited sum over states, and an approximate geometry. We find a value of specific rotation angle of 30° at 633 nm and 100° at 355 nm (Figure 14.12).

Our computed ADF specific rotation angle values range from 97° at 355 nm to 27° at 633 nm. These are closely comparable to the B3LYP values, 104.7° at 355 nm and 38.0° at 633 nm, respectively, reported by Wilson et al. [15] (who used a much larger (pVTZ) basis), and Gaussian B3LYP/6-311+G (d,p)//B3LYP/6-31G(d) values, 86.9° at 355 nm and 29.9° at 633 nm. These consistent calculations can be compared with gas-phase data from CRDP [11]; (1R,5R)-α-pinene has specific rotations of +187.0° at 355 nm and +46.0° at 633 nm as expected for the mirror image of the values for the (1S, 5S) form of the molecule discussed above.

Zanasi et al. [17] provide a consistent view of the ORD of α-pinene, shown in Figure 14.13, and advocate computational modeling of the specific rotation at several wavelengths, much more likely to give reliable guidance than a single value at a particular wavelength. However, there are exceptions to this conjecture. When optical rotation is small, molecular vibrations can have serious effects, even changing the sign of the optical rotation. This counterintuitive behavior will be discussed below. At some point a compound with high electronic chirality will have optical rotation large enough to overcome vibronic effects (Figure 14.13).

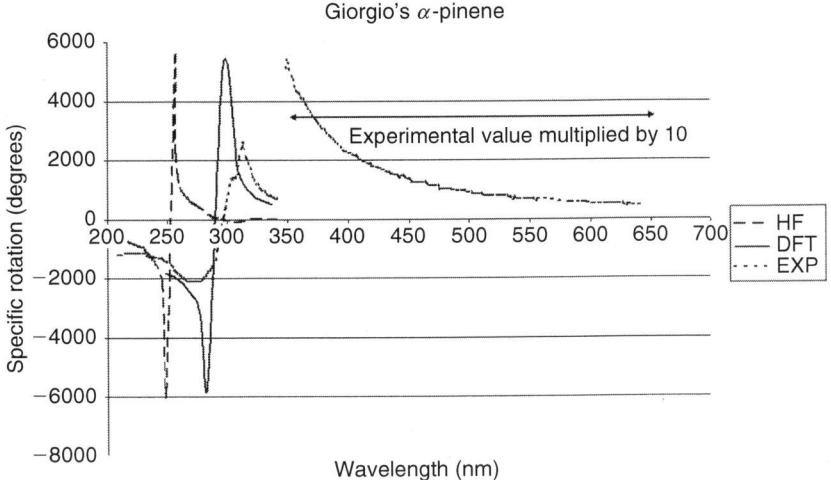

FIGURE 14.13
ORD curve for α-pinene. The vertical axis is the specific rotation [α]; the horizontal axis is the wavelength in nanometers. The solid line for wavelengths shorter than 350 nm is from TD-B3LYP calculations and the dashed line is obtained by HF calculations in the same basis. The dotted line is from their experimental observation. The signal at short wavelengths is much stronger than the signal at longer wavelengths. (From Zanasi, R., Viglione, R.G., Rosini, C., et al., *J. Am. Chem. Soc.*, 126, 12968, 2004. With permission.)

Ways to Improve CD and ORD Calculations

Use GIAOs

As Jensen [18] points out, computation of magnetic properties includes operators which refer to a gauge origin. This can be taken as the molecular center of mass, but this does not remove all difficulty. The gauge error depends on the distance between the centroid of the wave function and the gauge origin, which can be substantial for a large molecule. London atomic orbitals or gauge including (AKA invariant) atomic orbitals (GIAOs) [19] solve this problem. With the vector potential $\mathbf{A}_A$ referring to nucleus A, the ordinary spatial AO φ is transformed to

$$\chi_A(r, R_A) = \exp\left(-\frac{i}{c}\mathbf{A}_A \bullet \mathbf{r}\right)\phi_A(r, R_A)$$

All matrix elements with the GIAOs depend only on differences in vector potentials.

Use Linear Response Theory

Excited states are hard to model reliably—we can use excitation energies accurate to about 0.2–0.5 eV to help interpret low-energy optical absorption and CD spectra, but the higher energy states needed in the ORD expressions are even harder to capture. It is worthwhile evading the second-order sum over states which appears in the Rosenfeld formulation [1,2] of the electric dipole–magnetic dipole polarizability.

$$\beta_{pq} = \frac{c}{3\pi h} \text{Im} \left[\sum_{k \neq 0} \frac{\langle 0|\mu_p|k\rangle \langle k|m_q|0\rangle}{\nu_{k0}^2 - \nu^2} \right]$$

Modern methods rely on response theory. The TDHF formulation describes the response of a single determinant of canonical Roothaan–Hall molecular orbitals to an external perturbation. In DFT, one uses the Kohn–Sham formalism and seeks the frequency-dependent change in the density

$$\rho^1(r, \omega) = \sum_i P_{ai}^1(\omega)\varphi_{ia}$$

Here the indices i and a define a single excitation of an electron from occupied i to virtual a. The key quantity is the first-order response of the Kohn–Sham density matrix P^1; define

$$X_{ai} = P_{ai}^1; \quad Y_{ai} = P_{ia}^1$$

The dipole moment perturbation

$$\mu_u^1(\omega) = \sum_{ai} D_{u,ai} \left(X_{ai} + Y_{ai} \right) = \mathbf{D}_u \left(\mathbf{X} + \mathbf{Y} \right)$$

The analogy for the magnetic moment is

$$m_u^1(\omega) = \sum_{ai} M_{u,ai} \left(X_{ai} - Y_{ai} \right) = \mathbf{M}_u \left(\mathbf{X} - \mathbf{Y} \right)$$

The matrices $\mathbf{D}$ and $\mathbf{M}$ are

$$D_{u,ia} = \langle i|\mu_u|a\rangle; \quad M_{u,ia} = \langle i|m_u|a\rangle$$

The basis is the orthonormal Kohn–Sham orbital set; this is one of the details that makes this a DFT variant on TD-single determinant methods. Autschbach et al. [20] work through the manipulations needed to obtain the tensor β, first by perturbation of the magnetic moment by an electric field, then by the perturbation of the electric moment by a magnetic field.

Required quantities include integrals of the density functional exchange–correlation energy density, which is the second way in which this process is linked to the constructs of DFT.

Use Correlation-Corrected Methods (Time-Dependent DFT, CCSD)

We have already mentioned that replacing the expressions of HF-Roothaan theory with DFT analogs transforms the linear response TDHF theory to TD-DFT theory. Notable implementations of TD-DFT-based methods for CD/ORD modeling have been accomplished by Autschbach and coworkers [20] as mentioned already, Frisch et al. [21] (GAUSSIAN03), Ahlrichs et al. [22] (TURBOMOL), Helgaker et al. [23] (DALTON), Amos [24] (CADPAC), and Crawford et al. [25] (PSI3). Coupled cluster methods have also been employed within linear response theory to estimate polarizability tensors generally and the rotational strength tensor in particular. These are implemented in PSI3 and DALTON. Ruud et al. [26] have used coupled cluster methods for optical rotation and Crawford [27] has reviewed the modern development of methods for optical rotation, which has blossomed in the past decade. Applications to optical rotation are so far not numerous, but the prospect is bright. While the best calculations seem to provide useful models of CD and ORD spectra in general, much attention has been focused on some rarer small but resistant systems.

Methyloxirane: A Hard Case

> Make everything as simple as possible, but not simpler.
> —attributed to Albert Einstein

In our illustrations we strive for simplicity, but sometimes simplicity is just not achievable. *S*-methyloxirane, one of the smallest chiral molecules, has attracted intense interest from chemical theorists. Although even our simplest calculations have had some success for camphor and pinene, we might anticipate that representing the excited states of this system would be more difficult. As we noted before, CIS treatment using a modest basis probably works best when there is some sort of electronically "soft" chromophore in the molecule. At least some excitations of the carbonyl group in camphor and of the twisted ethylene fragment in the pinenes can be represented within the valence manifold. There is no such chromophore in methyloxirane, and excitations from the saturated ring must include not only CO and CC σ^* orbitals but Rydberg (ns, np, with principal quantum number $n \geq 3$) levels as well.

The experimental CD spectrum for *S*-methyloxirane is shown in Figure 14.14. The vertical axis defines the CD signal. Notice that the signal is small—less than 10 units—compared with our first examples. The gross features of the spectrum include a positive excursion at about 7.2 eV (ca. 170 nm) and a negative excursion about 7.8 eV (ca. 160 nm). There is

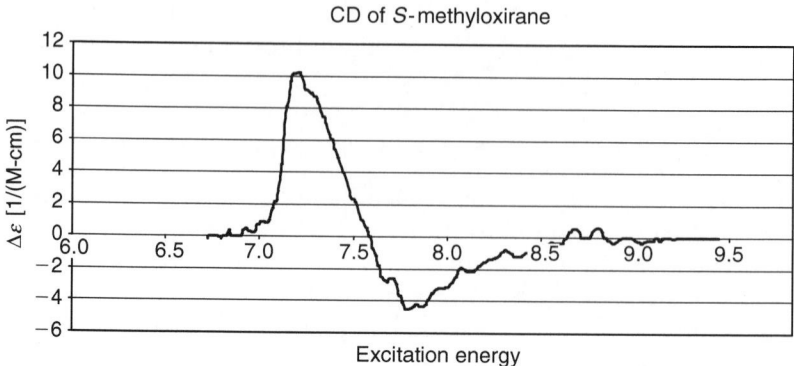

FIGURE 14.14

VUV CD S-methyloxirane. The vertical axis is the difference in extinction coefficient for left and right circularly polarized light; the horizontal axis is excitation energy in electron volts. (From Carnell, M., Peyerimhoff, S.D., Breest, A., et al., *Chem. Phys. Lett.*, 180, 477, 1991. With permission.)

considerable structure in the leading red edge of the CD signal and in both the absorption and CD spectra at higher energy.

We show below the simulated CD spectra obtained for the S-enantiomer using PCLOBE with the STO-4G basis, pi-shifted as we did before to model $\pi \rightarrow \pi^*$ transitions. It is barely noticeable in the minimum basis simulated spectra, but there is a weak transition from HOMO $\rightarrow$ LUMO predicted to be at 190 nm in the pi-shifted spectra followed at shorter wavelength by a strong negative band. The rotational strength of this weak transition in the minimum basis representation is opposite in sign to the low energy band in the observed spectrum. Clearly the minimum basis is failing to represent what is happening in the actual spectrum (Figure 14.15).

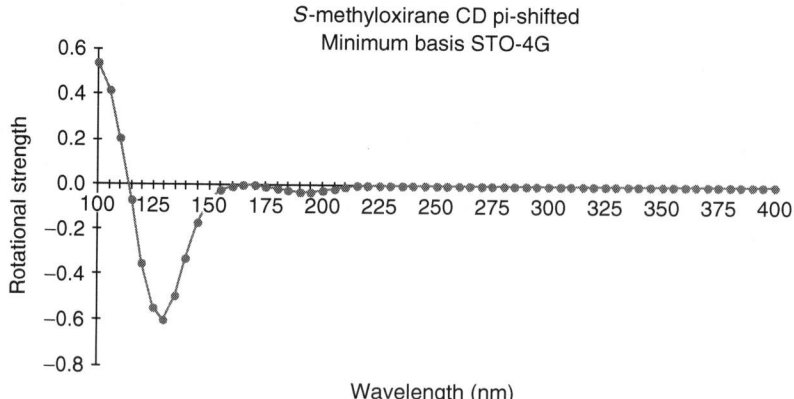

FIGURE 14.15

Simulated CD spectrum of S-methyloxirane. The vertical axis is rotational strength (scaled); the horizontal axis is wavelength in nanometers. The calculation uses an STO-4G minimum basis and 75×75 CIS. Note: Incorrect sign of small band on the red edge of the spectrum.

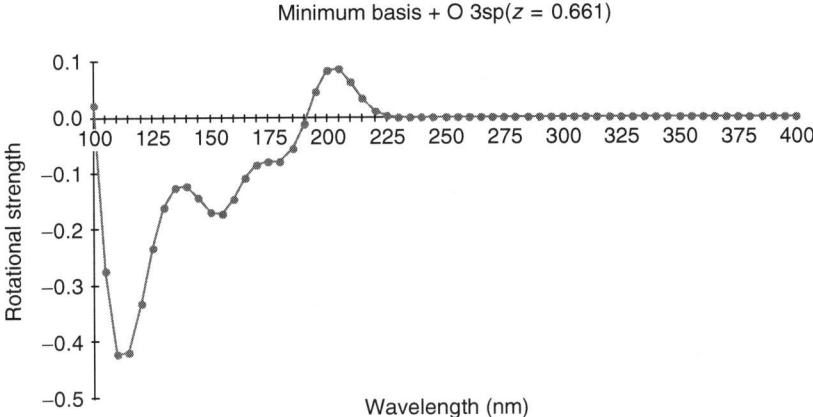

FIGURE 14.16

Simulated CD spectrum of *S*-methyloxirane. The vertical axis is rotational strength (scaled); the horizontal axis is wavelength in nanometers. The calculation uses an STO-4G minimum basis augmented by a 3sp shell on the O atom. The addition of the 3sp Rydberg orbitals on the O atom produces the correct (positive) sign for the lowest energy band.

Peyerimhoff [28] makes the point that low-lying Rydberg states are not well represented by molecular orbitals optimized for the ground state. Her use of orbitals optimized for triplet states improved her description of the *S*-methyloxirane system. She also added special diffuse orbitals to the basis to describe the Rydberg region [30]. We follow her example by adding a diffuse 3sp shell on the O atom of the epoxide ring. These s and p functions were forced to be orthogonal to the O atom 1s function, so they are not diverted to the variational improvement of the O core. Choosing an 3sp scale factor ζ of 0.661 shifted the low energy positive feature of the S-methyloxirane above 200 nm and enhanced its intensity considerably. The introduction of Rydberg orbitals to the basis makes a greater impact for methyloxirane than for the pinene species, perhaps a consequence of its smaller size and lack of a chromophoric center (Figure 14.16).

ADF's SAOP model, with an ATZP basis including diffuse functions on every atom, predicts a strong positive signal about 7 eV (180 nm), a few tenths of an eV to the red of the observed band (Figure 14.17).

The negative feature computed to lie at about 7.6 eV is shifted to the red of the experimental feature which is close to 7.8 eV. At higher energies the match to the experimental spectrum becomes qualitative but is still informative.

Optical Rotatory Dispersion for Methyloxirane

In PCLOBE we approximate the ORD by a sum over states, and these states are systematically shifted to the blue owing to our small basis. We would predict that our calculation would underestimate the specific rotation in the experimentally accessible range of 355–633 nm (Figure 14.18).

Indeed it appears that a realistic description of the ORD of methyloxirane does require accurate location of the excited states—a demanding task

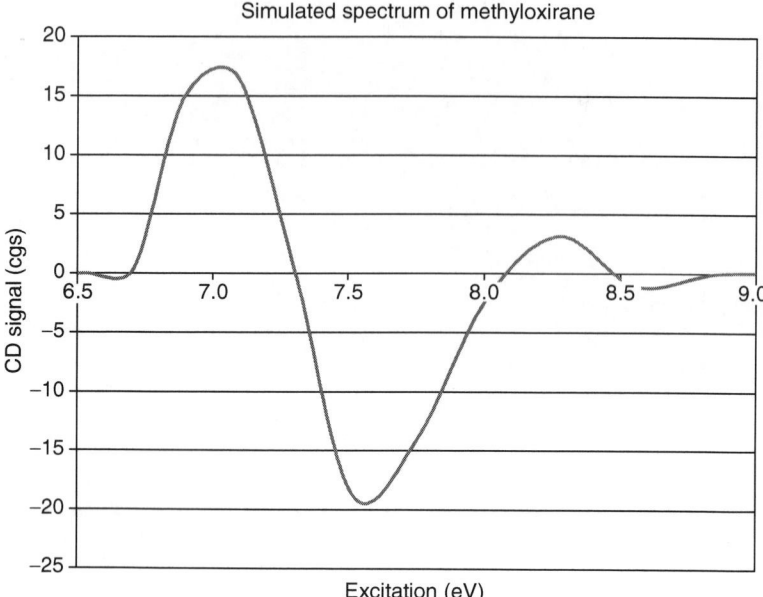

FIGURE 14.17
CD spectrum of *S*-methyloxirane according to density functional theory, obtained with the Amsterdam Density Functional software suite. The vertical axis is the circular dichroism in cgs units $\times 10^{40}$; the horizontal axis is excitation energy in electron volts. ADF results are from the SAOP/ATZP model.

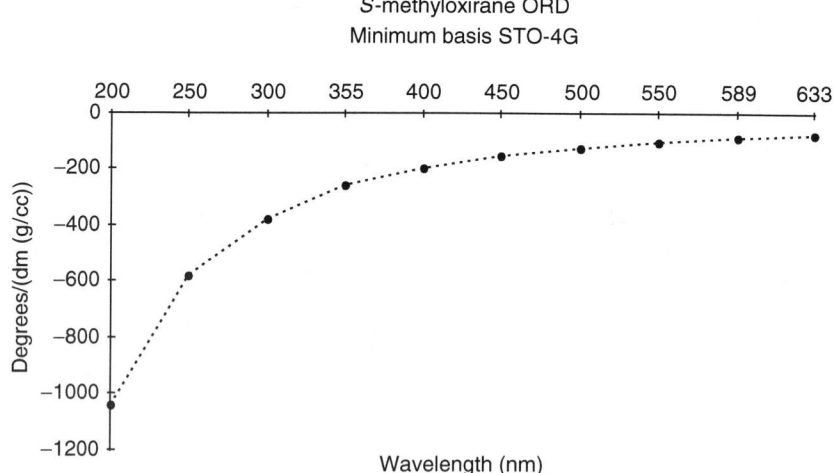

FIGURE 14.18
The calculated ORD spectrum of *S*-methyloxirane. The vertical axis is specific rotation; the horizontal axis is wavelength in nanometers. The calculation uses the STO-4G basis in a 75×75 CIS treatment of excitations and the dipole length formula.

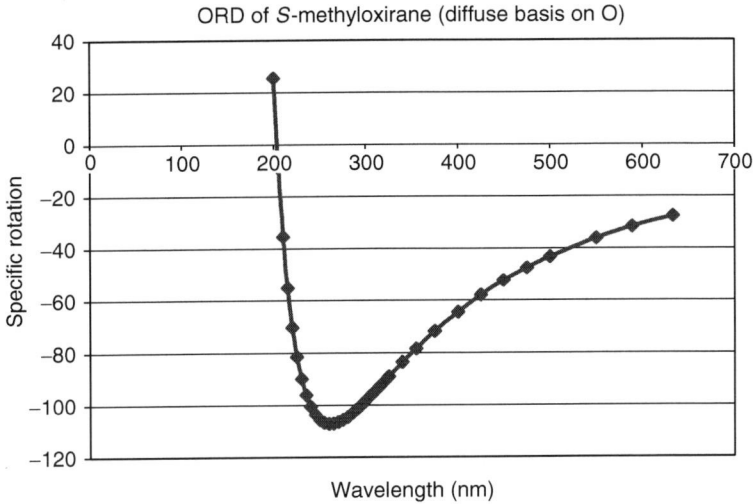

FIGURE 14.19
The calculated ORD spectrum of *S*-methyloxirane. The vertical axis is specific rotation; the horizontal axis is wavelength in nanometers. The calculation uses the the STO-4G basis augmented with a 3sp shell on the O atom in a 75×75 CIS treatment and the dipole length formula.

owing to their Rydberg nature. Adding the 3sp Rydberg orbitals on the O atom produces a Cotton effect near 200 nm similar to what is shown in Crawford's ORD curves [26] (Figure 14.19).

The large negative CD peak near 140 nm should determine the sign of the ORD curve at 633 nm; the small negative value of $[\alpha]_{633} = -8.39°$ is reasonable. The ORD curve remains negative down to 355 nm, but an upturn for a Cotton effect before 200 nm is obtained. In this small basis set, these results are far from quantitative, but the O 3sp Rydberg orbitals improve the minimum basis result. We might hope that more powerful modern methods will provide an improved general description. On the other hand we can maintain the usefulness of the simple model by augmenting the minimum basis for key atoms by diffuse 3sp Rydberg functions for systems where the lowest energy transitions are likely to have substantial Rydberg character.

Experimental Data and Advanced Calculations for Methyloxirane

The optical rotation of methlyoxirane at 589 nm, the experimentally convenient wavelength is small. This must be partly due to the high energy of the electronic transitions in this saturated molecule, which lacks any chromophore. The main Cotton effect pole is calculated by several methods [31] to be near 174 nm. When the optical rotation is small, other small effects and corrections can be important. Within the electronic structure model, choice of basis sets and details of the correlation model can affect the prediction. For nonrigid molecules, conformational changes, large

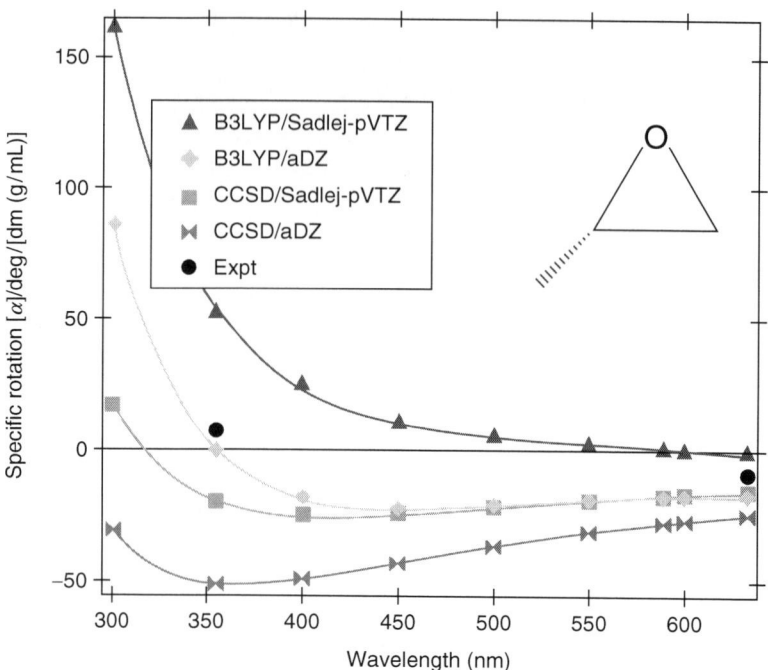

FIGURE 14.20

The computed ORD for S-methyloxirane. The two black dots indicate experimental values. Although the curve in Figure 14.19 indicates an upturn when the 3sp Rydberg shell is added to the STO-4G basis, the results here show the curve should turn up at longer wavelengths. Crawford has discussed reasons for the better results with B3LYP in his review. (From Crawford, T.D., *Theor. Chim. Acc.*, 115, 227, 2006. With permission.)

amplitude vibrations, and even much smaller changes in molecular geometry can have a substantial effect. Medium effects can be large, as is shown by Wilson et al. [15] who compared ORD curves in various media with the gas-phase data they obtained by CRDP.

Prof. Crawford has kindly provided additional results obtained since the original publication [31] for the multi-wavelength ORD curve of S-methyloxirane (propylene oxide) (Figure 14.20).

The two black dots are the values from gas-phase CRDP experiments for S-methyloxirane documented by Wilson et al. [15]. We see that the TD-DFT curve obtained with the B3LYP functional and a rather modest diffuse-augmented double-zeta basis agrees closely with the experimentally established values. This seems to be a case of getting a right answer for a wrong reason since the agreement is lost upon enhancing the basis or shifting the correlation model to CCSD. Crawford [27] maintains that the DFT results are a consequence of a systematic underestimate of the energy of the methyloxirane Rydberg transition; this puts the upward turn in the ORD curve at longer wavelength than it should be. CCSD (in its equation-of-motion variant) predicts the energy of that transition within 0.05 eV.

The nonrigidity of methyloxirane has been the subject of two interesting studies. Ruud and Zanasi [32] estimated the anharmonic zero-point vibrational corrections by TD-B3LYP with correlation-consistent polarized valence double- and triple-zeta basis sets. Keeping in mind the critique of Crawford [27] and Tam et al. [31] of the DFT description of methyldioxiran's lowest energy transition, they added their DFT estimates of vibrational effects to the best available CC3 coupled cluster estimate of the rotation angle. This correction, about 50° at 355 nm, added to the coupled cluster value of the electronic ORD at this wavelength, produced a net value of about 25°. These authors attribute the change in sign primarily to the almost-free rotation of the methyl group. Ruud and Zanasi's work reinforces the suspicion that changes in structure can have a disproportionate impact on the total optical rotation, especially for nonrigid systems. Careful and strict geometry optimization of a system is not an adequate response to this problem. Typically the differences between optimized structures obtained with various models are very small—smaller than the excursions executed during zero-point vibrations (Figure 14.21).

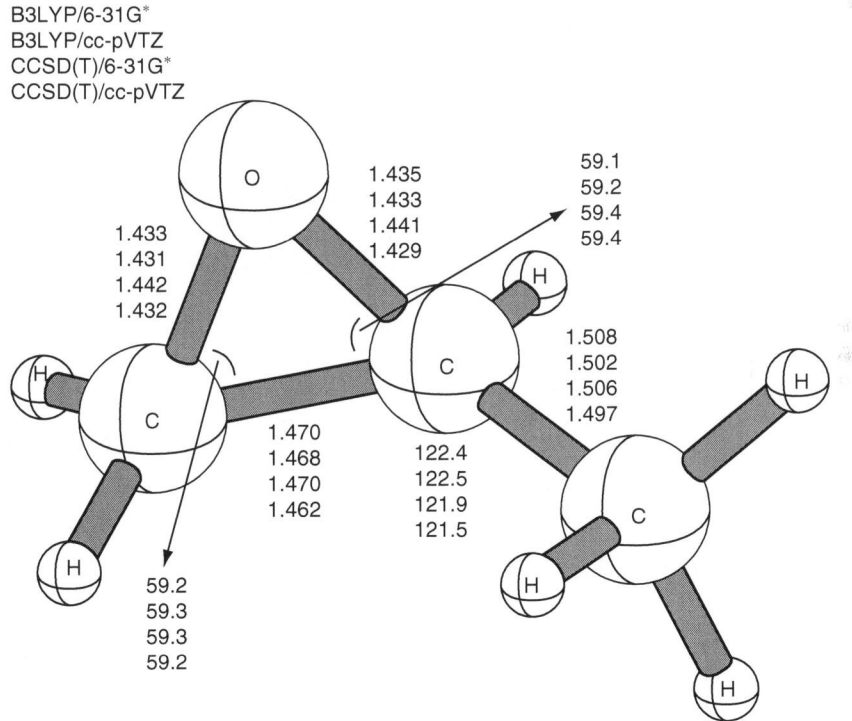

FIGURE 14.21

The slightly different optimized geometries of *S*-methyloxirane using B3LYP and CCSD(T) with 6-31G* and cc-pVTZ basis sets. When the optical rotation is small, changes in geometry may have a large effect on the ORD curve. (From Tam, M.C., Russ, N.J., and Crawford, T.D., *J. Chem. Phys.*, 121, 3550, 2004. With permission.)

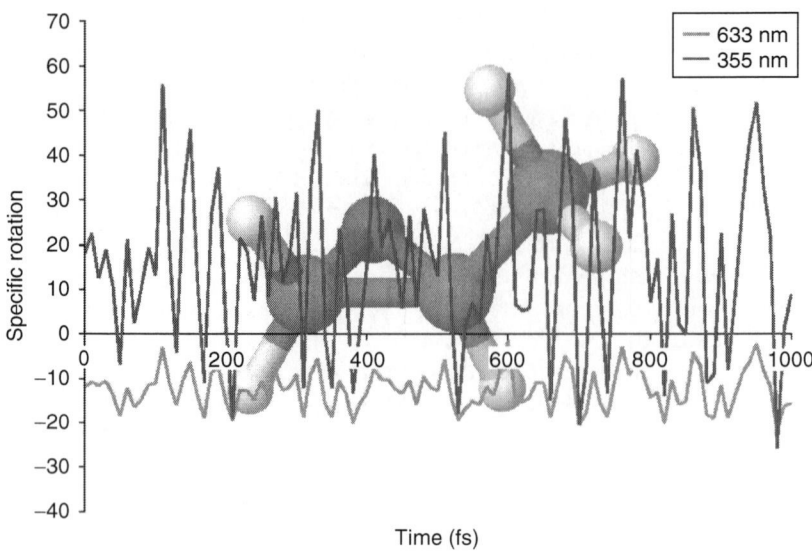

FIGURE 14.22
CD computed by Abrams at a sequence of geometries defined by molecular dynamics (see text) in collaboration with Crawford.

The range of vibrational motions can be described directly—if at considerable computational expense—by following the dynamics. Micah Abrams [33] generated a time series of structures by following a trajectory for 1000 fs (femtoseconds). Choosing structures at 10 fs intervals he computed the specific rotation at each point. As Figure 14.22 shows, the rotation at 589 nm averages about $-10°$ and the rotation at 355 nm averages about $+20°$. The root-mean-square variance in the computed angles is about $3°$ for 589 nm but much larger—more than $10°$, surely—and also occasionally makes excursions into negative values. Small shifts in geometry—which could come about from slightly different model chemistries or even different criteria for optimization—could give erratically varying results for the specific rotation angle.

A Brighter View

Dwelling on this hard case might have shaken the reader's confidence in the ability of even very computationally and theoretically advanced methods to match the CD and ORD behavior of molecules well enough that the computations can be a guide to interpretation of observations. We want to provide some examples to show that even rather approximate applications of the theories and models of CD/ORD can be of considerable help to experimental work.

TABLE 14.3

Performance of Model Chemistries in Various
Basis Sets for a Reference Set of Small- and
Medium-Sized Organic Molecules

Model	MAD
HF/6-31G(d)	35.3
HF/aug-cc-pVDZ	35.5
B3LYP/6-31G(d)	25.6
B3LYP/6-311++G(2d,2p)	18.7
B3LYP/aug-cc-pVDZ	20.1
B3LYP/aug-cc-pVTZ	23

Stephens et al. [34,35] defined a set of test cases for which sodium D-line rotations are known, and observed that DFT—specifically B3LYP—performs better than HF. Within B3LYP, a very large basis improved the mean absolute deviation (MAD) between computed and measured values to less than 20° (Table 14.3). (These statistics omit two serious outliers.)

This encouragement is reinforced by the ADF description of the same test set of molecules. Since the MAD can be reduced to about 20° by a suitable basis within TD-DFT, it appears that if the specific rotation is as much as 50°, one can be confident that the predicted sign is correct. The gradient-corrected BP86 functional seems only a bit better than the LDA (Vosko–Wilks–Nusair) functional, and SAOP seems to have a modest beneficial effect. The two large basis sets Vp (extensively polarized, not augmented with diffuse functions) and Vd (thoroughly augmented, but only modestly polarized) seem quite similar in performance. Since the quality of Vp results is at least comparable to Vd results, we infer (with some reticence) that a high level of polarization is as effective as adding diffuse functions. This may be an expression of the tendency for basis functions on atoms in a globular molecule to take part in the description of several atoms, especially when they are sufficiently diffuse. In ADF, the tendency of large basis sets to suffer linear dependence is recognized, and a countermeasure—projecting out the dependent combinations of basis functions—can be employed (Table 14.4).

Specific Examples

High symmetry molecules have enduring appeal to synthesists and theoretical analysts owing to their beauty and simplicity: examples of highly sought synthetic targets include cubanes, propellanes, icosahedranes, and even the highly strained tetrahedranes. An especially interesting set of high symmetry molecules also display chirality, owing to the absence of a plane of symmetry. C_2—and C_3—symmetric chiral molecules are of special interest in catalysis [36,37], but most such molecules are pursued simply for their aesthetic appeal. Farina and Morandi [38] have reviewed the

TABLE 14.4

Performance of ADF with Various Functionals
and in Various Basis Sets for a Reference Set
of Small- and Medium-Sized Organic Molecules

Model	MAD
LDA/IV	51
GGA/IV	43
LDA/Vp	23
GGA/Vp	19
LDA/Vd	25
GGA/Vd	24
SAOP/Vd	20

Note: Mean absolute deviation (MAD) of specific rotation angle $[\alpha]_D$ in a training set according to various pure density functional models. In this table LDA refers to a Vosko–Wilks–Nusair functional with the GGA is the Becke–Perdew 86 functional. IV is a DZ quality basis with polarization by one set of d functions on heavy atoms CNO and one set of p functions on each H atom. Vp is a TZ quality basis extensively polarized with two sets of d and two sets of f functions on heavy atoms CNO and three sets of p and two sets of d functions on each H atom. Vd is a TZ basis with d and f polarization sets on CNO and p and d sets on H atoms, augmented by diffuse s, p, and d sets on CNO and two diffuse p sets on H atom.

collection of such molecules known in the 1970s, which achieved C_2, C_3, D_2, and D_3 symmetry. Nakazaki [39] has extended the set, constructing systems with symmetry as high as T_d. New systems are reported regularly; for example, Kozhushkov et al. [40] described a family of 4,7,11-triheterotrishomocubanes which are both chiral and highly symmetrical—C_3 in the aza-hetero case they studied, shown in Figure 14.23. In the course of their B3LYP/6-31+G(d) geometry optimization they found a C_1-symmetric species very close in energy. We must assume this unsymmetric structure has one of the pyramidal NH bonds antiparallel to the other two. Symmetry suggests that only these two species, C_3- and C_1-symmetric, respectively, could be obtained. For the (all-R) species without substituents they computed values of the specific rotation with the TD-B3LYP/6-31+G(d,p)//B3LYP/6-31+G(d,p) model.

The two values computed for the (all-R) species were

$$[\alpha]_D^{25} = -214 \text{ and } [\alpha]_{435}^{25} = -436 \text{ for } C_1$$
$$[\alpha]_D^{25} = -172 \text{ and } [\alpha]_{435}^{25} = -342 \text{ for } C_3$$

Experimental values, measured for species in solution, were -28 and -475 for the all-R methyl-substituted species. The authors considered these

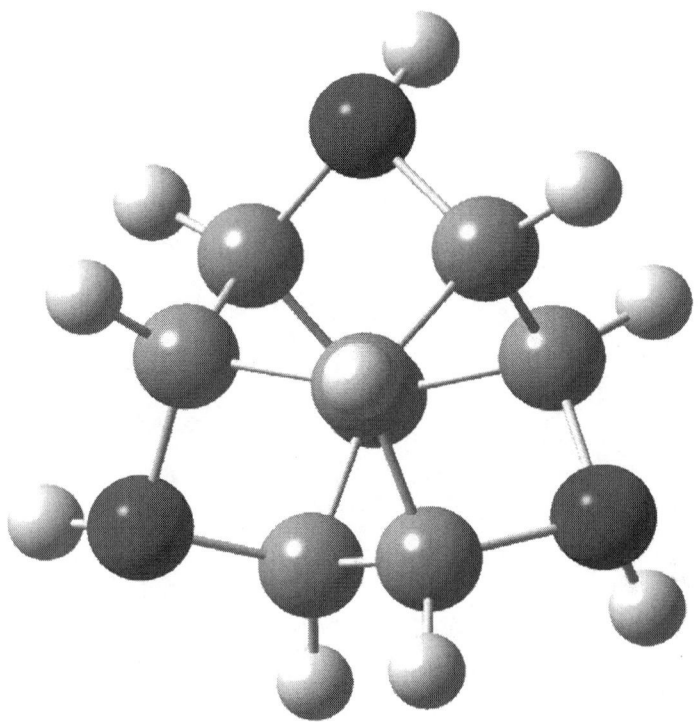

FIGURE 14.23
Structure of 4,7,11-triazatrishomocubane.

estimates "rather similar" for the two structures and also "rather close" to values for trioxatrishomocubane, the D_3 species shown in Figure 14.24.

The x-ray structure shows near-perfect D_3 symmetry, and the ORD measurements yielded

$$[\alpha]_D^{25} = -173 \text{ and } [\alpha]_{435}^{25} = -608$$

When judging what might be usefully similar values of the optical rotation, one should keep in mind the substantial effects of solvent. In this case, it seems quite unlikely that the sign of the computed rotation could be wrong. Our ADF-based SAOP/ATZP calculations of the ORD yielded $-204°$ at 589 nm and $-441°$ at 435 nm (Figure 14.25).

In what must have been a pure jeu d'esprit, Crawford et al. [41] have brought to bear the modern techniques of description of CD/ORD on the molecule "triangulene"—systematically, trispiro[2.0.0.2.1.1]nonane. This is a saturated hydrocarbon. Lacking chromophores it poses a severe test for time-dependent DFT, at least those variants that use a functional optimized for energies of atomization from the ground state. B3LYP is based on just

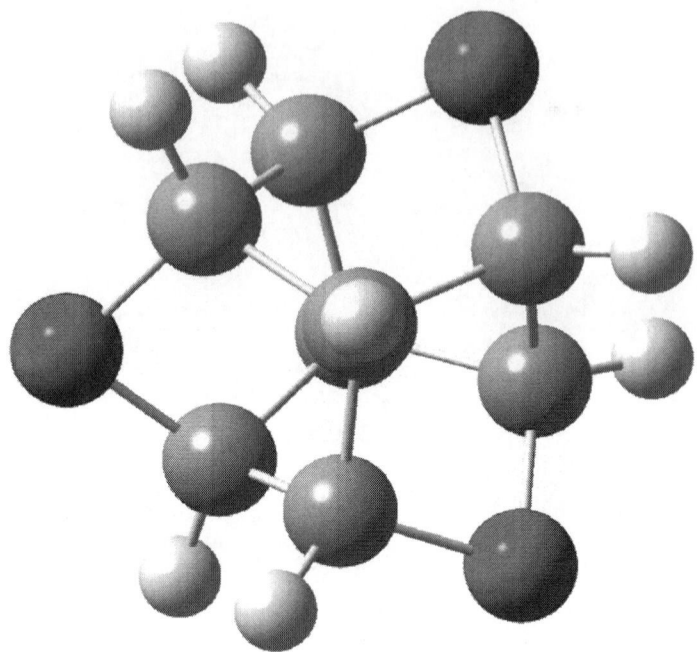

FIGURE 14.24
Structure of 4,7,11-trioxatrishomocubane.

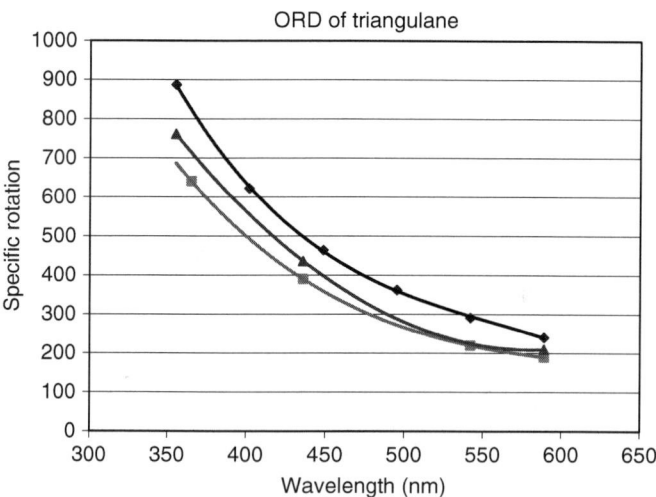

FIGURE 14.25
The ORD curves calculated for triangulane. The vertical axis is specific rotation; the horizontal axis is wavelength in nanometers by several methods. The points designated by squares are experimental values; triangles refer to values GAUSSIAN-03 with the density functional MPW in the 6-311+G(d,p) basis. Diamonds refer to points calculated in ADF with the SAOP model and the ATZP basis. Curves are polynomial fits to the data. (From Crawford, T.D., Owens, L.S., Tam, M.C., et al., *J. Am. Chem. Soc.* 127, 1368, 2005. With permission.)

such a functional. Crawford et al. employed their own CCSD-linear response theory and a further approximation called CC2 to the coupled cluster expansion, as implemented in DALTON. The specific rotation rises sharply from about 200° to 600° as the wavelength passes from 589 to 355 nm, approaching a transition with a positive Cotton effect. The outcome is unambiguous: TD-DFT overestimates the specific rotation by about 15%, while CC2 cuts the error to about 8%. The CCSD estimates are in error by no more than 10°—its values are almost impossible to distinguish from the observed values in the published graph. The flaw in DFT is its underestimate of the energies of Rydberg excitations: the first transition, estimated by CCSD (EOM) to lie at 7.21 eV, is placed by TD-B3LYP/aug-cc-pVDZ at 6.24 eV. Choosing the mpw1pw91/aug-cc-pVDZ//B3LYP/6-31G(d) model shifts the transition to 6.44 eV, and SAOP/ATZP locates that transition at 6.51 eV, still seriously shifted to the red.

The Phenomenon of Magnetic Circular Dichroism

Circular dichroism is a property associated with chiral systems, and its observation requires separation of those systems' optical isomers. The fundamental dissymmetry imposed by the nuclear framework on the electronic distribution is the cause of optical rotation and circular dichroism. However, all molecules can exhibit "induced" optical activity if placed in a dissymmetric environment. A magnetic field defines such an environment (Figure 14.26).

All that is necessary to convert a CD instrument to an MCD instrument is a powerful magnet. The Faraday effect attends the alignment of a magnetic field with the direction of propagation of a plane polarized beam of light. The medium, distorted by the magnetic field, rotates the plane of polarization according to

$$\theta = VH$$

Here θ is the extent of rotation (degrees), H is the strength of the magnetic field, and V is the Verdet constant. When the field is oriented perpendicular to the direction of propagation of the light, a different but related effect occurs and is called the Cotton–Mouton effect. In both cases some rotation of the plane of propagation occurs and in addition the beam acquires an ellipticity; that is, the magnetic field induces circular dichroism.

The Cotton–Mouton effect is quadratic in the magnetic field strength and so offers more sensitivity, but the spectral band shapes are quite complex for this effect and difficult to interpret. When the light beam is parallel to the magnetic field lines the spectral bands are simpler to interpret although this orientation, called the Faraday effect, is only linear in field strength. Because the Faraday effect orientation produces simpler band shapes it is preferred and commonly used in the majority of modern measurements. Note also

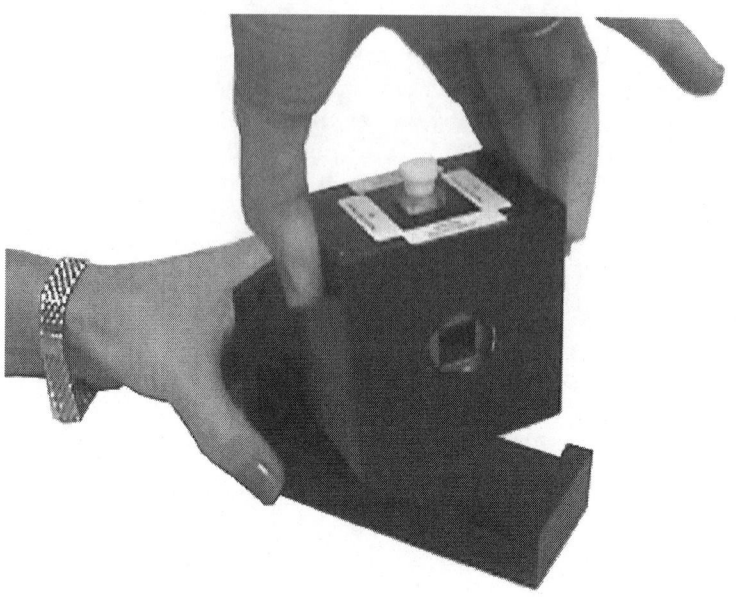

FIGURE 14.26

The conversion of a CD spectrometer to an MCD spectrometer requires only the provision of a magnetic field. Although very high axial magnetic fields can be obtained with superconducting magnets for use in measuring MCD, the OLIS Corp. offers a small compact permanent magnet which is sufficient to convert a CD spectrometer to measure MCD. In principle, any sample can be made optically active due to a strong magnetic field, but in practice MCD is more easily measured for electronically soft chromophores such as delocalized pi systems, porphyrins, indoles, ketones, and magnetic metal atoms such as Fe, Co, and the lanthanides. The strong MCD band of indole at 290 nm is easily measured with the OLIS permanent magnet using a 1 cm cell path.

that in the use of a superconducting solenoid it is more convenient to pass the light beam through the center of the coil.

Three phenomena contribute to MCD spectra. If the system under study has a degenerate excited state, the *A*-term arises as a consequence of Zeeman splitting of those degenerate levels. The *A*-term has a characteristic symmetric sigmoidal dispersion form. For a system with a paramagnetic ground state, population shifts among the Zeeman sublevels produce an asymmetric sigmoidal *C*-term. (This signal is generally strongly temperature dependent, becoming more intense at low temperatures when these populations differ considerably.) *B*-terms arise from field-induced mixing of states and appear in all systems with or without state degeneracies.

Theory of MCD

While important models for MCD were already presented by Caldwell and Eyring [5], the fundamental theory of MCD is primarily due to Buckingham and Stephens [42]. The perturbation analysis leads to three types of signals: *A*, *B*, and *C*:

$$A = \frac{3}{d_A} \sum_{a,j} \mathrm{Im}(\langle a|m|j\rangle\langle j|m|a\rangle)(\langle j|\mu|j\rangle - \langle a|\mu|a\rangle)$$

$$B = \frac{3}{d_A}\mathrm{Im}\left[\begin{array}{l} \sum_{k\neq J}(\langle a|m|j\rangle\langle k|m|a\rangle - \langle a|m|j\rangle\langle k|m|a\rangle)\dfrac{\langle j|\mu|k\rangle}{\hbar\omega_{KJ}} \\[2ex] \quad + \sum_{k\neq A}(\langle a|m|j\rangle\langle j|m|k\rangle - \langle a|m|j\rangle\langle j|m|k\rangle)\dfrac{\langle k|\mu|a\rangle}{\hbar\omega_{KA}} \end{array}\right]$$

$$C = -\frac{3}{d_a}\sum_j \mathrm{Im}(\langle a|m|j\rangle\langle j|m|a\rangle\langle a|\mu|a\rangle)$$

The uppercase indices A, J, K label states which may be degenerate; their components are labeled by a, j, k. The degeneracy index of the ground state is d_A. Here m refers to the magnetic dipole and μ to the electric dipole. These are vector-valued quantities. PCLOBE only treats closed-shell ground states and so does not treat C-terms. The working expressions for A and B terms appear below:

$$A(a \to j) = (1/2)\sum_s\sum_k \mathrm{Im}[\langle a|\mathbf{m}|j_k\rangle \times \langle j_k|\mathbf{m}|a\rangle \bullet \langle j_s|\boldsymbol{\mu}|j_k\rangle]$$

$$B(a \to j) = \sum_\lambda\sum_{k\neq j}\mathrm{Im}\left[\langle a|\mathbf{m}|j_\lambda\rangle \times \langle k|\mathbf{m}|a\rangle \bullet \frac{\langle j_\lambda|\boldsymbol{\mu}|k\rangle}{(E_k - E_j)} + \langle a|\mathbf{m}|j_\lambda\rangle\right.$$
$$\left.\times \langle j_\lambda|\mathbf{m}|k\rangle \bullet \frac{\langle k|\boldsymbol{\mu}|a\rangle}{(E_k - E_0)}\right]$$

For almost all stable organic molecules there is no issue of ground state degeneracy, and no C-term appears. A-terms do appear it there are nearly-degenerate excited states, but they can be modeled as a closely spaced pair of opposite signs. Intuitively accessible expressions for the B-term have been given by Helgaker.

$$B(n \to j)$$

$$= \frac{1}{6}\varepsilon_{\alpha\beta\gamma}\mathrm{Im}\left[\sum_{k\neq n}\frac{\langle k|m_\gamma|n\rangle\langle n|\mu_\alpha|j\rangle\langle j|\mu_\beta|k\rangle}{\hbar\omega_{kn}} + \sum_{k\neq j}\frac{\langle j|m_\gamma|k\rangle\langle n|\mu_\alpha|j\rangle\langle k|\mu_\beta|n\rangle}{\hbar\omega_{kj}}\right]$$

The Levi–Civita tensor assigns proper signs. We include the definition of a D-term as well, defined as the electric transition moment.

$$D(n \to j) = \langle n|\boldsymbol{\mu}|j\rangle \bullet \langle j|\boldsymbol{\mu}|n\rangle$$

Experimental data are reported as the ratios of the A-, B-, and C-terms to D. This is very helpful in canceling corrections for dielectric dispersion in the ratios since the dielectric constant can vary greatly over a range of wavelengths/frequencies. Experimentally the B/D and A/D ratios can also be obtained from measured spectra [43].

PCLOBE Modeling of Rings

Aromatic ring compounds have attracted considerable attention, since they lend themselves to very simple descriptions. Michl's particle-on-a-ring representation of the MCD of cyclic pi compounds [44–49] is the prime example. His model can be extended with refinements to the very significant porphyrin rings and substituted indoles so prominent in biological systems.

Although there are many cyclic pi-electron compounds, we need to treat noncyclic molecules and systems in which sigma bonds participate in the spectra. The simplest ab initio treatment suitable for these systems is to use a minimum basis set to describe the occupied orbitals for sigma as well as pi orbitals. We have already seen that even the description of the transitions of such compounds may require diffuse Rydberg orbitals. Let us investigate whether the simple treatment using a minimum basis will be able to describe MCD spectra. Bicyclic ring systems such as serotonin and melatonin (among many other indoles) [50–53] display a large positive MCD peak at about 285 nm. We can use the PCLOBE program to characterize the MCD spectrum of the parent indole, shown below.

The experimental spectrum has long been known. It is dominated by the positive feature ca. 287 nm, and shows strong signals at 280 and 276 nm as well. We take note of the broad negative feature below about 265 nm (Figure 14.27).

To illustrate the computation of B-terms, we employ Shillady et al.'s [53] adaptation of the perturbation theoretic equations, built into PCLOBE. As in the CD and ORD calculations, the energy differences can be recovered from CIS calculation of low-lying excited states. No new matrix elements are required. As usual we build the structure with CHEMSITE [10] and conduct a small 75×75 CIS/STO-4GL representation of the excitation spectrum. To compensate for the CIS overestimate of excitation energies, Del Bene and Jaffe's CNDO/S [54] method scaled certain one-electron integrals describing $p\pi$–$p\pi$ interactions. PCLOBE scales the CIS $\pi \rightarrow \pi^*$ excitation energies by a factor of 0.661 to fit the lowest energy transition in the benzene spectrum at 262.53 nm [55]. This produces excellent UV–visible transition energies for

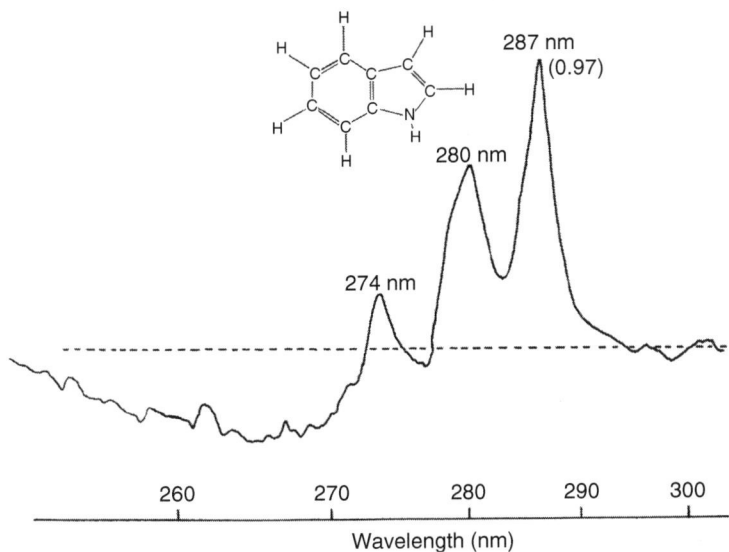

FIGURE 14.27

The MCD spectrum of indole in *n*-hexane. The trace was obtained by use of a Beckman DU monochrometer, with the slow scan of wavelength monitored by an Ithaco Dyantrac lock-in voltmeter with a time constant of 125 s. The magnetic field was 10 kG, the path length was 1 mm, and the concentration of indole was 1.71×10^{-3} M. The vibronic structure is evident. The large positive peak has signal strength of 0.97 in degrees-cm^2/(dmole gauss). (From Sprinkel, F.M., Shillady, D.D., and Strickland, R.W., *J. Am. Chem. Soc.* 97, 6653, 1975. With permission.)

indole. Furthermore, the computed signs and relative magnitudes of the 285 and 260 nm MCD bands of indole are in excellent agreement with experiment (Figure 14.28).

This approximate calculation suggests that there will be very strong MCD signals in the vacuum UV (below 200 nm). Predictions of this model however are likely to be unreliable at high energy (Figure 14.29).

Formaldehyde MCD

Carbonyl compounds have also been extensively studied using MCD. A rare spectrum of the $n \to \pi^*$ transition of formaldehyde near 320 nm was recorded by Seamans et al. [56] (Figure 14.30). On the basis of the formaldehyde data, the authors developed a guide to the interpretations of MCD spectra of other carbonyl compounds.

The perfectly C_{2v}-symmetric structure of formaldehyde shows no CD or MCD. This is what is expected from the analysis of Seamans et al. [56] who ascribed the intensity of the 320 nm band to vibronic effects. Here we present a computed MCD spectrum of a formaldehyde structure in which the carbonyl oxygen atom is displaced by 0.02 Å out of the plane of the molecule. With this model of distortion effects on the MCD spectrum, the 320 nm band is calculated to be a negative B-term. In this (nonaromatic)

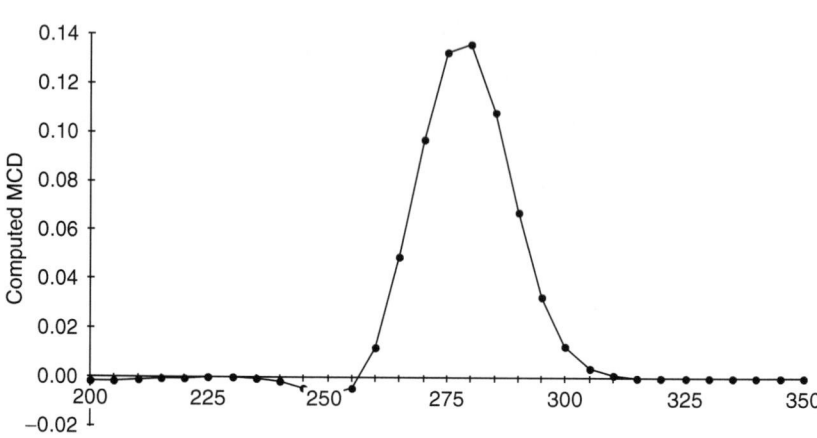

FIGURE 14.28
The simulated MCD spectrum of indole. The vertical axis is the MCD signal (see text); the horizontal axis is wavelength in nanometers. The calculation uses an STO-4G basis set with a 75×75 CIS treatment. The lowest energy transition wavelength has been shifted by a factor derived from fitting the first $\pi \rightarrow \pi^*$ transition of benzene. The spectrum is in qualitative agreement with the experimental spectrum but the vibronic splitting is missing since the computation is at a fixed geometry.

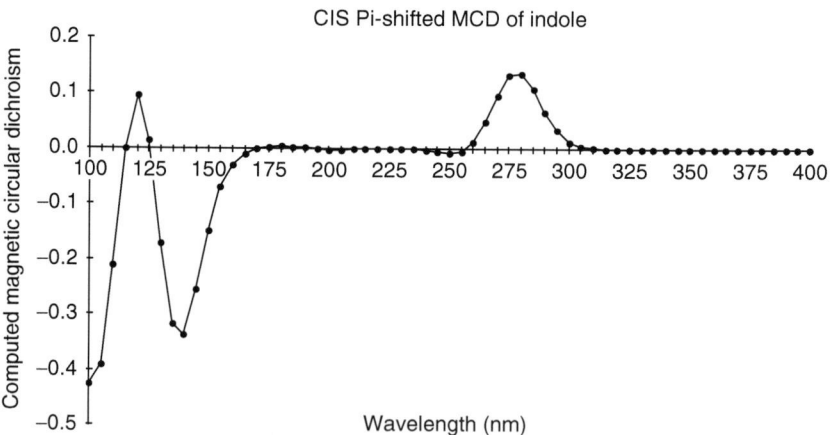

FIGURE 14.29
The simulated MCD spectrum of indole. The vertical axis is the MCD signal (see text); the horizontal axis is wavelength in nanometers. The calculation uses pi energy scaling (see text). The simulation shows predicted (as yet undocumented) VUV bands. While VUV-CD spectra are rare and MCD spectra are also rare, VUV-MCD spectra are extremely rare!

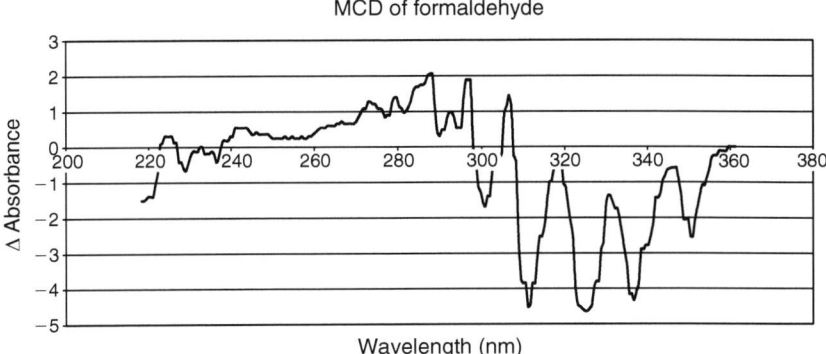

FIGURE 14.30
The vibronic MCD spectrum of formaldehyde measured in perfluorohexane at 20°C with a 49.5 kG axial field, unknown concentration. Vertical axis in 2×10^{-5} absorbance units. Clearly this spectral feature is vibronic in nature. (Redrawn by digitizing the original spectrum from Seamans, L., Moscowitz, A., Barth, G., et al., *J. Am. Chem. Soc.* 94, 6464, 1972. With permission.)

system no scaling was used on the CIS energies. Direct tests show that it does not matter whether the 0.02 Å distortion is above or below the sp^2 plane due to the vector product matrix elements in the B-terms. This outcome is a consequence of the fact that the virtual orbital enters into the magnetic dipole integrals in Stephens' formulas in a quadratic way. The second contribution to the B-term is

$$\sum_{k \neq A} (\langle a|m|j\rangle\langle j|m|k\rangle - \langle a|m|j\rangle\langle j|m|k\rangle)$$

Let us represent the π^* orbital (originally of symmetry B_1 in C_{2v}) as

$$|j\rangle_{a'} \approx |j\rangle_{b_1}^0 \pm \lambda |\varphi\rangle_{a_1}$$

The symmetry species b_1 and a_1 merge to species a' as the symmetry descends from C_{2v} to C_s. We can recognize that in the product of matrix elements above there will be a term linear in λ which will average to zero in the course of a vibration, and also a term quadratic in λ which will persist. This is not the case for the circular dichroism signal for achiral systems which break symmetry upon vibrations; these systems' rotational strength are linear in the magnetic dipole matrix integrals.

We see that the simple STO-4GL basis with only a 24×24 valence shell CIS treatment produced a qualitatively correct spectrum (Figure 14.31).

We present output from PCLOBE here with explanations for the case of formaldehyde. This molecule has simple but useful symmetry and offers

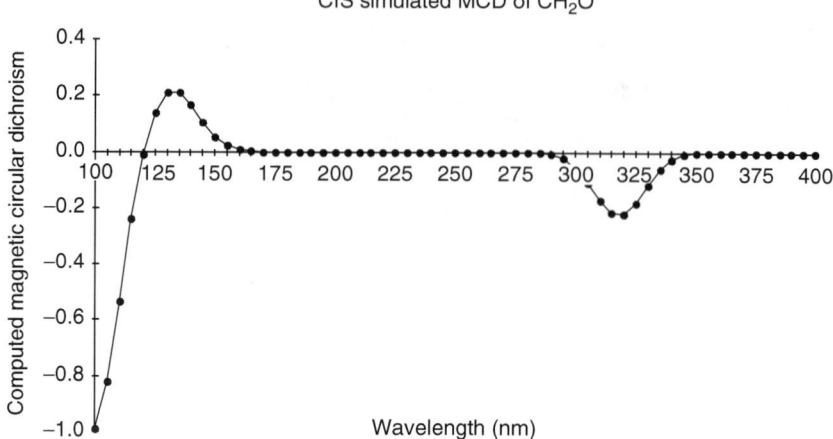

FIGURE 14.31

The simulated MCD spectrum of formaldehyde. The vertical axis is the MCD signal (see text); the horizontal axis is wavelength in nanometers. The calculation uses an STO-4G basis and a 75×75 CIS treatment of the excited states. As before, no pi-shift is needed for the carbonyl transition. Vibrational fine structure is evident in the experimental spectrum but the overall band envelope is negative. The bands in the VUV-MCD have yet to be measured.

examples of several significant types of electronic transitions: $\sigma \rightarrow \sigma^*$, $n \rightarrow \pi^*$, $n \rightarrow \sigma^*$, and $\pi \rightarrow \pi^*$. The choice of axes defining molecular plane in the C_{2v} point group is arbitrary. We will choose the same labels of x and y axes as did Seamans et al. [56] so that we can use their analysis of the carbonyl MCD spectra (Table 14.5).

We have already optimized the structure using the C_{2v} symmetry constraint in the optimization with the STO-4G basis set. The coordinates are not quoted here—but keep in mind that we have displaced the O atom by 0.02 Å out of the plane so that we can obtain some estimate of the effect of the nonplanar vibration of the carbonyl group. We quote the SCF energy and the orbitals' energies and coefficients. Using the minimum basis simplifies interpretation since we can identify each basis function as an atomic orbital.

TABLE 14.5

C_{2v} Character Table Used by Seamans et al. [83]

C_{2v}	E	C_2	σ_v	σ'_v		
A_1	1	1	1	1		z
A_2	1	1	−1	−1	R_z	
B_1	1	−1	1	−1	R_y	x
B_2	1	−1	−1	1	R_x	y

Alpha-spin Orbitals for Eight Filled Orbitals by Column

Ao#	Atom	Orbital	1	2	3	4	5	6	7	8
1	C	1s	−0.001	0.994	−0.117	−0.181	0.000	−0.026	−0.001	0.000
2	C	2s	0.010	0.027	0.215	0.650	0.000	0.089	0.004	0.000
3	C	2px	0.000	0.000	0.002	−0.001	0.000	0.004	−0.620	0.000
4	C	2py	0.000	0.000	0.000	0.000	0.568	0.000	0.000	0.253
5	C	2pz	0.007	0.000	0.136	−0.245	0.000	0.475	0.006	0.000
6	H	1s	−0.001	−0.006	0.030	0.214	0.220	−0.126	0.001	0.351
7	H	1s	−0.001	−0.006	0.030	0.214	−0.220	−0.126	0.001	−0.351
8	O	1s	−0.995	0.000	−0.226	0.106	0.000	0.107	0.000	0.000
9	O	2s	−0.025	−0.005	0.778	−0.462	0.000	−0.571	0.001	0.000
10	O	2px	0.000	0.000	−0.004	−0.002	0.000	−0.013	−0.648	0.000
11	O	2py	0.000	0.000	0.000	0.000	0.479	0.000	0.000	−0.850
12	O	2pz	0.006	−0.001	−0.236	−0.166	0.000	−0.681	0.014	0.000
MO type			CORE	CORE	A_1	A_1	B_2	A_1	B_1	B_2

Virtual orbitals

Ao#	Atom	Orb	9	10	11	12
1	C	1s	−0.002	−0.212	0.000	0.088
2	C	2s	0.011	1.556	0.000	−0.751
3	C	2px	−0.824	0.009	0.000	−0.020
4	C	2py	0.000	0.000	−1.224	0.000
5	C	2pz	0.008	−0.396	0.000	−1.336
6	H	1s	−0.006	−1.032	0.927	−0.193
7	H	1s	−0.006	−1.032	−0.927	−0.193
8	O	1s	0.000	0.048	0.000	−0.142
9	O	2s	−0.001	−0.324	0.000	1.101
10	O	2px	0.801	−0.002	0.000	−0.017
11	O	2py	0.000	0.000	0.359	0.000
12	O	2pz	−0.012	0.263	0.000	−0.949
MO type			B_1	A_1	B_2	A_1

The first column is dominated by an O1s orbital with the second column almost totally a C1s orbital. Column 8—the highest occupied MO of B_2 symmetry in C_{2v}—is the O2py lone-pair orbital mixed in an antibonding way with the local methylene CH_2 fragment orbitals, i.e., C2py oriented to bond with the out-of-phase combination of H1s orbitals. (Compare to orbital 6, which is the B_2 bonding combination of the O2py lone pair and the local orbitals of the CH_2 fragment.) Column 7 is the B_1-symmetric π orbital composed of the in-phase (bonding) combination of C2px and O2px orbitals, while column 9 is the out-of-phase combination, the B_1-symmetric π^* orbital; according to the orientation chosen by Seamans et al. [56] the x–z plane contains the π and π^* orbitals, in columns 7 and 9, respectively. The slight deviation of the O atom from planarity lowers the symmetry to C_s and allows mixing of A_1 and B_1 orbitals which are otherwise perfectly orthogonal σ and π.

Conveniently, the output matrix lists the first eight occupied orbitals together and the last four virtual orbitals as a second group. Single excitations from the occupied orbitals into the virtual orbitals—written as a linear combination of two determinants with a singlet spin function—are not usually good representations of excited states. In the CIS method, we find a variationally optimized linear combination of excitations from (at least some) occupied orbitals to (some) virtual orbitals. In the jargon of the program, NBS is the "number of basis functions" and CORE is the number of 1s orbitals. In the "frozen core approximation" the energetically isolated CORE MOs would not be used in the CIS model of excited states. HOMO(8) and LUMO(9) are the orbital numbers of the highest occupied and lowest unoccupied molecular orbitals.

CIS Output: Excited States

We have already seen PCLOBE's report defining the CIS space. There are 24 single excitations from the 6 occupied valence MOs to the 4 virtual orbitals defined by the minimum basis. The CIS states have also been quoted previously; we give brief reminders of their character.

```
                  SINGLET State Energy Levels
     (pi->pi* Excitations are scaled by 0.661 to fit benzene)

E(1)=0.14333780 au   3.90042 ev   31459.01 1/cm   317.9 nm (mostly CIS excitation #1)
   The wavelength adjusted to fit benzene pi-pi*=480.9 nm

E(2)=0.34625696 au   9.42213 ev   75994.62 1/cm   131.6 nm (mostly CIS excitation #9)
   The wavelength adjusted to fit benzene pi-pi*=199.1 nm

E(3)=0.45673642 au  12.42843 ev  100242.06 1/cm    99.8 nm (mostly CIS excitation #5)
   The wavelength adjusted to fit benzene pi-pi*=150.9 nm

E(4)=0.48607506 au  13.22678 ev  106681.14 1/cm    93.7 nm (mostly CIS excitation #13)
   The wavelength adjusted to fit benzene pi-pi*=141.8 nm

E(5)=0.67669186 au  18.41372 ev  148516.70 1/cm    67.3 nm (mostly CIS excitation #4)
   The wavelength adjusted to fit benzene pi-pi*=101.9 nm
```

Only singlets are reported here although an option to do triplet states, once available, has been suppressed. Next the oscillator strengths are computed with the dipole length expression. See the discussion below.

```
There are 4 transitions less than 15.0 ev.

State Energy=3.9004 ev        0.143338au       317.87 nm
Osc. Strength(f)=0.00000
M=0.00819    Mx=0.00000    My=-0.00819                Mz=0.00000 in Debyes
Transition Direction Angles ( 90.0, 180.0, 90.0) in degrees.
*** Wavelength as fitted to benzene pi-pi*=480.9 nm ***

State Energy=9.4221 ev        0.346257 au      131.59 nm
Osc. Strength(f)=0.01226
M=0.58566    Mx=0.58532    My=0.00000                 Mz=-0.02005 in Debyes
Transition Direction Angles ( 2.0, 90.0, 92.0) in degrees.
*** Wavelength as fitted to benzene pi-pi*=199.1 nm ***
```

```
State Energy = 12.4284 ev       0.456736 au      99.76 nm
Osc. Strength(f) = 0.30101
M = 2.52718    Mx = -0.06400    My = 0.00000              Mz = -2.52637 in Debyes
Transition Direction Angles ( 91.5, 90.0, 178.5) in degrees.
*** Wavelength as fitted to benzene pi-pi* = 150.9 nm ***
```

```
State Energy = 13.2268 ev       0.486075 au      93.74 nm
Osc. Strength(f) = 0.00001
M = 0.01058    Mx = 0.00000    My = 0.01058              Mz = 0.00000 in Debyes
Transition Direction Angles ( 90.0, 0.0, 90.0) in degrees.
*** Wavelength as fitted to benzene pi-pi* = 141.8 nm ***
```

Use of dipole velocity AO matrix elements are particularly convenient since a common integral treatment produces both the electric and magnetic transition integrals. The evaluation of two center angular momentum integrals for magnetic transitions is simple, requiring only the two-center dipole velocity results and overlap integrals. These advantages apply only to STO basis functions but can be well approximated with STO-NG functions. The oscillator strength defined by the dipole length is also computed as a measure of quality of the wave function. For an exact wave function the length and velocity forms should agree. Details are provided by Richardson, Shillady, and Bloor [29].

Transition moment (length) components T_α are computed for the (x,y,z) directions

$$T_\alpha(n) = \sum_{ij} C_i^j(n)\sqrt{2}\langle i|r_\alpha|j\rangle = \sum_{ij} C_i^j(n)\sqrt{2}\sum_{ab} A_{ia}A_{jb}\langle a|r_\alpha|b\rangle, \alpha = x, y, z$$

Then the formula by La Paglia [57] is used for the oscillator strength f_n.

$$f_n = (2/3)(E_n - E_0)\left(T_x^2 + T_y^2 + T_z^2\right) \text{ (dipole length)}$$

The dipole velocity form of the transition moment is

$$T_\alpha = C_i^j(1)\sqrt{2}\langle i|\partial/\partial\alpha|j\rangle, \alpha = x, y, z$$

The associated expression for the oscillator strength is

$$f_n = (2/3)[1/(E_n - E_0)]\left(T_x^2 + T_y^2 + T_z^2\right) \text{ (dipole velocity)}$$

Here all quantities appear in atomic units.

The classical interpretation is that the oscillator strength is the fraction of the electron which moves during a transition. A value between 0.1 and 1.0 for f_n indicates an intense transition. If f_n is between 0.001 and 0.1, the transition will likely be apparent in a UV–visible spectrum but if f_n is less than 0.001 the transition may be unobservable. Weakness may be due to a symmetry prohibition; if so, the transition acquires intensity only by vibration. However, a transition may have a very weak electric transition but be strongly allowed as a magnetic transition. This can yield a strong CD and/or MCD peak, as with the carbonyl $n \to \pi^*$ transition. Using the fact that the transition moment has Cartesian components it is also possible to estimate the direction of motion executed by the electron during the transitions. Defining the magnitude of the transition moment as T_n

$$T_n = \sqrt{(T_{nx} \times T_{nx}) + (T_{ny} \times T_{ny}) + (T_{nz} \times T_{nz})}$$

the direction cosines define the angles α, β, and γ as

$$\alpha = \cos^{-1}(T_{nx}/T_n)$$

$$\beta = \cos^{-1}(T_{ny}/T_n)$$

$$\gamma = \cos^{-1}(T_{nz}/T_n)$$

The oscillator strengths can be used to simulate the UV–visible absorbance (A) spectrum by draping a Gaussian over each at the transition wavelength and using a bandwidth parameter w.

$$A(\lambda) \propto \sum_n f_n \exp\left[-w(\lambda - \lambda_n)^2\right]$$

The direct CIS calculation predicts a very weak transition near 320 nm. This is the symmetry forbidden $n \to \pi^*$ transition, now bearing a small transition moment because the O atom is positioned 0.02 Å above the plane of the CH_2 group.

Once the CIS coefficients are available, the next order of business is to compute the arrays holding the electric (μ_x, μ_y, μ_z) and magnetic (m_x, m_y, m_z) transition moments over the CIS state functions. This, in principle, would require an elaborate sum of integrals involving the atomic orbital basis functions—and then hence involving the primitive Gaussian-lobe functions making up the representation of each basis function. We avoid a part of this cumbersome summing, using electric and magnetic integrals developed for Slater orbitals; we assume that the Gaussian expansion produces a faithful mimic of the STO.

```
*************************************************************************
      Computing Optical Activity Parameters and
      Faraday Effect Buckingham-Stephens Parameters
      using Single-Excitation Configuration Interaction,
      See P.J. Stephens, J. Chem. Phys. v52, p3489 (1970),
      F.S. Richardson, D.D. Shillady and J.E. Bloor,
          J. Phys. Chem. v75, p2466 (1973), and
      F.M. Sprinkel, D.D. Shillady and R.W. Strickland,
          J. Am. Chem. Soc., v97, p6653 (1975)
*************************************************************************
```

```
The Center of Mass is at Xm=0.020139 Ym=0.000000 Zm=1.119905
```

We do not use GIAOs, so the center of mass needs to be specified as the gauge origin.

Atomic Coordinates (bohr)				Orbital Scaling		
x	y	z	n	l	exponent	
−0.02014	0.00000	−1.11991	1	0	5.6727	
			2	0	1.6083	
			2	1	1.5679	
−0.02014	1.74973	−2.22015	1	0	1.2400	
−0.02014	−1.74973	−2.22015	1	0	1.2400	
0.01766	0.00000	1.12047	1	0	7.6579	
			2	0	2.2458	
			2	1	2.2266	

Here we insert the ground state (#1) which is orthogonal to all the CIS states.

```
The Configuration Interaction States

State No. 1 = 0.00000 ev.

1.000000x  1   1,  0.000000x  8  9,  0.000000x  8  10,  0.000000x  8  11,
0.000000x  8  12,  0.000000x  7  9,  0.000000x  7  10,  0.000000x  7  11,
0.000000x  7  12,  0.000000x  6  9,  0.000000x  6  10,  0.000000x  6  11,
0.000000x  6  12,  0.000000x  5  9,  0.000000x  5  10,  0.000000x  5  11,
0.000000x  5  12,  0.000000x  4  9,  0.000000x  4  10,  0.000000x  4  11,
0.000000x  4  12,  0.000000x  3  9,  0.000000x  3  10,  0.000000x  3  11,
0.000000x  3  12,  0.000000x  0  0,  0.000000x  0   0,  0.000000x  0   0,
```

The CIS states (#2...) make no reference to the ground state.

```
State No. 2 = 3.90042 ev.

 0.000000x  1   1,   0.977406x  8  9,  −0.002688x  8  10,   0.000000x  8  11,
−0.000748x  8  12,   0.000000x  7  9,   0.000000x  7  10,  −0.000204x  7  11,
 0.000000x  7  12,   0.000000x  6  9,   0.000000x  6  10,  −0.000035x  6  11,
 0.000000x  6  12,  −0.211352x  5  9,   0.000725x  5  10,   0.000000x  5  11,
 0.000471x  5  12,   0.000000x  4  9,   0.000000x  4  10,  −0.000032x  4  11,
 0.000000x  4  12,   0.000000x  3  9,   0.000000x  3  10,   0.000063x  3  11,
 0.000000x  3  12,   0.000000x  0  0,   0.000000x  0   0,   0.000000x  0   0,

State No. 3 = 9.42213 ev.

 0.000000x  1   1,   0.000000x  8  9,   0.000000x  8  10,  −0.000465x  8  11,
 0.000000x  8  12,   0.008412x  7  9,  −0.002565x  7  10,   0.000000x  7  11,
−0.001305x  7  12,  −0.993958x  6  9,   0.002199x  6  10,   0.000000x  6  11,
 0.001686x  6  12,   0.000000x  5  9,   0.000000x  5  10,   0.000363x  5  11,
 0.000000x  5  12,  −0.060377x  4  9,   0.000083x  4  10,   0.000000x  4  11,
 0.000199x  4  12,  −0.091185x  3  9,   0.000272x  3  10,   0.000000x  3  11,
 0.000314x  3  12,   0.000000x  0  0,   0.000000x  0   0,   0.000000x  0   0,

State No. 4 = 12.42843 ev.

 0.000000x  1   1,   0.000000x  8  9,   0.000000x  8  10,  −0.178362x  8  11,
 0.000000x  8  12,   0.933589x  7  9,  −0.000162x  7  10,   0.000000x  7  11,
−0.002066x  7  12,   0.008472x  6  9,  −0.066372x  6  10,   0.000000x  6  11,
 0.272088x  6  12,   0.000000x  5  9,   0.000000x  5  10,   0.098494x  5  11,
 0.000000x  5  12,  −0.002412x  4  9,  −0.043566x  4  10,   0.000000x  4  11,
−0.014025x  4  12,   0.000255x  3  9,  −0.025459x  3  10,   0.000000x  3  11,
 0.075077x  3  12,   0.000000x  0  0,   0.000000x  0   0,   0.000000x  0   0,
```

```
State No. 5 = 13.22678 ev.

  0.000000x  1   1,      -0.211291x  8  9,   -0.000030x  8  10,    0.000000x  8  11,
 -0.001302x  8  12,       0.000000x  7  9,    0.000000x  7  10,   -0.025455x  7  11,
  0.000000x  7  12,       0.000000x  6  9,    0.000000x  6  10,    0.000080x  6  11,
  0.000000x  6  12,      -0.977089x  5  9,    0.001740x  5  10,    0.000000x  5  11,
  0.000130x  5  12,       0.000000x  4  9,    0.000000x  4  10,    0.000143x  4  11,
  0.000000x  4  12,       0.000000x  3  9,    0.000000x  3  10,    0.000155x  3  11,
  0.000000x  3  12,       0.000000x  0  0,    0.000000x  0   0,    0.000000x  0   0,
```

. . . .
States above 15.0 ev are suppressed; the total is 25.
. . . .

```
State No.          25 = 47.67262 ev.

  0.000000x  1   1,       0.000000x  8  9,    0.000000x  8  10,   -0.047836x  8  11,
  0.000000x  8  12,      -0.099169x  7  9,    0.000959x  7  10,    0.000000x  7  11,
 -0.000111x  7  12,      -0.000539x  6  9,   -0.042044x  6  10,    0.000000x  6  11,
  0.047390x  6  12,       0.000000x  5  9,    0.000000x  5  10,    0.062454x  5  11,
  0.000000x  5  12,      -0.000258x  4  9,   -0.060674x  4  10,    0.000000x  4  11,
 -0.020211x  4  12,       0.001384x  3  9,    0.370998x  3  10,    0.000000x  3  11,
  0.915549x  3  12,       0.000000x  0  0,    0.000000x  0   0,    0.000000x  0   0,
```

The slowest step in the program is the evaluation of the six matrices over electric and magnetic transition moments with CIS state functions.

The (x,y,z) Dipole Velocity over Slater-type basis functions,
with the corresponding angular momentum matrix elements.
The origin of coordinates is centered at (xm,ym,zm) for both operators.

The DVx-matrix has been computed

The Lx-matrix has been computed

The DVy-matrix has been computed

The Ly-matrix has been computed

The DVz-matrix has been computed

The Lz-matrix has been computed

Next the (x,y,z) components of the electric and magnetic moments are computed in the basis of MO$_s$.

X-polarized Magnetic Transition Moments

X-polarized Electric Transition Moments

Y-polarized Magnetic Transition Moments

Y-polarized Electric Transition Moments

Z-polarized Magnetic Transition Moments

Z-polarized Electric Transition Moments

We suppress the details of the six large matrices defining the magnetic and electric moments in the basis of CIS States necessary for the final CD and MCD parameters.

```
Mx Interstate Transition Moments have been computed
My Interstate Transition Moments have been computed
Mz Interstate Transition Moments have been computed

Ex Interstate Transition Moments have been computed
Ey Interstate Transition Moments have been computed
Ez Interstate Transition Moments have been computed
```

Rotational strengths

Now we can construct the rotational strengths R_{0j} for each individual transition and evaluate all the terms in Stephens' MCD equations.

```
The R-term Transition Matrix
         0.00000      0.00000      0.00000      0.00000      0.00000
         0.00000      0.00000      0.00000      0.00000      0.00000
         0.00000      0.00000      0.00000      0.00000      0.00000
         0.00000      0.00000      0.00000      0.00000      0.00000
         0.00000      0.00000      0.00000      0.00000

The D-term Transition Matrix
         0.00008      1.31211      0.10581      0.00019      1.05231
         1.09175      2.35779      1.01869      0.03279      0.00012
         0.05283      3.18103      1.46242      1.24423      2.96297
         1.38039      1.22316      0.02067      0.20267      0.13346
         0.59870      0.60021      0.01013      0.79556

The Dipole Velocity Oscillator Strengths (f#)
         0.00000      0.04688      0.00499      0.00001      0.07348
         0.07990      0.19206      0.08391      0.00273      0.00001
         0.00465      0.30951      0.14364      0.12547      0.30804
         0.15050      0.14233      0.00242      0.02417      0.01623
         0.07790      0.10420      0.00178      0.14383

Computed Optical-Activity Parameters
    E = 3.900 ev, Lambda = 317.9 nm, Reduced-R = 0.000, G-factor = 0.000
    E = 9.422 ev, Lambda = 131.6 nm, Reduced-R = 0.000, G-factor = 0.000
```

Optical activity parameters

The rotational strengths R are all zero here because there is still a mirror plane left in the molecule. The D-term is defined by the electric dipole strength of each transition, previously called f; it is needed for the standard report of the MCD parameters, particularly the (B/D) ratio. Note the oscillator strengths are computed here in the dipole velocity form instead of the (in principle) equivalent dipole length.

****

Computed Faraday Effect Parameters assuming a Singlet ground state

(B/D) ratios summed over degeneracies in bohr magnetons per (1/cm)

0.0000338	−0.0000281	0.0000759	0.0001054	0.0000305
−0.0000411	−0.0000319	0.0000626	−0.0001569	−0.0001620
−0.0003168	−0.0000873	0.0001393	0.0000410	0.0000286
−0.0000271	−0.0000222	0.0001302	0.0001096	−0.0000796
−0.0000057	0.0000157	−0.0001181	0.0000099	

NOTE: the signs of the MCD bands use the Schatz
Convention that the sign of the MCD of H2O is NEGATIVE at
200 nm. Since Theta(H) =Consts. (A - B)H for H-field,
the SIGN OF THE B-Bands is MINUS (B/D) shown above;
D is always positive.

E = 3.900 ev, Lambda = 317.9 nm, (A/D) =0.00000000, (B/D) = 0.00003375

E = 9.422 ev, Lambda = 131.6 nm, (A/D) =0.00000000, (B/D) = −0.00002806

MCD Parameters

Finally, we come to the computed MCD parameters. In the Schatz convention [58] the sign of the *B*-terms as expressed in Stephens' formulas is reversed. The reversal produces a computed negative value for the MCD of water, as is found experimentally. Here, the first band of formaldehyde is found to be negative in agreement with the spectral feature recorded by Seamans et al. [56].

Now we can realize the convenience of using the ratio [43] of (A/D) and (B/D) so that the ratios are expressed in simple units, Bohr magnetons per wave number.

It is particularly interesting that no pi-shift is needed for the $n \rightarrow \pi^*$ transition and the simulated spectrum is in qualitative agreement with the experimental spectrum of Seamans et al. [56]. In this case, the pi-shifted spectrum is useless since the $n \rightarrow \pi^*$ transition is shifted far to the red. The question arises here as to when one should use the pi-shift. Recalling the success with the predicted CD spectrum of D(+)-camphor it would appear that the pi-shift is probably always needed except for the case of the $n \rightarrow \pi^*$ transitions and perhaps any other case of excitation of a nonbonded orbital. Apparently, the CIS is adequate for loosely bound electrons but all other cases can benefit from the pi-shift technique.

Given the computed parameters, the MCD spectrum can be simulated by a superposition of Gaussian functions.

$$[\Theta(\lambda)] \propto \sum_n (-B_n/D_n)e^{-((\lambda-\lambda_n)/15)^2}$$

PCLOBE writes MCD values computed at 5 nm increments (with and without the pi-shift) to the file *mcdspectra.txt* for later analysis and visualization. For preliminary inspection, simple coare line printer graphs are included in the PCLOBE output.

Use file cdspectra.txt for spreadsheet data.

****

See Sprinkel, Shillady and Strickland J. Am. Chem. Soc. v97, p6653 (1975)
for singlet MCD treatment used here.   Note - (B/D) = +MCD.
****

                        (Vertical Autoscale)
        Qualitative MCD Spectrum (B/D)x1.0d+4 Bohr Magnetons per (1/cm), Pi-Scaled

     0.25+---------+---------+---------+---------+---------+---------+---------+

     M   +   ++   +

     a   +   +++   +

     g   +   +++   +

     n   +   ++++   +

     e   +   +++++   +

     t   +   +++++   +

     i   +   +++++   +

     c   +   ++++++   +

         +   ++++++   +

   A 0.0+-------------------------------------W-------------------------+

     -   +----   -------   +

     B   +----   ------   +

         +----   --------   +

     T   +----   ------   +

     e   +----   ------   +

     r   +----   ------   +

     m   +----   -----   +

     s   +----   ----   +

         +----   ----   +

    -0.25+---------+---------+---------+---------+---------+---------+---------+

         100   150   200   250 nm 300 350   400   450

                        (W denotes a weak band)

(Vertical Autoscale)

Qualitative MCD Spectrum (B/D)x1.0d+4 Bohr Magnetons per (1/cm), Pi-Scaled

```
      0.25+---------+---------+---------+---------+---------+---------+
 M       +++        +++                                             +

 a       +++        +++                                             +

 g       ++++       ++++                                            +

 n       ++++       +++++                                           +

 e       ++++       +++++                                           +

 t       ++++       ++++++                                          +

 i       ++++       +++++++                                         +

 c       ++++       +++++++                                         +

         ++++       +++++++++                                       +

 A   0.0+---------------------------------------------------------------+

 -       +          -----------                                     +

 B       +          -----------                                     +

         +          -----------                                     +

 T       +          -----------                                     +

 e       +          -----------                                     +

 r       +          -----------                                     +

 m       +          ---------                                       +

 s       +          ---------                                       +

         +          ---------                                       +

     -0.25+---------+---------+---------+---------+---------+---------+
        100       150       200      250 nm 300  350       400      450
```

(W denotes a weak band)

Use file mcdspectra.txt for spreadsheet data
**** NORMAL FINISH OF PCLOBE ****

The STO subroutines in PCLOBE for the calculation of CD and MCD spectra have provided essential guidance making qualitative assignments of spectral bands [30,50,58,59]. It is remarkable that a modest investment in computational power can provide insight into the complex phenomena of optical activity and chirality as well as magnetically induced optical activity.

MCD—A Challenge to Modern Theory

The theoretical foundation for analysis of magnetic circular dichroism was laid by Buckingham and Stephens [42], Schatz et al. [58], and Stephens [43]. We have quoted their formulas, which require sums of matrix element products referring to all states of the system.

$$B(n \rightarrow j) = \frac{1}{6}\varepsilon_{\alpha\beta\gamma} \, \mathrm{Im} \left[\sum_{k \neq n} \frac{\langle k|m_\gamma|n\rangle \langle n|\mu_\alpha|j\rangle \langle j|\mu_\beta|k\rangle}{\hbar\omega_{kn}} + \sum_{k \neq j} \frac{\langle j|m_\gamma|k\rangle \langle n|\mu_\alpha|j\rangle \langle k|\mu_\beta|n\rangle}{\hbar\omega_{kj}} \right]$$

Coriani et al. [60] replace this by a form

$$B(a \rightarrow j) = i\varepsilon_{xyz} \left[\mathop{\mathrm{Lim}}_{\omega \rightarrow \omega_{rk}} (\omega - \omega_k)[\![\mu_z; m_y, \mu_x]\!]_{0,\omega} \right]$$

The unfamiliar notation refers to the response function, the poles of which contain the necessary information. Linear equations replace the unwieldy sums, as is described in reports by Jørgensen and coworkers [61,62]. Coriani et al. [60] computed B-terms for ethylene and p-benzoquinone. The smaller ethylene system permitted an exploration of the correlation-consistent sequence of basis sets defined by Dunning, and also a full-valence MCSCF study. Two features predominate the electronic absorption spectrum: the $X\,^1A_g \rightarrow 1\,^1B_{1u}$ and $1\,^1B_{3u}$, which are often characterized as a Rydberg and a $\pi \rightarrow \pi^*$ transition, respectively. There are, however, three significant B-terms in the MCD spectrum. This innocent-appearing system is unfortunately complicated by vibronic interactions and is sensitive both to the flexibility of the basis and the level of description of correlation. Despite such clouds of complication, experiment and theory agree that the leading feature of the MCD spectrum has a positive sign. This is also what we find at a glance at the schematic line-printer graph in the PCLOBE output (below). This is the outcome of the pi-shifted spectrum obtained using the STO-4GL basis and a molecular geometry optimized within C_s symmetry. The simpler calculation captures the leading feature of the spectrum, though it cannot describe the second negative band realistically.

```
           (Vertical Autoscale)

Qualitative MCD Spectrum (B/D)x1.0d+4 Bohr Magnetons per (1/cm), Pi-Scaled

     0.25+---------+---------+---------+---------+---------+---------+

  M     +                                                        +

  a     +                                                        +

  g     +          ++                                            +

  n     +          ++++                                          +

  e     +          ++++                                          +

  t     +          +++++                                         +

  i     +          ++++++                                        +

  c     +          +++++++                                       +

        +          ++++++++                                      +

  A  0.0+----W-----------W-W--------------------------------------+

  -     +------                                                  +

  B     +------                                                  +

        +------                                                  +

  T     +-----                                                   +

  e     +-----                                                   +

  r     +-----                                                   +

  m     +-----                                                   +

  s     +-----                                                   +

        +-----                                                   +

   -0.25+---------+---------+---------+---------+---------+---------+
        100       150      200      250  nm 300 350     400      450
```

                    (W denotes a weak band)

It is characteristic of modern formulations that the sum-over-states formulas produced by second-order perturbation theory are replaced by the energy derivatives which are part of response theory. Coriani et al. [60] investigated a derivative formulation of the B-term which also avoids the sum-over-states difficulty.

$$B(n \to j) = \frac{1}{2} \varepsilon_{\alpha\beta\gamma} \mathrm{Im} \left[\frac{\partial S_{nj}^{xy}}{\partial B_z} \right]$$

The central quantity is

$$S_{nj}^{xy} = \langle n|\mu_x|j\rangle \langle j|\mu_y|n\rangle$$

A coupled cluster approach has been defined, and expressions for the Verdet constant and the MCD B-term have been developed. We look forward to applications of this promising model to systems of greater chemical interest—and perhaps less challenging—than those described in first applications.

Most recently, Seth et al. [63] have applied time-dependent density functional theory to the computation of the MCD B-term for ethylene as well as a number of larger molecules. They employed the Becke–Perdew functional in a VTZ basis to obtain structures, and employed the SAOP model for estimates of spectra. SAOP has been proven to describe electronic excitations, polarizabilities, and related properties better than calculations with functionals which do not display correct long-range asymptotic behavior. In contrast to the very costly coupled cluster calculations, the TD-DFT calculation can treat large heteroaromatic pi systems. Apparently, the sum-over-states which is a necessary part of this calculation converges rapidly. SAOP requires a large (at minimum VTZP) basis, but seems to be converged if ca. 50 excited states are included in the sum. Furan, thiophene, and selenophene, for which experimental data are available, are described; computed MCD curves are in qualitative agreement with reported spectra. More impressive, the method is capable of describing the reversal in sign which is observed as N-methylpyrrole is converted to 2,5-dimethylpyrrole (Figure 14.32).

This is a most promising development in the theory of magnetic circular dichroism. Yet, the value of the simplest calculation cannot be dismissed. Here is the PCLOBE version of the leading edge of the MCD spectra of N-methyl and 2,5-dimethylpyrrole (Figure 14.33).

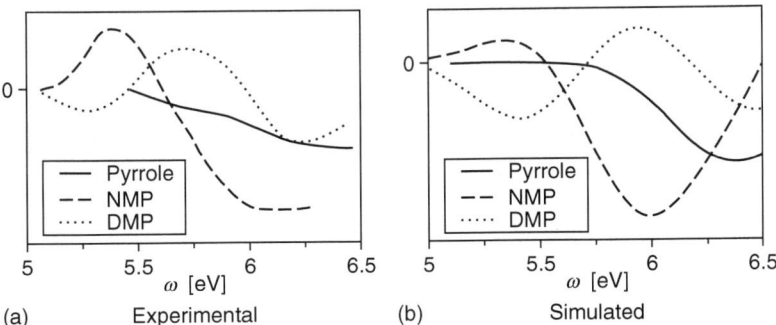

FIGURE 14.32
Simulated spectra of N-methylpyrrole and 2,5-dimethylpyrrole. The vertical axis is proportional with the MCD rotation angle (no scale defined); the horizontal axis is excitation energy in electron volts. Part a: Experimental MCD spectra. Part b: simulated MCD spectra based on TD-DFT. (From Seth, M., Autschbach, J., and Ziegler, T., *J. Chem. Theor. Comp.*, 3, 434, 2007. With permission.)

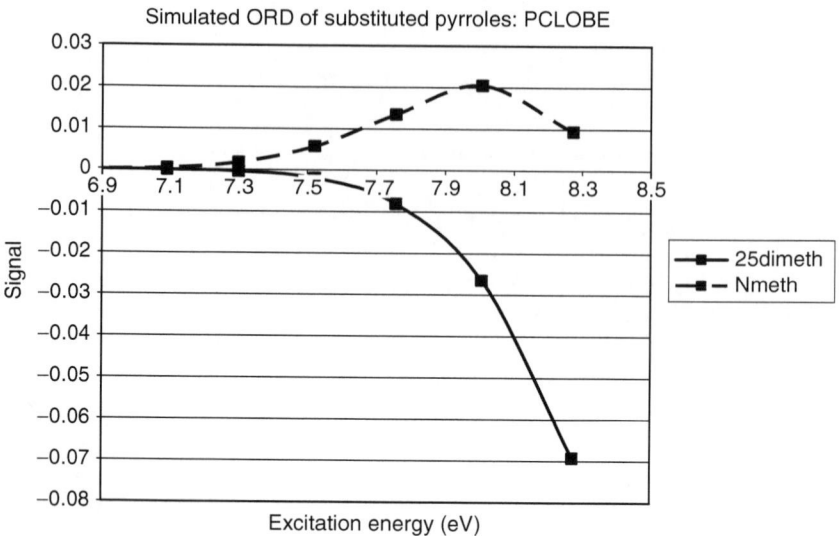

FIGURE 14.33

Computed MCD spectrum for *N*-methylpyrrole and 2,5-dimethylpyrrole. The vertical axis is the MCD signal; the horizontal axis is excitation energy in electron volts. The calculation uses an STO-4G basis, 75×75 CIS excited states, and the dipole length formula.

Seriously blue-shifted, and incapable of dealing with higher energy excitations, this approximate model captures the sign reversal in the low-energy spectra.

Note on Units for Optical Activity

The optical rotation calculation is conducted in ADF, GAUSSIAN, and PCLOBE in conveniently sized atomic units, but the rotational strength is reported by most programs in cgs units. The experimental observations of specific rotation are reported as angles, normalized by the experimental concentration of the chiral material and the path length. Here we wish to establish the connections among these quantities.

The rotational strength tensor **R** is composed of products of components of the dipole transition vector

$$\langle k|er|l \rangle$$

and the magnetic transition vector

$$\langle j|\boldsymbol{\mu}_m|0 \rangle = \sqrt{2}\langle l|e(\mathbf{r} \times \mathbf{p})|k \rangle = \left(\frac{e\hbar}{\sqrt{2}mc}\right)\langle l|(\mathbf{r} \times \boldsymbol{\nabla})|k \rangle$$

In atomic units, the mass and absolute charge of the electron have unit values as does the reduced form of Planck's constant. (The derived quantity c, the speed of light, is about 137 in the atomic unit system.) The magnetic moment is expressed more commonly in Bohr magnetons, which have values 9.274×10^{24} J/T in SI units and 9.274×10^{20} erg/gauss in the cgs system. The dipole factor is sometimes expressed in debyes, a unit outside either system but convenient for molecular values. Its unit value is defined as the dipole established by two (cgs) electrostatic unit (esu) charges separated by 1 Å, or 1×10^{18} statcoulomb-cm. The statcoulomb (or Franklin) is 3.3356×10^{10} coulomb. One atomic unit for a dipole (plus and minus absolute charges of an electron separated by 1 bohr) is 2.5417463 debye.

Rotational strengths are sometimes reported as Debye magnetons, which have a convenient range of values, but more strict use of cgs units would require the conversion factor

$$1 \text{ unit rotational strength (atomic unit system)} = 471.4 \times 10^{-40} \left(\frac{\text{esu} - \text{cm erg}}{\text{gauss}} \right)$$

as quoted by Pedersen and Hansen.* The observed optical rotation in radians per unit path length is defined by

$$\phi = \left(16\pi^3 N\gamma\right)\left(v^2/c^2\right) \sum_j \beta_{0 \to j}$$

where

$$\beta_{0 \to j} = \left(\frac{c}{3\pi h}\right) \frac{\text{Im}\{\langle 0|\mathbf{m}_e|j\rangle \cdot \langle j|\boldsymbol{\mu}_m|0\rangle\}}{v_j^2 - v^2} = \left(\frac{c}{3\pi h}\right) \frac{R_{0 \to j}}{v_j^2 - v^2}$$

In this expression, N is the concentration of the chiral agent in molecules/cm^3 and γ is a solvent correction dependent on the refractive index n.

$$\gamma = \left(\frac{n^2 + 2}{3}\right)$$

Multiplying φ by $(1800/\pi)$ yields the rotation angle in degrees per decimeter. Conversion to a molar rather than number concentration basis and division by that concentration and γ produces the generally reported specific rotation α:

$$[\alpha]_\lambda^T = \phi \frac{1800}{\pi N\gamma}\left(\frac{N_a}{M}\right) = \left(\frac{1800}{\pi}\right)\left(\frac{16\pi^3}{c^2}\right)\left(\frac{N_a}{M}\right)\left(\frac{c}{3\pi h}\right)\left(\frac{e^2 \hbar}{2mc}\right) \sum_j \frac{v^2}{v_j^2 - v^2} R_{0 \to j}$$

* Pedersen, T.B. and Hansen, Aa., *Chem Phys Lett*, 1995, 246:1.

Here M is the gram molecular weight. The specific rotation depends on the wavelength of observation and the temperature, and has units finally

$$[\alpha]_{589}^{25} = (\text{Degrees}/(\text{decimeter } (g/cc))).$$

With standard cgs values of electron charge, the Bohr radius, and the speed of light

$$q_e = 4.803204404 \times 10^{-10} g^{1/2} cm^{3/2}/s \text{ (esu)},$$

$$a_0 = 0.5291772108 \times 10^{-8} cm,$$

$$c = 2.99792458 \times 10^{10} cm/s \text{ [87}^{\text{th}} \text{ Ed. CRC]}$$

and assuming unit molar concentration, we obtain

$$[\alpha]_{589}^{25} = \left[\frac{2155237.437}{M}\right] \sum_j \left(\frac{\nu^2}{\nu_j^2 - \nu^2}\right) R_{0 \to j}.$$

Summary and Conclusions

In the past 20 years, electronic structure modeling has become capable of accurate description of chemical structures and energetics; nuclear magnetic resonance shielding and coupling; and optical rotation, circular dichroism, and most recently magnetic circular dichroism. Modeling has become an essential guide to interpretation of this broad range of phenomena. Essentially all research programs can profit from the perspective theory and software afford.

Acknowledgments

We have benefited from communications with Dr. Jim Cheeseman of Gaussian Inc. regarding the constants in the optical rotation calculation. Thanks to Prof. Dan Crawford for contributing results for *S*-methyloxirane; it is with much appreciation that we acknowledge Professor Micah Abrams for supplying his elegant depiction of methyloxirane dynamics. Prof. Sigrid Peyerimoff's use of Rydberg orbitals for the calculation of rotational strengths of *S*-methyloxirane was helpful. Mrs. Julie DeSa Lorenz supplied the picture from the OLIS Corp. Finally, we are especially grateful for the excellent diagrams for D(+)-camphor in the 1972 text by Crabbe. MCD theory continues rapid development. We are grateful to Professor Frank Neese for descriptions of new theory especially applicable to transition metal systems [64].

References

1. Moscowitz, A., *Adv. Chem. Phys.* 4:67, 1962.
2. Rosenfeld, L., *Z. Phys.* 52:161, 1928.
3. Condon, E.U., *Rev. Mod. Phys.* 9:432, 1937.
4. Atkins, P.W. and Friedman, R., *Molecular Quantum Mechanics*, 4th Ed., Oxford University Press, Oxford, UK, 2005.
5. Caldwell, D.J. and Eyring, H., *The Theory of Optical Activity*, Wiley, New York, 1971.
6. Kauzmann, W., *Quantum Chemistry*, Academic Press, New York, 1957.
7. Ford, T.A., MOSCOW Quantum Chemistry Program Exchange No. 107, Indiana University, Bloomington, 1957.
8. Szabo, A. and Ostlund, N.S., *Modern Quantum Chemistry*, Macmillan, New York, 1982.
9. Davidson, E.R., *J. Comput. Phys.* 17:87, 1975.
10. CHEMSITE, CHEMSW 4771 Mangels Blvd, Fairfield, CA, http://www.chemsw.com/.
11. Djerassi, C., *Optical Rotatory Dispersion*, McGraw-Hill, New York, 1960.
12. Crabbe, P., *Optical Rotatory Dispersion and Circular Dichroism in Organic Chemistry*, Holden-Day, San Francisco, 1965.
13. Mason, M.G. and Schnepp, O., *J. Chem. Phys.* 59:1092, 1973.
14. Budavari, S., Ed., *The Merck Index*, 12th Ed., Merck, Whitehouse, NJ, p. 1281, 1996.
15. Wilson, S.M., Wiberg, K.B., Cheeseman, J.R., et al., *J. Phys. Chem. A* 109:11752, 2005.
16. Robin, M.B., Basch, H., Keubler, N.A., et al., *J. Chem. Phys.* 48:5037, 1968.
17. Zanasi, R., Viglione, R.G., Rosini, C., et al., *J. Am. Chem. Soc.* 126:12968, 2004.
18. Jensen, F., *Introduction to Computational Chemistry*, Wiley, New York, p. 248, 1999.
19. Bak, K.L., Jorgensen, P., Helgaker, T., et al., *J. Chem. Phys.* 100:6620, 1994.
20. Autschbach, J., Ziegler, T., van Gisbergen, S.J.A., et al., *J. Chem. Phys.* 116:6930, 2002.
21. Frisch, M.J., Trucks, G.W., Schlegel, H.B., et al., Gaussian 03, Revision C.02, Gaussian Inc., Wallingford, CT, 2004.
22. Ahlrichs, R., Bar, M., Haser, M., et al., *Chem. Phys. Lett.* 162:165, 1989.
23. Helgaker, T.U., Almlof, J., Jensen, H.J.A., et al., *J. Chem. Phys.* 84:6266, 1986.
24. Amos, R.D., CADPAC 5: The Cambridge Analytic Derivatives Package, issue 5, Technical report, University of Cambridge, UK, 1992.
25. Crawford, T.D., Sherrill, C.D., Valeev, E.F., et al., PSI-3.3.0, Virginia Tech., Blacksburg, VA, 2007.
26. Ruud, K., Stephens, P.J., Devlin, F.J., et al., *Chem. Phys. Lett.* 373:606, 2003.
27. Crawford, T.D., *Theor. Chim. Acc.* 115:227, 2006.
28. Carnell, M., Peyerimhoff, S.D., Breest, A., et al., *Chem. Phys. Lett.* 180:477, 1991.
29. Richardson, F.S., Shillady, D.D., and Bloor, J.E., *J. Phys. Chem.* 75:2466, 1971.
30. Peyerimhoff, S.D., Private communication, 2007.
31. Tam, M.C., Russ, N.J., and Crawford, T.D., *J. Chem. Phys.* 121:3550, 2004.

32. Ruud, K. and Zanasi, R., *Angew. Chem.* 44:3594, 2005.
33. Abrams, M., Thanks to Prof. Abrams for this figure.
34. Cheeseman, J.R., Frisch, M.J., Devlin, F.J., et al., *J. Phys. Chem. A* 104:1039, 2000.
35. Stephens, P.J., Devlin, G.J., Cheeseman, J.R., et al., *J. Phys. Chem. A* 105:5356, 2001.
36. Gibson, S.E. and Castaldi, M.P., *Chem. Comm.* 3045, 2006.
37. Moberg, C., *Angew Chem. Int. Ed. Eng.* 37:248, 1998.
38. Farina, M. and Morandi, C., *Tetrahedron* 30:1819, 1974.
39. Nakazaki, M., *Topics Stereochem* 15:199, 1984.
40. Kozhushkov, S.I., Preuss, T., Yufit, D.S., et al., *Eur. J. Org. Chem.* 2590, 2006.
41. Crawford, T.D., Owens, L.S., Tam, M.C., et al., *J. Am. Chem. Soc.* 127:1368, 2005.
42. Buckingham, A.D. and Stephens, P.J., *Ann. Rev. Phys. Chem.* 17:399, 1966.
43. Stephens, P.J., *Chem. Phys. Lett.* 2:241, 1968.
44. Michl, J., *J. Am. Chem. Soc.* 100:6801, 1978; ibid. 6812; ibid. 6819.
45. Michl, J. *Tetrahedron* 40:3845, 1984.
46. Waluk, J. and Michl, J., *J. Org. Chem.* 56:2729, 1991.
47. Fleischhauer, J., Howeler, U., and Michl, J., *Spectrochim. Acta A* 55:585, 1999.
48. Fleischhauer, J., Howeler, U., and Michl, J., *J. Phys. Chem. A* 104:7762, 2000.
49. Fleischhauer, J. and Michl, J., *J. Phys. Chem. A* 104:7776, 2000.
50. Sprinkel, F.M., Shillady, D.D., and Strickland, R.W., *J. Am. Chem. Soc.* 97:6653, 1975.
51. Barth, G., Records, R., Bunnenberg, E., et al., *J. Am. Chem. Soc.* 93:2545, 1971.
52. Holmquist, B. and Valee, B.L., *Biochemistry* 12:4409, 1973.
53. Shillady, D.D., Castevens, C.M., Trindle, C., et al., *Biophys. Chem.* 105:471, 2003.
54. Del Bene, J. and Jaffe, H.H., *J. Chem. Phys.* 48:1807, 1968.
55. Harris, D.C. and Bertolucci, M.D., *Symmetry and Spectroscopy: An Introduction to Vibrational and Electronic Spectroscopy*, Oxford University Press, New York, p. 388, 1978.
56. Seamans, L., Moscowitz, A., Barth, G., et al., *J. Am. Chem. Soc.* 94:6464, 1972.
57. La Paglia, S.R., *Theor. Chim. Acta* 8:185, 1967.
58. Schatz, P.N., McCaffery, A.J., Suetaka, W., et al., *J. Chem. Phys.* 45:722, 1966.
59. Baldwin-Boisclair, S.D. and Shillady, D.D., *Chem. Phys. Lett.* 58:405, 1978.
60. Coriani, S., Jorgensen, P., Rizzo, A., et al., *Chem. Phys. Lett.* 300:61, 1999.
61. Olsen, J. and Jørgensen, P.J., *J. Chem. Phys.* 82:3235, 1985.
62. Hettema, H., Jensen, H.J.A., Jørgensen, P., et al., *J. Chem. Phys.* 97:1174, 1992.
63. Seth, M., Autschbach, J., and Ziegler, T., *J. Chem. Theor. Comp.* 3:434, 2007.
64. Ganyuskin, D. and Neese, F., *J. Chem. Phys.* 128:1, 2008 (forthcoming).

Index

A

Absorption band, 418
Absorption spectrum, 401, 420
ADF specific rotation angle, 432
Adiabatic and hybrid correlation
  functionals
  Becke's thermochemistry series, 286
  Kohn–Sham orbitals, 284–285
  PBE exchange–correlation functional,
    288–293
Alpha-spin Orbitals, 455
Amsterdam density functional (ADF)
  program
  deMon2k program, 319–327
  SAOP model, 316–319
  treatment of VWN exchange-
    correlation functional, 327–330
Angular momentum
  coupling of, 344
  of electron motion, 348
  imaginary Hermitian operator, 71
  with respect to gauge origin, 364
AO-to-MO transformation, 161
Atomic charges
  Mulliken charges, 110
  of nitrogen atom, 106
Atomic orbitals
  for atom, 119–120
  coefficients of local p–pi, 13–14
  contraction of lobes, 120
Atomic 2pz function, lobe representation
  of, 120
Atomic unit, 122
Atomization energies, 376, 378

B

Basis functions
  location and scale factor for, 78
  orthogonalization of, 78–79
  weighting coefficients of, 41

Benzene
  molecular orbitals for, 13–14
  $pi \rightarrow pi^*$ transition of, 15
Beryllium atom
  double-zeta expansion, 81
  Koopmans ionization energy, 84
  secular equation, 82
  vacant orbital, 83
BH molecule dissociation, and VB-CI
  program, 175–179
B3LYP-DFT, pyrrole, 411
B3LYP method, 235
Bond index, 106–107
Bonding and antibonding orbital
  Fock energies for, 393
Box model, *see* Particle-in-a-box model
Boys-Reeves algorithm, 165
Brillouin's theorem, 59
Brillouin theorem, 218
Brueckner orbitals, 222
trans-Butadiene
  molecular orbitals for, 6
  wave functions, pi electrons, 7

C

Camphor
  CD spectrum of, 424–425, 427
  structure of, 423
Canonical bonded diagrams,
  164–165
Carbon
  bonding in, 110
    by natural bond order analysis,
      111–113
  chemical shifts, 374
  isotropic shift, 353
Carbon monoxide
  bonding of, 401
  dipole moment of, 108
  orbital properties of, 109
  pi antibonding level, 394

Use of PCLOBE

I. Copying files

1. Insert the PCLOBE compact disk into your CD drive and locate its three files. One is a zipped file of illustrative examples and another is the "setup.exe" file.

2. Transfer the zipped file to your computer and extract the compressed files into a new folder, which you could name "PCLOBEXAMPLES"

3. Double click on "setup.exe" on the CD—this will install PCLOBE on your own drive. You will find the program in your programs list; you may construct a shortcut so to start PCLOBE from your desktop.

II. Running PCLOBE

1. Start (click on) PCLOBE either from the shortcut or from the programs list. If all goes well, you will see a Visual Basic display with selection buttons to define PCLOBE calculations.

2. Define a job, using structures supplied in PCLOBEXAMPLES
   a. Click the Browse button at the top left of the display panel.
   b. Find and click the "name.XYZ" file of your choice (open it).
   c. Select the desired basis from the list at center left.
   d. Enter an integer to define the gradient threshold ("3" $= 10^{-3}$). Selecting the default "0" defines a single-point energy calculation. This is required for certain tasks, and enforced by the interface. A gradient threshold of 4 or 5 will invoke calculation of vibrational frequencies. The default of 30 steepest descent iterations is best reset to "99" for geometry optimization.

3. Click on "Launch PCLOBE" to start the program. You should immediately see a black foreground output window showing progress through the task.

4. When the run finishes, click on the "View Output" button at the lower right part of the Visual Basic panel to look at the total output. If you wish to retain the output, save it with a unique file name.

Use of the VISTA operating system

Windows VISTA requires that you explicitly allow PCLOBE to read, write, and delete temporary files by setting permissions within the folder.

Special requirements on input files

Some options require special care that input files be prepared in accordance with certain conventions. For instance, the point group symmetry option requires that the input structure possesses the specified symmetry, and requires that the high order symmetry axis coincide with the "z" Cartesian direction. The examples for this input with names xxxsym.XYZ obey these requirements. Consult the "READpclobe.txt" file for more detail and an exception with the non-conventional Cnvx symmetry.

Use of the Rydberg basis requires that the file "rydberg.dat" contain a non-zero value of scaling for each atom to which a 3sp shell is to be added, remember to reset the value to "0.0" for the next run if the Rydberg orbitals are not desired (applies only to CIS in a STO-4G basis).

Preserving data written to placeholder files

From some tasks, PCLOBE produces data files labeled "uvspectra.txt," "cdspectra.txt," "mcdspectra. txt," and "ordspectra.txt." These are intended for use by graphics programs. The file labeled "lobe. xyz" is intended to provide input data to an external molecular modeler such as RASMOL. The file "lobe.draw" is intended for the external plotting program MOLEKEL. If information in these files is to be retained, they should be saved with more informative names.

An environmentally friendly book printed and bound in England by www.printondemand-worldwide.com

PEFC Certified

This product is
from sustainably
managed forests
and controlled
sources

www.pefc.org

This book is made entirely of sustainable materials; FSC paper for the cover and PEFC paper for the text pages.

#0060 - 021015 - C0 - 234/156/27 [29] - CB - 9780849384066